Environmental Engineering

Series Editors: R. Allan • U. Förstner • W. Salomons

Marion Hersh

Mathematical Modelling for Sustainable Development

With contributions from Ileana Hamburg

With 72 Figures

Springer

DR. MARION HERSH
University of Glasgow
Centre of Systems and Control
Department of Electronics and
Electrical Engineering
Glasgow G12 8LT

United Kingdom

E-mail: m.hersh@elec.gla.ac.uk

ISSN 1431-2492
ISBN 10 3-540-24216-3 **Springer Berlin Heidelberg New York**
ISBN 13 978-3-540-24216-1 **Springer Berlin Heidelberg New York**

Library of Congress Control Number: 2005907717

Springer is a partof Springer Science+Business Media
springeronline.com

Printed in Germany

Cover design: E. Kirchner, Heidelberg
Production: A. Oelschläger
Typesetting: Camera-ready by the Author
Printing: Stürtz AG, Gemany
Binding: Stürtz AG, Germany

Printed on acid-free paper 3o/2132/AO 5 4 3 2 1 0

Foreword

Many people are convinced that Sustainable Development and Mathematics are completely unrelated. Sustainable Development, in its role of a value laden imperative for polluting and over-consuming societies, seems to be totally unconnected to mathematical reasoning and ignorant of the values behind its symbols. Still, they are not only connected: they need each other.

Mathematics needs Sustainable Development. When science was gradually reinvented in European medieval societies, it was legitimised as contributing to the disclosure of God's divine creation. The conflicts that emerged became well known as a result of the clash between Galileo and the Church. Science found a new legitimacy through recognition that it was a powerful force against superstition. In the Enlightenment the argument was pushed forward by attributing Progress to the advancement of science: science could produce a better world by promoting rationality. In our modern society, science has become intimately linked to technology. Science for its own sake unfortunately rarely has positive outcomes in terms of research grant applications. Meanwhile, science and technology, and the progress they are supposed to produce, meet with wide scale scepticism. We all know of the current global problems: climate change, resource depletion, a thinning ozone layer, space debris, declining biodiversity, malnutrition, dying ecosystems, global inequity, and the risk of unprecedented nuclear wars. Science has to engage with these problems or lose its legitimacy.

Sustainable Development needs Mathematics. Although Sustainable Development might seem to be completely normative at first sight, we have never met anyone who outright disagrees with the necessity of Sustainable Development. There seems to be pretty much consensus that something has to be done. However, there is disagreement on the nature and speed of the required actions, just as there is disagreement on the various interests that would be affected. Many of the actions that are needed for Sustainable Development, such as the reduction of carbon dioxide emissions, require a high degree of global consensus. This consensus can only be created by solid facts. The facts will be clear when it is too late: when our country (the Netherlands) is flooded by climate change induced sea level rise, when inequity has caused unprecedented conflict, or when biological resources have totally disappeared as a result of the collapse of ecosystems. Mathematics is crucial for getting the facts clear. Without mathematics, anthropogenic (human generated) climate change would not have been recognised, just as the dynamics of ecosystems would be incomprehensible. We need mathematics to build global consensus just as we need it to search for solutions.

In this book a wide range of mathematical modelling techniques are described. The book is mathematically sound and it is written in a way that makes the material accessible for practical researchers without a very strong mathematical background. All the methods are illustrated by examples and case studies relevant to sustainable development. Different parts of the mathematical modelling cycle are discussed: modelling a system, calibrating and validating a model and using a model for decision making. We all know that models are never perfect. Therefore it is important to explicitly take into account the uncertainties of the various model components. This is especially true for models of complex environmental and social problems, where often only qualitative and imprecise information is available. Therefore, a large part of the book is devoted to fuzzy set and systems techniques. These techniques can be used to analyse and provide predictions of the behaviour of the underlying system(s), and also to produce measures of the certainty or accuracy of the results. This is very essential information for all policy makers.

May 2005

Arnold Heemink,
Dirk Jan Peet
Karel Mulder
Delft University of Technology

Dedication

For my good and supportive friend, Michael Johnson

Sustainable Development: Role and Scope of the Book

1 The Role and Scope of The Book

Many of the problems facing the world today are the result of unsustainable development. Global warming and the hole in the ozone layer are probably amongst the best known, but by no means the only examples. However discussion of sustainable development has tended to focus on environmental or sustainability aspects of sustainable development and ignored the equally important developmental side. Problems here include inequitable development, extreme poverty and starvation and resource wars, of which oil is currently most often the focus. However water has also been an object of dispute, and may replace oil in the near future as the focus of conflict, as water shortages become more extreme.

This gives rise to the question of what can be done to avoid environmental and social catastrophe. To do so will require action at a number of different levels, including by individuals in their own lives and by politicians and decision makers at the local, national and global levels.

In order to understand the problems better and make good decisions, appropriate analysis tools are required. Some of these tools are presented in this book. They are based on a number of different mathematical and computational techniques.

Previous analysis has tended to use 'soft' approaches, which do not require a knowledge of mathematical or computational techniques. These approaches can often be considered to be complementary to the techniques presented here. However the use of mathematical and computational methods can be advantageous, due to the complexity and interactive nature of many of the problems involved and can, for instance, support decision making and making trade-offs in complex problems. However many researchers are unfamiliar with the range of analytical, mathematical and computational methods that could be applied in this area. Therefore they are not able to take advantage of the full range of available methods

An Unsustainable Approach to End-of-Life Technology Disposal

in their research or analysis. This book aims to fill this gap by providing both a basic introduction and advanced technical details of some of the available mathematical and computing methods, as well as illustrating their use through case studies and examples.

The methods presented here are aimed specifically at sustainable development and the case studies and examples are all in this area, but they have a wide range of other application areas, including in economics, medicine and control systems.

The techniques presented include:

- Systems theory and methodologies for structuring complex sustainable development problems to make it easier to obtain a solution to them.
- Optimisation and decision making techniques to support policy formulation and other decision applications.
- Fuzzy sets and techniques for representing and analysing the mixture of quantitative and qualitative, uncertain and imprecise data occurring in environmental, developmental and social problems.

The *intended audience* of this book is students, academics and industrial and other practitioners, particularly in the areas of engineering, science, mathematics and economics, who are interested in applying mathematical methods to sustainable development. A *further audience* consists of non-specialists with an interest in sustainable development. The book is organised into three self-contained parts to allow readers to focus on areas of particular interest or study them non-sequentially. Each part contains appropriate background as well as more advanced material. There are numerous examples, both to clarify concepts and techniques and to illustrate their relevance and application to sustainable development. Each part also contains a longer and more detailed case study. In addition to the references cited in the text, each part has a section of additional reading materials to enable readers to explore the subject further. To allow the book to be used as a textbook for final year undergraduates or MSc students without disturbing the specialist reader, exercises are presented in the final chapter of each part. The preliminary pages of the book include a section on notation which presents the mathematical notation used in the book.

To prepare readers for the technical material in Parts I-III and, in particular, for the examples of its application to various problems in sustainable development, Chapter 1 provides an overview of some of the important issues in sustainable development. This includes references and additional reading. Other issues in sustainable development are introduced later the book. However, this book does not aim to cover *all* the issues of relevance to sustainable development, but should give readers a basic understanding of sustainable development. Therefore students, in particular, probably could benefit from an introductory course or self-study on sustainable development before using this book.

A number of useful books and articles are listed in the references and additional reading section after Chapter 1. Several introductory texts are listed in the Bibliography Section at the end of this Preface. The *State of the World* publications which are produced each year by the WorldWatch Institute provide useful discussions of a number of important topics in sustainable development.

Bibliography

Baker S (2005) Environment. Routledge
Elliott JA (2005) An introduction to sustainable development. Routledge
Gupta A (1998) Ecology and development in a third world. Routledge
Weaver P, Jansen L, Grootveld van G, Spiegel van E, Vergragt P (2000) Sustainable technology development. Greenleaf Publishing

Author's Affiliation

Marion Hersh, Department of Electronics and Electrical Engineering, University of Glasgow, Glasgow G12 8LT, Scotland.

Contents

Notation

Understanding the notation used is crucial for understanding mathematical and other technical literature. Unfortunately mathematical and scientific notation has not been standardised, though there is a degree of standardisation in some fields, but rarely across fields. Thus, for instance the symbol ‘u’ is generally used to represent an input in the control literature, but ‘x’ is often used for an input in the systems literature. This section lists some of the notation used in this book. As far as possible the notation traditional to a given field is used in sections relating to that field. Therefore the same symbol may be used with different meanings in different sections of the book.

General Notation

$\sum_{j=1}^{n} x_j$	summation	$x_1 + x_2 + ... + x_n$
$\prod_{j=1}^{n} x_j$	product	$x_1 x_2 ... x_n$
$x_1 \vee x_2$	max	the maximum of x_1 and x_2
$\bigvee_{i=1}^{n} x_i$	max (of n variables)	the maximum of x_1 to x_n
$x_1 \wedge x_2$	min	the minimum of x_1 and x_2
$\bigwedge_{i=1}^{n} x_i$	min (of n variables)	the minimum of x_1 to x_n

$\forall x$	for all	the given expression holds for all x
$A \Rightarrow B$	implies	A implies B i.e. if A holds, then so does B
$\begin{bmatrix} x_1 \\ x_2 \\ \vdots \\ x_n \end{bmatrix}$	column vector	column vector with elements x_1 to x_n
$[x_1, x_2, ..., x_n]$	row vector	row vector with elements x_1 to x_n
$\begin{bmatrix} x_{11} & x_{12} & \cdots & x_{1n} \\ x_{21} & x_{22} & \cdots & x_{2n} \\ \vdots & \vdots & \ddots & \vdots \\ x_{m1} & x_{m2} & \cdots & x_{mn} \end{bmatrix}$	matrix	matrix with m rows and n columns
^{T}	transpose	the transpose of a matrix replaces rows by columns.
$[a, b]$	closed interval	$\{x: a \le x \le b\}$
(a, b)	open interval	$\{x: a < x < b\}$
$I = [0, 1]$	unit interval	$\{x: 0 \le x \le 1\}$
$I_{\pm} = [-1, 1]$	interval	$\{x: -1 \le x \le 1\}$
$\mathcal{R}$	real line	$\{x: 0 \le x < \infty\}$
$\mathcal{N}_n$	set of n integers	$\{1, 2, ..., n\}$
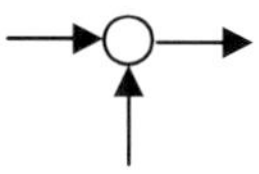	summing junction	inputs are added together unless otherwise indicated by a minus sign

$\therefore$	therefore

Set Theory Notation

$A_1 \cup A_2$	set union	all the elements that are in set A_1 *or* set A_2
$\bigcup_{i=1}^{n} A_i$	set union (n sets)	all the elements in any of the sets A_1 to A_n
$A_1 \cap A_2$	set intersection	all the elements that are in *both* set A_1 *and* set A_2
$\bigcap_{i=1}^{n} A_i$	set intersection (n sets)	all the elements in all of the sets A_1 to A_n
$x \in A$	element of	x is a member (element of) the set A
$A \subset B$	set inclusion	all the elements of set A are also elements of set B
$\varnothing$	null set	set with no elements

Fuzzy Set Theory Notation

$\mu(x)$	fuzzy set
$(\mu(x), \nu(x))$	augmented fuzzy set
$(\mu(x), \nu(x), \pi(x))$	third order augmented fuzzy set
$(\mu(x), \gamma(x))$	intuitionistic fuzzy set
$(\mu(x), \gamma(x), \pi(x))$	intuitionistic fuzzy set with intuitionistic index $\pi(x)$
$\mathcal{F}$	fuzzification mapping

$\mu_P o \mu_Q$	max min composition	$\max_{y \in Y} \min[\mu_P(x,y), \mu_Q(y,z)]$

Calculus Notation

$\int_{t_1}^{t_2} f(t)dt$	integral	integral of function f(t) between limits t_1 and t_2.
$\frac{dx}{dt}$, $\dot{x}$, x'	first derivative	derivative (gradient) of x with respect to t
$\frac{d^2x}{dt^2}$, $\ddot{x}$, x''	second derivative	second derivative of x with respect to t.
$\frac{d^nx}{dt^n}$, $x^{(n)}$	nth derivative	nth derivative of x with respect to t.
$\frac{\partial x}{\partial t}$	partial derivative	partial derivative of x with respect to t.
$\frac{\partial^n x}{\partial t^n}$	nth partial derivative	nth partial derivative of x with respect to t.
$\frac{\partial^2 f}{\partial x_i \partial x_j}$	partial derivative	partial derivative of f with respect to x_i and x_j.
$\underline{\nabla} f(\underline{x})$	grad	vector - partial derivatives of f with respect to x_1 to x_n
$\Phi(t_1, t_2)$	state transition matrix	matrix of transition between states at t_1 and t_2
$\Phi'(s)$	frequency domain state transition matrix	Laplace transform of state transition matrix

Decision Theory Notation

A	set of alternatives	
F	family of criteria	
$g_i(.)$	ith criterion function	
c(a, b)	concordance index of two options	
a S b	outranking relation	option a performs as least as well as b on all criteria
C(a S b)	concordant criteria	subset of criteria which agree with outranking relation
C(a Q b)	discordant criteria	subset of criteria which disagree with outranking relation
C(a P b)	indifferent criteria	subset of criteria which agree with outranking relation
a D b	dominance relation	option a performs better than b on all criteria
$a \succ b$	preferred	option a is preferred to b
$a \prec b$	less preferred	option a is less preferred than option b
$a \succeq b$	preferred or indifferent	option a is preferred or indifferent to option b
$a \preceq b$	less preferred or indifferent	option a is less preferred or indifferent to option b
$a \sim b$	indifferent	option a is indifferent to b

Conventions

A scalar is represented by a small letter, a vector by an underlined small letter and a matrix by a capital letter.
A time domain variable is represented by a small letter and a frequency domain variable or transform of a time domain variable by a capital letter.
An element of a set is represented by a small letter and a set by a capital letter.

List of Examples and Case Studies

Part I

Part II

Part III

1 Sustainable Development: An Overview

1.0 Learning Objectives

The main aim of this chapter is to give readers an overview of some of the main issues in sustainable development. This will support the presentation of mathematical methods and their application to problems in sustainable development in Parts I-III. Specific learning objectives for the chapter include the following:

- Understanding of the main definitions of sustainable development.
- Obtaining an overview of the main social, developmental and environmental issues underlying sustainable development in order to support application of the methods presented in later chapters to problems in sustainable development.
- Understanding of some the barriers to achieving sustainable development and the tradeoffs that may be required between developmental and environmental goals.
- Understanding of the precautionary principle.

1.1 Introduction

As discussed in the introduction to the book, unsustainable development is at the root of many of the serious problems facing the world. Although the focus of sustainable development is often the environment, unsustainable development has also resulted in serious social and developmental problems. Thus solutions are required to a range of problems, including global warming and the hole in the ozone layer, extreme poverty, with nearly half the world's population living on less that two dollars a day, resource wars, increasing water shortages and inequitable development with less than a fifth of the world's population consuming more than four fifths of its resources.

This chapter presents an overview of the main issues in sustainable development in order to prepare readers to apply some of the mathematical and other techniques presented in Parts I-III. It is divided into eight main sections. The chapter is introduced by a discussion in section 1.1 of the main definitions of sustainable development, of which the Brundtland one (WCED 1987) is probably the most commonly used. It is concluded by a summary of the chapter in section 1.8. Sections 1.2 and 1.3 consider the social and developmental aspects of sustainable development and sections 1.5 and 1.6 the environmental aspects. Section 1.4 can be considered to form a bridge between the 'environmental' and 'social' parts, as it considers some of the social consequences of environmental problems. The remaining section, 1.7, discusses the assimilative capacity of the natural environment and the role of the precautionary principle in trying to prevent further environmental and social damage.

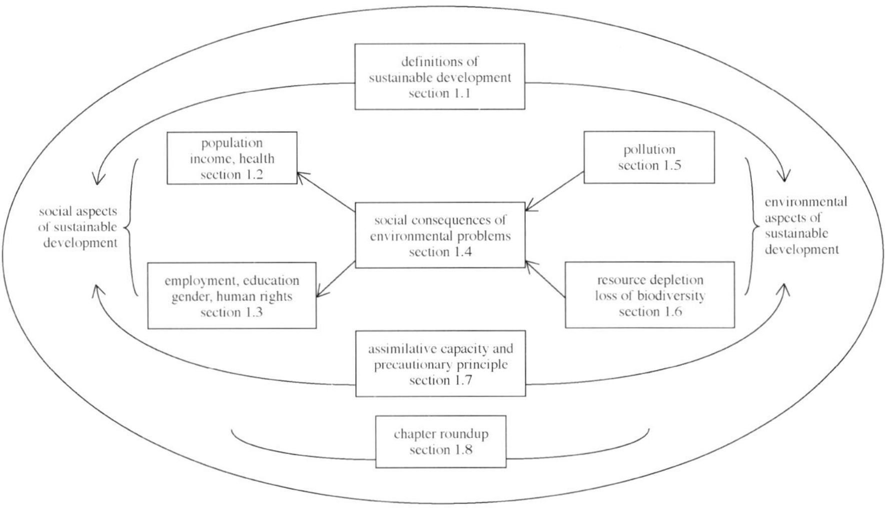

Layout of Chapter 1

1.1.1 Sustainable Development: Some Definitions

Increasing concern about the effects of human economic and industrial activities on the environment has led to a growing interest in sustainable development. This has been variously defined (Holmberg and Sandbrook 1992; Pearce et al 1989). The most generally accepted definition is that produced by the World Commission on Economic Development (WCED 1987), often called the Brundtland Commission after its chair, according to which sustainable development is *'development that meets the needs of the present without compromising the ability of future generations to meet their own needs.'* This definition is generally understood to bring together the following three factors:

- The development needs of humanity
- Protection and conservation of the natural environment
- Maintaining the ability of future generations to meet their own needs.

Fig. 1.1 illustrates the relationship between these factors. Another related approach to sustainable development involves the three sometimes conflicting policy goals of environmental integrity, economic efficiency and equity (including equity between generations) (Young 1992).

The Brundtland definition focuses on the satisfaction of *human needs* rather than human wants or the protection of the environment or other species (though they are included), but does not clarify what is meant by 'needs'. It also considers the ethical issues of equity within and between generations or balancing the needs of present and future generations and the needs of different groups in a given generation respectively. It should be noted that equity within a generation generally involves the redistribution of resources to the poorer i.e. the so called 'developing'[1] countries and to poorer people within countries (Kirkby et al 1995).

The Earth Summit in Rio in 1992 was proposed by the WCED to review progress in the five years after the launch of its report (WCED 1987), to promote follow-up arrangements and to set benchmarks. It achieved the *Rio Declaration* on legal principles for sustainable development, *Agenda 21*, and watered down, but still useful framework conventions on climate and biodiversity. Despite its success in involving a large number of people and raising awareness, Rio did not result in concrete measurable commitments in the areas requiring actions or, with few exceptions, additional resources for development. Other than Denmark, Norway, Sweden and the Netherlands, the industrialised countries have not kept their pledges of allocating 0.7% of GDP for development assistance and the average over the countries in the Organisation of Economic Cooperation and Development (OECD) is about 0.2% (Brundtland 2002).

There are a number of threats to sustainable development, including environmental degradation, inequitable and inadequate development in many parts of the world and global insecurity (Kirkby et al 1995). The necessity of resolving both the environmental and developmental crises is

---

[1] The terms 'developing' and industrialised countries will be used here to denote the groups of countries sometimes referred to as 'south' and 'north', 'third' and 'first world' or 'developing' and 'developed'. It should be noted that none of the terms in common use, including those used here, are totally satisfactory. In some places the terms low income and high income countries will also be used.

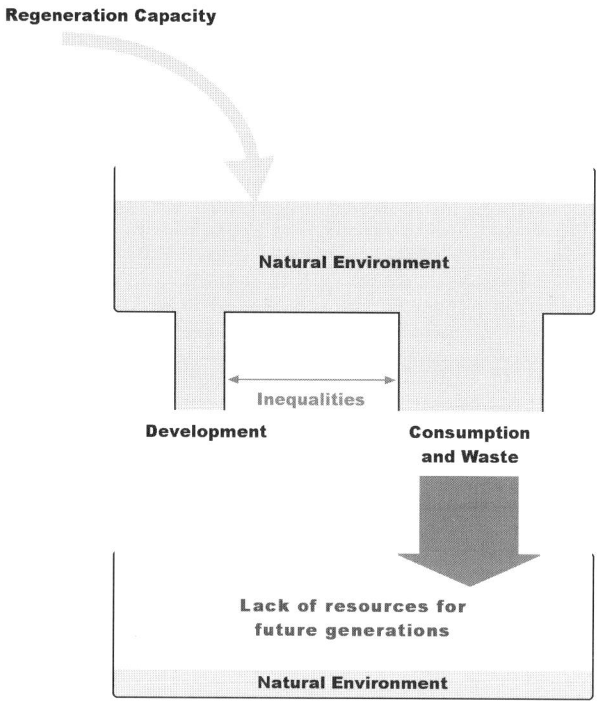

Fig. 1.1 Relationship between the components of sustainable development

generally recognised as a prerequisite for sustainable development, but the need for the resolution of conflict and world peace is less often considered in this context.

1.2 Population, Income and Health

1.2.1 Introduction

This is the first of two sections which discuss the social and developmental aspects of sustainable development. One of the main themes of this section is inequality in access to resources, particularly between, but also within countries. The discussion is introduced by the United Nations Millennium Declaration (section 1.2.2), since the nature of its objectives gives a good picture of the world's current state of development. The inequalities discussed include in access to and use of resources (section 1.2.3) in gross domestic product or national income (section 1.2.4) and life expectancy and health (section 1.2.5). Another important theme considered in section 1.2.3 is the interaction between population and resource consumption and the fact that it is the much smaller populations of the industrialised world which consume the majority of the world's resources.

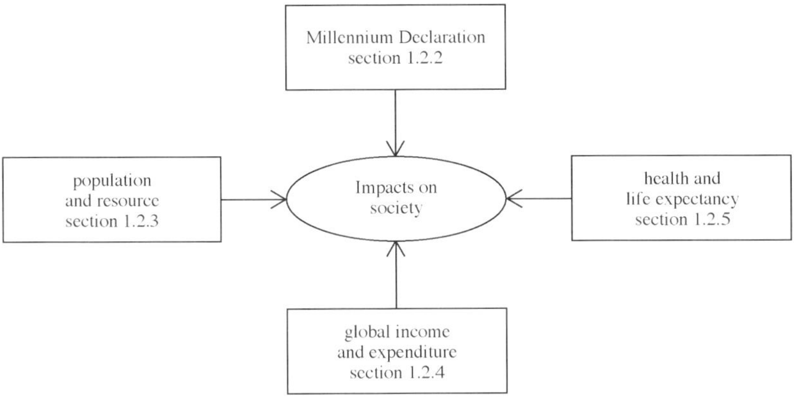

Layout of Section 1.2

1.2.2 The Millennium Declaration

The 189 nations of the United Nations adopted the Millennium Declaration in 2000 (WWW1 2005) with the aim of producing a substantial improvement in living conditions. The *Millennium Declaration* includes the following social objectives:

- Halving the proportion of the world's people living in extreme poverty, suffering from hunger and lacking access to clean drinking water.
- Reducing maternal mortality by 75%.
- Reducing mortality rates for children under five by two thirds.
- Universal completion of primary education and gender equality in access to education.
- Halting and then reversing the spread of HIV/AIDS, malaria and other major diseases.

There are also economic and environmental objectives. The nature of the social objectives and the fact that the World Bank considers them 'ambitious' (World Bank 2004) and that it is still open to question whether they will be met by 2115 (Bellamy 2004) illustrates the very serious nature of the social problems facing the world. Many donor countries have made additional commitments which would add 13.9 billion euros to annual aid if fulfilled, but an estimated additional 22-45 billion euros is required to enable the poorest countries to achieve the Millennium Development

Goals. There are also issues of how aid should be used to be most effective.

Other important social issues not in the Millennium Declaration objectives include employment, access to resources and equity i.e. equality of opportunities and access to resources both between and within countries. The global situation with regards to all these issues will be discussed in this and the following sections. Due to delays associated with data collection, the most recently published global data available at the time of writing was generally for 2002. Therefore, unless otherwise stated, it should be assumed that data is given for 2002.

Early ideas about development were often based on the belief that economic growth measured by gross domestic product would lead to development and almost magically reduce poverty and inequality. By the 1980s little growth had occurred in the 'developing' countries and there was no evidence that growth on its own could reduce poverty and inequality. The International Monetary Fund implemented so-called structural adjustment programmes aimed at reducing public expenditure and giving greater scope to market forces. The resulting cuts in spending on education, health and food subsidies had a very negative effect on development, with poorer people and particularly poor women, suffering the worst impacts, without even leading to economic growth. In the 1990s it was finally recognised that economic growth on its own would not lead to human development, but that development can lead to growth (Bellamy 2004).

1.2.3 Population and Resources

There are a number of different perspectives on population and the environment. Although there is a body of opinion that population growth is not a problem and will probably contribute to economic development and higher living standards (Simon 1980), it is more commonly held that fast growing populations have reached and exceeded the earth's capacity to feed them (Ehrlich 1968; Paddock and Paddock 1975) or provide sufficient resources and absorb the waste generated by human activity (Loh and Wackernagel 2004). The *ecological footprint* is a measure of the total area required to meet all a country's resource needs, absorb its waste and provide space for its infrastructure. The global ecological footprint exceeded the earth's surface area by 21% in 2001 (Loh and Wackernagel 2004).

Global population has grown very fast from about one billion in 1804 to an estimated 6.394 billion in September 2004 (WWW2 2005). Although population is still growing, the rate of growth peaked at 2.1% in the 1960s and has since dropped to 1.7% in the mid 90s (Rowley and Holmberg 1995) and to just under 1.3% in 2002 (Engelman et al 2002). This drop in population growth is largely due to women having fewer children on average, following a reduction in infant mortality rates and the increasing availability of modern means of contraception. However the continuing fall in mortality rates has not yet been balanced by an equivalent drop in birth rates, so that population is continuing to grow, although the rate of growth has dropped and populations in some of the countries of Central and Eastern Europe are now slowly falling (World Bank 2004). Even at this reduced rate of growth an additional 77 million people are still being added to the population each year, equivalent to 10 New York Cities. Population in the 48 poorest countries in the world is projected to triple by 2040 (Engelman et al 2002). This will significantly reduce the likelihood of strategies for poverty reduction being successful.

The focus of concern about *overpopulation* is now on environmental rather than food security consequences (Brown and Jacobson 1986), but is still focused on the 'developing' rather than the industrialised countries. However, it is the industrialised countries and the oil exporting countries, such as the United Arab Emirates and Kuwait, with their much smaller populations, which use the greatest proportion of the world's resources, four fifths of which are consumed by a fifth of the world's population (Kirkby et al 1995). 95% of the projected 3 billion growth in population by 2025 is expected to occur in the 'developing' countries. However, the *projected emissions* of the estimated 114 million increase in the population of the USA in the next 50 years are likely to be about the same as those of the estimated 1.2 billion increase in the population of Africa in this period i.e. about ten times as much for each person (Engelman et al 2002). The *ecological footprint* of each person in the USA, United Arab Emirates or Kuwait is about 32 times that of each person in Afghanistan, about 16 times that of each person in Bangladesh and Nepal and about 14 times that of each person in the Democratic Republic of Congo, Ethiopia and Mozambique (Loh and Wackernagel 2004, 2001 data). Water consumption per person per year in the USA in 46 times that in Ethiopia (Postel and Vickers 2004) and electricity use in the USA is 152 times that in Nigeria (Gardner et al 2004, data circa 2000) Thus overpopulation is not just an issue of total numbers, but of the quantity and type of resource use and unequal access to resources.

The high income countries are more *urbanised* than the low income ones, such as sub-Saharan Africa, where a third of the population lives in cities compared to more than three quarters (78%) of the population in the European Monetary Union. Movement from rural to urban areas is continuing throughout the world, with higher rates in the low income countries where a small proportion of the population is living in towns and cities. In the period 1990-2002, 17% of the population moved from rural to urban areas in East Asia and the Pacific, 12% in sub-Saharan Africa and 5% in the European Monetary Union (UN 2003). There is also some movement from urban to rural areas in the industrialised countries leading, for instance, to repopulation of some of the rural areas of Scotland (Stockdale et al 2000). However this movement out of cities is not (yet) strong enough to be reflected in regional or even national statistics, with a few exceptions. The percentage changes in urban populations of all except a very small number of countries, such as Georgia and Andorra, are higher and in many cases much higher than rural growth rates. In a number of countries the rural population is declining. For instance the rural populations of China and Europe have been estimated to be decreasing by 0.8% and 0.5% respectively (UN 2003).

Populations in the industrialised countries are *ageing* with 16.8% of the population of the European Monetary Union over 65 (World Bank 2004). On the one hand this is a consequence of welcome increases in life expectancy, whereas on the other it may result in associated social problems as the percentage of the population of working age drops. This may lead to attempts to increase the pensionable age. While some workers may personally wish to extend their working lives, in general this is likely to be very unpopular. Most trade unions are strongly opposed to any statutory increase in the retirement age, but would support increased flexibility for employees to decide when they retire.

There are considerable *inequalities* both within and between countries with regards to access to the world's resources and many of the statistics are very depressing. For instance eleven million children under five die each year in the 'developing' countries (UNDP 2004), though this is a reduction of three million compared to the figure at the end of the 1980s (Timberlake and Thompson 1990). An estimated 100 million children live and work on the streets of the world's cities, 40 million in Latin America, 25-30 million in Asia, and 10 million in Africa, due to war, poverty, violence, family disintegration and AIDS (KIT 2002). Although women in the poorer countries are the main providers of childcare, education and health services, as well as agricultural products, food, water and fuel (Kirkby et al 1995), their work is frequently both unpaid and not included in national accounting systems and economic measures such as gross national product.

Resource consumption depends on patterns of resource use and consequently on lifestyles, industrial and agricultural structures and consumption patterns, as well as population levels. Thus both changes (i.e. reductions in the industrialised countries) in resource use and stabilisation or reduction in population will be required for sustainable development. Stabilising and reducing population requires understanding of the interconnected social, cultural and economic, rather than the biological causes of population growth. These include the low status of women, high infant mortality and the lack of old age security (Lappé and Shurman 1995). Economic and social improvements in the position of women, including literacy and education, access to food and resources and the freedom to make decisions, generally lead to small families and, in combination with family planning programmes and reduced infant and child deaths, could lead to population stabilisation or even reduction (UNICEF 1991). Access to family planning could also save the lives of about 150,000 of the half a million women who die each year from causes related to pregnancy and giving birth. It could also significantly improve the nutritional health and quality of life of children, particularly in the 'developing' countries (UNICEF 1992).

1.2.4 Global Income and Expenditure

The United Nations System of National Accounts uses *gross domestic product* (GDP) as a measure of economic activity. It is the value of goods and services produced by labour and property located in a given country. Calculation of GDP is generally based on expenditures rather than income components as the data is more reliable (WWW3 2005). It is frequently used to compare the economic standing and relative prosperity of different countries. However GDP is inadequate as a measure of prosperity or the state of sustainable development for the following two main reasons:

- It includes the value of all goods and services obtained via monetary transactions, whether they have a positive or negative impact on society and the environment. For instance GDP includes the costs of cleaning up after oil spills and the value of all cigarettes sold. Clearly prosperity would be increased more by avoiding oil spills and smoking being phased out (in a way that does not infringe the rights of existing smokers), but this is not reflected in GDP.
- It excludes the value of goods and services that are not obtained by monetary transactions, such as housework. Clearly housework is necessary both to support other activities and of value in itself. A society in which no housework was taking place would be very unpleasant to live in.

Gross world product is defined similarly to GDP, but on a global scale and with similar problems as a measure of prosperity. Gross national income (GNI) is a measure of income from production wherever in the world it occurs (WWW4 2005). It therefore differs from GDP in including income from production which occurs outside the country. Despite their inadequacies, comparison of GDP or the related indicator gross national income still give a reasonable picture of the inequalities between countries. The USA has the largest value of GDP of nearly 10.4 trillion dollars, followed by Japan with nearly four trillion dollars. Guinea-Buisseau has the smallest value of GDP of 203 million dollars. In terms of gross national income per capita, Ethiopia and Burundi are the poorest nations, both with an income of 100 dollars per person and Luxembourg is the richest nation with nearly 39.5 thousand dollars per person, nearly 400 times as much. The average gross national incomes per capita for sub-Saharan Africa and South Asia are only 450 and 460 dollars per person respectively, compared to 20,320 dollars for the European Monetary Union (World Bank 2004), a difference of a factor of one forty-fourth.

Nearly two thirds (64%) of gross world product comes from services, with industry contributing most of the remaining third (32%) and agriculture 4%. However the transition from agriculture and manufacturing to services is more complete in the high income countries, with an average 71% of GDP from services, compared with only 46% in the low income countries (World Bank 2004). The low income countries still obtain nearly a quarter of their income (24%) from agriculture, compared with only 3% in the high income countries. It seems likely that the low income countries will follow the development path of increasing the percentage of income from services and reducing that from agriculture. While this will probably contribute to a strategy of increasing GDP, there are likely to be a number of unevenly spread associated social impacts, with increases in poverty, unemployment and hunger in some parts of the world. Service exports are dominated (39.2%) by computer, information, communications and other commercial services, but this trend is much stronger in the low than the high income countries (54.6% compared to 40.9%) particularly in South Asia (70.3%) (World Bank 2004), probably due to service industries being increasingly sited in areas of cheap labour.

There has been some discussion of the relationship between economic growth and sustainability. Another important issue is the relationship between economic growth and quality of life. Some indicators, such as the human development index, which is based on income per capita, average life expectancy, educational participation and literacy levels, show that economic growth is leading to satisfaction of human needs. For instance educational participation in the 'developing' countries is improving and life expectancy in the 'developing' countries has increased from 46 to 62 years since 1960, despite setbacks in some countries due to AIDS.

However broader indicators of quality of life give a more mixed message. For instance the index of sustainable economic welfare includes corrections to GDP, such as costs from income inequality, pollution, resources depletion and some types of medical care, and benefits from unwaged labour. It showed an increase in welfare with increases in GDP until some point in the 1970s, followed by stagnation or decline, despite continuing GDP growth (Michaelis 2002).

Many of both the industrialised and 'developing' countries are devoting a considerable proportion of their expenditure to the military (frequently called 'defence'), with average figures of 10% in both cases. This is diverting resources which could be used more constructively in other areas such as health, education, conflict resolution and employment. *Military spending* generally leads to the creation of *fewer jobs* than the same amount of other types of spending, as well as a different type of job mix which is more concentrated in higher skilled and higher paying occupations. Therefore an expansion of the military budget tends to widen the rising gap between highly and less educated workers (Bell et al 2004). This is a particular problem in the 'developing' countries where, despite evidence of the importance of *education for development* (Bellamy 2004), average spending on education is less than a third (3% of total expenditure) of average military spending. This compares with 12% of total expenditure going to education in the industrialised countries. In South Asia more than three times as much (17% of total expenditure) is allocated to the military as health (2%) and education (3%) combined. Despite the much higher proportion of total expenditure devoted to health in the 'developing' than the industrialised countries (11% compared to 3%)

A Youth Alert! peer education team visits a secondary school in Africa

(Bellamy 2004), expenditure per person on health is still about one hundred times as much in the high income countries ($2841) as in South Asia ($22) and sub-Saharan Africa ($29) (World Bank 2004; WHO 2003).

The 'developing' countries currently have a '*debt*' of 2.06 trillion dollars to the richer countries (Guttal 2000). Only 400 billion dollars (about 19%) of this amount was borrowed originally. The remainder has arisen from the very high rates of compound interest, of over 20% per year between 1973 and 1993 (Shah 2004). As a result 18% of their global income from the export of goods and services is being paid out to service external debt by the 'developing' and former eastern bloc countries. In Latin America and the Caribbean just over a *third* (34%) of external income is being lost in this way. In most of the 'developing' countries the costs of servicing public and publicly guaranteed debt are greater than income from official development assistance (aid) and there is consequently a net transfer of funds from the poorer to the richer countries. This is despite some debt adjustment and increases in overseas aid. Thus East Asia and the Pacific receive 0.4% of gross national income in aid, but have to pay out 2.3% to service debt, South Asia obtains 1.0% of gross national income (GNI) in aid, but pays out 2.4% of GNI for servicing debt and Latin America and the Caribbean receive 0.3% and pay out 4.2% (World Bank 2004). Other figures show that the 'developing' countries spend 13 times as much on debt repayment as they received in grants in 1998, an increase from nine times as much in 1996 (Shah 2004; Jubilee 2000). The balance of funds transfer between the International Monetary Fund (IMF) and sub-Saharan Africa changed in 1997 in favour of the IMF with a net transfer from Africa to the IMF of more than one billion dollars in 1997 and 1998 (Jubilee 2000).

Therefore many of the poorer countries would benefit more from a cancellation of debt than an increase in aid or would even benefit if aid was cancelled at the same time. Cancelling their debt would have further advantages in terms of reducing dependency, removing the so-called structural adjustment conditions imposed by the lenders, helping to *redress the economic balance* between the poorer and richer countries and improving the self and external images of the poorer countries. However the greatest contribution to eliminating poverty and supporting sustainable development would be given by cancelling all the external debt of the 'developing' and transition countries and increasing aid, but targeting it at grassroots organisations and the education of girls, which is generally agreed to have a major effect on development (Bellamy 2004; World Bank 2004). A number of proposals have been made for debt reduction following the tsunami in Asia at the end of 2004 and may lead to some changes in the (short-term) balance of debt, but are unlikely to lead to debt cancellation.

Pensions are going to become even more important as populations age. However public expenditure on pensions varies greatly from less than 1% of GDP in many countries of sub-Saharan Africa to over 10% in many European countries, with the highest figure of 17.5% in Italy. Pensions are considerable less than income in nearly all countries and are less than a fifth of average income in the Russian Federation (18.3%), Georgia (12.6%) and Argentina (18.7%) (World Bank 2004). Although pensioners unusually receive close to twice the average income in Benin, Burkina Faso and Togo, they are still affected by the generalised poverty in these countries. Poverty amongst pensioners is a problem even in the richer countries, with average pensions only a third of per capita incomes in the USA and around half (48.5% and 57.4% respectively) in the Netherlands and Finland respectively (World Bank 2004).

1.2.5 Health and Life Expectancy

Life expectancy and infant mortality vary considerably around the world. In particular there are very significant differences between the situations in the industrialised countries of the north and sub-Saharan Africa. Japan has the highest life expectancy of 81 years and Sweden the lowest figure for infant mortality of 2.8 infant deaths per 1000 live births. Life expectancy is also increasing faster in the industrialised countries in both real and percentage terms, with a increase of seven years compared to only two years in sub-Saharan Africa in the period 1970-2002. People in the industrialised north have *life expectancies* approximately twice those in sub-Saharan Africa (78 years compared to 46 years) and *under-five mortalities* which are only a fraction of those in sub-Saharan Africa (seven deaths per 1000 live births compared to 174) (Bellamy 2004). This very basic inequality needs to be corrected through measures such as universal access to clean water, adequate sanitation and good quality nutrition, the eradication of poverty and improved education, particularly for girls.

Infectious diseases continue to pose serious health problems in many 'developing' countries and 20 infectious diseases, including tuberculosis (TB), malaria and cholera, re-emerged or spread and 30 previously unknown deadly diseases, including HIV, hepatitis C and Ebola, appeared for the first time in the last quarter of the twentieth century (Gardner 2002). Pneumonia, TB, diarrhoea, malaria, measles and HIV/AIDS, which account for 90% of deaths from infectious diseases, are all preventable. Child deaths from diarrhoea were reduced by half between 1990 and 2000 due to improved nutrition, increased access to safe water, increased breastfeeding and oral rehydration therapy based on giving children an inexpensive solution of water, salt and carbohydrates. Deaths from TB have also reduced, again in part due to an inexpensive treatment programme (Gardner 2002).

However, *deaths from HIV/AIDS* have increased sixfold from just over half a million in 1990 to more than three million in 2000. Four fifths of these deaths are in sub-Saharan Africa, where most people cannot afford drug treatments. Globally 1.2% of the adult population has HIV/AIDS, whereas the figure for sub-Saharan Africa is 9% (Bellamy 2004). In addition to the tragedy of the deaths of large numbers of young people, there have been devastating social and economic impacts. The number of orphans and children who have lost one parent is projected to double to 26 million by 2010. According to UN estimates economic growth per person in half the countries of sub-Saharan Africa is falling by 0.5-1.2% a year due to AIDS (Gardner 2002). Although this may seem a small percentage and, as discussed in section 1.2.1, economic growth does not necessarily resolve the problems of poverty and inequality, it is a significant percentage of current GDP growth rates in sub-Saharan Africa (2.6%, World Bank 2004) and some economic growth will be one of many measures required to combat poverty in this area.

Poor nutrition and smoking are both significant causes of serious health problems. Half the world's population suffers from poor nutrition according to the World Health Organisation, with 1.2 billion people not getting enough to eat, 1.2 billion people overfed and at least two billion people (with some overlap with the first two groups) lacking adequate vitamins and minerals in their diets (Gardner and Halweil 2000).

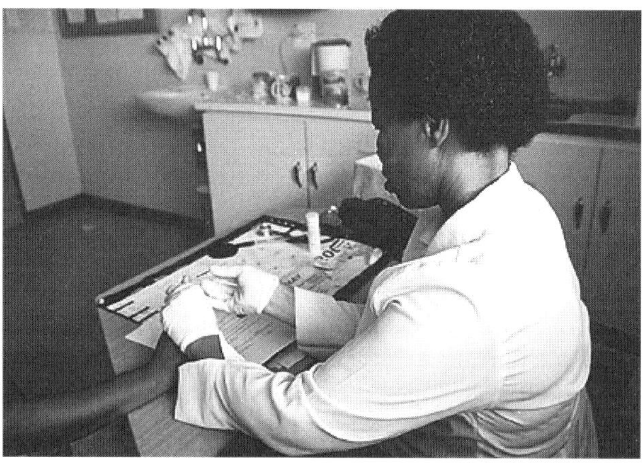

Voluntary HIV counselling and testing centre in Africa

Although there was a reduction by 10% in the number of hungry people in the 'developing' countries between 1980 and 2000, particularly in Asia and Latin America, the number of *underweight* children in sub-Saharan Africa has nearly doubled to 36%. However there are still severe problems in South Asia, where 44% of children are underweight. In some of the countries of sub-Sahara Africa the situation is even more serious, with nearly half the children in Somalia, Ethiopia and Nigeria underweight. Hunger is not restricted to the 'developing' countries either, and 10% of households in the USA are hungry, close to being hungry or worried about having enough to eat (Gardner and Halweil 2000). Poor nutrition can affect education and result in a 5-10% loss in learning capacity as well as delays in starting school. Children in areas with high iodine deficiency in Spain and the Philippines have on average three years less schooling than other children (Gardner and Halweil 2000).

About a third of adults in Europe and nearly two thirds (61%) in the USA are *overweight*. Obesity increased by 10-40% in most European countries and by half in the USA in the 1990s, largely due to sedentary lifestyles and the easy availability of cheap fatty and sugary foods (Gardner 2002). Overweight in children seems to be a particular problem in Central and Eastern Europe, where over 10% of children under five are overweight in several countries. Over a fifth of this age group are overweight in the Ukraine and Albania (World Bank 2004). This may be due to a combination of poor diet due to poverty and an unsuitable diet due to affluence. Overweight is also a problem in the poorer countries. Several Latin American countries, such as Colombia and Brazil have comparable proportions of overweight people (43% and 31% respectively) to the industrialised countries (Gardner and Halweil 2000). Cancer, diabetes and heart disease increased globally in the 1990s, partly due to an increase in life expectancy, but also to poor diets, lack of exercise and smoking. Adult onset diabetes increased fivefold from 30 to 143 million between 1985 and 1998, largely due to increases in the number of overweight people, and this figure could double to 300 million by 2025. According to 1999 UN estimates five billion people, more than 80% of the global population, have some degree of iron deficiency. This increases the risk of anaemia and cognitive impairment amongst the two billion most severely affected, particularly women and children in the poorer countries (Gardner and Halweil 2000).

It is generally *poverty* and *inequitable land distribution* rather than food shortages that lead to hunger. 80% of malnourished children in the developing countries in the 1990s were living in countries with food surpluses. A study in Brazil found that additional income earned by mothers was 20 times more likely to improve child survival than the same income earned by fathers (World Bank 2001). This is largely because women generally use all their earned income for household food and other needs, whereas men often devote up to a quarter of their income to entertainment and non-household expenses (Gardner and Halweil 2000). In addition women produce more than half the world's food and more than 80% in the rural areas of Africa, Asia and Latin America. However women (in the 'developing' countries) have little or no access to land ownership, credit, agricultural training, education and social privileges (Gardner and Halweil 2000).

Over and undereating are having severe effects on productivity in both the industrialised and 'developing' countries. Overeating is also putting a burden on health services. A study by the Physicians Committee for Responsible Medicine in the USA estimated the total annual medical costs related to meat eating at between \$29 and \$61 billion (Barnard et al 1995). The direct and indirect costs of severe overweight in the USA have been calculated at more than \$118 billion annually. This is more than twice the \$47 billion of direct and indirect costs attributed to smoking. Poverty reduction and subsidised food and free health care and education could have a significant effect on reducing under-nutrition, whereas education about nutrition could contribute to changing eating habits in the richer countries (Gardner and Halweil 2000).

While fewer people are now smoking in the high income countries, the prevalence of smoking has increased in the poorer countries and the percentage of men who smoke in the low and high income countries is comparable at just over a third of the population (37% and 36% respectively). However considerably fewer women smoke in the low income than high income countries, 7% compared to 21%. Globally nearly half the male population (46%) and just over a tenth (11%) of the female population smoke (World Bank 2004).

1.3 Employment, Education, Gender and Human Rights

1.3.1 Introduction

This is the second section on the social and developmental aspects of sustainable development. It continues the theme of inequalities, in particular in income and access to education (section 1.3.3), on the basis of gender (section (1.3.4) and due to unemployment or underemployment (section 1.3.2). These inequalities have led to serious poverty in much of the world, with nearly half the world's population living on less than the UN minimum to meet basic necessities (section 1.3.2). Another important issue covered is human rights and the value of a human rights approach to development (section 1.3.5). The education of girls is highlighted as one of the major factors that leads to development and the continuing gender based inequalities in access to education, income and parliamentary representation are noted (section 1.3.4). Discussion of female genital mutilation in section 1.3.6 is used to highlight the potential conflicts between women's rights and local cultures and the fact that action to end abuses of women's rights should be led by local people and in harmony with the local culture.

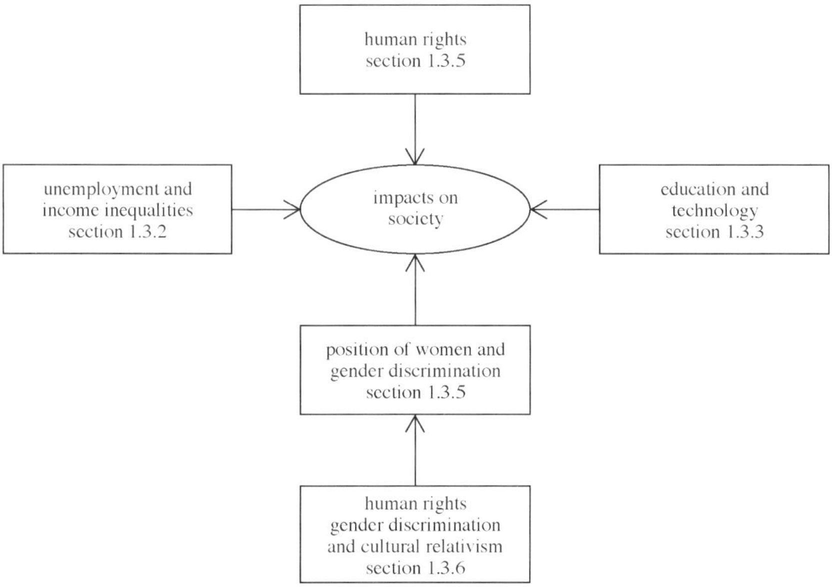

Layout of Section 1.3

1.3.2 Unemployment and Income Inequalities

Rapid economic growth in the poorer countries in the 1990s has been accompanied by both rapid population growth and a decline in the percentage of the total population living in poverty. However, the net result has been an increase in the absolute number of people living in *poverty*. In 2001 there were 1.1 billion people, mainly in China (0.89 billion) living on less than one dollar a day and 2.7 billion people, nearly half the world's population, living on less than two dollars a day, (World Bank 2004), which the UN considers the minimum required to meet basic needs (Gardner et al 2004). Galloping inflation is a serious problem in many countries, with rates of 43% in sub-Saharan Africa, 52% in Latin

America and the Caribbean and a staggering 111% in the republics of the former Soviet Union, the Central and Eastern European countries and the Baltic states (World Bank 2004).

Waged employment is the primary source of income for most people in modern economies. The global labour force (which includes people available for work as well as those in employment) has grown from 1.2 billion people in 1950 to an estimated 2.9 billion in 1998 (Renner 2000). An estimated additional 30 million jobs each year will be required for the next half century due to population growth in the 'developing' countries. Globally at least 150 million people were *unemployed* at the end of 1998 and another 900 million were *underemployed*, involuntarily working less than full time or working full time, but not earning a living wage. There are also 'discouraged' workers who have given up hope of ever finding a job, at least 9 million in the European Union i.e. at least half the number of the 18 million officially unemployed. At the same time many workers are working many hours of overtime, equivalent to two million jobs in the European Union (Renner 2000). In the USA 28% of workers are earning on or less than the official poverty level and real hourly earnings for production and non-supervisory workers declined 9% between 1978 and 1997 (Renner 2000).

Unemployment rates vary considerably between countries from 1.3% in Azerbaijan and 1.8% in Thailand to 31.9% in the Federal Republic of Macedonia and 33.8% in Namibia. Regional unemployment rates vary between 3.7% in East Asia and the Pacific to 11.1% for Central Asia and Central and Eastern Europe, with rates of 9.2% for Latin America and the Caribbean and 9.8% for the European Monetary Union (ILO 2003). There is currently no aggregate data for sub-Saharan Africa and the Middle East and North Africa, but many of the countries in these regions for which there is data, including Algeria (29.8%), Botswana (15.8%), Jordan (13.2%), South Africa (29.5%), have high unemployment rates. Although there is no aggregate figure for South Asia, several of the countries in this region for which there is data, such as Bangladesh (2.3%) and India (6.1%) apparently have relatively low unemployment rates (ILO 2003).

Although this data is based on sample surveys of the labour force or households, the data in different countries may not be totally compatible due to local cultural and other differences and differences in patterns of employment and understanding of unemployment between countries. Thus the apparent figure of 2.3% unemployment in Bangladesh may mean something very different in real terms than the same figure would in Germany (where unemployment is currently 8.6%) . It is often particularly difficult to measure employment and unemployment in agriculture in the developing countries due to strong seasonal effects. Women are frequently underrepresented in unemployment statistics, due to discrimination and the structural and social barriers which prevent them actively seeking work, as well as their greater responsibility for caring for children and dependent relatives.

The percentage of *unemployed youth* is generally higher than unemployment in the population as a whole. About 60 million people worldwide between the ages of 15 and 24 are unable to find work. This figure could increase significantly as the 43% of under-15s in the population in Africa and 26% in China get older (Renner 2000). Although child labour is decreasing, it still involves 11% of children in the 10-14 age group globally and 28% in sub-Saharan Africa (ILO 2003).

The informal sector is an important source of employment in sub-Saharan Africa and Latin America, where it provided nearly 60% of jobs in 1990. This has both advantages and disadvantages. The advantages are that it provides jobs for semiskilled and unskilled workers, is a major source of income for women, is more likely to adopt appropriate technologies and use local resources and has a major role in recycling and reusing waste materials. However the very significant disadvantages are the frequently poor working conditions, very low wages and lack of unemployment benefits (Renner 2000).

A number of factors have had an effect on the *availability of employment*. They include technology development, economic transition in the Central and Eastern European countries and the republics of the former Soviet Union and the focus on increasing labour productivity (items produced per person) rather than energy, materials and capital productivity. Technology development and the increased ease of movement of money, technology and machinery across borders have led to an increasing casualisation and contracting out of work, with a growth in temporary and part time jobs, subcontracting and outsourcing work, particularly to 'developing' countries with cheap labour. This has led to a reduction in job security and the bargaining power of the trade unions (Renner 2000).

Casualisation has affected *women workers* disproportionately. Although part time work has advantages for some women, as women still have the main responsibility for child care, part time workers frequently have much poorer pay and terms and conditions of employment than full time ones. Work continues to be gendered, with women's work generally low waged or unpaid and low status (Hersh 1999) and 'male' skills generally conferring higher status and considered to be of greater value than 'female' skills (Phillips and Taylor 1980). In addition to the gendering of jobs, there are significant differences in the sectoral distribution of employment for men and women. Over 40% more women than men in the low and middle income countries and nearly a third more in the high income countries are employed in services. Much smaller percentages of women than men are employed in agriculture and industry (World Bank 2004).

According to an index of *conditions of employment* and regulation of hiring and firing (with 100 indicating the most and 0 the least protection) the high income countries perform less well than the low and middle income countries (with an index of 44 compared to 55). The greatest protection (an index of 62) is in Latin America and the Caribbean and the USA has very little protection with an index of only 22 (World Bank 2004). This indicates that an increase in national prosperity does not necessarily benefit workers in terms of improved conditions of employment or an increase in job security.

A continuation of current trends will lead to a *polarisation* of the *work force* with a small group of highly skilled employees in secure, well paid jobs, probably working substantial overtime in highly stressed conditions, and other workers, often women, having periods of unemployment, irregular insecure jobs or being totally excluded from the labour market (Renner 2000). Craft manual skills are being lost. The number of unskilled jobs in manufacturing in western industrial countries declined by one fifth between 1970 and 1994, whereas the number of skilled manufacturing jobs has remained steady. However the number of people without recognised skills and qualifications is growing (Renner 2000).

Economic transition has resulted in a rapid rise in unemployment from nearly zero to about 10%, accompanied by a drop in real wages and dramatic increases in income inequality in most of the countries of Eastern Europe and the former Soviet Union. Real wages dropped by about 58% in Russia between 1989 and 1996 and wages are often paid months in arrears. The East Asia economic crisis in 1997 added 10 million people to the unemployment rolls and returned a significant part of the population to poverty, due to very limited unemployment benefits. 'Restructuring' to remove 'excess' labour in state and collective enterprises could lead to severe problems in several other Asian countries. For instance 3.5 million workers were made unemployed in China in 1998 alone and another 50 million job losses could follow (Renner 2000).

Labour productivity has been increasing since the industrial revolution and capital, *energy and materials* are steadily *replacing labour*. For instance in the USA labour productivity more than tripled in the period 1950 to 1996, whereas capital productivity (output per unit input of building and equipment) reduced by 14% and materials and energy productivity increased by only 24% and 15% respectively. In Japan labour productivity in manufacturing industry has increased nearly seven times since 1960 and in France 4.5 times. Some of the job losses in industry have been taken up by the service sector (Renner 2000). However the figures clearly show an unsustainable (in both human and environmental terms) focus on reducing inputs of jobs, while ignoring inputs of material, energy and capital. Protecting both employment and the environment would require a shift to increasing energy and materials productivity. Designing products to last longer and be easier to repair and upgrade could increase energy and materials productivity, lead to a shift in employment from manufacturing to repairing and upgrading and lead to a net increase in the number of jobs available. Shifting some taxation from pay roll costs to energy and materials could change the balance of costs between jobs, energy and materials, with a consequent increase in jobs and a reduction in energy and material use. This type of tax shift has been introduced in some countries in a limited form, but energy intensive industries have been exempt. For this approach to be effective this exemption would have to be removed (Renner 2000).

In addition to inequalities between countries there are also very significant *inequalities* between people *within countries*. Although data is available for the percentage share of income or consumption by the poorest and richest fifth (and tenth) of the population for most countries, the most recent year for which this data is available varies between 1983-85 and 2002. Therefore the situations in a number of countries may have changed significantly and international comparisons should be applied cautiously. In Lesotho, Namibia and Sierra Leona the poorest fifth of the population have less than 2% of total income or consumption (1.5%, 1.4% and 1.1% respectively), while the richest fifth enjoy over 60% (66.5%, 78.7% and 63.4% respectively). Income inequality is a particularly severe problem in many other countries in sub-Saharan Africa and some countries in South America, with the poorest fifth of the population having less than a twentieth and the richest fifth more than half of total income or consumption. The least unequal state based on distribution of income or consumption between the richest and poorest fifths of the population is Japan, where the poorest fifth of the population enjoys 10.6% of total income, and the richest fifth 35.6%, followed by a number of countries across Europe, with the richest fifth of the population enjoying 'only' 3.5-4.5 as much income or consumption as the poorest fifth of the population (World Bank 2004). Therefore even the least unequal countries are far from income parity.

1.3.3 Education and Technology

As well as its value for personal development and in its own right, *education* is important for *access to employment* and other *opportunities* and for encouraging development (Bellamy 2004). However much of the world's population does not have access to even basic education. Over half the population of sub-Saharan Africa and just over a quarter of the population in other low income countries does not even complete primary school. A number of countries in sub-Saharan Africa, such as Chad, Rwanda and Mozambique, have primary school completion rates of less than one in four. Poverty and illiteracy are closely related and the proportion of the population living on less than one dollar a day declines with the illiteracy rate, showing that education is an important tool for combating poverty. (World Bank 2004). Student teacher ratios vary greatly from 14 in the European Monetary Union to 45 in sub-Saharan Africa. This factor of three difference is likely to have a significant effect on the quality of schooling in sub-Saharan Africa compared to the European Monetary Union. Expected years of schooling vary between three for boys and one for girls (1990/1) in Mali and three and two for boys and girls respectively in Niger and Burkina Faso (1990/1) to 17 for both sexes in Australia and 16 and 17 for boys and girls respectively in the United Kingdom (2001/2) (World Bank 2004).

Science and technology capacity are becoming increasingly important for development, but vary significantly globally. For instance in the period

Deafblind children at Kwale School for the Deaf in Kenya planting trees

1990-2001 the high income countries had an average of one researcher in research and development per 300 people, whereas there was one per 6300 people in South Asia and one per 1700 people in East Asia and the Pacific. The highest ratio of researchers to head of population was in Japan followed closely by Sweden, both with one researcher for less than 200 people, and the lowest ratio was one per 500,000 in Senegal (UNESC0 undated). This difference in capacity is translated into differences in income, with Japan the second richest county in terms of per capita GNI and Senegal one of the poorest. High technology exports are nearly a quarter (23%) of the total of manufactured goods in the high income countries, but only 2% in the Middle East and North Africa and 4% in South Asia and sub-Saharan Africa (World Bank 2004).

There is an increasing *digital divide* between (and within) countries. In the European Monetary Union there is close to one computer for every three people (317.5 per 1000 population), whereas nearly 150 people have to share each computer in South Asia (6.8 per 1000 population) and about 84 in sub-Saharan Africa (11.9 per 1000 population). The figures for internet use in the 'developing' countries are slightly higher, but show a similar divide between the third (33.1%) of the population in the European Monetary Union who use the internet and only just over one in 70 people (0.14%) in South Asia and just over one in 60 people (0.16%) in sub-Saharan Africa who are internet users (ITU 2003ab).

1.3.4 The Position of Women and Gender Discrimination

There is now considerable research evidence that expanding basic education has a strong positive effect on economic development and that *educating girls* has an even stronger effect (Bellamy 2004). Increasing the number of girls enrolled in primary school generally increases gross domestic product per capita (Bellamy 2004) and countries that have lower education levels for women than men have reduced income, slower growth and increased costs of development (Dollar and Gatti 1999). Educating women also increases the probability their children will go to school and this likelihood increases with the mother's level of education. Research also shows that literacy and language skills gained by girls at school lead to improved health outcomes for them, their children and even their grandchildren (Bellamy 2004). Each year of extra education for mothers in 'developing' countries can reduce their children's under-five mortality by between five and 10% (Herz et al 1991). Improvements in women's education have been responsible for 43% of the reduction in child malnutrition in 'developing' countries in the last 25 years (Smith and

Haddad 2000). The sub-Saharan African countries with the highest levels of schooling for girls, Botswana, Kenya and Zimbabwe, also have the lowest levels of child mortality in this region, despite having greater poverty than some of their neighbours (UNFPA 2000). Education for girls also reduces the likelihood they will be trafficked or exploited as labourers. There is also increasing recognition that reducing poverty requires the empowerment of women (Bellamy 2004).

However there are still *serious inequalities* within as well as between countries in *access to education by women and girls*. While the ratio of female to male enrolments in primary and secondary school in 2001/2 was close to or just over 100% in the high income countries, Latin America, the Caribbean and Central Asia, it was only 81% in South Asia and 91% in the Middle East and North Africa (World Bank 2004). In Afghanistan female primary school enrolment in 1997-2000 was still only 15% and in Djibouti and the Congo only 28% and 32% with corresponding male figures of 58%, 37% and 33% respectively. Globally 65 million girls and 56 million boys are currently not receiving any schooling. 83% of the girls not receiving any schooling live in sub-Saharan Africa, South Asia, East Asia and the Pacific (Bellamy 2004).

Illiteracy rates are still worryingly high and there are significant differences in literacy for men and women with 83% of the male, but only 71% of the female population over 15 years old able to read and write. The lowest female adult literacy rate is 42% in South Asia, followed by just over 50% in the Middle East and Africa. The corresponding male literacy rates are all between two thirds and three quarters. Only the richer countries, Central and Eastern European countries and the republics of the former Soviet Union have female literacy rates approaching 100%.

Although the gender gap in primary school enrolment dropped from 86% to 92% during the 1990s, only 75% of girls complete primary school compared with 95% of boys. The greatest problems are in sub-Saharan Africa, where the number of girls out of school rose from 20 million in 1990 to 24 million in 2002 (Bellamy 2004). This is probably one of the reasons that sub-Saharan Africa is the poorest and most disadvantaged area of the world on most indicators.

Increasing attention to girls' education in the 1990s has resulted from the joint effects of the children's rights and women's movements. However there are still barriers to the schooling of girls, including beliefs about the effect of education on a girl's chances of marriage, worries about the school or the journey to it being unsafe and beliefs that the costs of sending girls to school are not economically justifiable. Some of these barriers can be removed by making schools safer, more accessible and free, increasing employment opportunities for women and providing

A Tanzanian Girl at School

information for parents. For instance abolition of school fees by the Kenyan government in 2003 led to another 1.3 million children enrolling in school, nearly half of whom were girls (Bellamy 2004).

Although *women* who get into *parliament* do not necessarily support policies that benefit women, equal representation of women is important both for improving their status and ensuring that women's issues get onto the agenda. Women are still seriously underrepresented in government, though the proportion of women is increasing. Globally only 15% of members of parliament were female in 2003, compared to 9% in 1987. Surprisingly Rwanda had the highest percentage at 49%, though this does not seem to have affected the position of women in Rwanda significantly, followed by Sweden with 45%. Yemen, the United Arab Emirates, Saudi Arabia and the Kyrgyz Republic had 0% women in parliament in 2003 (World Bank 2004).

The global *labour force* consists of only 40.7% *women*, a slight increase from 39.1% in 1990, though they make up 49.6% of the population. Only 28.6% of the labour force in the Middle East and North Africa and a third (33%) in South Asia is female. The highest percentage of women in the labour force occurs in Central and Eastern Europe and Central Asia (World Bank 2004). The percentage of *unemployed women* is significantly higher than that of unemployed men in a number of countries, for instance four times male unemployment in Egypt, three times male unemployment in Syria and about one and a half times male unemployment in the European Monetary Union, Uruguay and Albania. However it is comparable in some other countries, such as Singapore and Japan and less in a few countries, including Romania, South Africa and the United Kingdom (ILO 2003). It should be noted, as discussed in section 1.3.2, that many employed women are working in part time or temporary jobs and that women's unemployment tends to be underreported. Women also experience more underemployment than men. In a number of countries relatively high percentages of women are working as unpaid family workers. This is generally in addition to the unpaid housework and childcare carried out by the majority of women. For instance, the lower unemployment rates for women than men in Romania disguise the fact that 29.1% of women compared with 10.4% of men are working as unpaid family workers (ILO 2003). Nearly three quarters (73.2%) of women's employment is as unpaid family workers in Bangladesh and about half in Cambodia (53.3%), Panama (50.1%), and Turkey (51.3%). The corresponding percentages for men are significantly smaller at 10.1%, 31.6%, 16.7% and 10.2% respectively (ILO 2003).

Most women worldwide, regardless of race, culture or income, are affected by the experience or threat of violence. *Violence against women* starts at birth, with sex selective abortions and the killing of baby girls, in a number of countries, including China, Taiwan, South Korea, India, Pakistan and some sub-Saharan African countries. An estimated 60-100 million women and girls are 'missing' from the population as a result (Watts and Zimmerman 2002). Up to one in two women has experienced domestic violence, most frequently by an intimate partner, (Engleman et al 2002), one in five women experiences rape or attempted rape during her lifetime (IMAP 2000), and at least one in three women is beaten, coerced into sex or otherwise abused in her lifetime, generally by a member of her family or someone known to her. The World Health Organisation has reported that up to 70% of women who are murdered have been killed by their male partners (AI 2004a). The Council of Europe considers domestic violence to be the major cause of death and disability for women aged 16 to 44 and to account for more deaths and ill health than cancer or traffic accidents. The Russian government estimates that 14,000 women were killed by their partners or relatives in 1999, but has not introduced a specific law on domestic violence (AI 2004a). The majority of women do not report rape and sexual violence due to shame, fear, lack of legal rights and gender inequalities (Engelman et al 2002).

South Africa has the highest incidence of rape in the world, with conservative estimates of a million rapes every year and that one of every two women will be raped during her life time. Rape is now so prevalent that insurance companies are selling rape policies, with benefits including HIV tests, anti-retrivoral drugs, the morning after pill and home security upgrades (Logan 2000). Many girls in South Africa experience sexual violence from teachers and male students and a number stop attending school as a result (HRW 2001).

An estimated 130 million women and girls have been subjected to genital mutilation and another two million girls experience it each year (Engelman et al 2002). There are 16 countries in which over 15% of women between the ages of 15 and 49 have been subjected to genital mutilation and over 90% of women in Egypt, Guinea-Bissau, Mauritania, Moldavia and Sudan have been genitally mutilated (Bellamy 2004). The extent of mutilation varies and there is some evidence that less than 15% (about 19.5 million) of the women who have been genitally mutilated have experienced the most severe type (Toubia 1994; Meniru et al 2000). Genital mutilation can have serious consequences for a woman's physical and mental health and can include painful urination, menstruation and sexual intercourse, urinary retention and a tendency to urinary and genital tract infections and pelvic inflammatory disease, particularly when the most severe procedures are carried out (Meniru et al 2000; Engelman et al 2002).

Girls and women are more likely than boys and men to be sold into slavery and forced prostitution. In 2000 up to 5000 young girls were killed by relatives for 'shaming' their families by having sexual relations or socialising with a person of the opposite sex. In some cases this 'dishonour' was being raped. 80% of suicides in Turkey are women, with similarly high percentages in China, Afghanistan and Iran. Sexual violence in all its forms is preventing women controlling their reproductive health and deciding whether and when to have children. In addition to its value in removing a very direct form of discrimination against women, ending violence against women is likely to lead to positive demographic change, as women are able to make choices about when and with whom to have children (Engelman et al 2002).

1.3.5 Human Rights

The *Universal Declaration of Human Rights* was adapted by the UN General Assembly in 1948. The wide range of rights listed includes the following (though this is not the full list): life, liberty, to be equal in dignity and rights; freedom from slavery, torture and arbitrary arrest; to recognition as a person in law; equality before the law and equal protection from the law; to a nationality; to marry and found a family; to work, free choice of employment and to join a trade union and freedom of thought, conscience, religion, opinion and expression. These rights are independent of 'race, colour, sex, language, religion, political or other opinion, national or social origin, property, birth or other status'. Due to the early date disability, age, sexual orientation and changes in gender status are not listed explicitly, though they could possibly be considered to be included under 'other status'. An amended version was adapted by the Council of Europe in 1971. It has since then been further amended and the most recent version entered into force in November 1998 (WWW5 2005) It has also become part of national legislation, for instance as the Human Rights Act 1998 in the UK.

A *human rights approach to development* is being promoted by the United Nations. It includes the following principles presented in a Statement of Common Understanding (Bellamy 2004):

- All programmes of development cooperation, policies and technical assistance should further the realisation of human rights as laid down in the Universal Declaration of Human Rights and other international human rights instruments.
- Human rights standards and principles from the Universal Declaration of Human Rights and other international human rights instruments should guide all development cooperation and aid programmes in all sectors.
- Human rights should determine the relationship between individuals and groups (rights holders) and state and non-state organisations with obligations (duty bearers). A human rights approach identifies rights holders and their entitlements and works to support their ability to make claims and that of duty bearers to meet their obligations.

Unfortunately *human rights violations* are widespread. Although the numbers of people experiencing serious human rights violations may be greater in some countries than others, there are no parts of the world which are immune from human rights violations.

In 2003 extrajudicial executions (by the official military, police forces or unofficial paramilitaries, often called death squads) were carried out in 47 countries, including France, Germany, the UK, Mexico, Iraq, Angola, Burundi, and Rwanda. People were 'disappeared' by state agents in 28 countries, including Spain, Croatia, Angola, Ethiopia, Colombia and Libya. Torture and ill-treatment were carried out by security forces, police and other state authorities in 132 countries across the world. Prisoners of conscience were held in 44 countries and people were detained without charge or trial in 58 countries (AI 2004b).

The death penalty is still permitted, contrary to the right to life, in 78 countries and in exceptional cases in another 14 countries (AI 2004c). At least 2,756 people were sentenced to death in 63 countries in 2003 and 1,146 prisoners were executed in 28 countries (AI 2004bc). There are almost certainly additional cases not known to Amnesty International, making the true figure higher. Limited records indicate that at least 726 people were executed in China, but the true figure may be closer to 10,000. 108 executions were carried out in Iran, 65 in the USA and 64 in Vietnam. Since 1990 eight countries have executed people who were under 18 years old at the time of the crime, though four of these countries have since raised the minimum age to 18. The USA has executed more child offenders than any other country (19 since 1990), (AI 2004c), but finally outlawed the execution of child offenders in 2005.

The term *refugee* is used for people who are outside the country of their nationality or former habitual residence due to a well-founded fear of persecution on the grounds of race, religion, nationality or membership of a particular social group and unable to return on this account (UNHCR undated). There were about 13.7 million refugees at the end of 2003, a decrease of 920,000 from the start of the year (UNHCR 2004a). Although the refugee population dropped for the second consecutive year, whether the number of refugees continues to fall will depend very much on what the global community does to try and prevent the crisis situations that force people to become refugees and how such situations are resolved when they occur, as well as the measures taken to integrate refugees into host communities. The fact that the average duration of crises involving refugees has increased from nine years in 1993 to 17 years in 2003 (UNHCR 2004b) is not a good sign. In addition to the nearly 14 million refugees there are over one million asylum seekers i.e. people applying for refugee status and 25 million people who have been displaced inside their own countries (AI 2004d).

The industrialised countries are looking for ways to get out of their humanitarian obligations to refugees, in particular by trying to return them to the countries they first travelled or fled to (AI 2004d), setting up lists of 'safe' countries from which claims for asylum are not accepted and returning people prematurely to their country of origin before it is really safe for them to do so. However 60% of refuges at the end of 2003 were hosted by Africa, Central and South-West Asia and the Middle East and only 25% in Europe. The country hosting the largest refugee population was Pakistan, with 12% of refugees, followed by Iran (UNHCR 2004a). A number of countries, including Australia, Denmark and the UK, have introduced legislation which undermines the rights of asylum seekers to a decent standard of living (A1 2004d), as a result of which a number of asylum seekers have become destitute. The United Nations Commission on Human Rights has concern for about 17.1 million people, 57% of whom are refugees. 43% of this population are under the age of 18 and 11% are under the age of five (UNHCR 2004a).

1.3.6 Human Rights, Gender Discrimination and Cultural Relativism

This section and the previous section have discussed some of the social consequences of unsustainable development, including the violence and suffering inflicted by or at the instigation of more powerful groups in society on less powerful groups. This gives rise to the question of 'why?'. A detailed answer to this question could well form the subject of another book and will therefore not be considered here. However, this section will discuss this question in the context of gender discrimination and violence against women with the aim of highlighting some of the important issues.

Violence against women is almost universal (Fischbach and Herbert 1997) and takes place across race, culture, class, nation and income with, nce, only 16 small-scale pre-industrial non-patriarchal societies having no reported domestic violence (Levinson 1989). However, although violence and discrimination against women are universal, the forms they take differ in different cultures and often also change over time. Even within a given society, factors such as race, class, sexuality and disability affect a woman's experiences, her ability to fight for justice and possibly also her interpretation of these experiences.

A number of different examples of violence against women are discussed in section 1.3.4, including domestic violence, rape and attempted rape, forced prostitution and sexual slavery, honour killings and female genital mutilation. Female genital mutilation, also called female genital cutting or circumcision, will be discussed in more detail in this section. This issue has been chosen for more detailed discussion due to the complex issues it raises relating to the human rights and cultural relativism. There has been considerable political discussion over the appropriate term to be used, for instance (Meniru 2000; Meyers 2000; Cook et al 2002).

Although generally associated with the 'developing' countries and Africa, in particular, female genital cutting is also practised in Europe and the USA. Traditional forms of genital mutilation are practice by immigrant women, though prohibitive legislation means that daughters of wealthier families are normally sent abroad for cutting (Beckett and Macey 2001). Forms of genital mutilation *practised by westerners* include cosmetic modifications in the UK and USA (Essen and Johnsdotter 2004) and the reduction of 'ambiguous genitals' in the USA to ensure that girls are able to attain an appropriate 'feminine identity', though the latter procedure is increasingly being contested (Meyers 2000). Clitoridectomy was practised in the late nineteenth and early twentieth century in the USA and parts of Europe to control women's sexuality (Meyers 2000). Other related practices include cosmetic face and body surgery in the industrialised countries, foot binding in China and body modification to give a small waist through very tight lacing and the removal of ribs in England in the nineteenth century.

Female genital mutilation is sometimes seen as a religious obligation, most frequently associated with Islam. However it is not practised in the Islamic states of Iran and Saudi Arabia (Meniru et al 2000) and is practised across a range of cultures and religions. Although a number of Islamic leaders support female genital mutilation, it is not required or sanctioned by the Koran (Sala and Manara 2001; Easton et al 2003) and is probably counter to the tenets of Islam (Muslim Women's League 1999). The *cultural rationales* for female genital mutilation can be classified broadly into control of women's sexual behaviour, gender identity, group cohesion (Meyers 2000) and ensuring marriageability, but there are differences in the specifics and not all these reasons are valid in all areas where genital mutilation is carried out.

The need to control women and their sexuality in particular is the most frequently expressed reason for female genital mutilation. Control or repression of women's sexual behaviour is considered particularly important in societies where family honour is considered to be dependent on female chastity and genital mutilation is considered to be a guarantee of the 'honour', integration and submission of women (Allag et al 2001). However many societies have different standards for the acceptable sexual behaviour of men and women and put much greater stress on female than male chastity. Gender cutting/mutilation for gender identity reasons is also practiced in the USA and is intended to ensure that individuals have appropriate genitals for their gender category (Meyers 2000). There are related issues of the cultural construction of the female body and what is considered aesthetically beautiful, in this case for the appearance of the genitals (Manderson 2004). The procedures involved and their health and other consequences may be more extreme in the more stringent traditional forms of female genital mutilation than in many of the procedures used in the USA and Europe, but the principle of modifying the body to meet male determined standards of beauty is the same.

In a number of societies female genital mutilation is a central component of the rights of passage from child to adult and is considered to contribute to or be essential for maintaining tribal identity (James 1998; Cook et al 2002). The details of how and whether a woman's genitals are cut may be a distinguishing feature of different cultural groups and acceptance of genital cutting is sometimes seen as affirming local traditions against western influence (Meyers 2000). Genital mutilation may be a prerequisite for women to be considered marriageable (Allag et al 2001). In societies where there are few opportunities open to women, the ability to get married may be necessary for (economic) survival. Marriageability is also related to the control of women's sexual behaviour with women who do not meet particular norms of 'chastity' being considered unmarriageable. There is sometimes also a relationship with the 'commodification' of women who are considered as property which can be traded from father to husband as long as certain conditions, such as virginity, are satisfied. Unfortunately, in addition to justifying a practice which impinges on the human rights and often also the health of the affected women, many of these explanations result from and have the effect of maintaining a situation in which women have a subordinate position and in which their dignity, autonomy and independence are compromised (Sala and Manara 2001).

The response of the *medical and legal profession* to the various procedures which modify or mutilate women's genitals and other parts of their bodies has not been consistent. Many 'developing' and industrialised countries have legislation which forbids female genital mutilation, but are more likely to prosecute for the practice of traditional procedures than cosmetic 'modifications' or the modification of 'ambiguous genitals' (Essen and Johnsdotter 2004). However there is some evidence that legislation is not effective and may prevent women presenting themselves for medical care (Meniru et al 2000). Medical licensing authorities and professional associations in a number of countries have forbidden medical practitioners to carry out female genital mutilation to avoid giving it legitimation and recognition (Cook et al 2002), though medical practitioners are clearly required to give appropriate care to women who have experienced female genital mutilation. Similar professional objections are considered to apply in the case of an adult woman who wishes to be reinfibulated (which involves the labia being sewn together to obstruct the vagina except for a small orifice) after having been deinfibulated to give birth, both due to the associated medical harm (Toubia 1994) and because it is a 'medically unnecessary socially contrived procedure' which should not given respectability in this way (Cook et al 2002). However the same arguments have not been applied to purely cosmetic plastic surgery and breast augmentation, although they are also 'medically unnecessary socially contrived procedures'.

Female genital mutilation involves power issues due to race, gender and colonisation and raises questions as to how human rights can be maintained in a local cultural context. There has been considerable discussion of female genital mutilation in the context of universal values and cultural relativism. For instance, people from cultures which practice female genital mutilation often consider criticism by outsiders as cultural imperialism (Steinstra 2000), and believe that the practice is required to ensure inclusion of girls in the society and cannot be understood by outsiders (Sala and Manara 2001). However, it is also important to recognise that culture is not monolithic, but complex, changing and situated in a historical context. This historical context includes women's history, as well as any historical opposition to female genital mutilation and male domination within traditional and other cultures (James 1998). Some of the conflict between women's rights (a universal value) and the local cultural requirements can be reconciled by women from the region who know the local cultural context taking the lead in action against female genital mutilation, as proposed by a number of women's groups in 'developing' countries (Steinstra 2000).

One example of successful action led by women within the existing culture resulted from the *Tostan* ('breakthrough' in the Wolof language) *initiative* to provide non-formal education and literacy programmes for rural Senegalese women based on their own perceptions of problems and learning styles (Easton et al 2003). Follow-up modules on human rights and women's health led to discussion of female genital mutilation and resulted in the village of Malicounda-Bambara making a collective declaration in front of journalists renouncing the practice. Despite the resulting controversy, two other villages followed and the local imam was convinced to support the village. Ten villages in the intermarrying community of Milicounda-Bambara were visited to tell the villagers what Milicounda-Bambara had done and why and to encourage them to tell their own stories. Care was taken to avoid graphic terms or condemn practitioners and to affirm personal relationships. In the end representatives of these ten villages plus the original three met together to make a declaration abandoning genital mutilation. The movement spread to other parts of Senegal and by 2001 over 700 communities had made declarations. The Tostan initiative was spread to Mali and Sudan, where a majority of women rather than about half as in Senegal are genitally mutilated, but in a cut-down version. Although there was some positive reaction and three villages seem to have abandoned the practice, by 2003 there had been no public written declarations. The following conclusions can be drawn from the Tostan experience (Easton et al 2003):

- Treating female genital mutilation as a *social custom* that can be abandoned through local mobilisation under supportive conditions seems to work better than treating it as a scourge to be eradicated.
- As female genital mutilation is based in collective social behaviour, *collective action* tends to be more successful than individual action.

- A *human rights agenda*, with a focus on women's rights and health concerns, seems to give the best framework for local action against the practice.
- *Local empowerment* is important, but local target setting may lead participants to prioritise other issues.
- The initiatives should be driven by women, and also involve men.

Another approach which is based in local traditions is the *'circumcision through words'* rituals created by Kenyan women and their families in which the transition to womanhood is celebrated and cultural teachings are passed on, but no cutting takes place (Meyers 2000). The international organization Women Living under Muslim Laws uses education to encourage women to analyse and reconstruct their own identities within their cultures, to show how women live very differently in different Islamic regimes and show the weakness of the link between Islamic scripture and the requirement for female genital mutilation. This allows women to develop new views of womanhood and to abandon genital cutting without renouncing their religious beliefs (Meyers 2000). Action on female genital mutilation should also be accompanied by wider efforts to improve reproductive health and health services (Snow 2001).

Despite the importance of maintaining cultural (and religious) traditions and the fact that the right to one's own culture can be inferred from the Universal Declaration of Human Rights, practices which violate human rights, or undermine the dignity of or lead to inequality or subordination for particular social groups, such as women, cannot be justified in terms of culture or religion. However change should be promoted within the existing culture, both because this is likely to be more effective and because to do otherwise would itself be a human rights violation. It is also important to examine critically the practices of one's own society, culture or nation and not to assume that human rights violations, injustice, inequality and discrimination are something practiced by other 'barbaric' societies. Unfortunately this is not the case and almost all societies and nations practice some form of human rights violation, injustice, inequality and discrimination, though the severity does vary. Many of the issues discussed in this section can be analysed further using the techniques presented in Chapter 5.

1.4 Social Consequences of Environmental Problems

1.4.1 Introduction

Section 1.4 can be considered to form a bridge between the two sections (1.2 and 1.3) on the social and developmental aspects of sustainable development and the two sections (1.5 and 1.6) on the environmental aspects of sustainable development. The section shows that the social and environmental aspects of sustainable development are not totally separate, but interact with each other. In particular the social consequences of water shortages (section 1.4.2) and pollution (section 1.4.3) and the security impacts of environmental problems (section 1.4.4) are discussed.

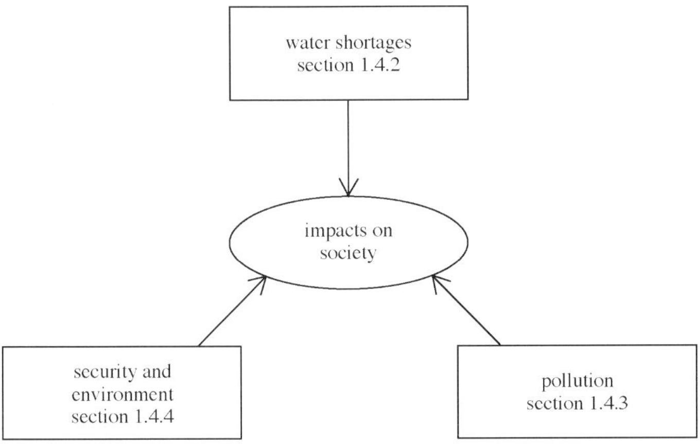

Layout of Section 1.4

1.4.2 Water Shortages

Water shortages and lack of access to clean water and sanitation can have serious consequences for health and development. For instance water born diseases kill an estimated 14,000 – 30,000 people every day (Gore 1992; Gardner 2002). 62% of mortality in sub-Saharan Africa in the early 90s was due to poor drinking water and sanitation (Steer and Wade-Gerry 1994) and half a million infants die in Asia and the Pacific each year from diarrhoeal diseases related to a lack of clean water and sanitation. Half of the world's population lacks adequate sanitation and a fifth lacks access to safe drinking water, statistics which have not improved significantly since the end of the 1980s (Gardiner 2002).

Shortage of sanitation is most acute in sub-Saharan Africa, where just over half the population does not have a safe water supply and 41% lacks adequate sanitation. It is therefore particularly worrying that rural sanitation provision in Africa decreased slightly at the end of the twentieth century and that there have only been minor improvements in urban water supply and sanitation (Gardiner 2002). However progress has been made in South Africa with access to a supply of safe drinking water extended to 8 million of the estimated 14 million people lacking it between 1994 and 2003 and the remaining 6 million people expected to have access by 2008 (Postel and Vickers 2004).

About a third of the world's population lives in regions of moderate to high *water stress* and it is estimated that two thirds will live in water stressed conditions by 2025 (WBGU 1999; UNEP 1999; Postel 2000). Most water stressed populations will be in Africa and South Asia, which are already experiencing extreme poverty and hunger. Currently 34 countries in Africa, Asia and the Middle East are classified as water stressed (Postel 2000). The Middle East and North Africa have only 1% of the world's freshwater flows for nearly 5% of the world's population and South Asia has 4% of freshwater flows for nearly 23% of the world's population (WRI 2000 2003).

Water is a fixed resource, but demand is increasing, despite reductions in some countries, for instance of 10% in the USA in the period 1980-1995. This is due to a reduction in water intensive industries, but possibly also to the adoption of water efficiency standards (Gardner 2002). The use of highly efficient drop irrigation technologies has also reduced water use in some countries (Gleick 2000). In large cities total municipal and industrial uses of water have grown by a factor of 24 in the last century and inefficient practices are leading to excessive water use in North America and much of Europe (Gardner 2002). The USA still has one of the highest per capita water usage rates in the world of 1932 cubic metres per person per year, despite irrigating only 11% of its cropland. This is approximately three times the water consumption per person in France and 46 times that in Ethiopia (only 42 cubic metres per person per year) (Postel and Vickers 2004). Annual demand for water for household and industrial uses in the 'developing' countries is projected to increase by 590 billion cubic metres between 1990 and 2020, equivalent to the flow of seven Nile rivers, and to increase from 13% to 27% of total water use (Postel 2000).

About 70% of water usage and up to 90% in developing countries is in agriculture (Postel and Vickers 2004). Therefore water scarcity can adversely affect development by reducing food supplies. About 40% of the world's food comes from the 17% of crop land that is irrigated, half of it in India, China, the USA and Pakistan. Although 60% of the *irrigation base* is less than 50 years old, its continuing productivity is already threatened, with a fifth of irrigated land damaged by salt, increasing numbers of rivers running dry for part of the year, and pressures to reduce water withdrawals due to damage to fisheries and aquatic habitats from excessive water use. However the greatest threat is the depletion of underground aquifers, as water scarce countries are increasingly tapping into ground water reserves to maintain or expand agricultural production. Nearly 10% of the world's grain harvest uses water that is pumped faster from wells than it can be replenished. The continuing availability of this part of the harvest and the avoidance of adverse consequences to the nations that consume it will depend on alternative sources of water being found (Postel 1999). This problem is widespread, including in the Middle East, much of the Western USA and Central and Northern China. Nine Indian states have growing water deficits totalling just over 100 billion cubic metres a year and China, the world's largest producer of grain, of 30 billion cubic metres a year (Postel 2000). Surface water irrigation efficiency is low in many countries, ranging from between 25 and 40% in India, Mexico and Pakistan to between 50 and 60% in Israel, Japan and Taiwan. Improvements in irrigation efficiency have been low in most regions due to the low cost of water for irrigation and the frequent lack of regulation (Postel and Vickers 2004).

Pressures from *competing water uses* may also reduce the water available for irrigation, with a consequent reduction in productive land and jobs, particularly for poor farm labourers. For instance, in China there will be a transfer of water from agriculture to industry as supplies drop, since industrial uses generate more jobs and 70 times more economic value. (Postel 2000). Cities currently are responsible for less than 10% of the world's freshwater withdrawals. However the majority of the world's cities with populations of at least 10 million people are in regions with mild to severe water stress, with demand outstripping available supplies. Increasing urban water demand will increase the pressure to transfer water to urban uses. Urban losses due to leakage and other factors are high and most urban water managers cannot account for 15-40% of their supplies. Taiwan loses nearly 2 million cubic metres of water every day due to leakages. Water losses in Spain are 24-34%, Albania up to 75%, Johannesburg 42% and Jordan 48%. However efficiency is possible and Singapore has managed to reduce losses nationwide to 5% and Copenhagen to 3% (Postel and Vickers 2004).

The other option followed in water scarce areas of increasing food imports also has serious disadvantages, as it creates dependency and indebtedness, reduces opportunities for other imports (Postel 2000) and has environmental costs due to the energy and emissions associated with

Fig. 1.2 Treadle pump in Africa

transporting food products. Thus more sustainable solutions will be required. These could include drip tank irrigation in combination with soil monitoring. There is also a need for making small scale irrigation schemes available to the tens of millions of the world's poorest farmers who do not have access to irrigation, as the technologies are too large for their small plots and too expensive. A successful example of small scale irrigation is the purchase of more than 1.2 million human powered treadle pumps (see fig. 1.2) in Bangladesh, where they have been used to increase the productivity of 2.5 thousand square kilometres of farmland (Postel 2000).

1.4.3 Pollution and Development

Another environmental problem with serious social consequences is pollution, with over a fifth of the world's population exposed to dangerous levels (Steer and Wade-Gerry 1994). *Pollution* has been defined as an 'undesirable state of the natural environment being contaminated with harmful substances as a consequence of human activities' (WWW62005) and is discussed further in section 1.5. Although the health effects and increased mortality due to pollution are experienced to a much greater extent in the 'developing' countries, its effects are also felt in the industrialised countries. For instance the life expectancy of workers in the oil, chemical and nuclear industries in the USA is 10 years less than the USA average (Mazzocchi 1990). Occupational exposure is also responsible for hundreds of thousands of cancer deaths each year (Stead and Stead 1996). Work related injuries and diseases are responsible for the deaths of over one million people each year and there are at least 250 million accidents and 160 million new cases of work related diseases each year (Brundtland 2002).

Sustainable development involves consideration of many different factors and sometimes requires tradeoffs to be made between them. For instance cash crops, other than cotton and groundnuts, may be beneficial to the environment, as they reduce the rate of soil erosion compared to food crops such as cereals and root crops (Repetto 1988). However they are highly damaging to development, as they provide little income to women, since cash cropping is largely a male activity in many countries (Brundtland 2002). Consequently increasing cash cropping is leading to a decline in the nutrition of women and children (Kennedy and Oniang'o 1990), as women generally devote their earned income to household food and other needs, whereas men spend up to a quarter of it on non-household expenses (Gardner and Halweil 2000).

1.4.4 Security and Environment

Global insecurity and high expenditure on armaments are problems of a lack of sustainability, both because of the devastating social and environmental consequences of war and the fact that arms spending diverts resources from development and environmental protection. However the link between the environment and security has only been recognised in the last 10 to 15 years (Brusaco-Mackenzie 2002). European historical and recent data show a relationship between occurrences of war and the strength of armed forces and between the total number of military personnel and the number of war casualties.

Military expenditure increased by 11% to 956 billion dollars from 2002 to 2003, but there was a significant reduction in the arms trade, with arms deliveries dropping from 42 billion dollars in 2002 to less than 29 billion in 2003. The high income countries have three quarters of military spending, but only 16% of the world's population. Nearly half of global military spending was in the USA, which was largely responsible for the 11% increase in global military spending (SIPRI 2004). Global military expenditure reached its peak in 1987 as a result of the enormous cold war military build-up and then declined in real terms from 1988-1996, but increased in 1997 for the first time since the end of the cold war. A sharp cut in Russian military expenditure led to a further decline in 1998. In real terms military expenditure in 2000 was about 40% lower than in 1987, but only 15% lower than in 1975 (SIPRI 2001). Unfortunately this seems to have set a trend of continuing military expenditures, with increases of 7% between 1998 and 2001 (SIPRI 2002) and further increases of 6.5% and 11% respectively in 2001-2 and 2002-3 (SIPRI 2004). These increases are at least partly motivated by the so-called 'war on terror' and possibly also by USA plans to extend its sphere of influence and economic activity, as indicated by the invasion of Iraq. This increase in global military spending raises serious concerns.

Although the high income countries are responsible for 75% of arms expenditure, it is the 'developing' countries which are the main purchasers of foreign arms, with nearly two thirds (63.9%) of arms transfer agreements in the period 1996-2003 and 53.1% of arms deliveries, a drop from 66.9% in the period 1996-9. Arms deliveries to developing nations were valued at 17 billion dollars in 2003. The USA has dominated arms sales with the 'developing' world in terms of both agreements and actual deliveries, with 6.2 billion dollars in transfer agreements and 6.3 billion dollars in deliveries in 2003. In 2003 Russia was the second arms trading nation with regard to agreements, valued at 3.9 billion dollars, and the United Kingdom the second nation with regard to deliveries, valued at four billion dollars. In the period 2000-2003 China made the most arms purchases of 9.3 billion dollars of arms transfer agreements, followed by the United Arab Emirates with 8.1 billion dollars worth (Grimmett 2004).

Arms races have reduced in importance, but spending on peace research and social programmes remains very small compared with military expenditure. Consequently it seems that many societies are geared towards war rather than peace. The unrestricted production and trade in arms, estimated as $3 trillion (in 1990 dollars) from 1945 onwards have contributed to extending conflicts. Although a number of arms control agreements have been made, they still allow the build-up of arms and the development of new weapons systems. Only a relatively small fraction of the existing stockpiles of nuclear weapons would be required to annihilate all life on earth.

Statistics on the number of *armed conflicts* vary slightly, according to the different definitions of the term armed conflict. However there is general agreement that the majority of conflicts are no longer between independent states, but internal to one state. In a number of cases, such as the conflicts in Chechnya and Kurdistan of Iraq, where independence is the main aim of one of the parties to the conflict, this party presumably considers that it should be treated as an independent state.

There were 36 armed conflicts in 28 countries in 2004, with armed conflict defined as a political conflict involving armed combat in which at least one thousand people have been killed, but including (non-politically motivated) resource conflicts and conflicts leading to state failure. Five states experienced more than one conflict in 2003. This in a downward trend from the peak of 44 conflicts in 1995. 31 of the conflicts are more than two years old, nearly two thirds (23) have been fought for more than 10 years and eight for over 25 years. The USA-led invasion of Iraq was the sole war between independent states, the USA and its coalition partners against Iraq. All the other conflicts were internal wars, involving government troops and armed opposition, insurgent or other groups (Project Ploughshares 2004), though, as indicated above, some of these groups were fighting for their right to be independent states. In 2003 more than four fifths (84%) of the conflicts were in Africa and Asia and more than a fifth (21%) of the countries involved in conflict were in the Middle East, though this figure has declined from 29% in 2002 (Project Ploughshares 2004). Mortality due to war or conflict, including non-combat civilian mortality which could be attributed directly to war or conflict, has been estimated as 50-51 million deaths for the period 1945-2000 and 130-142 million deaths for the whole of the twentieth century. The second figure includes deaths in German concentration camps (Leitenberg 2001). Mortality due to national political decision (Hobsbawm 1996) i.e. including causes such as genocide, starvation and deaths in prison camps as well as conflict, have been estimated at 214-226 million deaths for the twentieth century (Leitenberg 2001).

The combination of population growth, resource depletion and environmental degradation are important factors which can lead to conflict. Legal or illegal resource exploitation contributed to a quarter of the armed conflicts in 2000, whether as a trigger, exacerbating factor or by financing their continuation. Many resource related conflicts are being fought in areas of great environmental value. In 'developing' countries the benefits of resource exploitation, such as mining and logging, generally accrue to a small business or government elite and foreign investors and the costs, including environmental devastation, expropriation of land and disruption of traditional ways of life, are paid by local populations. This has led to violent conflict, for instance in Nigeria's Niger delta and some of the provinces in Indonesia, involving small scale skirmishes, sabotage and major human rights violations (Renner 2002).

In *resource based conflicts* much of the violence is directed against civilians to try to intimidate or drive out local populations in order to obtain access to resources (Renner 2002). This may include making an area uninhabitable by the indiscriminate spread of landmines, shelling houses and hospitals, chopping off limbs and systematic sexual violence. Young boys are often turned into child soldiers and young girls into sex slaves for the fighters and fighters may be forced to commit atrocities, often against their own relatives, to traumatise them and prevent them being accepted back into their communities (Kaldor 1999).

The availability and easy plundering of natural resources are a requirement for resource based conflicts, but not sufficient in themselves. Political, social, economic and military factors are also important. These include poor governance, the rise of private security forces and the proliferation of small arms and light weapons. The demand for consumer products and commodities fuels resource wars by making illegal resource exploitation lucrative. The globalisation of trade has made it very easy for warring groups to access key markets. Major international companies are contributing to perpetuating resource based conflicts by buying hot commodities such as conflict diamonds, providing revenues to governments at war and operating in countries with repressive or illegitimate governments (Renner 2002).

Environmental problems which can cause conflict include *overexploitation* of water resources, soil erosion, desertification and deforestation. There are conflicts or potential conflicts in a number of regions over water use. For instance water scarcity in the Nile Basin is causing dehydration, disease, hunger and the potential for conflict. The Nile passes through nine nations, but only two of them, Egypt and the Sudan, have made an agreement about its management (Brusaco-Mackenzie 2002). The threat of famine, generally resulting from inadequate matching of food supplies with local needs rather than an inadequate total food supply, often leads to population displacement and can affect the whole region. It has been suggested that food security may replace military security as the principal concern of many governments (Brown and Flavin 1999).

In many cases environmental factors interact with political, social and economic ones to increase instability and create a situation of potential conflict. However the environment has generally been considered a 'soft' subject in foreign policy on which agreement is possible even in circumstances of political tension or war. For instance the Geneva Convention on Long Range Transboundary Air Pollution (1979) was negotiated and came into force after ratification by all signatories during the Cold War (Brusaco-Mackenzie 2002).

1.5 Pollution

1.5.1 Introduction

This is the first of two sections on the environmental aspects of sustainable development and, in particular, air and water pollution. The earth can be considered to consist of a number of ecosystem*s*, which have a certain capacity for regeneration. For instance the atmosphere's composition is able to regenerate itself, but the rate of regeneration depends on the current state of the atmosphere, the rate pollutants are deposited and the type of pollutants (Dasgupta 1995). However problems can occur when the stresses on these ecosystems increase, for instance through high levels of pollution. Many environmental problems involve the degradation of renewable natural resources or their use at rates greater than they can be renewed. Many of the impacts of pollution are long term and wide ranging. Two particularly serious problems, which will be discussed in section 1.5.2, are global climate change due to anthropogenic (human generated) emissions of carbon dioxide and depletion of the stratospheric ozone level mainly due to emissions of chlorofluorocarbons (CFCs).

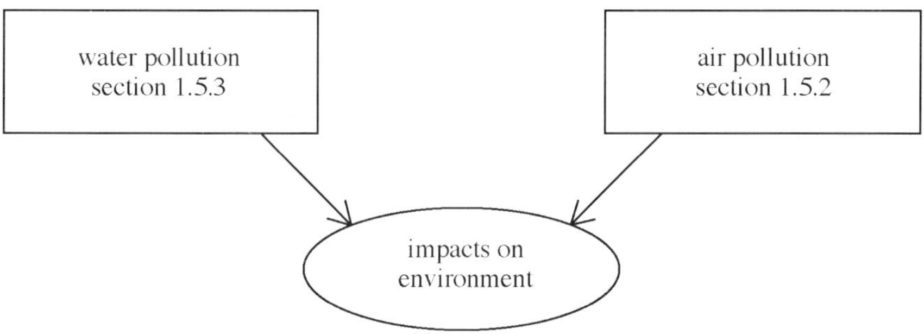

Layout of Section 1.5

1.5.2 Air Pollution

Environmental threats include air, water and soil pollution. There are a number of different types of *air pollutants*, including suspended particulate matter (SPM), lead, sulphur dioxide, carbon monoxide and nitrogen oxides. Sources of *SPM* include incomplete fuel combustion and vehicle exhaust gases, particularly from diesel engines. Health effects include increased incidence of respiratory diseases, such as asthma, bronchitis and emphysema, and increased mortality among children and elderly people.

Fine particulates can also carry heavy metals, many of which are poisonous and/or carcinogenic, into the lungs. High concentrations of *lead* affect the circulatory, reproductive, nervous and kidney systems and may cause hyperactivity and reduced learning ability in children. Although a long and increasing list of countries have banned leaded fuel and 80% of petrol (gasoline) sold is unleaded (OECD and UNEP 1999), vehicular emissions are still the main source of airborne lead. However their contribution has decreased from 90% (Brown 1990) to 74% in the mid-90s. Lead use in paint, batteries and other sources had been reduced and waste incineration and waste-water treatment technologies have been improved, leading to a reduction of nearly two thirds (64%) in total global atmospheric emissions of lead between 1983 and the mid-90s. Despite this reduction, the average person has 500-1000 times as much lead in their body as their pre-industrial revolution ancestors (McGinn 2002). Mercury is another highly toxic metal. The main sources of mercury are coal burning and solid waste disposal. Half of the world's annual mercury emissions from human activities come from Asia, largely because China and India burn about one third of the world's coal. Inorganic mercury is poorly absorbed, but the organic form methyl mercury is both very toxic and easily absorbed. Bacteria in polluted waters readily convert inorganic to organic mercury. A seventh of a teaspoon of mercury is sufficient to contaminate a one kilometre square lake for a year and about two thousand metric tons are emitted each year (McGinn 2002).

Local health risks due to *sulphur dioxide* include increased incidence of respiratory diseases, but are probably less serious than those due to lead and SPM. At high concentrations sulphur dioxide may also increase the death rates of people with heart and lung diseases (Conservation Foundation 1992). *Nitrogen dioxide* can harm the respiratory system and increase susceptibility to viral infections. Sources include motor vehicles, the burning of fossil fuels with a high sulphur content, nonferrous ore smelting, petroleum refining and power plants (Grossman 1994; Kane 1995). Nitrogen oxides and sulphur dioxide emissions are easily converted in the atmosphere into sulphuric and nitric acids. These acids mix readily with water in the air and have led to the *acidification* of lakes and streams, a reduction in the diversity and number of fish in some areas, damage to trees, crops and buildings and forest destruction (Graedel and Crutzen 1989; Kane 1994). Emissions of both sulphur and nitrogen oxides have more than doubled since 1950 (Kane 1994), but started levelling off in the 1990s. However this may be countered by growth in China, where low quality coal powers nearly three quarters of the economy (Kane 1995; Sawin 2004).

Truck Emitting Diesel Soot

Smog is produced when traces gases in the lower atmosphere, of which ozone is the most abundant, react with solar radiation to produce potentially dangerous reactive gases (Graedel and Crutzen 1989). Smog is a major problem in many urban areas in all countries, but is particularly severe in the 'developing' countries where 1.2 billion people are exposed to excessive low atmosphere air pollution (Steer and Wade-Gerry 1994). It can cause eye and lung problems and damage to foliage and crops. It has also reduced visibility in many national parks in the USA (Stead and Stead 1996). The burning of fossil fuels by vehicles, industries and power stations, the burning of agricultural wastes, forest fires and the emissions of millions of inefficient bio-fuel cookers have resulted in a giant and growing cloud of brown haze across much of Asia (Greenpeace 2002).

Many of the most *serious consequences* of *air pollution* are experienced *globally* rather than locally and after a time span of several decades, rather than immediately. These effects include global and regional climate change due to greenhouse gases and acid deposition due to emissions of sulphur and nitrogen dioxide (Grossman 1994). The *greenhouse effect* is due to gases which allow visible light radiation to reach the earth's surface, but do not let infrared heat radiation emitted by the earth's surface pass out. There is a natural greenhouse effect, which maintains the earth's surface at temperatures which can support life. It has only relatively recently been generally recognised that *anthropogenic* (due to human activity) gas emissions are responsible for a further greenhouse effect. This is due to emissions of additional greenhouse gases, such as water vapour, carbon dioxide, ozone, methane, nitrous oxide and chlorofluorocarbons, as well as (photo)chemically active gases such as carbon monoxide, nitrogen oxides and sulphur dioxide, and aerosol emissions. The concentrations of both naturally occurring and anthropogenic greenhouse gases in the air are very small compared with the concentrations of oxygen and nitrogen (Krause et al 1995).

Carbon dioxide is both the most prominent trace gas and the one whose concentration is increasing the most rapidly (Roodman 1995). Globally emissions rose by 1.7 billion metric tons from just under 21.3 billion metric tons in 1990 to just under 23 billion metric tons in 2000 (World Bank 2004). The increase is due to a mixture of industrial activity and the loss of vegetation, including through deforestation. The high income countries produce just over half (51.3%) of emissions, though they house less than 16% of the world's population. The *relative difference in emissions* is even greater on a per capita basis, with average emissions in 2000 of 12.4 metric tons of carbon dioxide per person per annum in the high income countries and only 0.7 metric tons per person in South Asia and 0.8 metric tons per person in sub-Saharan Africa (World Bank 2004).

The oil producing countries have the highest per capita emissions of carbon dioxide, of 21.9 metric tons per person in Kuwait followed by 21 metric tons per person in the United Arab Emirates. This is closely followed by the USA with annual emissions of 19.8 metric tons per person. The USA is the country with the highest total emissions of just over 5.6 billion metric tons annually, which is more than a quarter of the world's total output of carbon dioxide emissions. The next largest generator of emissions is China, with a total figure of just under 2.8 billion metric tons per year. However per capita emissions in China are only about a tenth of those in the USA or Kuwait, at 2.2 metric tons per person (World Bank 2004).

Emissions in the third largest producer, Russia, dropped by just over a quarter (27.7%) from just under 2 billion metric tons in 1990 to just over 1.4 billion metric tons in 2000. Unfortunately this probably has much more to do with the collapse of industry following the demise of the Soviet Union than a 'greening' of industrial and other activity. The lowest figure is 0.0 metric tons per capita in Burundi and Cambodia, with several other countries in sub-Saharan Africa having emissions of 0.1-0.2 metric tons per capita (World Bank 2004). Unfortunately this is an indication of poverty rather than the use of very clean energy. The fact that emissions in the whole of sub-Saharan Africa remained approximately stable, rising seven million metric tons to 478.8 metric tons in the 1990s (World Bank 2004) is unfortunately more a sign of lack of development than the greening of activities. Emissions in Latin America and the Caribbean, the Middle East and North Africa and South Asia are rising at a much faster percentage rate than in the rest of the world (41% , 63% and 59% between 1990 and 2000). Current emissions of each of these three areas are less than half a billion metric tons and therefore comparable with the growth in emissions in the second largest emitter, China, at nearly 0.4 billion metric tons and considerably less than the growth in emissions (0.79 billion metrics tons) of the largest emitter, the USA (World Bank 2004). Clearly any (significant) growth in emissions is worrying. A range of measures, including increased energy efficiency, energy conservation and the increasing use of clean renewable sources of energy will be required in both the 'developing' and industrialised countries to allow social and economic development in the 'developing' countries without further stressing the natural environment.

There is now *broad scientific consensus* that anthropogenic emissions of 'greenhouse' gases have led to climate change and associated consequences, such as sea level rise. The *UN Framework Convention on Climate Change* was signed at the 1992 Earth Summit and came into force in March 1994, with the objective of stabilising atmospheric concentrations of greenhouse gases at levels that will avoid 'dangerous anthropogenic interference with global climate', but not prevent economic development. The 1997 *Kyoto Protocol to the UN Framework Convention on Climate Change* committed industrial and former Eastern bloc nations (called Annex B nations) to collectively reducing their greenhouse gas emissions by 5.2% below 1990 levels during 2008-12, with specific targets for each country. This Protocol became binding on its 128 parties, including Russia, but not the USA, which has withdrawn, in February 2005. However only a few countries are on course to meet their Kyoto commitments (Dunn and Flavin 2002). Although opinions differ on the extent of emissions reductions that will be required to prevent further climate change, it seems likely that much more stringent reductions than those in the Kyoto Protocol will be required. One of the problems is that atmospheric levels of carbon dioxide are cumulative and therefore a reduction in emissions does not immediately translate into a reduction in atmospheric concentration.

The Intergovernmental Panel on Climate Change (IPCC) has concluded that *globally averaged surface temperature increased* by 0.6 ± 0.2 °C over the twentieth century and that for the range of scenarios considered the globally averaged surface air temperature is projected to increase by 1.4 to 5.8 °C by 2100 compared to 1900, with greater variation in the higher than the lower latitudes. *Globally averaged sea levels* are predicted to *rise* 0.09 to 0.88 metres by 2100 (Ahmad et al 2001). Other projections give a much

Evidence of global warming drowned here!

higher sea level rise of 0.5 to 1.5 metres in the next few decades and possibly several metres in the longer term (Krause et al 1995). Even a moderate sea level rise would threaten the coastal settlements in which half the world's population lives. The extent of climate warming and changes in precipitation will also vary by region and depend on future greenhouse gas concentrations. Both global temperatures and carbon dioxide levels have risen significantly since 1800 and the ten warmest years ever recorded occurred since 1980 (Roodman 1995). Many physical and biological systems have already been affected by regional climate changes, including shrinkage of glaciers, thawing of permafrost, (Ahmad et al 2001), lengthening of the mid to high latitude growing season, declines of some plant and animal populations and earlier flowering of trees. The area of oceans with water temperatures greater than 26°C has increased by a sixth in the last 20 years and sea levels are believed to be the highest in 5000 years (Linden 1994).

Although there may be beneficial effects, such as increasing agricultural productivity at high latitudes in some regions, the *net impacts of climate change* in most parts of the world will be *negative*. These impacts will be felt over varying time scales of decades for the restoration of slightly disturbed ecosystems to many centuries for equilibration of the climate system and sea level. Thus short term human activities will lead to a chain of events with long term consequences, that cannot easily be reversed, if at all. The nature of the impacts will also be affected by the rate of climate change and there are considerable uncertainties in the assessment (McCarthy et al 2001).

Modelling and other studies by the IPCC have given the following (amongst other) *projected adverse impacts* (Ahmad et al 2001):

- A general reduction in potential crop yields in most tropical and sub-tropical regions for most projected increases in temperature.

- A general reduction, with some variation, in potential crop yields in mid-latitude regions for increases in average annual temperature of more than a few degrees centigrade.
- Decreased water availability in many water scarce regions, particularly in the sub-tropics.
- An increase in the number of people exposed to water borne diseases, such as cholera, vector borne diseases, such as malaria, and deaths from heat stress.
- A widespread increase in the risk of flooding from increased heavy rainfalls and sea level rises.
- Increased energy demand for space cooling due to higher temperatures in the summer.

The resulting warmer oceans will lead to increased evaporation of moisture and a more humid and wetter climate, but a reduction in precipitation in the mid-latitude continental regions of North America and Eurasia. Extreme weather conditions with storms, floods and avalanches are likely to become more frequent. Warming of several degrees could lead to a major redistribution of cropping zones and changes in farming practices. Heat stress could adversely effect the world's rice growing regions with disastrous effects on rural subsistence farmers. Global warming would allow forest growth at higher altitudes and latitudes, but will probably occur faster than forests can migrate, causing a rapid dieback of existing forests. Climate change could affect both mid-latitude and tropical forests, where most of the world's biological species are found. Uncertainty in the likely effects and the time span over which they will occur makes planning difficult (Krause et al 1995).

The stratospheric (upper atmosphere) ozone layer protects the earth from the sun's ultraviolet radiation. Stratospheric *ozone depletion* is mainly a result of *emissions of chlorofluorocarbons* (CFCs) used in air conditioning, refrigerants, propellants and solvents. Ozone depletion is likely to produce an increase in skin cancer, a reduction in immune system efficiency and potentially severe effects on the food chain (Graedel and Crutzen 1989). It has taken a long time for the effects of CFCs to be established and, though they have now been banned, the development of substitutes without environmental impacts has not been immediate. This illustrates the problems in finding technological solutions even when effort is directed to doing so (Braun 1995). The tendency has been to replace CFCs with related compounds, such as hydrochlorofluorocarbons (HCFCs), with some but considerably less ozone destroying potential, or hydrofluorocarbons (HFCs), which contribute to global warming. This is partly a consequence of the fact that the different environmental conventions act separately, sometimes with counterproductive results (French 2002). The ban on producing and importing all HCFCs with the exception of those used as refrigerants in equipment manufactured before 2020 will not now come in until 2015. This exemption, which covers an estimated 10% of HCFCs will be removed in 2030 and there was an earlier date for prohibition on producing and importing HCFC-14b, which contributes to about a third of HCFC use (WWW72005).

1.5.3 Water Pollution

There are three main types of *pollutants in freshwater*:

- *Pathogens in raw sewage*, which cause a variety of debilitating and sometimes fatal diseases.
- *Heavy metals and synthetic organic* compounds from industry, mining and agriculture, which contaminate drinking water and accumulate in aquatic organisms.
- *Excessive nutrients* in sewage, agricultural runoff and industrial discharge, which cause high rates of algae and bacterial growth and eventually deplete water oxygen content.

Despite the dumping of toxic chemicals and other waste products, water resources in the industrialised countries have improved over the last 30 years. This has largely been as a result of public concern for safe water and the loud outcry over dumping incidents (Stead and Stead 1996). However a very high percentage of *raw sewage* is still being discharged without adequate treatment in the 'developing' countries, particularly in rural areas where only a just over a third (36.8%) of the population has access to improved sanitation (WRI 2004). Physical, biological and advanced sewage treatment can dramatically reduce the concentration of pathogens, but many of the 'developing' countries do not have the resources to do this. Human pathogens can be totally removed by adding chlorine to treated water, but the chlorine reacts with organic chemicals to form carcinogenic chlorinated hydrocarbons, and therefore the procedure is controversial (Grossman 1994).

Heavy metals and *toxic chemicals* discharged by industry, mining and agriculture are difficult to remove from drinking water and accumulate in fish and shellfish. They also accumulate in the bottom sediment which is released slowly over time. The following are examples of some of the health and environmental effects:

- Lead causes convulsions, kidney and brain damage, anaemia, cancer and birth defects.

- Nickel causes gastrointestinal and central nervous system damage and cancer.
- Arsenic induces vomiting, damage to the liver and kidneys and cancer.
- Mercury contributes to irritability, depression, kidney and liver damage and birth defects.

Sources of *heavy metal pollution* include smelting and heavy metal processing and the discharges of compounds used in paints, plastics, batteries and tanning and pulp and paper plants. In the 'developing' countries mining rather than industry is the main source of heavy metal discharges. Contamination of river and lake water by sewage or industrial discharges has increased the concentration of organic carbon that can be used by bacteria. This increases their demand for dissolved oxygen and reduces the dissolved oxygen available for fish. The oxygen available to the fish population has also been reduced by algal growth resulting from excess nitrogen and phosphorus and the decay of dying algae (Grossman 1994).

The largest quantity of *organic water pollutants* is produced by China, 6.2 million kilograms per day in 2000, an amount which has almost doubled since 1980. However the fact that Chinese emissions have decreased by over 1.4 million kilograms a day since 1995 is encouraging. The other big polluters are the USA (nearly 2 million kilograms per day), India (nearly 1.6 million kilograms per day), Japan (just over 1.3 million kilograms per day) and Russia (nearly 1.5 million kilograms per day) (WRI 2004).

1.6 Resource Depletion and Loss of Biodiversity

1.6.1 Introduction

This is the second of the two sections on the environmental aspects of sustainable development. Three main issues are covered: resource depletion, waste generation and loss of biodiversity, including through habitat destruction. Particular problems are likely to result from high energy consumption and the associated emissions of the greenhouse gas carbon dioxide (section 1.6.3). In addition to its value in its own right, biodiversity is important for maintaining the security of food supplies (section 1.6.4). An important cause is habitat loss, including extensive deforestation, particularly in the tropics, and loss of wetlands (section 1.6.4).

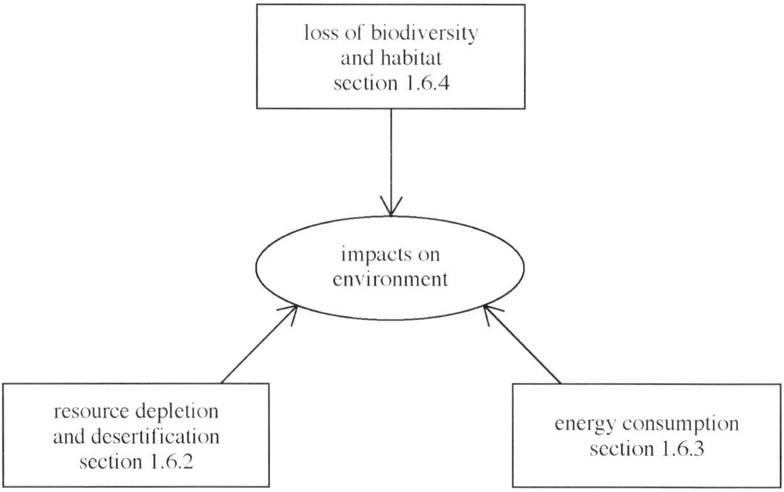

Layout of Section 1.6

1.6.2 Resource Depletion and Desertification

Natural resources are the basic source of material, financial and psychological wealth, but are being very poorly protected and have high rates of depletion. Environmental problems in the industrialised countries are increasingly a consequence of *consumption* rather than production patterns, since there has been some degree of regulation of industry (Michaelis 2002). It has been estimated that critical natural resources would last less than a decade if by 2030 average global consumption reached the level of average USA consumption in the late 1980s (Frosch and Gallopoulos 1989). Increased consumption is resulting from a combination of reduced production costs and increases in average global income, of nearly a third between 1970 and 1995.

Desertification is probably the most widely recognised form of resource degradation and may affect as much as one third of the world. Causal factors include the removal of trees, badly managed irrigation systems, overuse of sensitive environments and climate fluctuations. Attempts to reverse desertification have generally been unsuccessful and there are also similar patterns of environmental degradation in areas with high rainfall which are thousands of kilometres away from deserts (Wilson 1989). Reserves of topsoil and cropland are also being consumed at unsustainable rates as a result of overcultivation, salination, wind, chemical fertilisers and urbanisation (Brown 1995). As discussed in section 1.4.2, another important and increasingly scarce resource is water.

The other side of increasing consumption and resource depletion is the problem of *waste generation*, since waste reduction at source, reuse and recycling are only taking place to a limited extent. Finding new landfill sites is a major problem in the USA. In addition there is the issue of disposing of *hazardous wastes*, two thirds of which are produced by the chemical industry (Stead and Stead 1996). Other significant hazardous wastes include heavy metals and petroleum refinery, medical and nuclear wastes. Worldwide 300-500 million tons of hazardous wastes are generated each year, with 80-90% produced by the industrialised countries. The Basel Convention on the Control of Transboundary Movement of Hazardous Wastes and their Disposal and its 1995 amendments prevent the movement of hazardous wastes from richer to poorer countries and could limit the flow of toxic chemicals and wastes. Although the ban was passed by consensus and is universally supported by the 'developing' countries, some of the industrialised countries, in particular the USA, are opposing ratification (McGinn 2002).

Nuclear waste presents a particular problem, due to extreme toxicity, though it is produced in considerably smaller volumes than chemical wastes. Nuclear waste will remain dangerously toxic for thousands of years and none of the 25 or so countries producing nuclear waste have yet found an acceptable solution for its safe treatment and long term disposal (Gore 1992). Nuclear power has sometimes been presented as the solution to energy generation without global warming. However the (currently) intractable problem of treating the resulting waste and public concern about the associated health and environmental hazards mean that it is not the answer and that replacing coal fired power stations with nuclear power would purely replace one set of problems with another.

1.6.3 Energy Consumption

Energy consumption is increasing, due largely to increasing car and electricity use, which have resulted in a 50% increase in *carbon dioxide emissions* between 1971 and 1995. *Road and air traffic* have grown rapidly, with the number of motor vehicles increasing by a factor of 2.6 between 1970 and 1995 and air traffic growing by a factor of six in this period (Michaelis 2002). Air traffic has been expanding at nearly two and half times average economic growth rates from 1960 and is expected to virtually double between 2000 and 2015. Air travel is the world's fastest growing source of greenhouse gases and generates more than 600 million metric tons of carbon dioxide annually, which is close to the carbon dioxide produced by all human activities in Africa. For journeys under 500 km air travel produces slightly more carbon dioxide per passenger (0.17 kg/km) than car travel (0.14 kg/km) and three times as much as rail travel (FOE undated).

Just over half of the world's primary energy supplies are consumed in the OECD countries (Michaelis 2002). The richest fifth of the world's population is responsible for 58% of total energy consumption, whereas the poorest fifth uses less than 4% of the world's energy. Electricity consumption per capita more than doubled between 1980 and 1985 in the 'developing' countries, but still remained at only a tenth of that in the industrialised countries (Michaelis 2002). There are about 2 billion people who have *no access to 'modern' fuels or electricity* and another two billion who cannot afford refrigeration or hot water (Maier 2002). About 75% of energy consumption worldwide and 90% in the industrialised nations comes from *non-renewable fossil fuels* (Michaelis 2002), which are being exhausted very fast. The cost of oil production has doubled every 15 years (Commoner 1990), as sources become less accessible and more costly in both economic and environmental terms. The large scale combustion of fossil fuels has led to the highest atmospheric carbon dioxide levels ever and the highest global average temperatures since measurements began.

There are great variations in energy use between countries. The USA has the highest total consumption of any one country of over two and a quarter billion metric tons of oil equivalent annually or nearly eight metric tons per person. Energy consumption in China is also very high at over 1.1 billion metric tons annually, but this is equivalent to only 0.89 metric tons per person. The highest per capita rate of nearly 11 metric tons is in the United Arab Emirates. On average the high income countries consume eleven and a half times as much energy per person as South Asia (5.42 metric tons compared to 0.469 metric tons) and over eight times as much per person as sub-Saharan Africa (0.693 metric tons). Globally energy consumption per person increased by 1.1% and total energy consumption by nearly 18% in the period 1990-2001. However the rates vary greatly between countries (World Bank 2004).

Global electricity production increased by nearly a third (32%) over the period 1990-2001 to 15.44 trillion kilowatt hours. Over half of this increase came from the high income countries, which already have high energy consumptions and only 16% of the world's population. Globally coal is still the largest source of electricity production at 38.8%. The contributions from most of the other main sources, hydropower (16.6%), gas (18.3%) and nuclear power (17.2%), are approximately equal, with a smaller contribution (7.4%) from oil. Over the period 1990-2001 the oil and hydropower contributions dropped by 3.9% and 1.5% of the total respectively and the contribution from gas increased by 4.4% of the total (World Bank 2004). The efficiency of conversion of energy from primary sources to applications is on average about 28% with considerable variations between different uses and countries. At most 35% of the energy from coal is converted into electricity (Sawin 2004). Thus there is considerable potential for improved energy efficiency with an associated reduction in emissions.

10.9% of energy is generated from renewables and waste and this figure is growing at 1.5% per annum. The low income countries generate a much higher percentage of energy from *renewables and waste* than the high income ones, for instance 57.6%, in Sub-Saharan Africa and 39.2% in South Asia, compared with only 3% in the high income countries. However many of the 'developing' countries are decreasing the proportion of their energy requirements generated by renewables and waste. For instance South Asia and East Asia and the Pacific decreased the percentage of energy generated in this way by 10% and 5% respectively in the period 1990-2001. This decrease was not counterbalanced by the very slight percentage increase of an average of 0.1% in the high income countries and in the greatest energy user, the USA, there was a decrease of 0.1% of the energy from renewables and waste over this period (World Bank 2004).

Energy was found to be such a *difficult topic* at the Earth Summit in Rio that governments preferred to ignore it and there is no energy chapter in Agenda 21. If the industrialised nations manage to stabilise their oil consumption, but the developing countries follow the same type of 'development' path, a tripling of global oil output would be required. This would be difficult if not impossible to achieve, since production capacity is nearly at its maximum. In 1994 China became a net oil importer, but the significance of this has been generally ignored. The International Energy Agency has calculated that a business as usual scenario will lead to a 65% growth in global energy demand and a 70% growth in global carbon dioxide emissions, with two thirds of this growth in 'developing' countries and 95% of the increase from fossil fuels (Maier 2002). The *Factor 10 Club* has called for a *tenfold improvement* in material and energy productivity in the industrialised countries over the next 30 to 50 years. They consider this would halve resource use, while allowing a doubling of economic activity and an improvement in global equity. Several governments have accepted this concept as a benchmark, but little progress has been made towards achieving it (Michaelis 2002).

1.6.4 Loss of Biodiversity and Habitat

There has been concern for the preservation of species and ecosystems for more than a century, but in the last few years the loss of biodiversity has been considered a threat to sustainable development. Loss of biodiversity is a particularly serious problem, as it is totally irreversible (Wilson 1989). The term *biodiversity* is generally used to indicate species diversity or the number of species of plants, animals and 'lower organisms' in a given area. Although quantitative data on extinction rates is generally not available, estimates based on observed rates of loss of natural habitats and empirical rules relating numbers of species to habitat area predict the loss of 2-8% of currently existing species over the next 25 years. Species have always died out, but current rates of extinction are 1000 times greater than ever before. Most of the decline is occurring in the tropics which are the home of 50-90% of all species (Acharya 1995), the source of 25% of pharmaceuticals (Linden 1989) and the home of most of the 75,000 plants with edible parts (Wilson 1989). The number of members of tropical species declined by about 65% and temperate species by more than 10% in the period 1970-2000 (Loh and Wackernagel 2004).

In addition to the value of biodiversity in itself, it has a value in maintaining *security of food supplies* and as a source of pharmaceutical products. There are about 200 domesticated food plants, each consisting of many cultivated varieties or cultivars, but their continuing productivity is maintained by a reservoir of genes in a much larger number of related wild species. The development of high yielding cultivars for intensive agriculture has led to the loss of genetic diversity, which could be a major threat to the long term security of global food supplies. In several cases the discovery of resistant genes in wild relatives of food plants has prevented major crop failures (Murray 1995). Lack of genetic diversity is also believed to be a major cause of the disastrous consequences of the 1845 Irish potato blight, where famine, fever and emigration led to a reduction in population from eight million in 1845 to five million in 1848. Measures for protecting biodiversity include (Murray 1995):

- A biological inventory, with a source catalogue of plant and animal species.
- The establishment of environmental management systems to give warning of problems and allow action to be taken.
- Compensation to 'developing' countries to preserve genetic material.
- The incorporation of previously ignored environmental costs, such as loss of biodiversity, into financial analyses.

40% of the world's known fish species live in freshwater, although freshwater is only 2.5 of the world's water and 99% of it is locked up in ice caps or underground. Populations of both temperate and freshwater species declined by about 50% between 1970 and 1995 and 91 species (out of 25,000) became extinct in the last century (Loh and Wackernagel 2004).

The *loss and fragmentation of natural habitats*, particularly in the tropics, is currently the greatest threat to biodiversity. Important natural habitats include forests, wetlands and coral reefs. Deforestation is occurring at a great rate, particularly in the tropics, where only one tree is planted for every 10 cut and 1.54 million hectares were cleared between 1980 and 1990 (Acharya 1995; Brown 1993; McNeill 1989; Postel 1994). It took only 40 years for Ethiopia's forest cover to decline from 30% to 1% of its land area and India's forest cover has declined from over 50% to about 14% since 1900.

Total forest cover globally is just under 38.5 million square kilometres or just under 30% of total land area. Globally 0.2% of forest cover or 95,000 km^2 of forest is being lost each year, though some reforestation is occurring, particularly in the higher income countries and to a limited extent in the Middle East and North Africa, Central Asia and Eastern Europe. About 7% of tropical forest cover was lost and temperate forest cover was increased by about 1% between 1990 and 2000 (FAO 2003). In addition to reforestation being insufficient to maintain the area of global forest, there is a need for forest cover in all parts of the world and forest in different places support different ecosystems, with tropical ecosystems being particularly rich in species (Acharya 1995). Therefore planting of temperate forests cannot balance the loss of tropical forests. Deforestation is most severe in sub-Saharan Africa, Latin America and the Caribbean, which lost a total of nearly one million square kilometres of forest cover in the period 1990-2000, with 8% of forest cover being lost in sub-Saharan Africa and 5% in Latin America and the Caribbean. This includes nearly a quarter of a million square kilometres in Brazil.

In addition to forest clearance for economic activity, such as logging and plantations, forests are being cleared for farming and raising livestock, though a cleared tropical forest can only be used for farming for three to five years and grazing for five to ten years. In addition farming patterns have changed. The extended fallow period in the crop rotations of traditional societies allowed forest regrowth, but has become rare due to increases in the number of cultivators and migrant farmers who lack knowledge of appropriate farming methods. The loss of forests also results in loss of habitats, water and watersheds for very large numbers of people and the loss of an important carbon dioxide sink (Acharya 1995).

Siberian White Crane

Swans on Agaya Lake

Wetlands, including marsh, swamp and open shallow water, are important breeding grounds for many species. Between a quarter and a half of the world's wetlands have been lost. Drainage for land development, construction of dams and dykes and hydrological alteration due to the depletion of underground aquifers and pollution are the main causes of wetland loss (Corson 1994). In addition to supporting many species, wetlands help to stem floods and erosion, refill groundwater aquifers and facilitate the settling or removal of organic matter and microbes, (Kolloru 1994). Coral reefs support an enormous variety of marine life and are under threat from pollution, sedimentation and over-exploitation.

Climate change and, in particular, rapid stepwise changes in the environment due to climate systems with positive feedback, is also a threat to biodiversity. Rapid changes in patterns of precipitation and temperature can lead to species extinction either directly or through invasion by non-native species which can flourish in the changed climate conditions (Murray 1995). Freshwater ecosystems are being degraded by increasing demands for food, fibre, energy and water. For instance increasing use of water for irrigation to grow cotton and rice in central Asia has stopped the flow from the Amu Darya and Syr Darya rivers into the Aral Sea. The area of this sea reduced by more than half between 1960 and 2000 and its salinity increased nearly fivefold. The number of bird and mammal species living in the river deltas both reduced by about 50% over this period (Loh and Wackernagel 2004).

1.7 Assimilative Capacity and the Precautionary Principle

1.7.1 Introduction

This section introduces the precautionary principle which requires a proactive approach to the consideration of the social and environmental risks associated with the introduction of new technology and extreme caution to be exercised in deciding whether to introduce new technology when the likely impacts are not clear.

1.7.2 Assimilative Capacity, Source and Sink Capacity

Ensuring that global society follows sustainable development paths will require mechanisms for the formulation of *policies to promote sustainability* and the choice of sustainable projects, technologies and processes. This postulates an approach to decision making in which environmental, social and developmental factors are integrated with political and economic approaches, rather than an environmental twist being added to existing approaches or environmental and social issues being ignored as frequently happens at present.

There are a number of different variables which can be used to give measures for the sustainability of ecosystems and human activity systems. Ecological footprints have already been discussed in section 1.2.3. Other concepts including assimilative capacity, source and sink capacity and carrying capacity.

The *source capacity* of an ecosystem, such as the planet, is its ability to provide raw materials. The *assimilative capacity* or *sink capacity* of an ecosystem is its ability to absorb a substance without degrading or damaging its ecological integrity, including both structural and functional integrity (Chadwick and Nilsson 1993). Thus source and sink or assimilative capacity give a measure of the human activities that can be supported by a given system. The term human activities has been used here rather than population, as it is not just the level of population that is significant, but also the level of resource use and waste generation.

Critical load is the level of exposure to one or more pollutants below which significant or unacceptable harmful effects do not occur according to present knowledge (Nilsson 1986). In practice it is difficult to establish critical loads due to limited knowledge, very low thresholds for some type of pollutants and problems of scale (Chadwick and Nilsson 1993). There are also problems in defining environmental harm or damage and determining what types of 'harm' are acceptable. This is a question of social choice and value judgements rather than a purely technical issue (Dethlefsen et al 1993).

The term *carrying capacity* has been used for both populations and ecosystems. It limits the growth and development of human and other population in ways that are dependent on the relationships between finite resources and the consumers of those resources (Monte-Luna et al 2004). The related term *appropriated carrying capacity* describes the resource flows and waste assimilation capacity appropriated per unit of time from the capacity by a defined population (Rees 1996).

1.7.3 The Precautionary Principle

The *precautionary principle* was formulated in response to failures of environmental management approaches based on the assimilative capacity concept to protect the North Sea. In practice it generally cannot be proved that there is a 'safe' level for a given pollutant and there have been significant failures in identifying both harmful effects and the pathways by which they occur. The precautionary principle was introduced internationally at the First International Conference on the Protection of the Sea in 1984. It attempts to remove the need to prove a *causal link* between specific emissions and observed environmental damage before action is taken to reduce the environmental damage caused by substances with a known 'hazard potential'. The precautionary principle requires *anthropogenic inputs* into the environment of manufactured substances or very large quantities of natural substances to be avoided, unless this is likely to lead to damage elsewhere. It may be necessary to prioritise action on different substances in accordance with their ability to cause harm. Although what is meant by 'harm' is largely subjective, there are also objective considerations, such as the extent of scientific knowledge about the effects of the substance, its toxicity, persistence in the environment and bioaccumulability, natural fluxes of the substance in the ecosystem and stochastic effects. Truly *anticipatory environmental protection* should try to identify the potential for environmental harm in advance of the use of substances which may be released into the environment (Dethlefsen et al 1993).

Significant *uncertainties* about probable environmental consequences complicate decision making and provide a further justification for the precautionary principle. For instance the chemical, physical and biological effects of putting a given waste in a landfill site can only be known approximately and depend on a range of factors, such as how the site is operated and managed. The results of a particular environmental assessment will generally depend on the relative significance assigned to different factors, as well as the methodology and assumptions used. In addition there may be errors or imprecision in measurements. For instance investigation of the hazards from radio-caesium was originally based on the assumptions that physical depth distribution over time was the only significant variable and gamma radiation to a person standing on the surface the only path of human exposure. This approach to evaluation of the hazards ignored the fact that the high proportion of the population who are meat eaters could be contaminated by eating sheep that had consumed irradiated vegetation, making chemical mobility and consequently the uptake of caesium by vegetation from the soil significant variables (Wynne 1993). It also ignored the effects of radioactive contamination of plants and other animal species.

In addition members of the public may have very different views on *'acceptable' risks* from the scientific community and may express concern about levels of risk which are considered acceptable by system designers. The difference between a lack of evidence of risk and a lack of risk is often ignored, with a lack of firm evidence of risk often used to belittle public concerns, for instance about genetically modified organisms. Acceptance of risk is often related to (perceptions of) the associated benefits. In general members of the public are more willing to accept or discount (uncertain) risks when the associated benefits are clearly apparent and valuable than when there are no obvious benefits. For instance mobile phones have been widely accepted and used, despite the possibility of health risks (Blettner and Berg 2000), whereas there is considerable opposition to genetically modified organisms (Gaskell et al 2004). Risk analysis raises questions as to the relationship between technical design and evaluation and ethical and other assessments.

1.8 Roundup of Chapter 1

This chapter has presented an overview of some of the main issues in sustainable development as a background to the mathematical and computing techniques which will be presented in the subsequent chapters. It introduced the idea of sustainable development as involving social and developmental issues, economic development and the natural environment. In some cases there may be (apparent) conflicts between social and/or developmental and environmental goals. For instance combating poverty in sub-Saharan Africa will require a wide range of measures, many of which will lead to increased consumption with associated pressures on the natural environment. However the application of systems methodologies of the type presented in Part I could be used to determine solutions which encourage development while putting minimal stress on the natural environment. In other cases all the different goals may be in harmony. For instance educating girls and women reduces their average number of children, leading to population reduction and a reduction in consumption and waste generation.

Two of the main issues of concern in the context of sustainable development are population growth and resource consumption. Population growth is often seen as a problem of the 'developing' countries and resource consumption of the industrialised countries. However, in practice the smaller population growth in the industrialised countries is having as much or more impact on resource consumption than the much higher population growth in the 'developing' countries. Therefore both stabilising or reducing populations in the industrialised countries and reducing their resource consumption will be required to allow the people of the 'developing' countries to consume slightly more in order to achieve development.

Education, particularly of girls, is a priority for both social and economic development, but many 'developing' countries are only devoting a small proportion of public expenditure to education. There are still considerable disparities in many 'developing' countries between access to education by girls and boys and over a quarter of the population in the 'developing' countries does not even complete primary school. Unemployment is another and growing problem throughout the world. There are still wide disparities between the life expectancies and infant mortalities in the high and low income countries.

The environmental problems discussed in this chapter are generally a direct result of the unsustainable development paths followed by the industrialised countries and then emulated by the 'developing' countries. Two of the main, but by no means the only, problems are global warming and ozone depletion. Resolving these and other environmental problems will involve change at both the individual and policy level and will require the industrialised countries to act first, so they do not continue to enjoy the benefits of unwise development, while expecting the 'developing' countries to pay the price in terms of restricted development and austerity measures.

Reflecting on the material presented in this chapter on social injustices and environmental crises is intended to make readers aware of the urgent need for action, as well as the complexity and interconnections of many of the issues. Gaining a better understanding of complex and interconnected issues with the view to obtaining appropriate solutions requires the use of systems methodologies, which will be discussed in Part I. Deciding on appropriate policy and other measures to promote sustainable development requires optimisation and decision making techniques which will be considered in Part II and handling the mixture of qualitative and quantitative, imprecise and uncertain techniques associated with problems in sustainable development requires appropriate techniques, such as the use of fuzzy sets, which will be discussed in Part III.

References and Additional Reading for Chapter I

References for Chapter 1

Acharya A (1995) Tropical forests vanishing. In: Brown LR, Kane H, Lenssen H Vital signs, Starke L (ed), Norton, New York, pp 116-119

Ahmad QK et al (2001) Summary for policymakers. In: McCarthy JJ, Canziani OF, Leary AN, Dokken DJ, White KS (eds) Climate Change 2001, impacts, adaptation and vulnerability, contribution of working group II to the third assessment report of the Intergovernmental Panel on Climate Change (IPCC). Cambridge University Press, pp.1-73

AI (2004a) A global outrage: global and UK statistics, Amnesty International, http://www.amnesty.org.uk/svaw/vaw/global.shtm

AI (2004b) Amnesty International media briefing, http://web/amnesty/org/library/Index/ENGPOL100152004

AI (2004c) Facts and figures on the death penalty. Amnesty International, http://web.amnesty.org/pages/deathpenalty-facts-eng

AI (2004d) Amnesty International Report 2004. Amnesty International

Allag F, Abboud P, Mansour G, Zanardi M, Quéreux (2001) Mutilations genitals rituelles féminines. La parole aux femmes. Gynécol Obstét Fertil 29:824-828

Barnard, ND, Nicholson A, Howard JL (1995) The medical costs attributable to meat consumption. Preventive Medicine 24(6):646-655

Beckett C and Macey M (2001) Race, gender and sexuality: the oppression of multiculturalism. Women's Studies International Forum, 24(3/4):309-319

Bell, C, Collins SD, Ginsburg HL, Malloy M (2004) The permanent war economy: real security or false promise. National Jobs for All Coalition, http://www.njfac.org/sr4.htm

Bellamy C (2004) The state of the world's children, 2004, girls, education and development. UNICEF, New York

Blettner M, Berg G (2000) Are mobile phones harmful?. Acta Oncologica, 39(8):927-930

Braun E (1995) Futile progress, technology's empty promise. Earthscan

Brown LR (1990) State of the world 1990. Norton, New York

Brown LR (1993) A new era unfolds. In: Brown LR State of the World 1993, Starke L (ed) Norton, New York pp 3-21

Brown LR (1995) Grainland shrinks. In: Brown LR, Kane H, Lenssen N Vital Signs 1995, Starke L (ed) Norton, New York, pp 15-21

Brown L, Flavin C (1999) State of the world 1999: a Worldwatch Institute report on progress towards a sustainable society, Starke L (ed), Earthscan, London

Brown L, Jacobson JL (1986) Our Demographically divided world, Worldwatch Paper no 74, Washington DC

Brundtland GH (2002) Our common future and ten years after Rio: how far have we come and where should we be going. In: Dodds F (ed) Earth summit 2002, a new deal. Earthscan, pp 253-263

Brusaco-Mackenzie M (2002) Environment and security. In: Dodds F (ed) Earth summit 2002, a new deal. Earthscan, pp 179-201

Chadwick M Nilsson J (1993) Environmental quality objectives. In: Jackson T (ed) Production strategies, developing preventive environmental management in the industrial economy, Stockholm Environment Institute, pp 29-40

Commoner B (1990) Making Peace with the Planet. New York, Pantheon

Conservation Foundation (1992) State of the Environment 1982. Conservation Foundation, Washington DC

Cook RJ, Dickens BM, Fathalla MF (2002) Female genital cutting (mutilation/circumcision): ethical and legal dimensions. International Journal of Gynecology and Obstetrics, 79:281-287

Corson WH (1994) Global environmental issues and sustainable resource management. In: Kolloru RV (ed) Environmental strategies handbook. pp 923-970. New York, McGraw Hill

Dasgupta P (1995) Optimal development and the idea of net national product. In: Goldin IA, Winters LA (eds) The economics of sustainable development. Cambridge University Press, pp 111-142

Dethlefsen VT, Jackson T, Taylor P (1993) The precautionary principle. In: Jackson T (ed) Clean production strategies, developing preventive environmental management in the industrial economy, Stockholm Environment Institute, pp 41-62

Dollar D, Gatti R (1999) Gender inequality, income and growth: are good times good for women?. Gender and Development Working Paper Series No. 1, Development Research Group, World Bank, Washington, p 21

Dunn S, Flavin C (2002) Moving the climate change agenda forward. In: Flavin C, French H, Gardner State of the world 2002, progress towards a sustainable society, Starke L (ed), Earthscan, London, pp 24-50

Easton P, Monkman K, Miles R (2003) Social policy from the bottom up: abandoning FGC in sub-Saharan Africa. Development in Practice, 13:445-458

Ehrlich PR (1968) The population bomb. Ballantine Books, New York

Engleman R, Halweil B, Nierenberg D (2002) Rethinking population, improving lives. In: Flavin C, French H, Gardner State of the World 2002, progress towards a sustainable society, Starke L (ed), Earthscan, pp 127-148

Essen B, Johnsdotter S (2004) Female genital mutilation in the West: traditional circumcision versus genital cosmetic surgery. Acta Obstet Gynecol Scand, 83:611-613

FAO (2003) State of the world's forests. Food & Agriculture Organization of UN, http://www.fao.org/documents/show_cdr.asp?url_file=/DOCREP/005/Y7581E/y7581e00.htm

Fischbach RL, Herbert B (1997) Domestic violence and mental health: correlates and conundrums within and across cultures. Soc Sci Med 45(8):1161-1176

FOE (undated) Aviation and global climate change. Friends of the Earth, http://www.foe.co.uk/resource/reports/aviation_climate_change.pdf

French H (2002) Reshaping global governance. In: Flavin C, French H, Gardner State of the world 2002, progress towards a sustainable society, Starke L (ed), Earthscan, London, pp 174-198

Frosch RA, Gallopoulos NE (1989) Strategies for manufacturing. Scientific American, September, 144-152

Gardiner R (2002) Freshwater: a global crisis of water security and basic water provision. In: Dodds F (ed) Earth summit 2002, a new deal. Earthscan, pp 289-304

Gardner G (2002) The challenge for Johannesburg: creating a more secure world. In: Starke, L (ed) State of the World 2002. Earthscan, pp 3-23

Gardner G, Halweil B (2000) Nourishing the underfed and overfed. In: Brown LR, Flavin C, French H (ed) State of the World 2000, Starke L (ed) Earthscan, pp 59-78

Gardner G, Assadourian E, Sarin R (2004) The state of consumption today. In: Halweil B, Mastny L, Assadourian E, Starke L State of the world 2004, Starke L (ed), WW Norton and Co, pp 3-21

Gaskell G, Allum N, Wagner W, Kronberger N, Torgersen H, Hampel J, Bardes J (2004) GM foods and the misperception of risk perception. Risk Analysis, 24(1): 185-194

Gleick PH (2000) The changing water paradigm: a look at twenty-first century water resources development, Water International, 25(1):129

Gore A (1992) Earth in the balance: ecology and the human spirit. Houghton Miflin, Boston,

Graedel TE, Crutzen PJ (1989) The changing atmosphere. Scientific American, September, 58-68

Greenpeace (2002) Earth summit has answer to Asian brown cloud. http://www.greenpeace.org/news/details?item_id=21937

Grimmett RF (2004) Conventional arms transfers to developing nations. 1996-2003. Congressional Research Service Report for Congress, http://www.fas.org/man/crs/RL32547.pdf

Grossman GM (1994) Pollution and growth: what do we know?. In: Goldin I, Winters LA (eds) The economics of sustainable development, Centre for Economic Policy Research, pp 46

Guttal S (2000) Keeping debtors in place: debt relief under the enhanced HIPC initiative. In: The transfer of wealth: debt and the making of a global south. Focus on the Global South, Bangkok, pp 35-43

Hersh MA (1999) New technology, gender and employment. 14th IFAC Congress, Beijing, China

Herz B, Subbarao K, Habib M, Raney L (1991) Letting girls learn: promising approaches in primary and secondary education. World Bank Discussion Paper No. 133, World Bank, Washington DC, p. 19

Hobsbawm E (1996) The age of extremes: a history of the world, 1914-1991. Vintage Books

Holmberg J, Sandbrook R (1992) Sustainable development: what is to be done? In: Holmberg J (ed) Policies for a small planet. Earthscan, London, pp. 19-38

HRW (2001) South Africa rights: girls endure sexual violence in school. Human Rights Watch, Interpress Service

ILO (2003) Key indicators of the labour market, (3rd edn). International Labour Organisation Database

IMAP (2000) Statement on gender-based violence. International Medical Advisory Panel, London, IPPF, www.ippf.org/imap/gbv2000/index.htm

ITU (2003a) World telecommunication development report. International Telecommunication Union

ITU (2003b) World Telecommunication Indicators Database. International Telecommunication Union

James SM (1998) Shades of othering: reflections on female circumcision/genital mutilation. Signs: Journal of Women in Culture and Society, 23(4):1031-1048

Jubilee 2000 (2000) IFM takes $1 billion in two years from Africa as official report reveals five new reasons to drop the debt. http://www.jubilee2000uk.org/jubilee2000/news/imf094.html

Kaldor M (1999) New and old wars: organized violence in a global era. Stanford University Press

Kane H (1994) Sulfur and nitrogen emissions resume rise. In: Brown L, Kane H, Roodman DM Vital signs 1994, Smith L (ed), Norton, New York, pp 64-65

Kane H (1995) Sulfur and nitrogen emissions fall slightly. In: Brown LR, Kane H, Lenssen N Vital Signs 1995, Starke L (ed), Norton, New York, pp 94-95

Kennedy E, Oniang'o R (1990) Health and production effects of sugarcane production in South-Western Kenya, Food and Nutrition Bulletin, 12

Kirkby J, O'Keefe P, Timberlake L (1995) The Earthscan reader in sustainable development, Earthscan, London

KIT (2002) Street children – overview. http://www.kit.nl/frameset.asp?/specials/html/sc_street_children.asp&frnr=1&

Kolloru RV (1994) Risk assessment and management, In: Environmental strategies handbook. McGraw-Hill, New York, pp 377-432

Krause F, Bach W, Kooney J (1995) A target-based, least cost approach to climate stabilization. In: Kirby J, O'Keefe P, Timberlake L (eds) The Earthscan reader in sustainable development. Earthscan, London, pp 65-75

Lappé FM, Shurman R (1995) The population debate. In: Kirby J, O'Keefe P, Timberlake L (eds) The Earthscan reader in sustainable development. Earthscan, London, pp 104-109

Leitenberg M (2001) Deaths in wars and conflicts between 1945 and 2000. Conference on data collection in armed conflict, Uppsala

Levinson D (1989) Family violence in cross cultural perspective. Sage, Newbury Park

Linden E (1989) The death of birth. Time, January 2, pp 32-35

Linden E (1994) Burned by warming. Time, March 14, p 79

Logan J (2000) What is happening to the women in this country is genocide. Jane Magazine

Loh J, Wackernagel M (2004) Living planet report 2004, WWF

McCarthy JJ, Canziani OF, Leary AN, Dokken DJ, White KS (eds) (2001) Climate change 2001, impacts, adaptation and vulnerability, contribution of working group II to the third assessment report of the Intergovernmental Panel on Climate Change (IPCC). Cambridge University Press

McGinn AP (2002) Reducing our toxic burden. In: Flavin C, French H, Gardner State of the world 2002, progress towards a sustainable society, Starke L (ed), Earthscan, pp 75-100

McNeill J (1989) Strategies for sustainable economic development. Scientific American, September, 155-156

Maier J (2002) Energy: fuel for sustainable development. In: Dodds F (ed) Earth summit 2002, a new deal. Earthscan, pp 305-315

Manderson L (2004) Local rites and body politics. International Feminist Journal of Politics, 6(2):285-307

Mazzocchi T (1990) Economics and the work environment, the other economic summit. Houston, TX

Meniru GI, Hecht BR, Hopkins MP (2000) Female circumcision: at our doorsteps and beyond. Prime Care Update Ob/Gyns 7:231-237

Meyers DT (2000) Feminism and women's autonomy: the challenge of female genital cutting. Metaphilosophy, 31(5):469-491

Michaelis L (2002) Sustainable consumption and production. In: Dodds F (ed) Earth summit 2002, a new deal. Earthscan, pp 264-277

Monte-Luna P del, Brook BW, Zetina-Rejón MJ, Cruz-Escalona VH (2004) The carrying capacity of ecosystems. Global Ecol Biogeogr, 13:485-495

Murray M (1995) The value of biodiversity. In: Kirby J, O'Keefe P, Timberlake L (eds) The Earthscan reader in sustainable development. Earthscan, London, pp 17-29

Muslim Women's League (1999) Female genital mutilation. http://www.mwlusa.org/publication/positionpapers/fgm.htm

Nilsson J (1986) Critical loads for nitrogen and sulphur. Nordic Council of Ministers, Copenhagen

OECD & UNEP (1999) Phasing lead out of gasoline: an examination of policy approaches in different countries. OECD and UNEP, Paris, p 3

Paddock W, Paddock P (1975) Famine 1975!. Little, Brown and Company, Boston

Pearce D, Markandaya A, Barbier EB (1989) Blueprint for a green economy. Earthscan, London

Phillips A and Taylot B (1980) Sex and skill: notes towards a feminist economics. Feminist Review 6:79-88

Postel S (1994) Carrying capacity, earth's bottom line. In: Brown LR State of the World 1994, Starke L (ed), Norton, New York, pp 2-21

Postel, S (1999) Pillars of sand. W.W. Norton and Company, New York, p 80

Postel S (2000) Redesigning irrigated agriculture. In: Brown LR, Flavin C, French H State of the world 2000, Starke L (ed), Earthscan, London, pp 39-58

Postel, S, Vickers A (2004) Boosting water productivity. In: Halweil B, Mastny L, Assadourian E, Starke L State of the world 2004, Starke L (ed),WW Norton and Co, pp 46-65,

Project Ploughshares (2004) Armed conflicts report 2004. http://www.ploughshares.ca/content/ACR/ACR00/ACR00.html

Rees WE (1996) Revisiting carrying capacity: area-based indicators of sustainability. Population and Environment, 17(3)

Renner M (2000) Creating jobs, preserving the environment. In: Brown LR, Flavin C, French H State of the world 2000, Starke L (ed), Earthscan, pp 162-183

Renner M (2002) Breaking the link between resources and repression. In: Starke L (ed) State of the world 2002. Earthscan, pp 149-173

Repetto R (1988) Economic policy reform for natural resource conservation. World Bank Environment Department Working Paper 4

Roodman DD (1995) Carbon emissions resume rise. In: Brown LR, Kane H, Lenssen N Vital Signs 1995, Starke L (ed) Norton, New York, pp 66-67

Rowley J Holmberg J (1995) Stabilizing population: the biggest challenge. In: Kirby J, O'Keefe P, Timberlake L (eds) The Earthscan reader in sustainable development. Earthscan, London, pp 115-122

Sala R and Manara D (2001) Nurses and requests for female genital mutilation: cultural rights versus human rights. Nursing Ethics 8(3):247-258

Sawin JL (2004) Making better energy choices. In : Halweil B, Mastny L, Assadourian E, Starke L State of the World 2004, Starke L (ed), WW Norton and Co, pp 24-43

Shah A (2004) Third world debt undermines development, the scale of the debt crisis. http://www.globalissues/TradeRelated/Debt/Scale.asp

Simon J (1980) Resources, population, environment: an oversupply of false bad news. Science, 208:1434

SIPRI (2001) SIPRI Yearbook 2001, Armaments, disarmament and international security. Oxford University Press

SIPRI (2002) SIPRI Yearbook 2002, Armaments, disarmament and international security. Oxford University Press

SIPRI (2004) SIPRI Yearbook 2004, Armaments, disarmament and international security. Oxford University Press

Smith, LC, Haddad L (2000) Overcoming child malnutrition in developing countries: past achievement and future choices. International Food Policy Research Institute, Washington DC, p 44

Snow R (2001) Female genital cutting: distinguishing the rights from the health agenda. Tropical Medicine and International Health, 6(2):89-91

Stead WE, Stead JG (1996) Management for a small planet, 2nd edn, Sage Publications

Steer A, Wade-Gerry W (1994) Environmental strategies for the developing world. In: Kolluru RV (ed) Environmental strategies handbook. McGraw Hill, New York, pp 851-904

Stienstra D (2000) Cutting to gender: teaching gender in international relations. International Studies Perspectives 1:233-244

Stockdale A, Findlay A, Short D (2000) The repopulation of rural Scotland: opportunity and threat, J. Rural Studies, 16:243-257

Timberlake L, Thompson L (1990) When the bough breaks: our children, our environment. Earthscan, London

Toubia N (1994) Female genital mutilation and the responsibility of reproductive health professionals. International Journal of Gynaecology and Obstetrics 46:127-135

UN (2003) World urbanization prospects: the 2001 revision, United Nations

UNDP (2004) Human development report 2004 cultural liberty in today's diverse world. United Nations Development Programme, (http://hdr.undp.org/reports/)

UNEP (1999) Global environment outlook 2000. Earthscan, London, also at http://www.unep.org/geo2000/

UNESCO (undated), Statistical yearbook, United Nations Education, Science and Cultural Organisation

UNFPA (2000) The state of world's population 2000. New York, p 23

UNHCR (undated) Convention and protocol relating to the status of refugees. United Nations Commission on Human Rights

UNHCR (2004a) 2003 Global refugee trends. United Nations Commission on Human Rights, http://www.unhcr.ch/statistics

UNHCR (2004b) Protracted refugee situations. UN Commission on Human Rights,http://www.unhcr.ch/cgibin/texis/vtx/home/opendoc.pdf?tbl=STATISTICS&id=40ed5b384

UNICEF (1991) The state of the world's children. UNICEF, New York

UNICEF (1992) The state of the world's children. UNICEF, New York

Watts C, Zimmerman C (2002) Violence against women: global scope and magnitude. The Lancet, 359:1232-1237

WBGU (German Advisory Council on Global Change) (1999) World in transition. ways towards sustainable management of freshwater resources, annual report 1997. Springer Verlag, Berlin

WCED (1987) Our common future, report of the world commission on environment and development. Oxford University Press

WHO (2003) World health report. World Health Organisation

Wilson EO (1989) Threats to biodiversity. Scientific American, September, 108-116

World Bank (2001) Engendering development: through gender equality in rights, resources and voice. Oxford University Press, pp 152-154

World Bank (2004) World development indicators 2004. The World Bank

WRI (2000) World resources 2000-01: people and ecosystems: the freying web of life. World Resources Institute

WRI (2003) World resources 2000-01: people and ecosystems: the freying web of life. World Resources Institute

WRI (2004) Earth trends, the environmental information portal. World Resources Institute, http://earthtrends.wri.org/

WWW1 (2005) http://www.un.org/millennium/declaration/ares552e.htm

WWW2 (2005) www.infoplease.com

WWW3 (2005) http://www.cftech.com/BrainBank/FINANCE/GDP.html

WWW4 (2005) http://en.wikipedia.org/wiki/Gross_National_Income

WW5 (2005) http://conventions.coe.int/treaty/EN/Treaties/html/005.htm
WWW6 (2005) http://www.thefreedictionary.com/pollution
WWW7(2005) http://www.epa.gov/ozone/title6/phaseout/hcfc.html
Wynne B (1993) Uncertainty and environmental learning, In: Jackson T (ed) Production strategies, developing preventive environmental management in the industrial economy. Stockholm Environment Institute, pp 64-84
Young M (1992) Sustainable investment and resource use. Parthenon, Park Ridge

Additional Reading for Chapter 1

Ayres RU (1993) Industrial metabolism. In: Jackson T (ed) Production strategies, developing preventive environmental management in the industrial economy. Stockholm Environment Institute, pp 165-199
Bellamy C (2004) The State of the World's Children 2004. The United Nations Children's Fund (UNICEF)
Brown LR (1994) Facing food insecurity. In: Starke L State of the world 1994. Norton, New York, pp 177-197
Brown LR (1995) World economy expanding faster. In: Brown LR, Kane H, Lenssen N Vital signs 1995, Starke L (ed) Norton, New York, pp 70-71
Buckingham-Hatfield S, Evans B (eds) (1996) Environmental planning and sustainability. John Wiley and Sons, Chichester
Commoner B (1989) Rapid population growth and environmental stress. In: Consequences of rapid population growth in developing countries. Proc. of UN Expert Groups Meeting, August 1988, United Nations, New York
Daly H (1991) From empty world economics to full world economics: recognizing an historic turning point in economic development. In: Goodland R, Daly H, Serafy S (eds) Environmentally sustainable development: building on Brundtland, The World Bank, Washington DC
Daly H (1993), The perils of free trade. Scientific American, November, 50-57.
DG VIII (1997) Preparation of guidelines in the water resources sector. October
Doh K, Son J (1991) Policy measures on recycling of waste in Korea, Joint UNIDO/UNDP/APCTT workshop on recycling of waste in Asia and the Pacific Region, Manila. European Communities, Luxembourg.
Douglis C (1993) Images of home. Wilderness, Fall, 12-22.
DTI (1991) Market barriers to material reclamation and recycling, UN Recycling Advisory Unit of the Environment DTI, Warren Springs Laboratory, Steverage
EC (2002) Corporate social responsibility a business contribution to sustainable development. European Commission Employment and Social Affairs, Luxembourg
Ehrlich PR, Ehrlich AH (1990) The population explosion. Simon and Schuster, New York

European Foundation (1998) Employment and sustainability digest report, European Foundation for the Improvement of Living and Working Conditions, Luxembourg

European Communities (1999) Employment and environment principles and recommendations from the European consultative forum on the environment and sustainable development.

GEC (1999) The politics of GM food risk, science and public trust. ESRC Global Environmental Change Programme (www.gecko.ac.uk)

Ginther K, Denters E, de Waart PJIM (1995) Sustainable development and good governance. Martinus Nijhoof Publishers

Jackson T (1991) Least-cost greenhouse planning: supply curves for global warming abatement. Energy Policy, 19(1):35

Jackson T (1993a) Hazardous futures. In: Jackson T (ed) Production strategies, developing preventive environmental management in the industrial economy, Stockholm Environment Institute, pp 105-124

Jackson T (1993b) Principles of clean production. In: Jackson T (ed) Production strategies, developing preventive environmental management in the industrial economy, Stockholm Environment Institute, pp 143-164

Jackson T, Costanza R, Overcash M, Rees W (1993) The 'biophysical' economy, aspects of the interaction between economy and environment. In: Jackson T (ed) Production strategies, developing preventive environmental management in the industrial economy, Stockholm Environment Institute, pp 3-28

Kagee A, Naidoo AV, Van Wyk S (2003) Building communities of peace: the South African experience: counseling and building communities of peace. International Journal for the Advancement of Counselling, 25(4):225-233.

King EM, Mason AD (2001) Engendering development: through gender equality in rights, resources and voice. World Bank and Oxford University Press, Washington DC, p. 88

Layard A, Davoudi and Batty S (eds) (2001) Planning for a sustainable future. Spoon Press

Lenssen N (1995) Nuclear waste still accumulating. In: Brown LR, Kane H, Lenssen N Vital Signs 1995, Starke L (ed) Norton, New York

Leyshon, A, Lee R,Williams CC (2003) Alternative economic spaces. Sage Publications

Margolis H (1996) Dealing with risk, why the public and the experts disagree on environmental issues. University of Chicago Press

Mills E, Wilson D, Johansson T (1991) Getting started: 'no-regrets' strategies for reducing greenhouse gas emissions. Energy Policy, 19(6), 526

OECD (1998) Eco-efficiency. Organisation for Economic Cooperation and Development, Paris

OECD (2000) Towards sustainable development indicators to measure progress. Rome Conference, Organisation for Economic Cooperation and Development

OECD (2001) Environmentally related taxes in OECD countries issues and strategies. Organisation for Economic Co-operation and Development

OECD (2002) Sustainable development strategies, a resource book. Earthscan

OECD (2004) Measuring sustainable development integrated economic environmental and social frameworks. Organisation for Economic Cooperation and Development

Ornstein R, Ehrlich P (1989) New world, new mind. Touchstone, New York

Petrick JA, Quinn JF (1994) Deforestation in Indonesia: policy framework for sustainable development, J. of Asian Business, 10(2):41-56

Postel S (1990) Toward a new 'eco'-nomics. World Watch, September/October, 20-28

Postel S (1993) Facing water scarcity. In: Brown LR State of the world 1993, Starke L (ed), Norton, New York, pp 22-41

Pradervand P (1991) The interaction of population and natural resources, Background Document for Social Services Division, IUCN, Gland, Switzerland

Shiva M (1995) The politics of population policies. The Earthscan reader in sustainable development, Earthscan, London, pp 110-114

Smith JB, Tirpak D (1988) The potential effect of a global climate change on the United States. Draft Report to Congress, October

Weenan van H, (1997a) Design for sustainable development concepts and ideas. European Foundation for the Improvement of Living and Working Conditions, Luxembourg

Weenan van H (1997b) Design for sustainable development guides and manuals. European Foundation for the Improvement of Living and Working Conditions, Luxembourg

Weiner J (1990) The next hundred years: shaping the fate of our planet. Bantam, New York

WRI (1992) World resources 1990-91. World Resources Institute, Oxford University Press, New York and London

Wynne B (1986) Uncertainty, technical and social. In: Brooks H and Cooper C (eds) Science Public Policy, Pergamon Press

Part I Holistic Approaches and Systems Methodologies

I.1 Introduction

Systems approaches have been gaining in popularity in a number of new fields, including social, economic, information and engineering. There are a number of different definitions, but they are all in agreement that one of the most important aspects of a system is the *relationship* between the *components*. Thus a system can be defined as an assemblage or combination of elements or parts (including methods, procedures or doctrines) that forms a complex or unitary whole, such as a transportation system or a system of organisation and management. A system consists of components or operating parts, attributes or properties of the components and relationships between the components and attributes. It is the relationship between the parts rather than the parts themselves that determines the behaviour of different systems. This allows the same principles to be applied to many different types of systems and systems in many different fields (O'Connor and McDermott 1997).

System properties which relate to the whole system, rather than the parts are called *emergent properties*, as they emerge from the interactions of the different components in the whole system (Checkland and Scholes 1999). These properties are often unpredictable, particularly when the system is complex. Understanding a system generally requires consideration of the whole system. Traditional approaches to investigating and understanding problems and situations have generally been based on analysis in the sense of decomposition into parts and examination of the parts. Systems approaches often carry out analysis in both directions i.e. *composition* or investigation of the complete system which can be built up from components, including the interactions between components, and *decomposition* or investigation of the component parts and subsystems comprising a given system.

Systems approaches have, at least in part, been motivated by the need to deal with (organised) complexity. They can provide tools for structuring complex situations, without losing a sense of the whole, thereby allowing tradeoffs to be made between the full range of complex interacting factors and interests. Complexity increases with the number of components in a system, the number of different types of components and the number of different (or different types of) relations between them. For systems with a (relatively) small number of components it is often possible to either solve the system equations exactly or obtain good approximations, whereas statistical methods can be applied to systems with very large numbers of components. However both these approaches tend to be unsatisfactory for systems with moderate or moderately large numbers of components (Weinberg 1975), but systems approaches can still be applied in this case.

Although many systems based approaches to *structuring complex situations* are based on hierarchy, systems are not necessarily hierarchical. There are a number of non-hierarchical systems approaches which focus on interconnections and interactions. Thus hierarchy is only one of a number of different approaches to dealing with complexity. However the general model of organised complexity is based on a hierarchy of levels of organisation. Each level is more complex than the one below, as it is characterised by emergent properties which do not exist at the previous level.

Another important pair of ideas in systems thinking is *communication and control* (Checkland and Scholes 1999). Control generally involves the imposition of constraints and takes into account at least two hierarchical levels, with the upper level imposing constraints on the lower one. All control processes depend on communication or the flow of information in terms of instructions or constraints.

As discussed in Chapter 1, sustainable development is concerned with a number of complex problems resulting from harmonising the sometimes conflicting requirements of human developmental needs, protection and conservation of the natural environment and maintaining the ability of future generations to meet their own needs (WCED 1987) or the sometimes conflicting policy goals of environmental integrity, economic efficiency and equity (Young 1992). Social, developmental and environmental systems all comprise a number of subsystems. Emergent properties are also of importance when considering the interactions between these subsystems. Indirect and long term (often unforeseen) effects are also likely to be significant. Control or regulation is clearly another very important idea in sustainable development, since moves towards sustainable development generally involve reducing or otherwise regulating undesirable and unsustainable impacts of human activity, while promoting positive developmental effects.

The systems features inherent in consideration of sustainable development make systems approaches particularly appropriate for representing, discussing and analysing both sustainable development as a whole and the various problems associated with it. In some cases systems approaches can be applied directly to obtain solutions to specific sustainable development problems, whereas in others it is more appropriate to apply them to structure problems and to clarify issues to allow other techniques to be used to obtain solutions.

Part I of this book is concerned with the application of systems approaches to sustainable development. It is divided into six chapters (chapters 2-7). Chapter 2 introduces a number of basic concepts in systems theory, including systems thinking, general systems and different types of systems (approaches), such as 'hard' and 'soft' systems. It concludes with a discussion of the application of systems approaches to sustainable design and production. Since systems theory is well developed and a number of different methodologies have been developed and applied, it illustrates what is possible, as well as the advantages and drawbacks of the different approaches with particular reference to sustainable development.

Chapter 3 presents the mathematical concepts underlying systems approaches from an engineering perspective and therefore contains a number of definitions. Amongst other things it clarifies the differences between feedback and feedforward and open and closed systems. This mathematical background prepares the way for a discussion of systems modelling using state space representations in chapter 4. State space representations give a useful and frequently used approach to system modelling. They are particularly common in a control context, but no means restricted to it. Some of the modelling principles in state space systems are illustrated by a simplified model of a regional system. The chapter closes with a detailed discussion of the model of the arms race between two nations due to Richardson (1960). This model illustrates the application of many of the state space techniques discussed in the chapter, as well as how state space system approaches can be used to draw conclusions about system stability and the practical implications of such conclusions.

Moves towards sustainable development generally require change. This occurs in a political and ideological context and is facilitated by the effective implementation of policies and understanding of the barriers to change. Thus chapter 5 discusses mental models, sense making and organisational structures and how understanding of the associated issues can be used to increase understanding of the organisational, psychological and social barriers to change, with the aim of overcoming them. It also considers the use of multi-loop action learning as a tool for achieving changes in behaviour and attitudes at the individual, organisational and societal levels. The ideas presented in this chapter are illustrated by a discussion of the systemic factors, including lack of a safety culture, which led to the catastrophic accident at the Union Carbide Plant in Bhopal in 1984.

Mathematical modelling approaches to understanding systems were discussed in chapter 4. Chapter 6 presents systems approaches based on structuring a problem and working through a number of stages, either sequentially or as an iterative process with backtracking to increase understanding of the problem, highlight important issues and either obtain a solution or clarify what is required in order to do so. The systems methodologies considered include systems engineering approaches, the SIMILAR Method, Checkland's Soft System Methodology and the Systems Failure Method. As discussed in the text, these methods have a number of similarities, as well as very important differences, and there can be a role for applying more than one method to an intransigent or important problem and comparing the solutions obtained.

Part I concludes with chapter 7, which discusses the application of the SIMILAR method and Checkland's Soft Systems Methodology to waste management in Glasgow and the Clyde Valley in Scotland. The case study illustrates the application of the two methods to increase understanding of a complex problem and suggest potential improvements in the situation, as well as the similarities and differences between the two approaches. Chapter 7 also contains tutorial exercises which can be used to give readers practice in some of the techniques presented in Part II, as well as to gain a deeper understanding of the different methods and methodologies and their advantages and drawbacks.

I.2 Learning Objectives

Readers should gain an understanding of systems approaches and their application to sustainable development. Specific learning objectives for Part I include the following:

- Understanding of the term system and the basic concepts in systems theory.
- Understanding of why systems approaches are applicable to problems associated with sustainable development, as well as their advantages and drawbacks.
- Understanding of the mathematical and engineering background to systems, including the difference between linear and non-linear,

feedback and feedforward, and open and closed systems and their differences from open and closed systems.

- Understanding of the main principles of state space approaches and the ability to apply them to problems in sustainable development
- Understanding of how systems approaches can be applied, for instance through the use of mental models, to clarify the barriers to implementing policies for sustainable development.
- The ability to apply a number of systems methodologies to sustainable development problems, including systems engineering methods, the SIMILAR Method, Checkland's Soft Systems Methodology and the Systems Failures Method, as well as the ability to choose the appropriate methodology or combination of methodologies.

2 Introduction to Systems Ideas

2.0 Learning Objectives

The main aims of this chapter are to introduce some of the systems concepts which will be developed further in later chapters and to present some of the systems approaches already being used in sustainable design and production. Specific learning objectives include the following:

- Understanding of what is meant by 'systems thinking'.
- Understanding of the differences between hard and soft systems.
- Understanding of the role and applications of general systems theory.
- Knowledge of some of the different systems approaches that have been applied to sustainable design and production, the differences between them and their advantages and disadvantages.

2.1 Introduction

This chapter introduces some of the basic concepts in systems theory. In particular the ideas of systems thinking and the differences between hard and soft systems are discussed in section 2.2. Section 2.3 considers general systems theory and section 2.4 presents systems approaches to sustainable design and production. The chapter is summarised in section 2.5

2.2 Systems Thinking and Hard and Soft Systems

Systems or *systemic thinking* involves consideration of the 'emergent' properties of the whole system, which generally do not hold or are not meaningful for the component parts. For instance a bicycle is a very flexible and environmentally friendly means of transport when its components are connected in a particular way, but the individual components, such as the wheels, the frame and the handlebars, cannot be used as a means of transport on their own (Checkland and Scholes 1999).

Systems thinking generally involves comparing constructed abstract wholes, often called systems models, with the perceived real world in order to learn about it to increase understanding or carry out some activities. There are two complementary traditions of systems thinking: *hard* or *engineering systems* (Blanchard and Fabrycky 1990; Chestnut 1967; Klir 1972; Sage 1992) approaches consider the world to be systemic, whereas *soft systems* (Checkland 1996; Checkland and Scholes 1999; NHS 1996) approaches consider the process of investigation itself to be a system. Systems thinking provides a methodology for investigating a wide range of problems based on the following two premises (Kramer and Smit 1977):

- Reality is considered in terms of *wholes*.
- A system is found within an *environment* with which it interacts i.e. open rather than closed systems are considered (see section 3.12).

The main properties and differences between the hard and soft systems approaches are stated in table 2.1. However both approaches can often give useful insight into problem situations and contribute to their resolution. The choice of appropriate methods and methodologies will depend largely on the nature of the problem.

Table 2.1 Comparison of hard and soft systems methodologies

Hard Systems	Soft Systems
System objectives are *precisely stated* at the start.	System objectives are allowed to *evolve* as understanding of the problem develops.
The system of interest is *already defined* and the problem is to define objectives and then meet them.	Consider *unstructured problems* in which defining the system of interest is an important part of the problem.
Only the *problem* being investigated is considered to be a system.	Both the *problem* and the *process of investigation* are considered to be systems.
Generally *quantitative* and *mathematical*.	Generally *qualitative* and *descriptive*.

2.3 General Systems

General systems theory aims to have a unifying effect and counteract the increasing fractionation of science and to prepare definitions and classifications which are likely to generate fruitful theories in the narrow sense (Rapoport 1972). It draws on the general tendency toward integration of the different natural and social sciences and the *similar problems and concepts* which have developed in widely different fields, generally totally independently of each other. For instance exponential growth laws hold for populations of bacteria, humans and animals and the number of publications in genetics or other areas of science (Bertallanfy 1968). General systems theory can be considered to be a general science of *wholeness*. It concentrates on the development of methodologies and approaches that are applicable to all or at least a wide range of different types of systems within a range of different disciplines. However, it also allows the development of methodologies for subclasses of systems. The aims of general system theory include the development of unifying principles for the individual sciences and the integration of scientific education (Bertallanfy 1968). General systems models can often be developed and applied when there is insufficient information about a system and its operation to obtain a detailed mathematical model. The general systems model can then be used to provide a basis for further study or more detailed analysis (Weinberg 1975).

General systems methodologies include the use of analogies between pairs of systems to derive equations or otherwise obtain additional information about one of the systems. The use of analogies will be most effective if it can be used to give a strong *isomorphism* (Rapoport 1972) or one-to-one mapping between specific entities. This requires two systems S_1 and S_2 to be defined, with S_1 the system of interest. The analogy is based on specific characteristic(s) of the two systems and requires both systems to have appropriate characteristics that can form the basis of the analogy. A transformation T or set of mappings is then defined between the main characteristics of the two systems. If the transformation T transforms the characteristic C_2 in the system S_2 into the characteristic C_1 in the system S_1, then the system S_2 can be used as a model of the characteristic C_1 in S_1. If this holds for all the significant characteristics of the system S_1, then, as illustrated in fig. 2.1, the system S_2 is a model of S_1 and can be transformed into S_1 using the transformation T.

The *applications of general system theory* include the following (Mesarovic and Takahara 1975):

- Study of systems with uncertainty or systems in new fields of work, where the development of a general systems model provides the basis for further study or more detailed analysis. General systems models require very few assumptions and can therefore be applied to, for instance, social and political systems, where more restrictive models give an unsatisfactory approximation.
- Derivation of a common approach or unification of the different more specialised branches of systems theory, for instance to help in model building.
- Provision of a means of studying and describing large scale and complex systems, with complexity resulting from the ways in which the relationships between the variables are described or the number of details considered. The use of a general systems model which is less structured and concentrates only on key factors may facilitate analysis. This differs from traditional approximation methods which omit less important system components, by considering the system as a whole, but from a less detailed viewpoint.
- Precise definition of concepts and interdisciplinary communication. For instance, the different understandings of the term adaptation used in psychology, biology and engineering can be expressed in terms of general systems theory and then compared.
- Provision of a precise language for multidisciplinary problems and interdisciplinary communications. This allows knowledge and concepts developed in one field to be applied in others or interdisciplinary problems to be studied without the artificial barriers of discipline specific language.
- Facilitating the synthesis and organisation of human knowledge in diverse, but related areas.

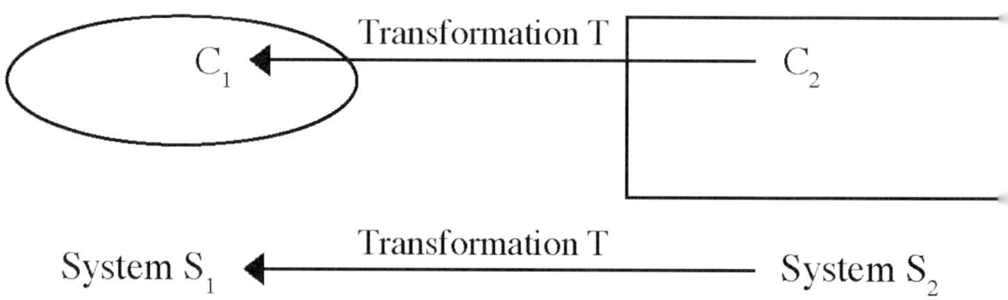

Fig. 2.1 Representation of a given system S_1 by a general system S_2

2.4 Systems Approaches to Sustainable Design and Production

2.4.1 Introduction

This section discusses systems approaches to sustainable design and production and some of the methodologies that have been developed to support these processes. A *systems approach to sustainable design and production* requires an understanding of the various processes involved and the materials used as well as the range of potential impacts on the environment, society and development, including equity factors. Environmental impacts can loosely be divided into two categories: *excess resource* (including energy) *consumption* and *excess waste* (including hazardous waste) *generation* beyond the source or sink capacities of natural ecosystems. *Social factors* include effects on the availability and conditions of employment, exposure to emissions and toxic substances and political and civil rights. *Developmental factors* include universal access to high quality education and health care, clean water and nutritious food, as well as a good quality of life. Development requires the appropriate use of technology, rather than necessarily a continuing spiral of technological development. It should also be noted that continuing economic growth is not sustainable (Meadows et al 1972 1992) and may not lead to positive developmental impacts for all sections of the population. *Equity issues* are very important in the context of development, since in the past development has occurred at highly differential rates and the very high rates of consumption of resources and generation of waste in the richer, more industrialised countries may have had a negative effect on development elsewhere.

Short-term strategies for making production more sustainable will probably require tradeoffs between the reduction of different types of environmental, social and developmental impacts. In the long term it will be necessary to ensure that natural source and sink capacities are not exceeded. Industrial production generally has impacts on the local and possibly also the global natural environment, as well as some social impacts. The extent of these impacts will depend on a number of factors, including the type of production, the size of the firm and the measures taken to reduce negative social and environmental impacts. However, it is generally only the larger and, in particular, multinational firms which have a significant impact on development in the long term and over a wider area, though the combined effect of a large number of small firms is also likely to be significant. Another very important consideration is *human centred design* and the *user friendliness* of green products and processes.

A related concept is *design for all* (which is also called universal design) and the aim of which is the design of general purpose products and devices so they can be used by as wide a range of different people as possible, including disabled people. The concepts of universal and sustainable design can be combined to give what can be called *eco-design for all* (Hersh 2001 2002ab). In some cases design for all may require products to be designed to be compatible with appropriate assistive technology devices rather than to be designed directly to be used by disabled people (Hersh 2002b).

As discussed in the introduction to Part I, systems approaches provide (holistic) techniques for structuring problems to ensure the consideration of factors that affect the whole problem and interactions between the components. Structuring of products and processes often involves division into a number of different *life cycle stages*, typically five or six. This allows the environmental, social and other impacts at each stage to be considered separately, which can often be helpful in deriving strategies for reducing negative impacts, as well as consideration of effects over the whole life cycle.

A commonly used *list of life cycle stages* is (Graedel and Allenby 1995; Rhodes 1995):

- raw material acquisition,
- material processing,
- manufacture,
- transport and distribution,
- use, reuse and maintenance
- disposal or recycling.

Example 2.1: Impacts of Computer Life Cycle Stages

An example of the use of life cycle stages to structure the representation of environmental impacts is given in table 2.2, which presents the main wastes produced at the six computer life cycle stages (Hersh 1998). An increasing ten-point scale indicates the importance or volume of each type of waste. Therefore the value of 2 taken by precious metal (waste) at the disposal and recycling life cycle stage shows that only relatively small amounts of precious metal wastes are produced at this stage, whereas the value of 9 for paper and card at the use and reuse life cycle stage shows that very large quantities of paper and card are consumed (and disposed of) at this stage. Thus, the structure based on six life cycle stages helps clarify which wastes are the most significant at each stage. However decision making on strategically targeted impact reduction also requires information

Table 2.2 Some of the wastes generated by the six computer life cycle stages

Life Cycle Stage	Water	Carbon Dioxide	Other Air Emissions	Hazardous Wastes	Lead Glass	Plastics	Iron	Other Metals	Paper/ Card	Rock/ Soil	Precious Metals
Material acquisition	4	4	4	2						7	
Material processing	3	3	3	2							
Manufacture	6	7	4	5	5	5	5	3			
Tranport & distribution		6	6						4		
Use & reuse		4	4			4			9		
Disposal/ Recycling				1	3	4	6	3	3		2

on the global significance of each type of waste, the type of problems it causes and whether and how this waste can be reduced without being replaced by other types of waste or generating other unsustainable activities.

2.4.2 Systems Methods for Sustainable Design and Production

In response to increasing pressure by organisational stakeholders, environmental management has become more strategically oriented. There are a number of different systems based approaches to integrating environmental issues into industry, such as design for the environment (DFE), life cycle analysis (LCA), total quality environmental management (TQEM), ecofusion and green supply chain management. There are also a number of national and international environmental standards, of which the most recent is the *International Standards Organisation (ISO) 14000 series* (Sarkis et al 1996, Hillary 2000). *DFE* philosophy requires consideration of environmental, health and safety criteria over the complete product life cycle in the design process (Allenby and Pugh 1992; Fiksel 1996; Lenox and Ehrenfeld 1995; Paton 1994). The properties of the materials used in the manufacturing process, such as their recycling and reuse capabilities and their (potential) long term impacts on the environment (Sarkis et al 1996), must all be considered. Other important considerations are the energy consumption and efficiency of the manufacturing process and ease of both assembly and disassembly. DFE requires environmental considerations to be introduced right from the start of product design or upgrading to ensure that their effects on the environment are understood before manufacturing decisions are finalised. A related concept, *product stewardship* (Bast and Korpalski 1996; Korpalski 1996), uses three main tools: DFE guidelines, product assessment and product stewardship metrics.

DFE tools range from simple information resources to complex software-based systems. However common features include checklists, design standards and guides and databases of chemicals and materials. Features considered to be of importance include:

- a database of environmental information on topics such as environmental legislation and policy, pollution and emissions, air and water quality, habitats and biodiversity.
- full integrability with computer-aided engineering and design or existing design tools.
- support for technical tradeoff studies in real time
- wide applicability.
- general compatibility with standards and current engineering design and management.
- the facility to provide general rankings or ratings of the environmental performance of different products and processes.

DFE focuses mainly on the design process, whereas LCA is targeted at analysis of the design and involves an examination of all aspects of product design from the preparation of input materials to end use. This includes evaluation of the types and quantities of both input materials, such as energy and raw materials, and product outputs, including atmospheric emissions, solid and aqueous waste and the end product. It also includes evaluation of ways of reducing the environmental impacts of products and processes.

LCA has four interrelated components: inventory analysis, impact analysis, life cycle costing and improvement analysis (Alting and Legarth 1995; Graedel and Allenby 1995; Keoleian and Menerey 1993), though life cycle costing is often omitted (Graedel and Allenby 1995). The inventory analysis stage obtains quantitative and qualitative information that fully describes all life cycle stages, whereas impact analysis accurately assesses the influences of the activities identified by the inventory on specific environmental properties and gives a priority ranking to the relative seriousness of any changes in these properties. Obtaining this priority ranking may involve subjective as well as scientific factors.

A number of different approaches can be used to obtain and present the results of overall impact assessment. One approach uses a series of *matrix displays*, with sources (specific activities or industries) and critical properties (specific impacts) as row and column headers. Part of a typical matrix display is illustrated in fig. 2.2. In the case of the environmental impacts over the computer lifecycle, the columns are labelled with the different environmental impacts, such as energy and resource consumption, and the rows with the different life cycle stages. A single cell shows the (relative) seriousness of a given type of environmental impact at a particular life cycle stage, as well as the degree of reliability of this information. The use of a graphical representation facilitates comparison of the impacts of different life cycle stages (Graedel and Allenby 1995). Inclusion of both the significance of the impacts and the reliability of the data is important, as the different types of data may have been collected in very different ways, including by estimation, and have very different degrees of accuracy. *Improvement analysis* applies DFE techniques to the results of the first three LCA stages to improve the environmental plan of the process, product or facility.

Ecofusion (Miyamoto et al 1996) involves the application of product list, composition and life cycle flow windows to environmental valuation of products described in terms of their composition and life cycle stages. It has three units: environmental product assessment, LCA and assembly/

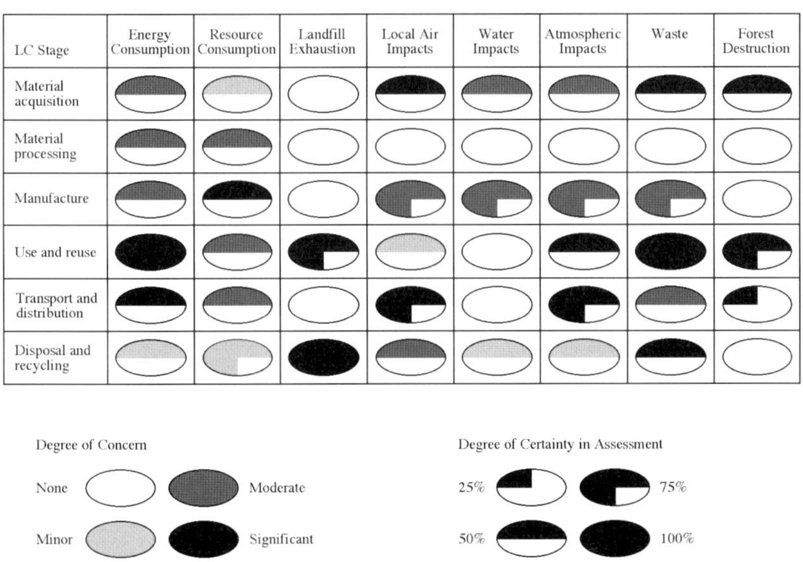

Fig. 2.2 Matrix display: environmental impacts of computer life cycle stages

disassembly evaluation, with associated techniques. *TQEM* (GEM 1992) focuses on the management and elimination of wastes and the continuous improvement of processes and systems. *Green supply chain management is* linked to the external relations of manufacturing firms and involves logistics planning (Sarkis et al 1996).

Industrial ecology provides a systems framework for environmental management that treats the industrial world as a natural ecosystem, which forms part of local ecosystems and the global biosphere (Lowe 1993 1996, Bourg and Erkman 2003). This approach can be used to achieve sustainable environmental performance by emulating characteristics of natural systems, such as eliminating the concept of 'waste' (as something that cannot be absorbed constructively elsewhere in the system), making each ecosystem element multifunctional and continually circulating and transforming materials. The approach allows for the assumptions of either unlimited resources and waste, limited resources and waste and no waste (Lowe 1993), with the second case i.e. limited resources and waste being the most realistic.

2.5 Roundup of Chapter 2

Chapter 2 introduced the concept of systems and some of the main ideas in systems theory. In particular systems have emergent properties, so that the behaviour of a system is more than the sum of its component parts and a system is situated in an environment with which it interacts. One of the categorisations of systems methodologies is into hard and soft. Hard systems approaches are quantitative, mathematical and based on clearly defined problems, whereas soft systems approaches are qualitative, descriptive and involve determining the problem to be considered. There are a number of different systems approaches to sustainable design and production, including life cycle analysis and design for the environment.

Chapter 3 continues the general introduction to systems theory in chapter 2 by presenting the mathematical basis of systems theory, which underlies the hard and engineering approaches to systems.

3 Engineering Mathematics Representations

3.0 Learning Objectives

The aim of this section is to present the main concepts in engineering mathematics representations of systems. Some illustrations of the direct applications of these concepts to mathematical modelling in the context of sustainable development are presented here. Many of the later chapters in both Part I, as well as Parts II and III will draw on these concepts. Specific learning objectives include the following:

- The ability to use mathematical notation to represent systems.
- Understanding of some of the main systems categories, including linear and non-linear, open and closed, open and closed loop, discrete and continuous time.
- Understanding of the relevance of systems models to sustainable development.

3.1 Introduction

This chapter will present some of the main concepts in engineering mathematics representations of systems, including linear and non-linear (section 3.2), discrete and continuous time (section 3.4), open and closed (section 3.12) open and closed loop (section 3.5), feedback (section 3.5) and feedforward (section 3.6), deterministic and stochastic systems (section 3.7) and time and frequency domain (section 3.8). The transfer function and impulse response are also introduced (section 3.13). These concepts will be drawn on later in the book, particularly in chapter 4 and the modelling stage of the SIMILAR Method in chapter 6. They are also essential for understanding the engineering systems literature. A summary of the chapter is presented in section 3.16.

The term engineering mathematics representation is used to indicate that simple functional expressions, often based on input/output relations, are generally used rather than topological or set relations. Many of the definitions and types of systems given in the subsequent subsections can be applied in both a hard and soft systems context, though, other than feedback, they are more commonly used in hard systems approaches.

A general system S can be defined as a *relation on a collection of abstract sets* or, in more mathematical terms, as (Mesarovic 1972)

$$S \subset X\{V_i: i \in I\} \quad (3.1a)$$

where X denotes a Cartesian or cross product and I is the index set. The Cartesian or cross product of two sets V_1 and V_2 is the set of pairs (v_1, v_2) where v_1 and v_2 are members of the sets V_1 and V_2 respectively. This definition can easily be extended to the Cartesian product of all sets V_i with i in the index set I. This gives an ordered set with jth term v_j where v_j is a member of the set V_j and j is in the index set. When the index set is finite, the expression (3.1a) can be rewritten as

$$S \subset V_1 \times V_2 \times \ldots \times V_n \quad (3.1b)$$

The components V_i, $i \in I$ of S are called the *systems objects*. Expressions (3.1a) and (3.1b) are very general and include the specific cases when the system can be expressed as a set of differential or difference equations, as well as the case when there is incomplete information and the system is described by a set of verbal statements. Mathematical system representations can be divided into *input-output* and *goal seeking*. In the case of single input single output systems, expression (3.1b) reduces to the simple case

$$S \subset V_1 \times V_{21}$$

which, putting $V_1 = X$ and $V_2 = Y$, can be rewritten as

$$S \subset X \times Y \quad (3.1c)$$

where X and Y are the input and output objects respectively.

Alternatively the system can be described by a set of rules that associates an output time function y(t) with an input time function x(t) as, for example,

$$y(t) = F[x(t)] \quad (3.2)$$

Input-output representations can be divided further into transfer function and state space representations. *Transfer function representations* give a relatively simple representation of the relationships between the system inputs and outputs, but are restricted to the case of time invariant linear systems. In the case of multiple inputs and outputs the representation is given by a matrix of transfer functions, with each entry representing the relationship between an input-output pair. *State space representations* give the relationship of the system inputs and outputs to the internal states of the system as well as the relationship between the inputs and outputs and can also be used for nonlinear and time varying systems. They are generally expressed in terms of a set of first order differential or difference equations and will be discussed further in chapter 4.

In goal seeking approaches the desired behaviour of the system is defined in terms of a goal seeking process, such as a cost function to be minimised. This includes optimal control. Thus the relationship between the system input(s) and output(s) is obtained indirectly from the goal seeking process, rather than stated directly, as in the case of input-output representations.

3.2 Linear Systems

One of the important distinctions is between *linear* and *non-linear* systems. A *linear* system satisfies superposition and scaling. The *superposition* property states that the output of a system with two (or more) inputs is the sum of the outputs of the individual systems i.e.

$$F(x_1 + x_2) = y_1 + y_2 \tag{3.3a}$$

Systems which have the *scaling* property are scale invariant. In a scale invariant system multiplication of an input by a constant factor has the effect of multiplying the output by the same factor i.e.

$$F[ax(t)] = ay(t) \tag{3.3b}$$

Therefore the output of a linear system with two (or more) weighted inputs is the weighted sum of the outputs of the individual systems.

$$F(a_1x_1 + a_2x_2) = a_1y_1 + a_2y_2 \tag{3.4}$$

The term linear is used as linear functions of one variable can be represented by a straight line. Linear functions of two variables can be represented by a plane. Systems which do not satisfy (3.4) are nonlinear. Most real systems are non-linear to some extent. However, in some cases it can be assumed that the system is approximately linear. Otherwise it is often possible to linearise a non-linear system about an equilibrium or steady state point. However the linearisation only holds close to the equilibrium point. When there is more than one equilibrium point, it will in general be possible to obtain a different linearisation about each equilibrium point.

Example 3.1: Production of Computer Monitors

Many industrial processes are assumed to be approximately linear with regards to the resources consumed and wastes generated. This means that, for instance, manufacturing 100 CDs will consume 100 times the materials and generate 100 times the waste of manufacturing one CD. An assumption of linearity makes it possible to inventory the different categories of environmental impacts of an industrial process and use the impact per unit product to calculate the total impacts for a given production line over a particular period.

For instance IBM used data from a life cycle inventory of PVC manufacturing to inform decision making about materials to provide electromagnetic shielding and the decorative finish for the top cover of a monitor (Brinkley et al 1996). IBM was able to use this data, since the manufacturing process is linear with respect to material input and material consumption is linear with respect to environmental impacts, such as emissions. Therefore, for instance, manufacturing 1000 covers requires 1000 times the weight of PVC required for one cover and processing 1000 kilograms of PVC resin emits 1000 times the weight of carbon dioxide and nitrogen oxides as processing one kilogram of PVC resin. The assumption of linearity is clearly appropriate in this case. However linearity could no longer be assumed if, for instance, the type of manufacturing process used depended on the number of monitor covers being produced.

3.3 Time Invariance

In a *time invariant* system the response due to an input is not dependent on the time of occurrence of the input, so that a *time shift* in the input signal causes an equal time shift in the output waveform i.e.

$$y(t-T) = F[x(t-T)] \text{ for any } T. \tag{3.5}$$

Systems which are not time invariant are time varying.

Example 3.2: Growth of Population with Limited Resources

Consider the following equation which is used in sociology to describe the growth of human populations with limited resources (Bertalanffy 1968).

$$\dot{x} = a_1 x + a_2 x^2 \tag{3.6a}$$

The system in (3.6a) is time-invariant as the coefficients a_1 and a_2 are constants rather than functions of time, but non-linear as the equation contains a term in x^2. There is thus an implicit assumption in this model that the limited resources remain approximately constant over time. However if the limited resources varied with time the system could become time-varying, as in equation (3.6b)

$$\dot{x} = a_1(t)x + a_2(t)x^2 \tag{3.6b}$$

Equation (3.6a) is an example of the logistic curve and is discussed further in section 4.4.

3.4 Discrete and Continuous Time Systems

In *discrete time systems* values of the variables are only given at discrete instants of time, whereas in *continuous time systems* the signals and variables are functions of a continuous variable, which is defined for all values of time. A discrete time signal only has information on the values of the variables at the sampling instants and does not have any information on their values between these time instants. In discrete systems the relations between the variables are generally defined by *difference equations*, whereas they are defined by *differential equations* in continuous systems.

Discrete time systems only produce outputs at discrete times. These time intervals are often regular, of size T. The value of the signal at the kth time interval is therefore x(kT). Since the sampling interval T is constant for a given signal or system, it is often omitted and the kth signal value written as x_k. A discrete time signal therefore consists of the set of

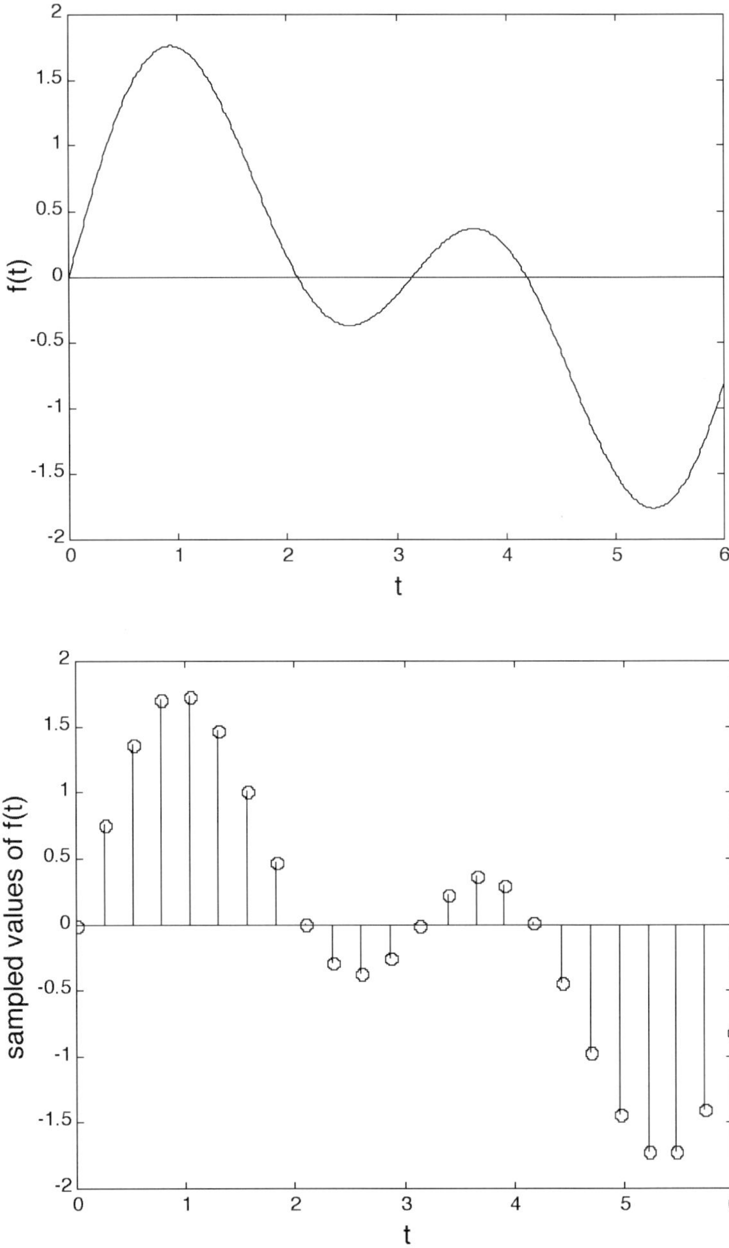

Fig. 3.1 Obtaining a discrete from a continuous time signal by sampling

values $\{x_k\}$for all integer values k.

Continuous time signals can be converted to discrete time signals using an *analog-to-digital converter* which *samples* the signal every T seconds to produce a sequence of values x(0), x(T), x(2T), ... called a *sampled data* or *discrete time signal.* These values are stored as digital words i.e. as an ordered arrangement of binary digits or bits. The term sampled data signal is often used for discrete time signals even if they are not obtained from sampling. T is called the *sampling interval.* The value of T will depend on the type of signal being sampled, for instance it may be 100 μs for a speech signal and 1 second for a central heating temperature signal. The sampling frequency or rate is 1/T Hz or 2π/T rad/s. Thus a speech signal has a sampling frequency of 10,000 Hz and a central heating temperature signal of 1 Hz.

According to the sampling theorem, the sampling frequency should be at least twice the highest significant frequency in the continuous signal. This will prevent confusion between the original signal and aliases of high frequency components and allow the continuous signal to be reconstructed from the discrete version. Any naturally occurring continuous signal can be converted into discrete time or digital form by sampling and then processed digitally. Sampling to give a discrete time signal is illustrated in fig. 3.1.

The discrete time signal can be converted back to a continuous time signal using a zero order hold, defined by

$$u(t) = u(nT) \quad \text{for } nT \leq t < (n+1)T \tag{3.7}$$

and illustrated in fig. 3.2.

Example 3.3: Global Population Growth

There is considerable concern about the (exponential) growth in human

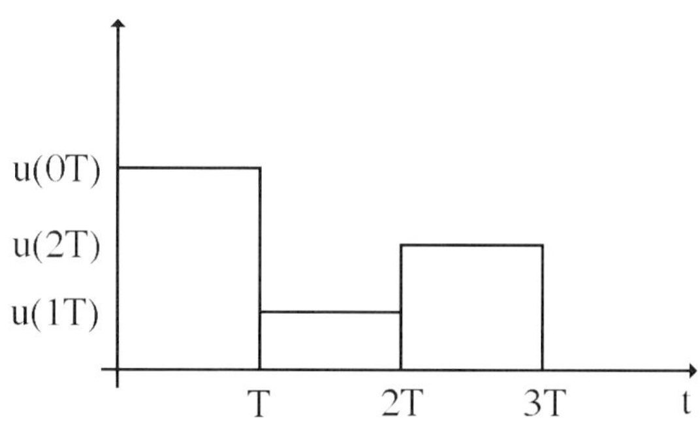

Fig. 3.2 Zero order hold

population and the likely effects of this increasing population on the carrying capacity of the planet and, in particular, on its source and sink capacities i.e. the ability to produce resources and absorb waste. Therefore population data is often extrapolated to try and determine probable future population levels at different points in time and whether and at what level a steady state will be reached.

Global or national population is a continuous function of time since events which lead to sudden increases or decreases in population, such as wars and plague, are generally localised and occur over a period of time. This results in smoothing of the global or national population function and avoidance of any discontinuities. However population census data is only

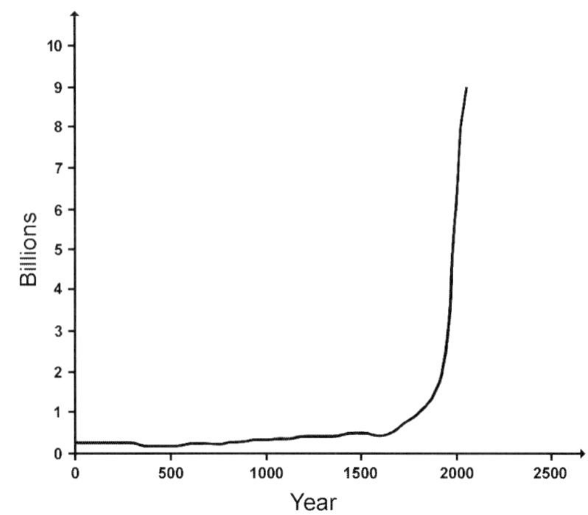

Fig. 3.3a Continuous time representation of population data

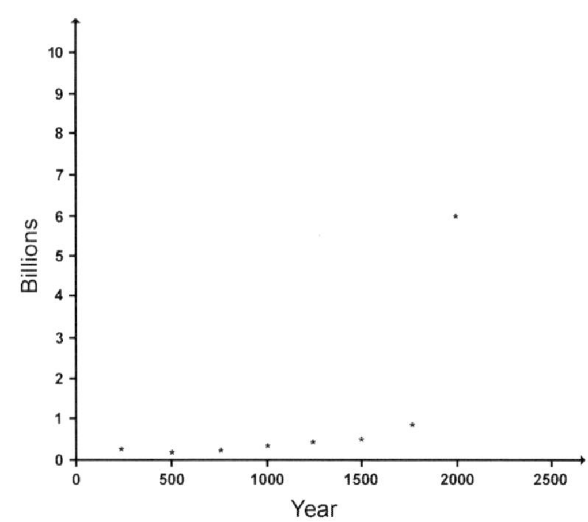

Fig. 3.3b Discrete time representation with sampling time 250 years

collected once every ten years in many countries. This is equivalent to discretisation with a sampling time of ten years. In the intervening period population is estimated annually or quarterly using data on births, deaths, immigration and emigration. When population data is plotted the digital data is often used to produce a continuous graph by interpolation between the points to give a smooth graph. Part of the reason for this is the small scale on the time axis. However the data could equally be presented in discrete form, either as a set of points or using a bar chart. Figs. 3.3a and 3.3b respectively show continuous and discrete time representations of population data.

3.5 Open and Closed Loop Systems

In *open loop* systems (fig. 3.4a) there is no mechanism by which the output is able to affect the input. In *closed loop* systems (fig. 3.4b) information about the output is fed back and used to update the input. Thus closed loop systems have the advantage over open loop systems of being able to use system information to *respond to disturbances* and *changes in the system parameters*. Many real systems have interactions between their elements which can be modelled using feedback.

In an open loop system the relationship between the output y(t) and the input r(t) is as follows

$$y(t) = G_p r(t) \qquad (3.8)$$

showing that there is a fixed relationship between the output and input.

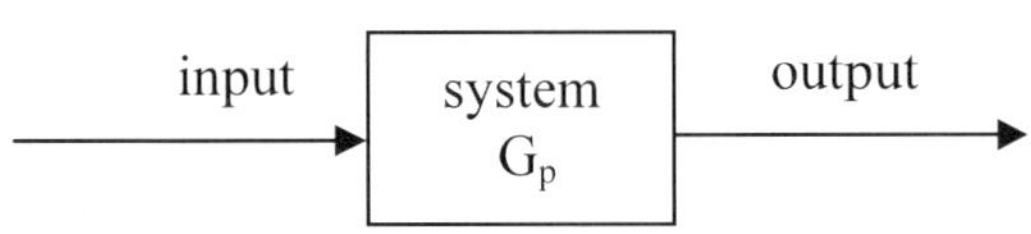

Fig. 3.4a Open loop system

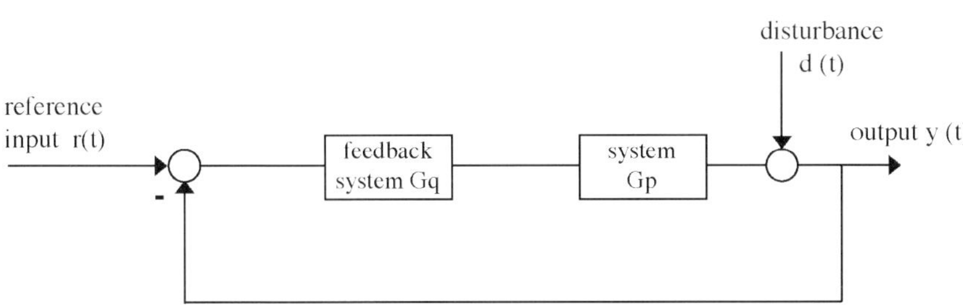

Fig. 3.4b Closed loop system with negative feedback

In a closed loop system with negative feedback, the relationship between the output y(t), reference input r(t) and disturbance d(t) can be obtained as follows:

$$y(t) = d(t) + G_pG_c[r(t) - y(t)]$$

$$\Rightarrow y(t) = \frac{G_pG_c}{1+G_pG_c} r(t) + \frac{G_pG_c}{1+G_pG_c} d(t) \qquad (3.9a)$$

Therefore in the feedback case the output is the sum of two terms due to the reference input and the disturbance respectively. Changing the gain G_c of the feedback system can be used to reduce the impact of the disturbance.

There are two different types of *feedback*: negative or balancing feedback and positive or reinforcing feedback. In *negative feedback systems*, which include many industrial control systems, output values are fed back and compared with a reference value and the resulting error signal is used to control the output behaviour, for instance to follow a particular trajectory. This type of feedback can be designed to keep the system stable and maintain it in a desired state or drive it to a goal.

In *positive feedback systems*, rather than the output being subtracted from the reference signal to give an error signal, it is added to it. This tends to reinforce existing behaviour rather than regulate it and can therefore lead to exponential growth and instability. Examples of positive feedback leading to instability include excessive consumption and resource use. However positive feedback can be used to reinforce desired as well as undesirable behaviour. For instance in education positive feedback acts as positive reinforcement which encourages further learning. Positive feedback is very rarely used in industrial control systems, but can occur in natural and socio-economic systems. In the case of positive feedback the output is given by

$$y(t) = \frac{G_pG_c}{1-G_pG_c} r(t) + \frac{G_pG_c}{1-G_pG_c} d(t) \qquad (3.9b)$$

The only difference from the negative feedback equation (3.9a) is the minus sign in the denominator. Feedback is particularly important when there are disturbances in the system or its behaviour is changing, as is generally the case with social and environmental and other real systems.

Feedback systems can also be further divided into the following four categories:

- Natural systems with natural feedback, such as skin temperature control through perspiration cooling.
- Natural systems with human-devised feedback control, such as increased agricultural production through the use of crop rotation, irrigation and organic compost, or the containment of a forest fire caused by lightning.
- Human-devised systems with human-devised feedback control, such as control systems for regulating industrial processes or propulsion control in rail transit systems.
- Human-devised systems with natural feedback, such as industrial production or energy consumption systems, with natural feedback control resulting from the limited availability of energy and other resources.

It is the combined natural and human-devised systems that are particularly important in the context of sustainable development. For instance, although natural feedback mechanisms are effective, there is a limit to the available natural resources and attempts to consume them at too high a rate will lead to instability and unsustainable development.

In the case of natural systems with human-devised feedback control, there are issues of the appropriate steady state or equilibrium level, as well as the environmental, social and developmental consequences of the control mechanisms used and the possible disruption of natural feedback mechanisms. Consideration of several other feedback loops may be necessary to determine the consequences of human modifications of natural feedback systems.

In addition although feedback control systems can be designed for systems with human elements, implementing this type of control is not as easy as for industrial processes. For instance, the control measures may require unpopular changes in behaviour. Some of the barriers to implementing feedback or other measures to achieve sustainability are discussed in chapter 5.

Example 3.4: Regulation of Traffic Growth

The feedback of information on traffic growth and the resulting state of road congestion has led to the construction of new roads to provide additional capacity. The increase in road capacity has then resulted in further traffic growth to fill this additional capacity. Overshoot occurs, so that the new road capacity is exceeded and renewed congestion recurs in a fairly short time span. A more sustainable approach would be to use negative feedback in combination with regulatory measures to track target levels of road use. Initially this approach could be used to reduce traffic levels and subsequently maintain them at an appropriate level. Actual traffic levels could be fed back and compared with desired levels and the extent of the difference used to determine the type and extent of the regulatory measures required. This is illustrated in fig. 3.5.

In the case of traffic regulation, regulatory measures should generally involve promotion of a modal shift from private car use to public transport, cycling and walking and encouragement for facilities to be sited near population centres to reduce the need to travel. However many local and national politicians have been reluctant to introduce more than token measures of this type due to concerns that they will be unpopular.

Feedback of information on the system output(s) can be used to *update the system behaviour* to make it conform to desired specifications. Therefore it is important that this information should be reasonably accurate. However a number of different factors can result in misleading feedback. For instance extreme events occur, but any predictions based on them could easily lead to serious errors. Unfortunately it is not always easy to identify extreme events and care has to be taken in the interpretation of data, not to confuse significant real shifts in data values

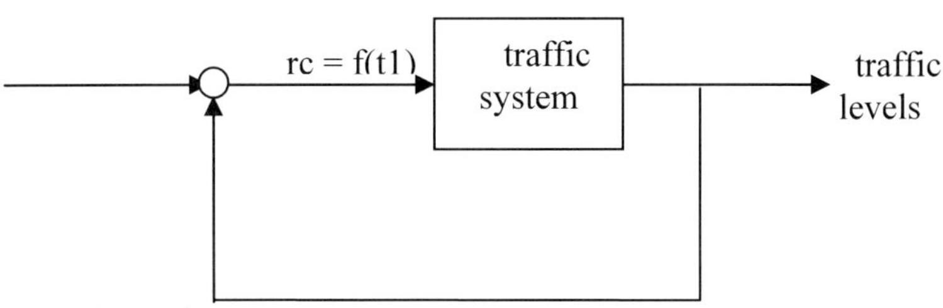

rc = road capacity
tl = traffic levels

Fig. 3.5a Positive feedback – road construction and traffic growth

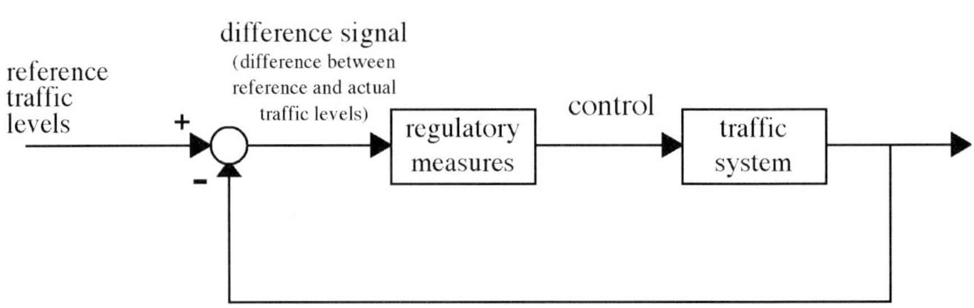

Fig. 3.5b Negative feedback – regulatory measures

with extreme events. To a certain extent this has occurred with climate change data. For a long time it was argued that observed changes, such as really high or low values of temperature and/or rainfall were due to natural causes rather than anthropogenic emissions.

Systems of many different types will experience limits on the extent to which change in a particular direction can occur, even in the presence of reinforcing loops i.e. positive feedback which boosts the response. This is due to the presence of negative feedback which provides *balancing loops*, which act to counter the reinforcing loop(s) and reduce the response for a given effort. This generally places limits on the final result and the extent to which change can occur in a given direction. In some cases this change may be desirable, whereas in others it is an unwanted direct or indirect consequence of system actions, often boosted by positive feedback.

For instance a particularly important case is the effect of human economic activity, supported by technological developments on the natural environment. Human economic activity, leads to resource consumption and waste generation. Despite the development of some green technologies, technological development acts as positive feedback or a reinforcing loop and contributes to growth in human economic activity, which leads to increased resource consumption and waste generation. However the ability of the planet to supply resources and absorb waste is limited.

Therefore resource depletion and the reduced sink capacity of the earth will (after a time delay) act as negative feedback or a balancing loop which will lead to reductions in human economic activity. However there is unlikely to be a smooth transition and the consequences of environmental collapse could be disastrous. It is therefore imperative that mechanisms are investigated to reduce resource consumption and waste generation before this point is reached.

3.6 Feedforward

In feedback (control) systems, information about system performance is fed back and used to update and improve system performance. In *feedforward* (control) systems a model of the system is constructed and used to obtain the input. In the engineering case the input is obtained from the model by calculating the *inverse dynamics*. In addition to its engineering applications, for instance in regulating industrial processes, feedforward has an important role in systems based on individuals and organisations, but not in natural and environmental systems. In human systems feedforward is based on the ability to *anticipate* the future. However, though most individuals and organisations can predict or guess the future to some extent, the ability to do this accurately is very limited and inaccuracies increase with the time span over which prediction takes place.

The successful use of feedforward can eliminate or reduce the effects of predictable disturbances. However it is critically dependent on the availability of an accurate model of the system. Thus an unsatisfactory response can occur for the following reasons:

- The system model is inaccurate. When the system of interest is non-linear, errors and the deviation from the desired path can grow very rapidly.
- The initial state of the system may not be accurately known, estimated or measured.
- Unlike in the case of feedback, it is not possible to correct for errors due to unmeasurable disturbances.

As illustrated in fig. 3.6, feedforward is generally used in combination with feedback to reduce the effects of system disturbances.

It should be noted that figure 3.6 is a frequency domain diagram which gives the relationship of the output to the input and disturbance in the frequency domain (see section 3.8). The associated time domain variables y(t), u(t) and d(t) could be obtained from them by inverse Laplace transformation and the relationship of the time domain output y(t) to the input u(t) and disturbance d(t) by inverse Laplace transformation of equation (3.10) below.

From fig. 3.6,

$$Y(s) = D(s) + G_p[-G_{FF}D(s) + G_c\{R(s)-Y(s)\}]$$

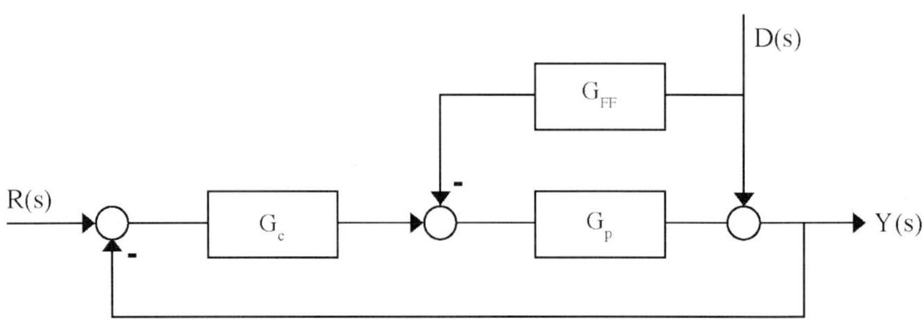

Fig. 3.6 Feedforward control

$$\Rightarrow Y(s) = \frac{G_pG_c}{1+G_pG_c}R(s) + \frac{1\text{-}G_pG_{FF}}{1+G_pG_d}D(s) \qquad (3.10a)$$

Therefore the effect of the disturbance D(s) on the output Y(s) can be eliminated by choosing

$$G_{FF} = G_p^{-1} \qquad (3.10b)$$

so that the numerator ($1\text{-}G_pG_{FF}$) of the disturbance term is zero. However, real systems are proper or strictly proper i.e. the order of the denominator is greater than or equal to that of the numerator. Therefore, when the system G_p is strictly proper from equation (3.10b) G_{FF} will have a higher order numerator than denominator and consequently not be realisable. This problem can be resolved and a realisable transfer function obtained by modifying G_{FF} by the addition of an nth order denominator term, with a repeated pole at $s = -\alpha$ where n is chosen to make G_{FF} strictly proper. This gives

$$G_{FF} = \frac{G_p^{-1}}{(s+\alpha)^n} \qquad (3.10c)$$

Problems can also occur when the system transfer function G_p is non-minimum phase i.e. has unstable zeros, since these become the poles of the feedback controller transfer function G_{FF} and would therefore make it unstable.

Feedforward can allow anticipation of future effects on the present, and this can lead to self fulfilling or self defeating prophecies, according to whether the feedforward is positive (reinforcing) or negative (balancing). For instance severe real shortages can be created by people purchasing additional goods in response to rumours of possible shortages (O'Connor and McDermott 1997). In the context of sustainable development feedforward can lead to unsustainable policies.

Example 3.5: Feedforward Control of Cooling Water

Many industrial and energy generation processes use very large quantities of cooling water. The most sustainable option is to operate a closed system in which the same water is repeatedly circulated, after purification if necessary. However many countries currently have regulations specifying the conditions to be met by water discharged into the environment, but do not require water to be reused. Therefore a considerable proportion of the cooling water used in industry is discharged to the environment (after appropriate processing).

Since many aquatic ecosystems are temperature sensitive, one of the important statutory conditions is on the temperature of the discharged water. This can be achieved by using feedback control to control the flow of water into a holding tank system from which the water is discharged into the environment. The feedback controller uses the error or difference in temperature between the discharged water and the desired outflow temperature (reference value) determined by the statutory requirements, as illustrated in figure 3.7a.

However, in many cases the inflow temperatures are subject to a wider range of variation than the feedback controller can satisfactorily regulate, leading to a rise in the outflow temperature. This additional temperature variation can be treated as a process disturbance which is easily measured. Thus, measurement of this temperature disturbance can be used in a feedforward controller to minimise its effects and recover good control of the outflow temperature, as illustrated in figure 3.7b. Therefore the use of feedforward in combination with feedback has improved the control of the outflow temperature compared with the use of feedback on its own, thereby avoiding damage to aquatic ecosystems through excessive temperatures.

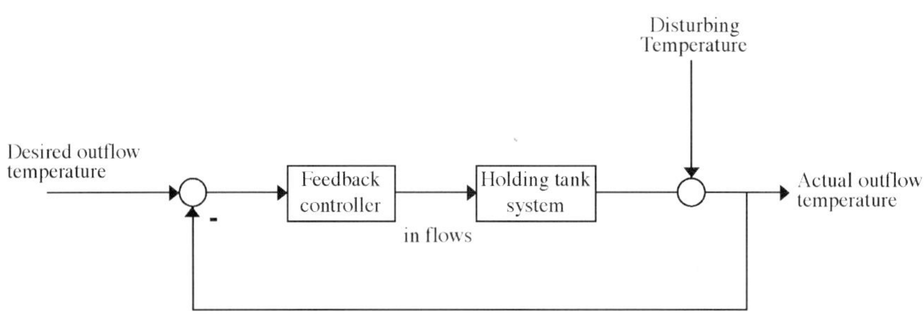

Fig. 3.7a Feedback control of outflow temperature of discharged water

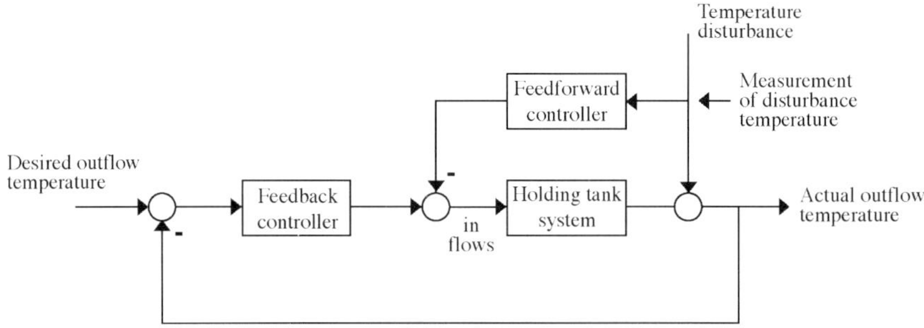

Fig. 3.7b Feedforward control of outflow temperature of discharged water

3.7 Deterministic and Stochastic Systems

Deterministic systems have determinate or precisely defined relationships between all output and input quantities, which can be expressed without the use of probabilities or statistical expressions. In *stochastic systems* it is not possible to express the relations between all pairs of output and input variables precisely and at least one of these relations requires the use of a conditional probability or stochastic relationship. Probabilistic systems can be further divided into simple (memoryless) and complex (sequential) systems. In simple probabilistic systems all the system variables are defined in terms of the instantaneous values of other variables, whereas in complex probabilistic systems some variables are defined in terms of a mixture of present and previous values of the other variables.

3.8 Time and Frequency Domain Systems

Most real systems can be modelled in terms of either *time or frequency domain variables* i.e. as either a function of time or frequency. Both representations are equally valid and which representation is most appropriate will depend on the context. *Transforms* can be used to move between the time and frequency domains. For instance Laplace or Fourier transforms can be used to convert continuous time domain variables and equations to frequency domain ones and z or delta transforms to move between the time and frequency domain for discrete time systems.

The *Laplace transform* of the function x(t) is given by

$$X(s) = \int_0^\infty x(t)e^{-st}dt \tag{3.11}$$

The *z-transform* X(z) of the sequence $\{x_n\} = \{x_o, x_1, x_2, ...\}$ is defined as

$$X(z) = Z(\{x_n\}) = x_0 + x_1 z^{-1} + x_2 z^{-2} + ... = \sum_{n=0}^{\infty} x_n z^{-n} \tag{3.12}$$

Laplace, z and other transforms have a number of *properties* (Bolton 1994; Craig 1964; Robinson 1968) which can be used to make it easier to calculate the transform. Although the exact mathematical form of these properties differs for the different transforms, the basic principles of properties such as the final and initial value theorems, time and frequency shifting, multiplication by the frequency variable in the frequency domain and convolution are similar for all the commonly used transforms. Closed mathematical expressions can be obtained for inverse transforms, but these are rarely used. Instead techniques such as partial fractions are used in combination with tables of transforms and the properties of the different transforms to obtain the inverse.

3.9 Causality, Dynamic and Instantaneous Systems

Systems can be classified according to whether the output depends only on current or also on past and/or future values of the input and output. A system is *causal* or *nonanticipatory* if its response to an input depends only on current and previous input and output values and not on future values. Causal systems have zero impulse response (see section 3.13) for negative t. In addition if two inputs to a causal system have the same values for $t \leq t_o$, then the outputs from this system will also take the same values for $t \leq t_o$.

A system for which the output is a function of the input at the present time only is *instantaneous* or *memoryless*. The output of a *dynamic* system depends on past or future, as well as present values of the input. If the system is causal and dynamic, its output depends only on present and past values of the input. Most real systems are causal and dynamic, but noncausal systems do occur, for instance in digital signal processing.

In many real situations cause and effect relationships are complex. Rather than a single well-defined cause, such as a particular event, leading to a well-defined single effect, such as another event at a slightly later time, there may be a complex web of interacting influencing factors or a chain of interacting causes and effects. In other cases, even when there are clearly defined causal relationships, the effect(s) may occur after a considerable time delay or at a considerable distance in space or be totally out of proportion to the original causes. All these factors can make it difficult to identify the original causes. Since systems approaches consider interactions, they can be an appropriate way to model more complex causal relationships, with the term cause used in the widest possible sense. The so-called *butterfly effect* is a good example of the complicated and unpredictable cause and effect interrelationships that occur in many real systems. This is named after the talk entitled 'Does the flap of a butterfly's wings in Brazil set off a tornado in Texas' given by the meteorologist Edward Lorenz. The talk was motivated by Lorenz's observations of computer models of weather patterns. He noticed that very small changes in initial conditions frequently resulted in totally different weather patterns (O'Connor and McDermott 1997).

3.10 Time Delays

Most systems have *time delays*. For instance there is a time delay before the full impact of the environmental effects of emissions from industrial processes are felt. Many of these effects are also cumulative and therefore will continue to be felt for a considerable period of time. Thus when, for instance, industrial processes are modified to reduce emissions, there will generally be a considerable time delay before the environmental effects of these reductions in emissions are realised. Consequently, although the first scientific papers suggesting that chlorofluorocarbons (CFCs) could destroy atmospheric ozone were published in 1974, it was only in 1985 that clear evidence was obtained of a deep hole in the ozone layer above the Antarctic. The 1985 measurements showed the effects of CFCs released in 1970, as it takes 15 years for a CFC molecule released at the earth's surface to reach the high stratosphere and contribute to breaking down the protective ozone layer. Such time delays can increase the difficulties involved in determining system effects (O'Connor and McDermott 1997).

Long time delays between the original causes and their effects are not uncommon in complex systems. In some cases the time delay is so great that it is very difficult if not impossible to reverse the action. Industrial chemicals are a case in point. For instance polychlorinated biphenyls (PCBs) have been used worldwide in electrical equipment and dumped in landfills and sewers for the last 40 years. PCBs move slowly through the soil and into groundwater. They are fat soluble and can accumulate in the body. Their concentration increases as they move up the food chain and is greatest in sea birds, mammals and human breast milk. They interfere with the human immune system and reproductive functions. Although their production has been banned in most countries since the 1970s, 70% of the PCBs produced are still in use in electrical equipment and the other 30% has already been released into the environment. Of this 30%, about 3.3% (i.e. 1% of the PCBs originally produced) has reached the oceans, where the PCBs are having an effect on marine life. 96.7% of the 30% released to the environment (29% of the PCBs originally produced) remain in the soil, lakes and rivers and are likely to be absorbed by living creatures for decades or even centuries (O'Connor and McDermott 1997).

3.11 Physical and Abstract Systems

Systems can be classified as physical or abstract. *Physical* systems have measurable variables and parameters, whereas *abstract* systems do not have measurable variables or parameters. Many of the important variables in sustainable development, including poverty and clean air, are abstract. However both these abstract variables can be described in terms of a number of physical variables, such as income or concentration of carbon dioxide and nitrogen oxides. Physical systems can be further divided into real physical and conceptual physical systems. In real physical systems the quantities really exist, whereas in conceptual physical systems they are only assumed. Real physical systems are always bounded, whereas abstract and conceptual physical systems can be unbounded. *Bounded* systems have a finite number of external quantities and a finite structure, whereas *unbounded* systems have an infinite number of external quantities or an infinite structure.

3.12 Open and Closed Systems

Systems can be categorised as open or closed according to whether they interact with their environments. *Open* systems allow an exchange of materials, energy and information between the system and its environment and a steady state, which may involve a high degree of order, can be achieved by continuous exchanges with the environment. Conventional physics has restricted consideration to *closed* systems, which are isolated from their environments, whereas more modern physical theory also considers open systems. Since a closed system does not exchange material and energy with its environment, only increasing disorder (entropy) is possible. According to the *second principle of thermodynamics* a closed system must eventually reach a time independent state of equilibrium with maximum entropy and minimum free energy. Energy is not required to maintain this equilibrium state and no energy can be obtained from it. For instance a closed reservoir contains considerable potential energy, but cannot be used to drive a motor. In a closed system a distribution of molecules of different velocities with fast high temperature molecules at one side and slow low temperatures molecules at the other side is improbable. The difference between open and closed systems is illustrated in fig. 3.8.

However many real systems, particularly biological systems, are open and exchange materials and energy with their environments. These

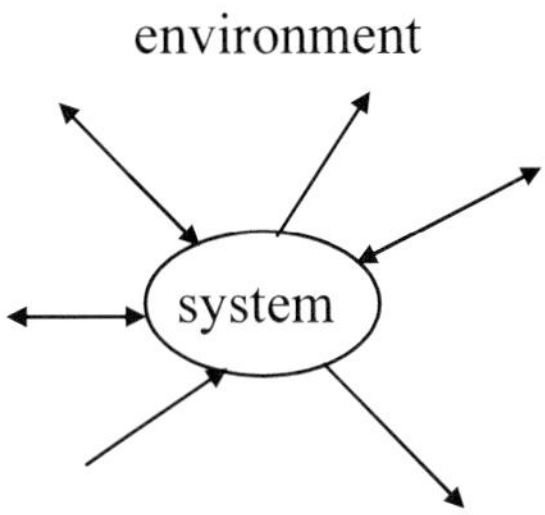

Fig. 3.8a Open system

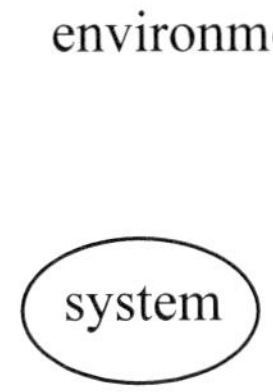

Fig. 3.8b Closed system

include living organisms, which take in, for instance, food, water and oxygen, transform them through respiratory, metabolic and other processes to produce energy and structural materials, such as muscle and fat and output carbon dioxide and various liquid and solid wastes. Such open systems, including living organisms, are in a steady state, but not a state of (static) equilibrium.

The application of physical theory to open systems has led to increased understanding of a number of physical and biological phenomena and to the development of the *principle of equifinality*. In a closed system the initial conditions totally determine the final state that can be obtained for a given system and a different final state can only be obtained if the initial conditions or the inputs are changed. According to the principle of equifinality in an open system the same final state can be obtained from different sets of initial conditions and in different ways. For instance a sea urchin can develop from a complete ovum, each half of a divided ovum or the fusion of two whole ova. Many open systems, including living organisms, show moves towards increasing order, heterogeneity and organisation. This is due to the import of materials from the environment, which may lead to a reduction in entropy. In an open system the total change in entropy (dS) is given by the sum of the changes in entropy due to the transfer of materials (d_1S) and irreversible processes in the system (d_2S) (Bertalanffy 1968) i.e.

$$dS = d_1S + d_2S \tag{3.13}$$

By the second principle of thermodynamics, the term d_2S is always positive. However d_1S can be positive or negative, with negative entropy arising, for instance, from the import of free energy or negative entropy. This leads to a trend towards increasing order in living organisms. If an open system reaches a steady state, this state will be determined by the system parameters and independent of the initial conditions. Thus the same final state can be reached from different initial conditions and after various disturbances.

In studying systems, appropriate choices have to be made about the system *boundary*. At the simplest level, obvious physical limits, such as a person's skin or the wall round a factory, will be taken as a boundary. However there may not always be an obvious physical boundary and, even when there is, it may be appropriate to draw the system boundaries either more widely or more narrowly. Systems are about interactions and interrelations, which are not necessarily limited by physical structures and boundaries. It has therefore been suggested (Ulrich 1968) that the system boundary should be taken where relations are less concentrated, so that most of the system relations will be inside the boundary. Choices about the boundary may also determine whether a particular system is considered to be open or closed. However in practice the boundaries of most real systems would have to be drawn excessively widely to make the system closed.

Example 3.6: Cities and Self-Sufficient Communities

A city can be considered a system, as its components, including people, animals, plants, buildings, infrastructure and communal organisations, interact in many different ways and the city as a whole has a number of distinct properties not held by its constituent parts and which emerge from their interactions. There is generally a movement of population between the city and its environment, with people coming in temporarily from the environment to live or to work and moving out either permanently or for short periods, for instance to go on holiday or on business trips. Cities also import a wide range of goods, including food, that cannot be produced in the city, and also produce a variety of goods which are exported. Many cities import electricity from a national or regional grid rather than producing sufficient energy to meet all their needs within the city. Therefore a city is clearly an open rather than a closed system.

However there is generally some ambiguity as to how the city boundary should most appropriately be drawn, largely due to the growth in urban population and the spread of many cities into urban conglomerations. In some cases a continuous urban spread is divided into separate towns or cities for administrative purposes or contains areas which identify as separate cities for historical or other reasons. Despite a degree of ambiguity, drawing the boundary sufficiently widely to include all exchanges of population and trade within the boundary would in general lead to a system which could no longer be meaningfully identified as a 'city'. For instance many cities import food products from all round the world and export goods to a large part of the world (though these practices are not sustainable due to the energy consumption and emissions resulting from transportation).

A 'self-sufficient' rural community is closer to a closed system than a city is. However, in practice, few communities are totally self-sufficient and most of them trade some goods or produce in both directions. Even when there is no population exchange and minimal trade with the outside world, changes in ideas and ideologies in the outside world and developments in technology are likely to have an influence, even if only over a very long time span. In some contexts it will be appropriate to treat such communities as closed systems, since the interactions with the environment will be minimal and relatively insignificant, whereas in others it will be necessary to use an open system model.

Example 3.7: Warkworth Penitentiary as an Open System

Fig. 3.9 shows Warkworth Penitentiary in Canada. From its appearance and the perspective of many of its inmates this is a closed institution. However there are clearly frequent exchanges with the environment i.e. everything outside the prison. All raw materials need to be brought in from outside, though some of them will be processed and, for instance, food prepared inside the prison. The prisoners are also brought into the prison from outside (the environment) and most of them are eventually released into the environment. Therefore this prison is an open system. This illustrates the difference between the use of the technical terms open and closed systems and the more colloquial uses of the terms open and closed.

Fig. 3.9 Warkworth Penitentiary, Canada

3.13 Transfer Function and Impulse Response

Linear time invariant systems have a particularly simple system description based on the *impulse response* in the time domain and the *transfer function* in the frequency domain. In addition the response of such systems to any given input can be determined by their response to a particular test input, the *impulse function* $\delta(t)$.

This follows from the convolution property of the impulse function, by which any function is unchanged by convolution with an impulse function i.e.

$$x(t) = x(t) * \delta(t) = \int_{-\infty}^{\infty} x(T)\delta(t-T)dT \tag{3.14}$$

The integral (3.14) can be considered as the limit of a weighted sum of delayed delta functions i.e.

$$x(t) = \lim_{\Delta t \to 0} \sum_{n=-\infty}^{\infty} x(n\Delta t)\,\delta(t - n\Delta t)\Delta t \tag{3.15}$$

Thus when the system is time invariant and linear, knowledge of the system response to $\delta(t)$ gives the system response to any input $x(t)$. The response of the system to the impulse $\delta(t)$ is called the *impulse response* $h(t)$. Using superposition and time invariance in (3.15) the response due to any input $x(t)$ is given by

$$y(t) = \lim_{\Delta t \to 0} \sum_{n=-\infty}^{\infty} x(n\Delta t)\,h(t - n\Delta t)\Delta t = \int_{-\infty}^{\infty} x(T)h(t-T)dT \tag{3.16a}$$

which can be written using (3.14) as

$$y(t) = x(t) * h(t) \tag{3.16b}$$

Therefore the output due to any input is given by the *convolution* of the input with the impulse response. Using the convolution theorem (according to which the transform of the convolution of two functions is equal to the product of the transforms) on (3.16b) gives

$$Y(\omega) = X(\omega)H(\omega) \tag{3.17}$$

where H(ω) is the transform of the impulse response and is generally referred to as the transfer function. Thus the transfer function of a single input single output (SISO) linear system is the ratio of the output transform to the input transform. Therefore, from (3.17) the transform of the output of a linear system is equal to the transform of its input multiplied by the transfer function i.e. in the frequency domain the output is obtained from the input by multiplication by the transfer function.

3.14 Sensitivity

The *sensitivity* of a system function, F(s), to changes in another plant function G(s), is defined as the ratio of the percentage change in F(s) to the percentage change in G(s).

$$S_G^F = \frac{\Delta F/F}{\Delta G/G} = \frac{G}{F}\frac{\Delta F}{\Delta G} \approx \frac{G}{F}\frac{\partial F}{\partial G} \tag{3.18}$$

Thus S_G^F is used to denote the amount by which F changes in response to changes in G. In theory the sensitivity function can be used to investigate the effects of changes, such as an increase in consumption, on, for instance, global climate change or resource depletion. However, in practice it is difficult to obtain equations which adequately describe social, developmental and environmental systems without excessive simplification. In many cases, such as climate change, there are a large number of interconnected equations, making it very difficult to obtain explicit expressions for the sensitivity functions.

3.15 Cascade of Systems

Systems can be combined or *cascaded* together, in either series or parallel. Fig. 3.10a shows the *block diagrams* for two systems in parallel.

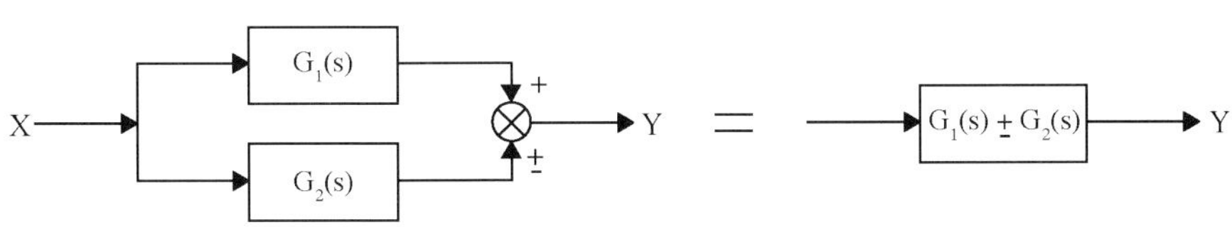

Fig. 3.10a Block diagram for two systems in parallel

The systems can be combined using block diagram algebra to give

$$Y(s) = X(s)G_1(s) + X(s)G_2(s) = X(s)[G_1(s) + G_2(s)]$$

so that the overall transfer function is

$$G(s) = \frac{Y(s)}{X(s)} = G_1(s) + G_2(s) \tag{3.19}$$

The result can similarly be demonstrated when the sign is negative. Fig. 3.10b shows the block diagrams for two systems in series.

Fig. 3.10b Block diagram for two systems in series

The systems can be combined using block diagram algebra to give

$$Y(s) = X(s)G_1(s)G_2(s)$$

so that the overall transfer function is

$$G(s) = \frac{Y(s)}{X(s)} = G_1(s)G_2(s) \tag{3.20}$$

Fig. 3.11 shows the block diagrams for two systems connected by a feedback loop.

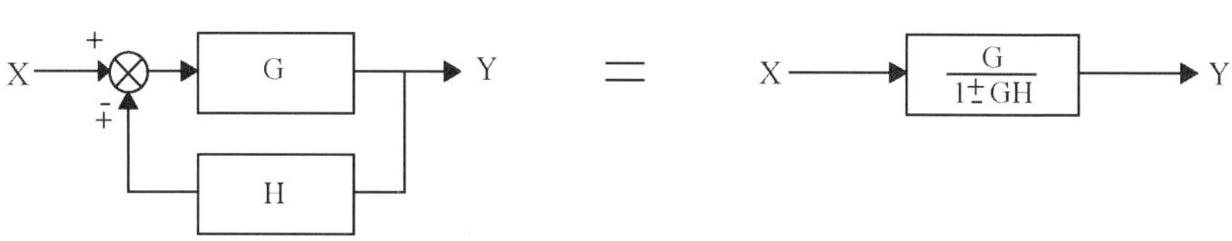

Fig. 3.11 Two systems connected by a feedback loop

The systems can be combined using block diagram algebra to remove the feedback loop to give

$$Y(s) = G(s)X(s) \pm G(s)H(s)X(s) = [1 \pm G(s)H(s)]\, X(s)$$

so that the overall transfer function is

$$G(s) = \frac{Y(s)}{X(s)} = \frac{G(s)}{1 \pm G(s)H(s)} \tag{3.21}$$

These expressions have been derived for continuous time systems, but they are equally valid for discrete time systems, with the only difference that the transfer functions would be functions of z not s.

3.16 Roundup of Chapter 3

This chapter has presented some of the main concepts in engineering mathematics representations of systems. These concepts are essential for understanding the engineering systems literature and will also be drawn on later in the book, particularly in chapter 4. They provide the foundations for the tools readers can use to construct their own mathematical models of problems in sustainable development.

Important concepts presented in the chapter include input and output, feedback, open and closed systems, open and closed loop systems, linear systems and discrete and continuous time systems. A closed loop system can be obtained from an open loop system by adding feedback i.e. feeding back information about the system output. In a closed loop control system this information can be compared with a reference input and used to update system performance. These concepts should not be confused with open and closed systems. An open system interacts with its environment, whereas a closed system is totally self-contained and does not interact with its environment. Linear systems are particularly useful as the output from the sum of two inputs is the sum of the outputs for each input separately (superposition) and multiplying the input by a constant just multiplies its output by a constant (scaling). Another useful distinction is between continuous time and discrete time systems. Most real systems are continuous time, but discrete time systems are useful due to the proliferation of digital computers. Continuous time systems are defined as a function of a continuous variable, whereas discrete time systems are only defined at discrete time instants and can be obtained from continuous time systems by sampling.

The mathematical concepts presented in this chapter will be drawn on in chapter four in the presentation and analysis of a particular type of mathematical systems representation, called state space systems.

4 State Space System Representations

4.0 Learning Objectives

This chapter introduces the main concepts in state space system representations and illustrates their application by examples, of a regional system and the arms race between two (groups of) nations. Specific learning objectives include the following:

- Understanding of the terms 'state space' representation, 'state variable' and 'state vector'.
- The ability to use state space representations to obtain models of systems of interest in sustainable development.
- The ability to obtain state space from transfer function representations and vice versa.
- Understanding of the importance of stability and how the system stability properties are obtained.
- The ability to apply state space representations appropriately, particularly in the sustainable development context, and an understanding of their advantages and limitations.

4.1 Introduction and Definitions

This chapter discusses the use of state space system representations. The basic concepts are introduced and some definitions are presented in section 4.1. Linear transformation of state space representations is discussed in section 4.2. The use of the state transition matrix to describe the transition between two system states is introduced in section 4.3, where an expression for the system output in terms of thc state transition matrix is also derived. Section 4.4 considers the solution of the state equation in the non-linear case, including linearisation about an equilibrium point and the form and solution of the equations in a number of special cases. The derivation of a transfer function from the state space representation and the use of realisations to obtain state space representations from the transfer function are considered in section 4.5. Section 4.6 presents an example of the application of state space techniques to modelling the arms race between two (groups of) nations. The chapter is summarised in section 4.7.

State variable representations of a system can be used to give a complete description of its *internal behaviour*, whereas a transfer function representation only gives external or input-output relationships. Modes corresponding to equal zeros and poles cancel out and do not appear in the transfer function representation, whereas the state space representation shows the full internal structure. A *state space* representation is based on a series of first order differential or difference equations plus an output equation.

A linear time-invariant system has the following state space representations:

$$\dot{\underline{x}} = A_s\underline{x}_t + B_s\underline{u}_t \tag{4.1a}$$

$$\underline{y}_t = C_s\underline{x}_t + D_s\underline{u}_t \tag{4.1b}$$

for a continuous time system and

$$\underline{x}_{t+1} = A_d\underline{x}_t + B_d\underline{u}_t \tag{4.2a}$$

$$\underline{y}_t = C_d\underline{x}_t + D_d\underline{u}_t \tag{4.2b}$$

for a discrete time system, with

$$\underline{x}_t = \begin{bmatrix} x_1(t) \\ x_2(t) \\ \vdots \\ x_n(t) \end{bmatrix}, \quad \underline{u}_t = \begin{bmatrix} u_1(t) \\ u_2(t) \\ \vdots \\ u_m(t) \end{bmatrix} \quad \underline{y}_t = \begin{bmatrix} y_1(t) \\ y_2(t) \\ \vdots \\ y_p(t) \end{bmatrix} \tag{4.3}$$

in both cases, where $\underline{x}_t$, $\underline{u}_t$ and $\underline{y}_t$ are the state, input and output vectors respectively. These vectors are continuous time vectors for continuous time systems and sampled data or discrete time vectors for discrete time systems.

There are two (sets of) equations, the state equations (4.1a) or (4.2a) and the output equation(s) (4.1b) or (4.2b) for continuous and discrete time systems respectively. The output equations have the same form in the continuous and discrete time case, but the state equations consists of a number of first order differential equations in the continuous time case and first order difference equations in the discrete time case. The subscripts 's' and 'd' in the matrices A, B, C and D will be omitted subsequently.

The components of the state, input and output equations do not necessarily represent the same kind of physical variables. The variables $x_1(t)$ to $x_n(t)$ are called *state variables* and the vector $\underline{x}_t$ the *state vector*. It is a vector in the mathematical and not the physical sense. A vector in the mathematical sense is an ordered arrangement of variables, whereas a vector in the physical sense is one variable which has both magnitude and direction, such as velocity. The state variables are *not* unique. The appropriate choice of state variables depends on the particular problem. State variables may represent real physical or abstract variables, but will not necessarily all represent the same type of variable. For instance the state variables in a given problem could include population levels, energy consumption and emissions of toxic waste.

A generalised operator notation could be used for both the continuous and discrete time cases to give a unified approach, but will not be discussed further here. Such a representation could also be used for state space representations based on other operators, such as the delta operator (Middleton and Goodwin 1990; Hersh 1993).

For multivariable systems A, B, C and D are matrices, whereas B and C are respectively a column and row vector and D a scalar for single input single output systems. The matrix D is often zero. The *order* of a system is defined as the minimum number of initial conditions that have to be specified at an initial time t_o to describe the behaviour of the system for later times when the input u(t) is given.

The matrix A is called the *system matrix*. Its eigenvalues determine the system dynamics. The term system poles is used equivalently to system eigenvalues. The *system poles* or *eigenvalues* are the roots of the denominator of the system transfer function. The positions of the system poles determine whether or not the system is stable, with the stability regions in the s (continuous time) and z (discrete time) planes shown in figures 4.1a and 4.1b respectively. When the poles are complex, the system has oscillatory behaviour. The matrix A is a square matrix, with dimension n equal to the *system order* i.e. the number of state variables or the number of system poles or eigenvalues. The dimensions of B, C and D are n x m (number of states by number of inputs), p x n (number of outputs by number of states) and p x m (number of outputs by number of inputs) respectively. The numbers m and p are respectively the numbers of inputs and outputs.

Many real systems are non-linear and/or time varying. Time varying linear continuous or discrete time systems can be represented by equations

(4.1) and (4.2) respectively with the matrices A, B, C and D functions of time A_t, B_t and C_t and D_t. The most general state representation is for non-linear time-varying systems, where the state equations have the general form

$$\begin{aligned}
\dot{x}_1(t) &= f_1[x_1(t), x_2(t), ..., x_n(t), u_1(t), u_2(t), ..., u_m(t), t] \\
\dot{x}_2(t) &= f_2[x_1(t), x_2(t), ..., x_n(t), u_1(t), u_2(t), ..., u_m(t), t] \\
&\vdots \\
\dot{x}_n(t) &= f_n[x_1(t), x_2(t), ..., x_n(t), u_1(t), u_2(t), ..., u_m(t), t]
\end{aligned} \tag{4.4}$$

for a continuous time system and

$$\begin{aligned}
x_1(t+1) &= f_1[x_1(t), x_2(t), ..., x_n(t), u_1(t), u_2(t), ..., u_m(t), t] \\
x_2(t+1) &= f_2[x_1(t), x_2(t), ..., x_n(t), u_1(t), u_2(t), ..., u_m(t), t] \\
&\vdots \\
x_n(t+1) &= f_n[x_1(t), x_2(t), ..., x_n(t), u_1(t), u_2(t), ..., u_m(t), t]
\end{aligned} \tag{4.5}$$

for a discrete time system. The output equations have the form

$$\begin{aligned}
y_1(t) &= g_1(x_1(t), x_2(t), ..., x_n(t), u_1(t), u_2(t), ..., u_m(t), t) \\
y_2(t) &= g_2(x_1(t), x_2(t), ..., x_n(t), u_1(t), u_2(t), ..., u_m(t), t) \\
&\vdots \\
y_p(t) &= g_1(x_1(t), x_2(t), ..., x_n(t), u_1(t), u_2(t), ..., u_m(t), t)
\end{aligned} \tag{4.6}$$

for both continuous and discrete systems. $x_1(t)$ to $x_n(t)$ are the state variables and $u_1(t)$ to $u_m(t)$ the inputs and are continuous time variables in the continuous time case and discrete time or sampled data variables in the

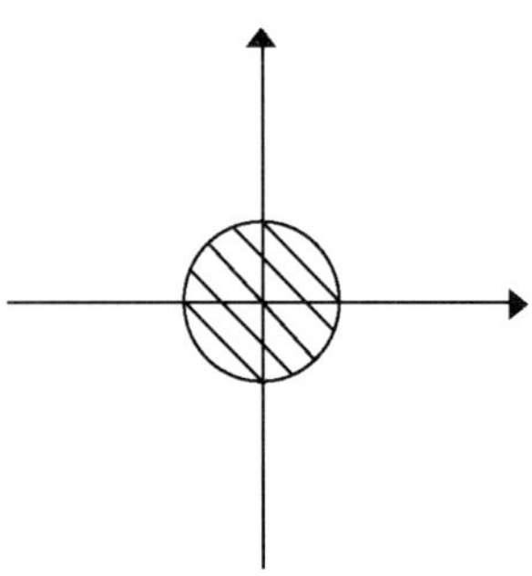

Fig. 4.1a z-plane stability region

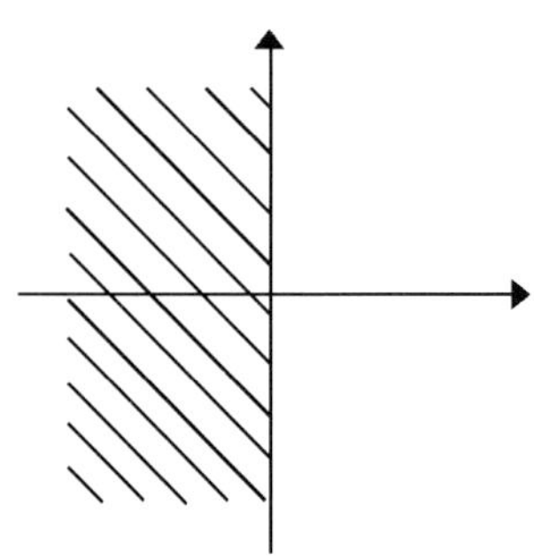

Fig. 4.1b s-plane stability region

discrete time case. The forms of the functions f_1 to f_n and g_1 to g_p depend on the particular system.

Example 4.1: A Regional Model

There have been a number of different approaches to developing global, regional and local models, as well as models of resource use, including (Forester 1969 1971; Meadows et al 1973 1974 1992). Although it is clearly useful to investigate resource depletion and population growth on a global basis, many relevant variables, such as education or quality of life, vary significantly in different parts of the world. Therefore the model considered here will be a regional one. In addition, most existing models do not include technology or technological development as a variable, although it has had a very significant impact. In order to obtain a satisfactory description a large number of variables would have to be included, of which the following is only a partial list: population, technological development, development of sustainable technology, availability and access to education, availability and access to food supply, availability and access to clean water, quality of life, standard of living, health care, pollution, resource availability and/or resource depletion. In addition there would be a number of inputs for the movement of materials, resources, ideas and people between the regional system and its environment. This would give a large and complicated model.

Therefore to obtain a simple model, only the following five state ($x_1, \ldots, x_5$) and four input ($u_1, \ldots, u_4$) variables will be used:

Population	x_1
Technological development:	x_2
Quality of life (including equity and positive development):	x_3
Environmental destruction, including pollution:	x_4
Economic activity:	x_5
Net influx of population i.e. immigration - emigration:	u_1
Scientific and technical creativity:	u_2
Pressures to reduce environmental impacts e.g. from legislation and public opinion:	u_3
Investment:	u_4

Many of the variables are abstract variables. In addition aggregation of variables, for instance to form the variables x_3 and x_4, makes it difficult to specify appropriate units for these variables. Therefore the mathematical expressions and graphs obtained from solving the equations (4.7) and (4.8) will be treated as descriptive of the evolution of the variables rather than giving numerical values.

These variables can then be used to obtain the following model, which is assumed to be linear in the first variable and non-linear in the others:

$$\dot{x}_1 = a_{11}x_1 + u_1 \tag{4.7a}$$

$$\dot{x}_2 = f_2(x_1, x_3, x_5, u_2) \tag{4.7b}$$

$$\dot{x}_3 = f_3(x_1, x_2, x_4, x_5, u_3) \tag{4.7c}$$

$$\dot{x}_4 = f_4(x_1, x_2, x_4) \tag{4.7d}$$

$$\dot{x}_5 = f_5(x_1, x_2, x_4, x_5, u_4) \tag{4.7e}$$

The non-linear terms in equations (4.7) can then be expanded to give the following model:

$$\begin{aligned}\dot{x}_2 = {} & \{a_{21}x_1[u(t)-u(t-c_1)] + [a_{21}c_1 - a^*_{21}x_1][u(t-c_1) - u(t-c_1-c_1a_{21}/a^*_{21})]\}x_3^r \\ & + a_{25}x_5[u(t)-u(t-c_5)] + [a_{25}c_5 - a^*_{25}x_5][u(t-c_5) - u(t-c_5-c_5a_{25}/a^*_{25})] \\ & + b_2 u_2\end{aligned} \tag{4.8a}$$

$$\begin{aligned}\dot{x}_3 = {} & a_{31}x_1[u(t)-u(t-d_1)] + [a_{31}d_1 - a^*_{31}x_1][u(t-d_1) - u(t-d_1-d_1a_{31}/a^*_{31})] \\ & + a_{32}x_2[u(t)-u(t-d_2)] + [a_{32}d_2 - a^*_{32}x_2][u(t-d_2) - u(t-d_2-d_2a_{32}/a^*_{32})] \\ & - a_{34}x_4^k + a_{35}x_5[u(t)-u(t-d_5)] \\ & + [a_{35}d_5 - a^*_{35}x_5][u(t-d_5) - u(t-d_5-d_5a_{35}/a^*_{35})] + b_3 u_3\end{aligned} \tag{4.8b}$$

$$\dot{x}_4 = x_1 x_2 x_4 \tag{4.8c}$$

$$\begin{aligned}\dot{x}_5 = {} & a_{51}x_1[u(t)-u(t-g_1)] + [a_{51}g_1 - a^*_{51}x_1][u(t-g_1) - u(t-g_1-g_1a_{51}/a^*_{51})] \\ & + a_{52}x_2[u(t)-u(t-g_2)] + [a_{52}g_2 - a^*_{52}x_2][u(t-g_2) - u(t-g_2-g_2a_{52}/a^*_{52})] \\ & - a_{54}x_4^p + a_{55}x_5[u(t)-u(t-g_5)] \\ & + [a_{55}g_5 - a^*_{55}x_5][u(t-g_5) - u(t-g_5-g_5a_{55}/a^*_{g5})] + b_4 u_4\end{aligned} \tag{4.8d}$$

In these equations when the effects of the other variables are ignored, the dependence of $\dot{x}_2$ on x_1 and x_5; $\dot{x}_3$ on x_1, x_2 and x_5 and $\dot{x}_5$ on x_1, x_2 and x_5 is piecewise linear as illustrated in fig. 4.2 for $\dot{x}_3$ and x_1, with $d_1 = 7$, $a_{31} = 3$ and $a^*_{31} = 7$.

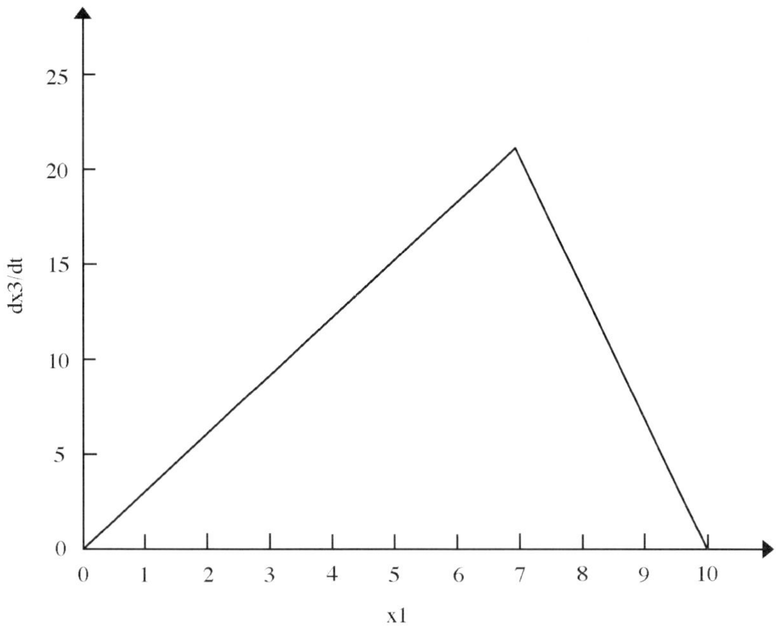

Fig. 4.2 Dependence of $\dot{x}_3$ on x_1

4.2 Linear Transformations

A *linear transformation* is a transformation of a linear system to give an equivalent state space representation with a different state vector and state, input and output matrices A*, B* and C*. The input and output vectors remain unchanged. A linear transformation is generally defined by a particular n-square matrix. Linear transformations can be applied to give state representations with nice properties such as a diagonal system matrix, A*. However after transformation it will generally not be possible to identify physical or abstract variables which correspond to the state variables.

A state space representation can be converted from state vector $\underline{x}$ to state vector $\underline{x}^*$ using the linear transformation

$$\underline{x}^* = P^{-1}\underline{x} \tag{4.9a}$$

$$\Rightarrow \underline{x} = P\underline{x}^* \tag{4.9b}$$

Replacing $\underline{x}$ in equations (4.1) by $\underline{x}^*$ using equation (4.9b) gives

$$P\underline{\dot{x}}^* = AP\underline{x}^* + B\underline{u}$$

$$\Rightarrow \underline{\dot{x}}^* = P^{-1}AP\underline{x}^* + P^{-1}B\underline{u}$$

$$\underline{y} = CP\underline{x}^* + D\underline{u}$$

$$\Rightarrow \underline{\dot{x}}^* = A^*\underline{x}^* + B^*\underline{u} \tag{4.10a}$$

$$\underline{y} = C^*\underline{x}^* + D^*\underline{u} \tag{4.10b}$$

$$\text{with} \quad A^* = P^{-1}AP, \quad B^* = P^{-1}B, \quad C^* = CP, \quad D^* = D \tag{4.10c}$$

Equations (4.10) have been derived for a continuous time system. In the case of a discrete time system equation (4.10a) is replaced by

$$\underline{x}_{t+1} = A^*\underline{x}_t^* + B^*\underline{u}_t \tag{4.10d}$$

but the form of output equation and expressions for the coefficients are the same as in the continuous time case i.e. as in equations (4.10b) and (4.10c), though the vectors in equation (4.10b) are now discrete time rather than continuous time.

The form and properties of the matrices A*, B*, C* and D* are determined by the choice of P. For instance the state space representation with the state matrix A diagonal can be obtained by choosing the matrix $P = [\underline{e}_1, \ \underline{e}_2]$ where the two vectors $\underline{e}_1$ and $\underline{e}_2$ are the system eigenvectors.

4.3 State Transition Matrix

The *state transition matrix* $\Phi(t, t_0)$ describes the transition between the state at two different times i.e.

$$\underline{x}(t) = \Phi(t, t_0)\underline{x}(t_0) \tag{4.11}$$

It is also the solution of the unforced ($\underline{u} = 0$) state equation i.e.

$$\underline{\dot{x}}_t = A\underline{x}_t \tag{4.12a}$$

$$\text{or} \quad \underline{x}_{t+1} = A\underline{x}_t \tag{4.12b}$$

The transition property of the state transition matrix can be expressed for both continuous and discrete time systems as

$$\Phi(t, t_0) = \Phi(t, t_1)\,\Phi(t_1, t_0) \tag{4.13a}$$

and the inverse matrix is given by

$$\Phi(t_1, t_0)^{-1} = \Phi(t_0, t_1) \tag{4.13b}$$

The following derivation is for continuous time systems, but similar techniques can be applied to discrete time systems. Solving equation (4.12a) gives

$$\Phi(t, t_0) = e^{A(t-t_0)} \tag{4.14}$$

This matrix can be used to solve the state equation (4.1a) to give (Friedland 1987)

$$\underline{x}(t) = \Phi(t, t_0)\underline{x}(t_0) + \int_{t_0}^{t} \Phi(t, \tau)B\underline{u}(\tau)\,d\tau \tag{4.15a}$$

$$= e^{A(t-t_0)}\underline{x}(t_0) + \int_{t_0}^{t} e^{A(t-\tau)}B\underline{u}(\tau)\,d\tau \tag{4.15b}$$

There are a number of different methods for obtaining the state transition matrix. However numerical difficulties may be encountered for systems which are not relatively simple. One of the simplest involves obtaining the associated Laplace transform $\Phi_s'(s)$ (or z transform $\Phi_z'(z)$) and then using inverse Laplace (or z) transformations to find $\Phi(t)$. $\Phi_s'(s)$ can be obtained by taking Laplace transformations of equation (4.12a) to give

$$sX(s) - \underline{x}(0) = AX(s)$$

and rearranging as

$$s\underline{X}(s) - A\underline{X}(s) = \underline{x}(0)$$

$$\Rightarrow (sI - A)\underline{X}(s) = \underline{x}(0)$$

$$\Rightarrow \underline{X}(s) = (sI - A)^{-1}\underline{x}(0) \tag{4.16a}$$

From equation (4.11), putting $t_0 = 0$, $x(t) = \Phi(t)x(0)$

$$\Rightarrow \underline{X}(s) = \Phi_s'(s)\underline{x}(0) \tag{4.16b}$$

Equating (4.16a) and (4.16b) gives

$$\Phi_s'(s) = (sI - A)^{-1} \tag{4.17}$$

$\Phi(t)$ can be obtained from (4.17) using partial fraction expansion, Laplace transform tables and the properties of Laplace transforms. However, as indicated above, numerical difficulties can be encountered for systems which are not relatively simple. It can be shown similarly that

$$\Phi_z'(z) = (zI - A)^{-1} \tag{4.18}$$

from which $\Phi(t)$ can be obtained using partial fraction expansion, z transform tables and the properties of z transforms. The value of $\Phi(t)$ (in the continuous time case) can then be substituted in equation (4.15a), which can be used to calculate the state vector at any time t. The system output can then be obtained from the state vector using equation (4.1b).

4.4 Solution of the State Equation in the Non-Linear Case

Although many real systems have some degree of non-linearity, a linear approximation can often be obtained by *linearisation* about a *stationary*, *steady state* or *equilibrium* point. Therefore (linear) state equations can be used to represent a wide range of different types of systems. Some systems have several stationary points, some of which may be stable and others unstable, and consequently a different linearisation about each one. In the steady state, the differential terms $\dot{x}_i$ (i=1...n) in equations (4.4) are zero, giving

$$f_1 = f_2 = \ldots = f_n = 0 \tag{4.19}$$

This set of simultaneous equations can be solved to obtain the values

$$x_1 = x_1^* \quad x_2 = x_2^* \quad \ldots \quad x_n = x_n^* \tag{4.20}$$

Equations (4.4) can be rewritten using the new variables x_i', i = 1...n, defined as

$$x_i = x_i^* - x_i', \ i = 1 \ldots n \tag{4.21}$$

to give

$$\dot{x}_i' = f_i'(x_1', x_2', \ldots, x_n') \quad i = 1 \ldots n \tag{4.22}$$

It will then be assumed that a Taylor series expansion of equations (4.22) can be obtained (Bertalanffy 1968) as

$$\dot{x}_i' = a_{i1}x_1' + \ldots + a_{in}x_n' + a_{i11}x_1'^2 + a_{i12}x_1'x_2' + a_{i22}x_2'^2 + \ldots \tag{4.23}$$

The general solution of this system of equations is

$$\begin{aligned}
x_1' &= k_{11}e^{\lambda_1 t} + k_{12}e^{\lambda_2 t} + \ldots + k_{1n}e^{\lambda_n t} + k_{111}e^{2\lambda_1 t} + \ldots \\
x_2' &= k_{21}e^{\lambda_1 t} + k_{22}e^{\lambda_2 t} + \ldots + k_{2n}e^{\lambda_n t} + k_{211}e^{2\lambda_1 t} + \ldots \\
&\vdots \\
x_n' &= k_{n1}e^{\lambda_1 t} + k_{n2}e^{\lambda_2 t} + \ldots + k_{nn}e^{\lambda_n t} + k_{n11}e^{2\lambda_1 t} + \ldots
\end{aligned} \tag{4.24}$$

where the k_{ij} are constants and the λ_i are the roots of the characteristic equation

$$\begin{vmatrix} a_{11} - \lambda & a_{12} & \cdots & a_{1n} \\ a_{21} & a_{22} - \lambda & \ddots & \vdots \\ \vdots & \ddots & \ddots & a_{(n-1)n} \\ a_{n1} & \cdots & a_{n(n-1)} & a_{nn} - \lambda \end{vmatrix} = 0 \tag{4.25}$$

Equations (4.24) give the solutions round the steady state point. When all the λ_i are either real and negative or have negative real parts if complex, then the x_i' approach zero over time and, using equations (4.21), the steady state is stable. If one or more of the λ_i is real and positive or has a positive real part if complex, the steady state will be unstable and the system will move away from it. If one or more of the λ_i is zero, then the system is critically stable and small changes or perturbations are likely to result in instability. When some of the λ_i are complex, the system will contain periodic terms, which will be damped if the real part is negative.

Several special cases can be obtained for state spaces of order one or two (Bertalanffy 1968). In the single state variable case, one of the simplest expressions is exponential growth or decay obtained by expanding the equation (4.22) as a Taylor series

$$\dot{x} = a_1 x + a_{11} x^2 + \ldots \tag{4.26}$$

and only retaining the first term, giving

$$\dot{x} = a_1 x \tag{4.27a}$$

with solution

$$x = x_o e^{a_1 t} \tag{4.27b}$$

When two terms are retained in the Taylor expansion (4.26), the solution is sigmoid and obtains a limiting value. It is called the *logistic curve* and has the following equation

$$x = \frac{a_1 C}{1 - K e^{-a_1 t}} \tag{4.28a}$$

where

$$C = \frac{-1}{a_{11}}, \qquad K = \frac{a_{11} x_o + a_1}{a_{11} x_o} \tag{4.28b}$$

The logistic curve (4.28) describes, for instance, reactions catalysed by the reaction products in chemistry and the law of Verhulst (Bertalanffy,1968) in sociology which describes the growth of human populations with limited resources. It is illustrated in fig. 4.3 for the values $a_1 = 1$, $a_{11} = -1$ and $x_o = 0.5$.

Another commonly occurring equation in morphology, biochemistry and physiology is the *allometric equation* obtained for systems with two state variables and only one term in the state equation i.e.

$$\dot{x}_1 = a_1 x_1 \tag{4.29a}$$

$$\dot{x}_2 = a_2 x_2 \tag{4.29b}$$

This pair of equations has the solution

$$x_1 = c_1 e^{a_1 t} \qquad x_2 = c_2 e^{a_2 t} \tag{4.30a}$$

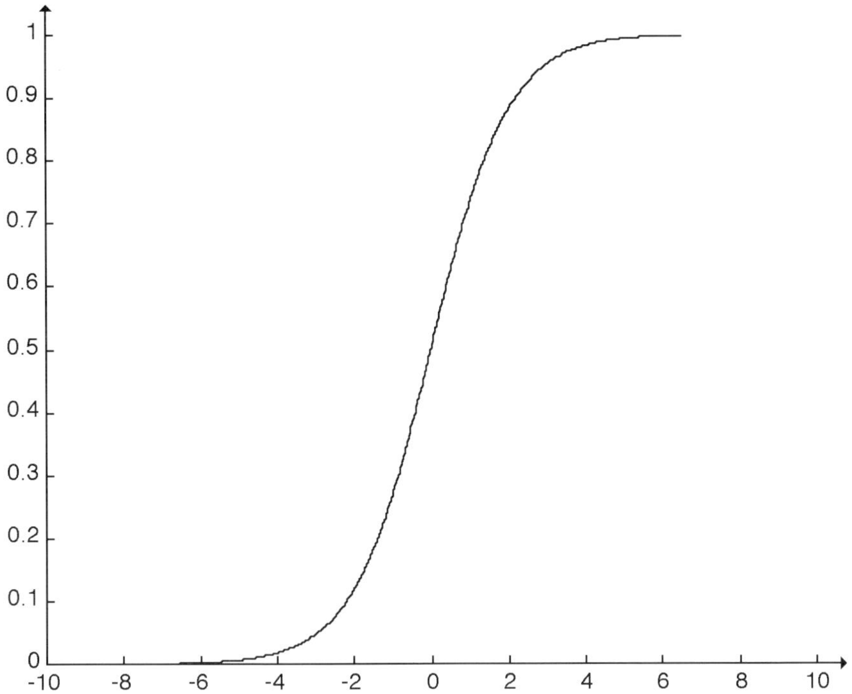

Fig. 4.3 Logistic curve

$$\text{or } x_1 = b x_2^{\alpha} \quad \text{with } \alpha = a_1/a_2 \text{ and } b = c_1/c_2^{\alpha} \tag{4.30b}$$

This implies that the relative growth rates of the variables x_1 and x_2 are proportional to each other for the part of the life cycle for which the allometric relation holds.

4.5 Relationship between State Space and Transfer Function Representations

A *unique transfer function* can be obtained from the state space representation. This is given by

$$G(s) = C(sI - A)^{-1}B + D \tag{4.31a}$$

in the discrete time case and

$$G(z) = C(zI - A)^{-1}B + D \tag{4.31b}$$

in the continuous time case.

The term *realisation* is used for a set of state space equations derived from or which give rise to the transfer function. As already stated, the state space representation is not unique. Therefore a number of different state space representations or realisations can be obtained from a given transfer function. A *minimal realisation* is one in which the state matrix A has least dimension, so that the corresponding system has the smallest possible number of state variables. To illustrate the different approaches to obtaining state space from transfer function representations, three different minimal realisations will be derived. These realisations will be obtained for continuous time single input single output systems. The forms of the realisations are similar for discrete time systems with $\dot{x}_i$ replaced by $x_{i.t+1}$ for i =1...n.

Consider a system with transfer function

$$G(s) = \frac{c_m s^m + c_{m-1}s^{m-1} + \ldots + c_0}{s^n + b_{n-1}s^{n-1} + \ldots + b_0} \tag{4.32}$$

Real systems are generally *proper* i.e. the order of the denominator of the transfer function is greater than or equal to that of the numerator ($n \geq m$). When these orders are equal i.e. n = m, the system is proper and

$$D = G(\infty) \tag{4.33}$$

Therefore $G(\infty)$ can be subtracted from G(s) to give a *strictly proper* transfer function

$$G'(s) = G(s) - G(\infty)$$

$$= \frac{a_{n-1}s^{n-1} + a_{n-2}s^{n-1} + \ldots + a_0}{s^n + b_{n-1}s^{n-1} + \ldots + b_0} \tag{4.34}$$

It should be noted that G'(s) will be equal to G(s) if G(s) is strictly proper and that the coefficient a_{n-1} is not necessarily non-zero i.e. the order of the numerator may be less than n-1. Realisations of minimal order will have order n and can be derived in a number of different ways (Friedland 1987).

Realisation 1:

Set

$$x_1 = y$$
$$x_2 = \dot{x}_1 + b_{n-1}x - a_{n-1}u$$
$$x_2 = \dot{x}_2 + b_{n-2}x - a_{n-2}u$$
$$\vdots$$
$$x_n = \dot{x}_{n-1} + b_1 x - a_1 u$$

so that

$$\dot{x}_n = b_0 x - a_0 u$$

Rearranging each of these equations, so that only the first derivatives are on the left hand side, gives the realisation

$$\dot{\underline{x}} = \begin{bmatrix} -b_{n-1} & 1 & 0 & \cdots & 0 \\ -b_{n-2} & 0 & 1 & \ddots & \vdots \\ -b_{n-3} & 0 & 0 & 1 & 0 \\ \vdots & \vdots & \vdots & \ddots & 1 \\ -b_0 & 0 & 0 & \cdots & 0 \end{bmatrix} \underline{x} + \begin{bmatrix} a_{n-1} \\ a_{n-2} \\ a_{n-3} \\ \vdots \\ a_0 \end{bmatrix} u \tag{4.35a}$$

$$y = [1 \;\; 0 \;\; \ldots \;\; 0]\, \underline{x} \tag{4.35b}$$

Realisation 2:

Divide the numerator and denominator by the highest power of s, s^n, so that

$$\frac{Y(s)}{U(s)} = \frac{a_{n-1}s^{-1} + a_{n-2}s^{-2} + \ldots + a_0 s^{-n}}{1 + b_{n-1}s^{-1} + \ldots + b_0 s^{-n}}$$

This equation can be solved by putting

$$Y(s) = [a_{n-1}s^{-1} + a_{n-2}s^{-2} + \ldots + a_0 s^{-n}]E(s) \tag{4.36a}$$

$$U(s) = [1 + b_{n-1}s^{-1} + b_{n-2}s^{-2} + \ldots + b_0 s^{-n}]E(s) \tag{4.36b}$$

Equations (4.36) can then be used to define the state variables as follows:

$$
\begin{aligned}
x_1 &= e(t) \\
x_2 &= \dot{x}_1 \\
x_2 &= \dot{x}_2 \\
&\vdots \\
x_n &= \dot{x}_{n-1}
\end{aligned}
$$

so that

$$\dot{x}_n = u - b_0 x_1 - \dots - b_{n-1} x_{n-1}$$

Rearranging these equations, so that only the first derivatives are on the left hand side, gives the realisation

$$\underline{\dot{x}} = \begin{bmatrix} 0 & 1 & 0 & \cdots & 0 \\ 0 & 0 & 1 & \ddots & \vdots \\ \vdots & \ddots & \ddots & \ddots & 0 \\ 0 & \cdots & 0 & 0 & 1 \\ -b_0 & -b_1 & -b_2 & \cdots & -b_{n-1} \end{bmatrix} \underline{x} + \begin{bmatrix} 0 \\ 0 \\ \vdots \\ 0 \\ 1 \end{bmatrix} u \qquad (4.37a)$$

$$y = [a_0 \ a_1 \ \dots \ a_{n-1}]\underline{x} \qquad (4.37b)$$

Realisations 1 and 2 have the negative coefficients of the denominator of the transfer function in the state matrix, in the first column and last row respectively. The coefficients of the numerator are in the input and output vectors $\underline{b}$ and $\underline{c}$ respectively.

Realisation 3:
When all the poles of G(s) are real and single, a realisation with the state matrix A diagonal can be obtained. G(s) can be expanded in partial fractions as

$$\frac{Y(s)}{U(s)} = \frac{k_1}{s - r_1} + \frac{k_2}{s - r_2} + \dots + \frac{k_n}{s - r_n}$$

where r_1 to r_n are the poles of G(s). The state variables can then be defined to satisfy

$$\dot{x}_1 = r_1 x_1 + u$$

$$\dot{x}_2 = r_2 x_2 + u$$
$$\vdots$$
$$\dot{x}_n = r_n x_n + u$$

This gives the realisation

$$\underline{\dot{x}} = \begin{bmatrix} r_1 & 0 & \cdots & 0 \\ 0 & r_2 & & \\ \vdots & \ddots & \ddots & \\ 0 & \cdots & 0 & r_n \end{bmatrix} \underline{x} + \begin{bmatrix} 1 \\ 1 \\ \vdots \\ 1 \end{bmatrix} u \tag{4.38a}$$

$$y = [k_1 \ k_2 \ \ldots \ k_n] \underline{x} \tag{4.38b}$$

This realisation differs from the other two realisations in having the poles on the diagonal and their coefficients in the output $\underline{c}$ vector, rather than the negative denominator coefficients in the state matrix and the numerator coefficients in the input or output vector.

4.6 Example: The Arms Race

A simple model of the arms race between two nations has been obtained by Richardson (1960). This can be written in state space notation as

$$\dot{x}_1 = -a_{11} x_1 + a_{12} x_2 + b_1 u_1 \tag{4.39a}$$

$$\dot{x}_2 = a_{21} x_1 + -a_{22} x_2 + b_2 u_2 \tag{4.39b}$$

where the state variables x_1 and x_2 and coefficients a_{ij} are defined as follows:

- x_1 and x_2: the arms expenditures of each of the two adversarial (groups of) nations.
- a_{12} and a_{21}: threat coefficients which cause the (groups of) nations to increase their arms expenditure.
- a_{11} and a_{22}: fatigue and expense coefficients, resulting from the burden of continuing the arms race.
- u_1 and u_2: grievances and other factors which lead to nations arming at a constant rate.

Therefore all four coefficients a_{ij} will be positive and the reduction aspect has been indicated by negative signs before the coefficients a_{11} and a_{22}. The output y is taken to be the total arms expenditure of the two nations or groups of nations. It is given by the sum of x_1 and x_2 i.e.

$$y = x_1 + x_2 \tag{4.39c}$$

The state and output equations can be written in matrix form as

$$\dot{\underline{x}} = A\underline{x} + Bu \tag{4.40a}$$

$$y = C\underline{x} \tag{4.40b}$$

with

$$A = \begin{bmatrix} -a_{11} & a_{12} \\ a_{21} & -a_{22} \end{bmatrix} \qquad B = \begin{bmatrix} b_1 \\ b_2 \end{bmatrix} \qquad C = [1 \quad 1] \tag{4.41}$$

Therefore the transfer function representation is given by

$$G(s) = C(sI - A)^{-1}B$$

$$(sI - A)^{-1} = \begin{bmatrix} s + a_{11} & -a_{12} \\ -a_{21} & s + a_{22} \end{bmatrix}^{-1}$$

$$= \frac{1}{s^2 + (a_{11} + a_{22})s + a_{11}a_{22} - a_{12}a_{21}} \begin{bmatrix} s + a_{22} & a_{12} \\ a_{21} & s + a_{11} \end{bmatrix}$$

$$\therefore G(s) = C(sI - A)^{-1}B$$

$$= \frac{1}{s^2 + (a_{11} + a_{22})s + a_{11}a_{22} - a_{12}a_{21}} [1 \quad 1] \begin{bmatrix} s + a_{22} & a_{12} \\ a_{21} & s + a_{11} \end{bmatrix} \begin{bmatrix} b_1 \\ b_2 \end{bmatrix}$$

$$= \frac{1}{s^2 + (a_{11} + a_{22})s + a_{11}a_{22} - a_{12}a_{21}} [s + a_{22} + a_{21} \quad s + a_{11} + a_{12}] \begin{bmatrix} b_1 \\ b_2 \end{bmatrix}$$

$$= \frac{b_1(s + a_{22} + a_{21}) + b_2(s + a_{11} + a_{12})}{s^2 + (a_{11} + a_{22})s + a_{11}a_{22} - a_{12}a_{21}}$$

$$\Rightarrow G(s) = \frac{(b_1 + b_2)s + b_1(a_{22} + a_{21}) + b_2(a_{11} + a_{12})}{s^2 + (a_{11} + a_{22})s + a_{11}a_{22} - a_{12}a_{21}} \tag{4.42}$$

$\therefore$ the system poles are given by

$$s = -0.5(a_{11} + a_{22}) \pm 0.5\sqrt{(a_{11} + a_{22})^2 - 4a_{11}a_{22} + 4a_{12}a_{21}}$$

$$= -0.5(a_{11} + a_{22}) \pm 0.5\sqrt{(a_{11} - a_{22})^2 + 4a_{12}a_{21}} \tag{4.43}$$

The case of unforced input i.e. $u = 0$ will be considered first. When $a_{12} = a_{21} = 0$, the system of equations (4.39) is decoupled into two separate equations, in which there are no threats or other pressures leading to an arms race. This gives the following output, which shows exponential decay

$$y(t) = K_1 e^{-a_{11}t} + K_2 e^{-a_{22}t} \tag{4.44}$$

where the values of K_1 and K_2 are determined by the initial conditions and the matrix C. In the more general case, the system has two real roots as $(a_{11} - a_{22})^2 + 4a_{12}a_{21} \geq 0$ (since all the coefficients a_{ij} are positive).

Therefore the system becomes unstable and exhibits exponential growth if one root is positive i.e. if

$$-0.5(a_{11} + a_{22}) + 0.5.\sqrt{(a_{11} - a_{22})^2 + 4a_{12}a_{21}} > 0$$

$$\Rightarrow (a_{11} - a_{22})^2 + 4a_{12}\, a_{21} > (a_{11} + a_{22})^2$$

$$\Rightarrow a_{12}\, a_{21} > a_{11}a_{22} \tag{4.45}$$

i.e. the product of the threat coefficients is greater than the product of the fatigue coefficients, which implies that the pressures toward an arms build up are greater than those resisting it. When inequality (4.45) does not hold the system is stable.

The particular case with the following values of A and B will now be considered:

$$A = \begin{bmatrix} -0.2 & 0.5 \\ 1 & -0.2 \end{bmatrix} \qquad B = \begin{bmatrix} 0.5 \\ 0.5 \end{bmatrix} \tag{4.46}$$

The characteristic equation is now

$$s^2 + 0.4s - 0.46 = 0 \tag{4.47}$$

$$\Rightarrow s = -0.2 \pm \sqrt{0.5} = 0.5071 \text{ or } -0.9071 \tag{4.48}$$

The frequency domain state transition matrix Φ'(s) will be calculated and then used to obtain expressions for the state vector and system output. It is given by

$$\Phi'(s) = (sI - A)^{-1}$$

$$= \frac{1}{s^2 + (a_{11} + a_{22})s + a_{11}a_{22} - a_{12}a_{21}} \begin{bmatrix} s + a_{22} & a_{12} \\ a_{21} & s + a_{11} \end{bmatrix} \tag{4.49}$$

Substituting coefficients from equation (4.46) in equation (4.49) gives

$$\Phi'(s) = \frac{1}{s^2 + 0.4s - 0.46} \begin{bmatrix} s + 0.2 & 0.5 \\ 1 & s + 0.2 \end{bmatrix}$$

$$= \frac{1}{(s - 0.5071)(s + 0.9071)} \begin{bmatrix} s + 0.2 & 0.5 \\ 1 & s + 0.2 \end{bmatrix} \tag{4.50}$$

The time domain state transition matrix can be obtained from the inverse Laplace transform of Φ'(s). First partial fraction expansion techniques are

Arms races can have unexpected dangers!

applied to each term in the matrix to give

$$\Phi'(s) = \frac{\Phi'_1(s)}{(s-0.5071)} + \frac{\Phi'_2(s)}{(s+0.9071)}$$

$$= \frac{1}{(s-0.5071)}\begin{bmatrix} 0.7071 & 0.5 \\ 1 & 0.7071 \end{bmatrix} + \frac{1}{(s+0.9071)}\begin{bmatrix} -0.7071 & 0.5 \\ 1 & -0.7071 \end{bmatrix} \tag{4.51}$$

Inverse Laplace transforms of equation (4.51) can be obtained from transform tables to give

$$\Phi(t) = \begin{bmatrix} 0.7071 & 0.5 \\ 1 & 0.7071 \end{bmatrix} e^{0.5071t} + \begin{bmatrix} -0.7071 & 0.5 \\ 1 & -0.7071 \end{bmatrix} e^{-0.9071t} \tag{4.52}$$

i.e. a weighted sum of the modes based on the two eigenvalues 0.5071 and -0.9071.

From equation (4.15a) with $t_0=0$

$$\underline{x}(t) = \Phi(t)\underline{x}(0) + \int_0^t \Phi(t-\tau)B\underline{u}(\tau)\,d\tau$$

$$= \begin{bmatrix} 0.7071 & 0.5 \\ 1 & 0.7071 \end{bmatrix}\begin{bmatrix} x_1(0) \\ x_2(0) \end{bmatrix} e^{0.5071t} + \begin{bmatrix} -0.7071 & 0.5 \\ 1 & -0.7071 \end{bmatrix}\begin{bmatrix} x_1(0) \\ x_2(0) \end{bmatrix} e^{-0.9071t}$$

$$+ \int_0^t \begin{bmatrix} 0.7071 & 0.5 \\ 1 & 0.7071 \end{bmatrix}\begin{bmatrix} u_1 \\ u_2 \end{bmatrix} e^{0.5071(t-\tau)}\,d\tau + \int_0^t \begin{bmatrix} -0.7071 & 0.5 \\ 1 & -0.7071 \end{bmatrix}\begin{bmatrix} u_1 \\ u_2 \end{bmatrix} e^{-0.9071(t-\tau)}\,d\tau$$

$$= \begin{bmatrix} 0.7071x_{10} + 0.5x_{20} \\ x_{10} + 0.7071x_{20} \end{bmatrix} e^{0.5071t} + \begin{bmatrix} -0.7071x_{10} + 0.5x_{20} \\ x_{10} - 0.7071x_{20} \end{bmatrix} e^{-0.9071t}$$

$$+ \begin{bmatrix} 1.3944u_1 + 0.9860u_2 \\ 1.9720u_1 + 1.3944u_2 \end{bmatrix}\left[1 - e^{0.5071t}\right] + \begin{bmatrix} 0.7795u_1 - 0.5512u_2 \\ -1.1024u_1 + 0.7795u_2 \end{bmatrix}\left[1 - e^{-0.9071t}\right] \tag{4.53}$$

The output y(t) can now be obtained by substituting $\underline{x}(t)$ from equation (4.53) and C from equation (4.41) in equation (4.40b) to give

$$\begin{aligned} y(t) &= x_1(t) + x_2(t) \\ &= [1.7071x_{10} + 1.2071x_{20} - 3.3664u_1 - 2.3804u_2]e^{0.5071t} \\ &\quad + [0.2929x_{10} - 0.2071x_{20} + 0.3229u_1 - 0.2283u_2]e^{-0.9071t} \\ &\quad + 3.0435u_1 + 2.6087u_2 \end{aligned} \tag{4.54}$$

This shows an unstable situation with exponential growth in arms expenditure, as would be expected from a system with one positive real root. This unstable behaviour could have been predicted without calculating the system poles by using inequality (4.45). In this case, using (4.46) for the coefficients of A,

$$a_{12}a_{21} - a_{11}a_{22} = 0.5 \text{ x } 1 - -0.2 \text{ x } -0.2 = 0.46 > 0 \tag{4.55}$$

so that inequality (4.45) holds.

Linear transformation of the equations (4.40) and (4.41) with (4.46) will now be carried out to obtain a state space representation with A* diagonal.

$$\dot{\underline{x}}^* = A^*\underline{x}^* + B^*u \tag{4.56a}$$

$$y^* = C^*\underline{x}^* \tag{4.56b}$$

This can be obtained using the transformation E = $[\underline{e}_1 \;\; \underline{e}_2]$, where the two vectors $\underline{e}_1$ and $\underline{e}_2$ are the system eigenvectors. The relationship between the coefficients e_{11} and e_{12} of the eigenvector can $\underline{e}_1$ be obtained by solving the equation

$$A\underline{e}_1 = \lambda\underline{e}$$

$$\Rightarrow \begin{bmatrix} -0.2 & 0.5 \\ 1 & -0.2 \end{bmatrix} \begin{bmatrix} e_{11} \\ e_{12} \end{bmatrix} = (0.5071) \begin{bmatrix} e_{11} \\ e_{12} \end{bmatrix} \tag{4.57}$$

The first row of equation (4.57) gives

$$\begin{aligned} -0.2e_{11} + 0.5e_{12} &= 0.5071\, e_{11} \\ \Rightarrow 0.5e_{12} &= 0.7071e_{11} \\ \Rightarrow e_{11} &= 0.7071e_{12} \end{aligned} \tag{4.58a}$$

It can be shown analogously that for the second eigenvector

$$e_{21} = -0.7071 e_{22} \tag{4.58b}$$

Normalising with the coefficients e_{12} and e_{22} equal to one gives the eigenvectors

$$\underline{e}_1 = \begin{bmatrix} 0.7071 \\ 1 \end{bmatrix} \quad \underline{e}_2 = \begin{bmatrix} -0.7071 \\ 1 \end{bmatrix}$$

$$\Rightarrow \quad E = \begin{bmatrix} 0.7071 & -0.7071 \\ 1 & 1 \end{bmatrix} \tag{4.59}$$

$$\Rightarrow E^{-1} = \frac{1}{1.4142}\begin{bmatrix} 1 & 0.7071 \\ -1 & 0.7071 \end{bmatrix}$$

$$= \frac{1}{1.4142}\begin{bmatrix} 1 & 0.7071 \\ -1 & 0.7071 \end{bmatrix}\begin{bmatrix} -0.2 & 0.5 \\ 1 & -0.2 \end{bmatrix}\begin{bmatrix} 0.7071 & -0.7071 \\ 1 & 1 \end{bmatrix}$$

$$\Rightarrow A^* = E^{-1}AE$$

$$= \frac{1}{1.4142}\begin{bmatrix} 0.5071 & 0.3586 \\ 0.9071 & -0.6414 \end{bmatrix}\begin{bmatrix} 0.7071 & -0.7071 \\ 1 & 1 \end{bmatrix} = \begin{bmatrix} 0.5071 & 0 \\ 0 & -0.9071 \end{bmatrix} \tag{4.60a}$$

$$B^* = \frac{1}{1.4142}\begin{bmatrix} 1 & 0.7071 \\ -1 & -0.7071 \end{bmatrix}\begin{bmatrix} 0.5 \\ 0.5 \end{bmatrix} = \begin{bmatrix} 0.6036 \\ -0.6036 \end{bmatrix} \tag{4.60b}$$

$$\text{and} \quad C^* = [1 \quad 1]\begin{bmatrix} 0.7071 & -0.7071 \\ 1 & 1 \end{bmatrix} = [1.7071 \quad 0.2929] \tag{4.60c}$$

This example has illustrated the application of state space techniques to a simple model of the arms race between two (groups of) nations. State space techniques have been used to obtain conditions in which the system is stable i.e. the arms expenditures of both sides remain stable, and to obtain expressions for the changes in arms expenditure of the two nations over time. A separable form of the state equation with the state matrix A diagonal was also obtained. This representation can be separated into two subsystems and therefore the system equations solved more easily.

However in this representation the two state variables have been transformed and therefore no longer represent the arms expenditures of the two nations.

4.7 Roundup of Chapter 4

This chapter has introduced the main concepts in state space representations. State space representations can be used for both continuous and discrete time systems. They consist of a set of first order equations, differential equations in the continuous time case and difference equations in the discrete time case, plus one or more output equations. State space representations can be obtained for both linear and non-linear systems. In the linear case the equations involve matrices with constant coefficients and are therefore particularly easy to manipulate. State space equations can be obtained for both single input single output (SISO) and multivariable (MIMO) systems. In the SISO case there is one output equation and in the MIMO case there is one output equation for each output.

The state vector is the vector of state variables i.e. the variables on which the description is based and which are differentiated or differenced in the set of first order equations. The dimension of this vector is equal to the order of the system. In the linear case the eigenvalues of the state matrix determine the properties of the system dynamics and, in particular, whether or not the system is stable. The state transition matrix can be used to obtain the transition between the system states at two different times.

In the non-linear case linearisations can be obtained around equilibrium or steady state points, if the system has such points. The logistic curve is obtained in the non-linear case by a Taylor expansion with only one term. It describes the growth of human populations with limited resources.

State space representations and analysis techniques can be applied to problems in sustainable development to obtain state space models and analyse them. This was illustrated in the chapter by application to a model of the arms race.

Chapter 5 has a more descriptive approach than chapters 3 and 4, since it considers mental models and sense making. However, though not applied in practice in chapter 5, some of the mathematical techniques presented in chapters 3 and 4 are applicable to the mental models in chapter 5.

5 Mental Models, Sense Making and Risk

5.0 Learning Objectives

The main aim of this chapter is to show how systems techniques, particularly those based on mental models, can be used to investigate the barriers to sustainable development inherent in many organisational structures, ideologies and ethos (the character or disposition which is inherent to an organisation). These concepts and techniques are illustrated by a discussion of the systemic factors which led to the Bhopal disaster. Specific learning objectives include the following:

- Understanding of the concept of mental models
- Understanding of how mental models and other systems techniques can be applied to increase understanding of both the barriers to sustainable development and the processes that can be used to encourage and develop it.
- Understanding of the concept of risk and the factors that affect perceptions of risk.
- Understanding of multi-loop action learning and the ability to apply it to problems in sustainable development.

5.1 Introduction

Discussion of sustainable development often focuses on mechanisms, such as the details of particular policies and whether or not they are sustainable or on what could be considered blueprints for sustainability. However change of all types, including moves towards sustainable development, generally occurs in a political and ideological context. Policies which have been decided on are often implemented with different degrees of effectiveness. Therefore it is useful to discuss the application of systems methods to increasing understanding of the organisational, psychological and social barriers to change and, in particular, change in the direction of sustainable development. This is the aim of this chapter.

It is set out as follows. Section 5.2 introduces mental models and discusses the use of feedback in learning, with an illustration relating to misinterpretation of ozone layer data. Multi-loop action learning, which can lead to changes in behaviour and/or attitudes at the individual, organisational and societal levels, is discussed in section 5.3. Section 5.4 considers risk, including the importance of considering the technical and social or political components together, and the precautionary principle. Safety cultures, including environmental and social safety cultures, are discussed in section 5.5. Some of the concepts and techniques in the chapter are illustrated by an example of the Bhopal accident in section 5.6 and the chapter is summarised in section 5.7.

5.2 Mental Models

One approach to improving understanding of individual and organisational mechanisms and decision making processes is through the use of mental models. *Mental models* can be considered to fulfil two distinct functions:

- To provide tools for researchers and others to analyse the processes by which individuals and organisations model the world.
- To provide a model of how individuals and organisations represent external reality to themselves. In some cases mental models fulfil a particular function rather than represent a particular real system.

Mental models may be held by individuals or by groups, organisations, cultures and nations. The mental models of individuals will be influenced to different extents by the accepted views of the surrounding society. In many cases individuals will incorporate these accepted views directly into their mental models, whereas others will reject these accepted views and construct mental models in opposition to them. However some people will be able to draw on and evaluate (possibly unconsciously) a number of different sources of views and information in drawing up their mental models. Mental models can take many forms and their contents are very varied. They may include ways of combining different concepts to construct more complex concepts. One approach to combining concepts involves treating them as characteristic functions which can be combined using a truth value (Johnson-Laird 1983). For instance the concept of an industrial process x having a greater rate of emissions of the greenhouse gas carbon dioxide than another industrial process y has the value *true* if x does indeed emit more carbon dioxide than y (under analogous conditions).

Although on the surface very different, mathematical and mental models have a number of similarities. For instance both types of models use a combination of assumptions and an internally self-consistent system of logic to obtain a model or representation of reality. The type of model will be mainly determined by the nature of the 'symbols' used and the ideas that can be expressed will largely depend on the *contextual tools*, including symbols, available for expressing them.

Some *symbols* have considerable social and political meaning. For instance, most societies have a number of significant symbols, such as flags and other devices and sometimes significant (symbolic) individuals, which have real symbolic importance to at least some groups in the society, and possibly the majority of the population. Since there is often a relationship between the types of symbols used and the ideas they can express, changes in symbols can sometimes have a (significant) effect on the mental models of the societies that use them and may even promote new ways of thinking and behaving. Thus, for instance the change from a constitutional monarch to a ceremonial president as head of state could have an effect on the class system, through the symbolic associations of a president as opposed to a monarch, though there is no real change in political power. However new symbols need to be recognised and accepted to gain (symbolic) meaning and purely changing symbols without making other changes will not be sufficient to achieve new sustainable models and ways of thinking. Changes in symbols may also result from changes in status. Thus, for instance, when a country gains independence, it generally takes a new flag. It should be noted that the symbols used in advertising are generally particularly powerful and that they are currently promoting high consumption lifestyles, with the product being advertised being made to stand for everything good in life. Therefore either a reduction or a change in the symbolic meaning of advertising will probably be necessary to reduce consumption.

Any restrictions on the set of symbols or other contextual tools that may be used will lead to restrictions on the different types of concepts that can be expressed with them. Thus the use of models based on a restrictive or sparse set of symbols could lead to a lack of imagination in proposing solutions to environmental, social and other problems and the consequent failure to consider certain types of solutions. However the elimination of certain types of symbols, as in the case of advertising, can have positive as well as negative consequences.

Limiting mental models can give rise to artificial barriers in the absence of real barriers, through the belief that something cannot or should not be done. Therefore the mental models on which the system structure rests are often the best leverage points for change. This is particularly important in the context of sustainable development, where the barriers to change are frequently political and sometimes also social rather than technical. In particular politicians and decision makers may have an inaccurate understanding of the causes of social and environmental problems. They may also have mistaken views on what types of risk the general public will accept as the price of 'progress' or what changes to lifestyle they will accept in order to prevent or alleviate negative environmental impacts. On the other hand the removal of symbols which represent concepts such as war could have very positive effects.

Mental models can be considered to form a system which is used to make sense of other systems. Thus they can be used by individuals and organisations to structure their experiences and try to understand them. They often form the basis of *sensemaking* (Weick 1995), which is about creating order and gaining retrospective understanding of experiences. Systems techniques can be applied to understanding and interpreting mental models and, in particular, to understanding how mental models can effect policy formulation and implementation. This is particularly relevant in the context of sustainable development.

New experiences are generally interpreted in terms of existing mental models. When something occurs which does not fit with the mental model there are a number of different possible responses, ranging from total collapse to inspired creativity, which has been defined as 'figuring out how to use what you already know in order to go beyond what you currently think' (Bruner 1983). In the case of collapse and an inability to respond, this will very occasionally be due to a situation which does not make sense. There are extreme situations or experiences, such as rape, holocaust, war and genocide, which may be edited out by suppressing the memories, because it is not possible to make sense of them. However problems in understanding a situation are more likely to be due to an inability to make sense of it in terms of existing mental maps.

Although mental models can be revised to take account of new evidence, it is more common for experience(s) to be *edited by selection and filtering* to make them conform better to existing mental models (O'Connor and McDermott 1997). There is also frequently a failure to either acknowledge or report evidence which does not fit existing models and therefore may not be believed (Weick 1995). The underlying thinking is frequently of the type 'It doesn't make sense to me, so it can't have happened!'. Thus sensemaking may involve considerable distortion of real situations in order to make them harmonise with existing mental models. Rewriting of experiences is likely to be a particular problem when the new data challenges existing mental models (preconceptions or philosophical, political, religious and other beliefs) on which an individual's life is based and changing these models would force a critical examination and the possible rejection of their life philosophy and/or way of life. It should be noted that individuals and organisations may also have a vested interest in rejecting experiences that do not fit their mental models. Thus, for instance many politicians will often ignore evidence that their policies are not working, due to a dogmatic belief in the policy and/or a vested interest in continuing to belong to the political party that espouses this policy.

Another common process in editing mental models is *generalisation* in which one experience is taken to represent a particular group. Generalisation can be useful in allowing knowledge, skills or techniques to be applied in different situations (O'Connor and McDermott 1997). However generalisations are often used to justify bigotry and prejudice and are at the basis of stereotypes. Systemic approaches can also provide tools for questioning 'universal' assumptions and 'truths', which it is believed that 'everyone' holds, but which are frequently false. These 'universal' assumptions are often based on generalisations from the behaviour and/or beliefs of dominant groups and are therefore often false for non-dominant and minority groups. Examples in Western Europe include the beliefs that everyone speaks English and that everyone celebrates Christmas. Negative generalisations and stereotypes are generally based on individual negative experiences with members of non-dominant and minority groups. For instance in Western Europe negative generalisations (by white people) about black people based on one bad experience are not uncommon, whereas similar generalisations about white people are much rarer.

Feedback and feedforward can be applied to mental models similarly to other systems. *First order* or *adaptive learning* uses negative feedback to change responses and move closer to the desired goals (O'Connor and McDermott 1997). It is the principle behind the control of many industrial processes and successful control strategies and can, for instance, be applied to reduce energy consumption. The same principles can be applied to individual or organisational decision making and environmental and other policy formulation. Thus adaptive learning with feedback can be used to identify and reject or modify solutions that have not worked (well) in the past and develop new solutions. This is particularly important in the context of sustainable development, where past mistakes have been repeated too often. For instance new roads are still being built as the response to traffic congestion. However, it has been amply demonstrated that this results in traffic generation (Pfleiderer and Dieterich; SACTRA 1994) which leads to further congestion, as the growth in traffic overfills the increased provision, as well as increasing emissions and energy consumption.

Adaptive learning frequently works well, but will not produce the desired results if the basic problem is with existing mental models, since it does not change them. In this case *double loop learning* with an additional loop with negative feedback which can effect changes to existing mental models is required. Positive feedback in the second loop should be avoided as it will act to reinforce existing mental models. Applying adaptive feedback to existing mental models can be particularly important in situations where they act to preserve the status quo. For instance, there has been considerable scapegoating in much of Europe of immigrants and refugees as responsible for problems, such as poverty, homelessness and unemployment. As well as leading to injustice and contributing further to the suffering experienced by all refugees and many immigrants, harsh policies towards refugees and migrants will do nothing to resolve these problems. However using negative feedback to revise mental models could remove the barriers to understanding the real causes of problems such as poverty and homelessness, which would be a first step towards resolving them (O'Connor and McDermott 1997).

Example 5.1: Measurement of the Ozone Layer

For a number of years NASA scientists failed to find evidence of a reduction in ozone concentration, because they had set their computers to reject very low readings which they believed could not occur. In response to published evidence of a deep hole in the ozone layer over the Antarctic they repeated the tests on the original data without rejection of very low readings and found that ozone concentrations had been falling steadily for seven years (O'Connor and McDermott 1997). This example illustrates both how mental models can lead to the misinterpretation of experimental data and the ability to update mental models in response to evidence. In this case the mental model was based on the erroneous scientific hypothesis that there is a lower limit on ozone concentrations.

The two loops are illustrated in fig. 5.1. The inner loop models the situation based on the original mental model with the assumption that there was a lower bound on possible ozone readings. The outer loop models the process by which the mental model of the ozone layer was updated. Measurements of the ozone layer are fed back in both loops. They are compared with the expected values in the inner loop and used to update the regulatory measures. The 'reference' expected values are generated by the mental model and when they are too low readings are not recorded in the measurement system.

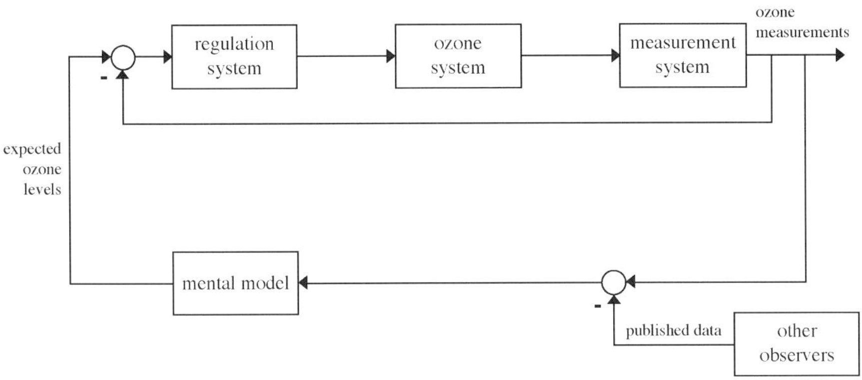

Fig. 5.1 Changing mental models: measurements of the ozone layer

In the outer loop the measured values are compared with the published data and the difference is used to update the reference expected ozone levels. The minimum expected value now no longer acts to filter out low values of ozone data and the measured ozone data are equal to the actual values.

5.3 Multi-Loop Action Learning

5.3.1 Introduction

Although it is not always immediately apparent what measures are required to promote sustainable development, it is often easier to identify appropriate measures than to implement then in practice. In many cases action will be required by a range of different groups and organisations and there may be structural, institutional and other barriers. Attitudinal barriers can be particularly difficult to overcome and may result in a lack of recognition of existing environmental and social problems, as well as a reluctance to take action, particularly if this would result in or require a change in lifestyles.

The use of double loop learning to achieve changes in existing mental models was discussed in the previous section. In this section this approach is taken further through multi-loop action learning. *Multi-loop action learning* is an extension to include quadruple loop action learning (Hersh 2004ab) of methods categorised (Nielson 1996) as single, double and triple loop action learning. The approach has generally been applied in the context of ethical behaviour in organisations, but is equally relevant to sustainable development. There is also a significant overlap between ethics and sustainable development with regards to both organisational ethos and action. Multi-loop action learning is illustrated in fig. 5.2.

One of the advantages of multi-loop action learning is that it provides a structure in which to consider where the barriers to change are occurring, for instance whether at the individual, organisational or wider societal level. This then indicates the level at which action will be required and where leverage is necessary. The different levels of action loop learning can be defined as follows (Nielson 1996; Hersh 2004ab):

- *Single loop action learning* is about changing behaviour rather than learning about ethics and changing values.
- *Double loop action learning* involves changes in values (generally of individuals) as well as behaviour.
- *Triple loop action learning* involves changes in the underlying tradition or ethos of the organisation, as well as changes in values and behaviour.
- *Quadruple loop action learning* additionally involves changes in the ethos or tradition of the surrounding society. Alternatively it involves changes in the underlying tradition with reference to the nature of the organisation in addition to its practice.

Example 5.2: Employment of Disabled People

These different levels of multi-loop action learning will be illustrated by the following example. Here the aim is to increase the proportion of disabled people in senior positions in an organisation.

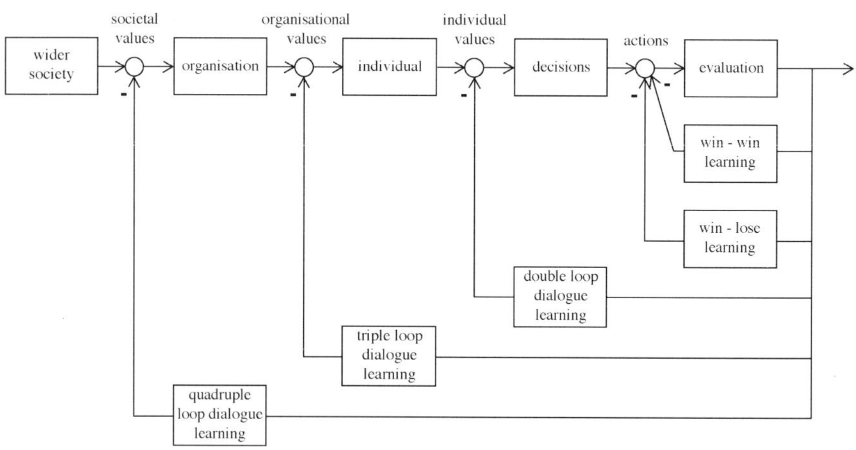

Fig. 5.2 Multi-loop action learning

- Single loop action learning could lead to measures to increase the proportion of disabled people recruited to senior positions, for instance due to fear of legal action on the grounds of disability discrimination, without any increase in awareness of the ethical responsibility to recruit more disabled people or a change in values.
- Double loop action learning could lead to a change in ethical values on the part of some individuals in the organisation with a recognition of the ethical responsibility not to discriminate against disabled people, in addition to practical measures. This ethical commitment is likely to make the practical measures more effective than they would be otherwise.
- Triple loop action learning could lead to a change in the ethos of the organisation with a recognition of both the value to the organisation and the ethical responsibility to employ more disabled people at a senior level. This could be accompanied by measures to overcome structural barriers and make the organisation attractive as a place of employment for disabled people.
- Quadruple loop action learning could lead to a change in the ethos of the wider society with an ethical commitment to the value of diversity in society and ensuring equality and elimination of discrimination for the diverse population. Disabled people would be considered one of many of the diverse groups which enrich society. Furthermore measures would be taken to remove structural barriers and make all environments attractive and accessible to the whole population, of which disabled people are a part.

A range of methods will be required to implement multi-loop action learning. Further information and examples of the methods given in the following sub-sections can be found in Nielson (1996).

5.3.2 Single Loop Action Learning

Single loop action learning methods have been divided into win-win and win-lose. In win-win methods there are some benefits to everyone. This is not the case in win-lose methods and therefore pressure of some type is required to implement them.

Win-lose methods include the following:

- Bottom-up forcing methods, such as whistleblowing (Hersh 2002c), refusal to cover-up unethical behaviour and avoiding the implementation of unethical policies and orders.

- Top down forcing methods, such as the boss producing and imposing ethical policies without consultation. However this is ethically questionable, in particular because it will result in a reduction of autonomy for workers.
- Win-lose negotiations, such as leverage building, good guy-bad guy and making extreme demands in order to then achieve a 'compromise'. However these methods may themselves be unethical, depending on how they are applied. For instance leverage building could have some similarities to blackmail.

Win-win methods include the following:

- Mutual gain negotiations in which there are sufficient benefits on both/all sides to agree a settlement. Ethical issues are sometimes included without being mentioned explicitly, particularly if they are part of the agenda of one party.
- Persuasion in which language and/or illustrations are used to convince the other parties to do something. People are sometimes persuaded against unethical behaviour or unsustainable activities by the fear of getting caught, the benefits of the ethical conduct or sustainable activities or the disadvantages of the unethical behaviour or unsustainable activities rather than concern for ethics or sustainability.
- Minimal peaceful coexistence negotiations aim to obtain an improvement of a bad situation rather than mutual gains.

Both win-win and win-lose methods can be effective in encouraging or suppressing unethical behaviour or unsustainable activities, but without leading to any learning, changes in values or greater understanding of ethics or sustainability. They are likely to lead to results more quickly than double and triple loop methods. Win-win behaviour may result in better relationships and lead to productive cooperations and could therefore give more scope for learning and changes in values through the subsequent application of double loop dialogue methods. However win-win approaches can result in learning for people who learn through doing. Bottom-up forcing is appropriate when it is not proving possible to find any other ways of dealing with unethical behaviour or unsustainable activities. However the other win-lose methods are themselves questionable in ethical terms and should therefore generally be avoided.

5.3.2 Double Loop Action Learning

Double loop methods are generally based on dialogue. They can be used to develop or maintain an ethical organisational culture which is committed to sustainability and can sometimes lead to the adoption of ethical values. They can sometimes have win-win outcomes, though this is not the (main) aim. They are appropriate when there is misunderstanding or lack of knowledge about what is ethical in addition to unethical behaviour, but should not be applied when the problem is purely unethical behaviour or unsustainable activities. They are also unlikely to work in environments which discourage dialogue and/or where there is pervasive low level pressure to unethical behaviour or unsustainable activities.

Double loop dialogue methods include the following:

- Socratic iterative dialogue
- Action-science dialogue
- Action-inquiry dialogue.

Each of these methods involves a number of stages or components which will be listed below, but will not be discussed in any detail. *Socratic iterative dialogue* consists of the following four components which are carried out sequentially:

1. An initial respectful and friendly approach.
2. The facilitator asks the other participants for a potential solution and helps them consider its advantages.
3. The facilitator helps the other participants consider the drawbacks of the potential solution and iteratively suggests other solutions to retain the advantages and reduce the drawbacks of previous solutions.
4. Continuing the process iteratively until the solution cannot be improved further.

Action-science dialogue has the following seven implementation rules (Argyris et al 1985), which are not necessarily sequential:

1. Combine advocacy with inquiry.
2. Illustrate your inferences with fairly directly observable data.
3. Make your reasoning explicit and publicly test for agreement at each inferential step.
4. Actively seek contrary data and alternative explanations.
5. Recognise that mistakes occur and that you can learn from them.
6. Actively investigate your impact on the learning context.
7. Design ongoing experiments to test competing views.

Action-inquiry dialogue has the following four components (Torbert 1987):

1. Framing: determining the frame or purpose of everyone's and not just the speaker's participation.
2. Advocating: what the speaker proposes in the frame.
3. Illustrating: using a concrete example to clarify what the speaker means.
4. Inquiring: how the others respond to the speaker's perspective and initiative.

All three dialogue methods can be used to change both values and behaviour, but require a certain degree of consensus from the dialogue participants for success. The iterative Socratic method is explicitly friendly and respectful and may be the most appropriate with less powerful people, people from cultures where there is little experience of advocacy and directness, and/or subordinates fearful of being misunderstood if they present different perspectives. Action-science is the most thorough as it explicitly uses new experimental and experiential data and multiple alternatives. It is therefore likely to take the longest. Action-inquiry is the most focused on the required action and therefore likely to be the fastest and most direct. It is also possible to combine elements of the different methods, for instance to add explicit friendliness to the action-science method or experimentation to action-inquiry.

5.3.3 Triple Loop Action Learning

There are a number of different *triple loop* methods, including the following:

- Woolman's friendly disentangling
- Friendly upbuilding
- Friendly reconstruction
- (Adversarial) deconstruction
- Experimental neopragmatism.

Woolman's (1774) *friendly disentangling* method has four main components:

1. Framing a 'we' fellowship relationship with others and seeking the cause of current problematic behaviour in biases of the embedded tradition.
2. Approaching others in a friendly way.

3. Asking for help in disentangling problematic behaviour from potential biases in 'our' embedded tradition system.
4. Working with those willing to experiment with alternative behaviours and values that are not based on the problematical biases of the tradition system.

The social tradition is both criticised and treated as a partner in the dialogue. The method can enable peaceful change, reform the biases in organisational traditions and facilitate change of ideas. It sometimes leads to win-win solutions. The *'we' fellowship relationship* is a central element of the approach. It is therefore unlikely to work when the people involved do not see a 'we' relationship or 'us versus them' behaviour is more important. It will also not work when the people in power present ethical and sustainability issues in terms of their self interests and ignore alternative perspectives. It is inappropriate when there is no significant negative bias in the system and the cause of unethical behaviour or unsustainable activities lies largely in individuals rather than the organisation.

The concept of upbuilding is derived from Kierkegaard's work (1944 1967). *Friendly upbuilding* is based on protecting and extending an existing ethical tradition in a destructive external environment and using it to solve problems. It combines the following three processes:

1. Explicit building from within an ethical environment as a way of framing and solving problems in a problematical environment.
2. Generation of ethical decisions informed by the tradition.
3. Leaving the tradition open to criticism, modification and further development.

Upbuilding can help organisations resist negative environmental pressures, and simultaneously develop their own traditions. It can facilitate peaceful change and enable ethical learning by individuals. Since it is based on an existing ethical tradition, it will not work when the tradition is insufficiently ethical or there is insufficient consensus about its ethical components. It may also be too conservative.

Friendly reconstruction (Gadamer 1989), (adversarial) deconstruction (Derrida 1981; Foucault 1972) and experimental neopragmatism (Rorty 1982 1991) are postmodernist approaches. *Postmodernism* is a particular period or perspective of western thought. It differs from earlier philosophies in not recognising a centre or single ideal purpose in life. Instead there is a recognition of the value of the similarities and differences of both one's own and other traditions. Friendly reconstruction tries to bridge differences, whereas (adversarial) deconstruction exposes negative biases and experimental neopragmatism aims at peaceful coexistence and small improvements through experimentation.

In *(adversarial) deconstruction* criticism of negative biases and oppressions is used to achieve and maintain difference and diversity. This is often accompanied by adversarial criticism of individuals representing the mainstream. *Friendly reconstruction* builds on tradition based positive biases and commonalities. It combines this with friendly criticism and dialogue about positive and negative biases in cultures and systems. *Experimental neopragmatism* experiments with win-win solutions in the absence of objective standards or principles. Rather than examining positive and negative biases, it is accepted that people from different traditions may have different needs and experiments are used to explore temporary win-win solutions which satisfy these different needs.

Each of the postmodernist approaches has advantages and disadvantages and the three methods can be combined. Reconstruction can bridge cross-cultural differences to solve problems, whereas deconstruction can enable people to act constructively by removing blame from individuals and neopragmatism can facilitate low risk incremental improvements. However reconstruction will not work when the divisive effect of negative biases is stronger than the unifying effect of positive ones and deconstruction can be dangerous for the critics of a powerful system. There may be no win-win experimental outcomes for neopragmatism or the process may break down.

It may be possible to successfully apply some of the triple loop action learning methods to quadruple loop action learning in some circumstances. Currently the methods used in achieving *quadruple loop action learning* are those of political campaigning, including letter writing, leafleting, lobbying and demonstrating. Unlike triple loop action learning, quadruple loop action learning often requires action by large numbers of people. Success may be dependent on understanding systemic mechanisms and applying appropriate leverage.

5.4 Risk

Most systems are characterised by uncertainty and risk. *Risk* can be defined as the probability of injury, damage or loss in a particular context over a given period of time (Sage 1992). It is particularly significant in the context of sustainable development, where, for instance, the effects of the implementation of new and existing technologies on the natural environment, health, social development, ecosystems and biodiversity are both significant and characterised by considerable uncertainty and risk. In many cases there is a high degree of uncertainty, making it difficult if not impossible to quantify the associated risks. The risk associated with an event can generally be considered to be a combination of the probability of the event and the significance of its impacts (Sage 1992). However it is often useful to characterise risky events by both these aspects separately, rather than to combine them.

Decision making about risk and what risks are considered acceptable will generally be affected by a number of different factors, including the following:

- The probability, type and severity of the likely consequences. There is often greater concern about events with significant impacts, but low probabilities, such as the risk of dying in a plane crash, where the number of deaths at one time is generally large, than in a car accident, where the statistical risk is much higher.
- Whether the risk has been assumed voluntarily. Most people are less concerned about risks they have assumed voluntarily, such as those involved in risky sports, than risks they are exposed to involuntarily, for instance due to living next to a nuclear power station.
- Whether the risk is to known or unknown individuals, with risks to known people generally weighted more highly.
- Whether the consequences are immediate or delayed.

Many risks have both a technical component and social and political consequences. The *technical component* relates to the likely consequences of risks, whereas the *social and political component* is concerned with attitudes to these consequences and whether or not they are considered acceptable. In practice it is not possible and generally not desirable to consider the technical and social and political consequences totally separately from each other. Particularly in the context of limited resources, views on the political and social acceptability of risks may determine which specific consequences are investigated. The process may also be iterative with attitudes changing or evolving as more technical information becomes available. It is sometimes assumed that some risks are technically complicated and can only be understood and evaluated by experts. However in practice the ability of the general public to assess such risks is generally greater than is assumed (ESRC 1999). There are also ethical issues associated with the imposition of risks on people who have not given informed consent (Martin and Schinzinger 2004).

When there is uncertainty about the risk, the *precautionary principle* should be applied. This is based on a willingness to take action in advance

Tell me it's safe

of firm scientific evidence to avoid potential damage and/or costs to society, the environment and/or future generations resulting from delay (Harremoës et al 2002; O'Riordan and Cameron 1994). However this principle has rarely been implemented and there is generally considerable resistance from governments and firms to not implementing particular technologies or not carrying out particular projects on account of their potentially damaging effects.

The existence of risk and the measures taken to reduce it may put limits on personal freedom and may require people to refrain from certain actions to reduce risks. Therefore the effects of risk on personal freedom are an important factor that should be considered in decision making about the introduction of a particular technology or industrial process. Some restrictions will be generally accepted, as the limiting effects are small and not out of proportion to the increase in safety. For instance most people will accept not wearing jewellery and tying back their hair when working on certain types of machinery. However other safety measures may raise concerns. For instance there is currently serious concern about proposals to introduce identity cards in the UK. Unacceptable intrusions on personal freedom can often be avoided by an appropriate choice of measures. However in some cases it will either not be possible to reduce the risk associated with a particular technology to an acceptable level or to do so without restricting personal freedom. This implies that in general this technology should not be implemented. Any associated benefits would have to be unusually significant to outweigh the loss of personal freedom. Equity considerations with regards to the distribution of risk also imply that safety measures should effect everyone equally and not target a particular group. For instance regulations about covering wounds and precautions in handling blood products avoid the ethical and other problems associated with compelling either particular individuals or the whole workforce to be tested for HIV status.

5.5 Safety Cultures

An important aspect of organisational culture is the attitude to risks. Some organisations have developed a culture which encourages safe practices and significantly reduces both the likelihood of accidents and the likelihood of injuries resulting from any accidents that do occur. The characteristics of a *safety culture* generally include the following (Fortune and Peters 1995; ShamRao 2001):

- Positive attitudes to criticism and other feedback, both from lower levels in the organisation and from outside the organisation.
- Consideration of boundaries and patterns of internal communication when designing processes, procedures and working practices.
- Encouraging concern for the consequences of organisational activities and the effects of individual actions.
- Encouraging involvement and commitment and successful conflict resolution.

The concept of a safety culture can easily be extended to sustainability or environmental and social safety. An *environmental safety culture* would encourage practices which consume resources wisely and do not generate large quantities of waste. There would be a high consciousness of the importance of environmental conservation and protection as well as the impact of human activities on the environment in all aspects of planning and decision making. A *social safety culture* would encourage practices that combat injustice and reduce inequality in society. In industry this would include a focus on employment issues, with good working conditions and high wages. It would probably lead to alternative forms of organisation rather than the traditional division into workers and management.

There have been a number of serious large-scale accidents in technological systems, including nuclear and chemical plants, such as Three Mile Island, Bhopal and Chernobyl, and aircraft crashes, including the Kegworth crash (Smith 1992). These accidents have generally been blamed on 'operator error', but deeper investigation has found the causes to lie in *systemic factors* resulting from the interaction of human and technical systems and a lack of attention to human factors (Meshkati 1991).

When accidents occur, human operators often provide a convenient scapegoat, as blaming them allows preservation of the status quo and does not require the high shutdown and retrofitting costs that would result if designs were found to be faulty or threaten those in charge by blaming management (Perrow 1986). Although operator error is often the triggering action which results in the occurrence of an accident, the causes of this error are generally found to be systemic, with these systemic factors leading to a set of circumstances which make it very difficult to avoid errors. In addition such systemic factors generally result in a situation which makes the occurrence of an accident highly probable and only waiting for a trigger, which need not be operator error, to occur.

5.6 Example: Human Factors and Lack of Safety Culture in Large-Scale Technological System Accidents

As an example of the systemic factors which lead to error and accidents, the well known and catastrophic accident at the Union Carbide Plant in Bhopal, India will now be considered. On December 3, 1984 operators became alarmed by a leak and overheating in a storage tank, containing methyl isocyanate, a toxic chemical used in pesticides. Within an hour the leak exploded, sending 40 metric tons of deadly gas into the atmosphere. As a result 3800 people were killed and more than 200,000 others injured (New York Times, Sept 12, 1990). Deaths continued at the rate of, on average, two people a day amongst those injured and the health of the survivors continued to deteriorate (Los Angeles Times, March 13, 1989).

By the late 1970s operations at the Bhopal plant had been transformed from mixing chemicals to make pesticides to the production of chemical ingredients. Although Union Carbide was aware of the hazards involved, it did not transfer all the safety mechanisms available. For instance its West Virginia plant used computerised instruments to control safety systems and detect leaks, whereas safety controls at Bhopal were all manual and workers were asked to detect leaks by seeing or smelling them (Martin and Schinzinger 2004). Although workers may develop considerable sensitivity to the presence of chemicals, leaks can only be detected in this way at much high concentrations than if computerised instruments are used, thereby increasing the risks.

The Indian government required the Bhopal plant to be operated entirely by Indian workers. Initially care was taken to ensure adequate training, but

Union Carbide relinquished its supervision of safety at the plant in 1982 for financial reasons, despite retaining general financial and technical control. At its last inspection in 1982 a team of US engineers warned of many of the hazards that led to the disaster. Safety practices eroded over the next two years and there was a move away from US safety standards to lower Indian ones. A number of other factors contributed to the reduction in safety standards, including labour relation problems, which led to a high turnover of employees, and internal management disputes. Many of the managers, including those at the time of the accidents, were not properly trained in the hazards and appropriate operating procedures for a pesticides plant (ICFTU 1985). Most employees had little technical education and inadequate training, including in the safety and health hazards of the toxic substances used in the plant.

Plant management was characterised by authoritative and sometimes manipulative styles and organisational rigidity. This rigidity was largely responsible for the lack of effective measures to deal with the five major accidents reported at the plant between 1981 and 1984. Organisational culture was also largely responsible for the lack of attention to warnings about safety problems, such as those resulting from the October 1982 release of methyl isocynatate, hydrochloric acid and chloroform, which spread to the surrounding community. All signs about operating and safety procedures were in English, though many of the operators only spoke Hindi (Meshkati 1991). Workers did not always wear safety gloves and masks, even when they experienced chest pains and vomiting, due to the high plant temperatures resulting from a lack of air conditioning (Martin and Schinzinger 2004).

The primary defence against gas leaks was a vent-gas scrubber designed to neutralise the gas. At the time of the 1984 accident it had been shut down due to a suspension of production of methyl isocyanate six weeks previously and was turned back on too late to be of any use. The second line of defence, a flare tower to burn off any escaping gas missed by the scrubber, was also inoperable, as a section of pipe connecting it to the tank was being repaired. Attempts by workers to minimise damage by spraying water 30.5 metres into the air were ineffective as the gas was escaping from a 36.6 metres high stack.

Human factors were largely ignored in both the design and operating stages. For instance the pressure gauge that should have indicated a build-up of chemicals round the relief valve was located away from the control room. It was supposed to be monitored manually and had no links to the control room or warning systems. The absence of an important panel from the control room meant that the leak did not appear on monitors and the lack of oxygen masks led to a mass exodus of control room operators as gas levels increased. Broken or malfunctioning gauges were common and made it hard for operators to interpret what was happening (ICFTU 1985). In addition operators were overloaded and found it difficult to check all the relevant parameters on about 70 different panels.

The socio-economic context in which the accident occurred was one of considerable inequalities and great poverty. As a result thousands of desperately poor migrant labourers were squatting in vacant areas around the plant, in the hope of gaining employment or trying to use the plant's water and electricity. The presence of these poor labourers greatly increased the numbers of deaths and injuries resulting from the escape of gas. This was further magnified by the lack of evacuation plans or information from Union Carbide or the Indian government about the dangers from the plant (Martin and Schinzinger 2004). This indicates the importance of wider systemic factors both in magnifying the effects of disasters and in leading to desperately poor people accepting risks which would not be acceptable elsewhere.

Analysis in terms of multiple-loop action learning indicates that there were problems in each of the loops. At the first loop level there were serious problems with safety and management practices, human factors issues and a lack of training. At the second loop level managers and workers in Bhopal seem not to have been strongly committed to safety, possibly due to a lack of knowledge. At the third loop or organisational level Union Carbide seems to have had an ethos which did not take safety issues and workers' welfare at their Indian plant (Bhopal) as seriously as at their US plants, totally ignored its responsibility to act on (serious) safety warnings and did not really understand the local context. At the fourth or societal level there seems to have been little concern that training, safety standards and management practices in Bhopal were deficient, particularly in comparison with US ones and an acceptance that many people in India were desperately poor, possibly combined with a tacit belief that the life of the average person in India was of little value. In many cases when there are problems at all four loop levels, leverage can be used to make changes in practices at the single loop level. This is unlikely to have been feasible here due to the nature of the practices that needed to be changed and the lack of people in the organisation with the knowledge and attitudes to want change. Therefore changes in practices would have probably required changes in the ethos of the organisation and possibly also of the surrounding society. The tragedy is that these changes did not occur before there was a major accident.

It is unlikely that the accident would have occurred if there had been a safety culture at Bhopal, since measures would have been taken to ensure that all the safety measures were operational, for instance by regular testing. In a safety culture the available safety measures would also have been more stringent than required, rather than insufficient and there would have been high quality bilingual safety training and materials, covering both accident avoidance and minimisation of damage in the case of an accident. Unfortunately the culture at Bhopal seems to have been the antithesis of a safety culture, what could be considered an 'unsafety' culture.

The example also illustrates the importance of applying systems approaches that consider all factors over the relevant time span, rather than either finding a convenient scapegoat or ascribing the cause to bad luck which led to the failure of all safety systems at one time. Although it is not possible to foresee all eventualities, a well designed system based on a safety culture has sufficient redundancy in safety critical components that failure of all safety systems at one time is almost impossible.

Systems analysis techniques can also be applied to a comparative study of major industrial accidents, including Bhopal, Three Mile Island, Chernobyl and a much earlier nuclear power plant accident at the National Reactor Testing Station in Idaho. One of the worrying results of such analyses is the presence of several common factors, including design errors, instrumentation errors and lack of attention to human factors, as well as a tendency to focus on the details of what happened rather than the underlying and often long term reasons for it (Meshkati 1991).

5.7 Roundup of Chapter 5

This chapter has discussed some of the techniques that can be used to investigate and analyse the barriers to sustainable development. Mental models provide tools to analyse the processes by which individuals and organisations model the world or obtain a model of how individuals and organisations represent external reality to themselves. There are some similarities, such as the use of symbols, but also significant differences, from mathematical models.

Double loop learning can change mental models by applying feedback from the real situation to examine and update the mental model in the light of the evidence. The updated mental model can then be applied to obtain action to resolve the problem based on the new mental model. Single loop feedback provides information to update actions, but filters it through existing mental models, which may be based on prejudice or inaccurate information.

Multi-loop action learning provides techniques to modify individual and/or organisational and/or societal behaviour and views. Single loop action learning is about changing behaviour rather than values, whereas double loop action learning involves changes in the values (of individuals) as well as behaviour. Triple loop action learning additionally involves changes in the underlying tradition or ethos of the organisation and quadruple loop action learning involves changes in the ethos or tradition of the surrounding society.

All organisations have a culture or ethos. This culture can either promote sustainability of make it more difficult. The concept of a sustainability culture is an extension of the idea of a safety culture in which sustainable design practices (see chapter 2) are encouraged.

The system concepts in chapter 2, mathematical techniques in chapters 3 and 4 and consideration of mental models and multi-loop action learning have provided the various tools required for the systems methodologies presented in chapter 6.

6 Systems Methodologies

6.0 Learning Objectives

This chapter introduces a number of systems methodologies and illustrates their application to sustainable development by a number of short examples. Specific learning objectives include the following:

- Understanding of the basic principles and the range of sustainable development applications of a number of systems engineering techniques, including network management, interaction matrices, tree representations and causal diagrams.
- The ability to apply the above techniques to sustainable development problems.
- Understanding of the differences between hard and soft systems approaches.
- Knowledge of the SIMILAR method and the ability to apply it to sustainable development problems.
- Knowledge of Checkland's Soft Systems Methodology and the ability to apply it to sustainable development problems.
- Knowledge of the Systems Failures Method and the ability to apply it to sustainable development problems.

6.1 Introduction

This chapter applies some of the theory and techniques presented in chapters 2, 3 and 4 to systems methodologies. The chapter is organised as follows. Section 6.2 discusses systems engineering and presents a number of different methods, including network management tools, interaction matrices, tree representations and causal diagrams. Hard systems approaches and, in particular, the SIMILAR method are considered in section 6.3. Sections 6.4 and 6.5 both present soft systems methodologies,

Checkland's Soft Systems Methodology and the Systems Failures Method respectively. A summary of the chapter is given in section 6.6.

6.2 Systems Engineering

6.2.1 Introduction and Basic Concepts

There have been a number of different approaches to defining *systems engineering*. However most of them have the following components (Sage 1992):

- Theories, tools and system management procedures.
- Information and knowledge organisation.
- Application of these tools and procedures to the analysis and interpretation of the impacts of proposed policies or systems for the resolution of real world problems, including consideration of the needs and values of stakeholders. The term *stakeholder* is used for an individual, group or organisation that is interested in and/or affected by a particular decision, activity, proposal or other system.

System design is a major aspect of systems engineering. It generally involves a creative process which results in the determination of design specifications or a design architecture for a product, process or system. Systems design should take account of the needs and requirements of appropriate stakeholders. It generally involves the following stages, often on an iterative basis:

- *Formulation* of the design problem in which requirements and objectives of the client group or relevant stakeholders are identified and design alternatives or options generated.
- *Analysis* of design alternatives in which the impacts of the design options are evaluated.
- *Interpretation and selection* in which the impacts of the design alternatives are compared with each other, using the requirements and objectives of appropriate stakeholders as the criteria. The most acceptable alternative is selected for implementation or further study in a later phase of the systems engineering life cycle.

Systems engineering methodologies often involve iterative hypothesis generation and the testing of alternatives. Analysis of the various steps involved can be used to divide the system engineering life cycle into seven phases, as follows (Sage 1992):

- Identification of requirements and specifications.
- Preliminary conceptual design.
- Logical design and specification of the system architecture.
- Detailed design, production and testing.
- Operational implementation.
- Evaluation and modification.
- Operational deployment.

6.2.2 Network Management

Network management tools, such as the *program evaluation and review technique* (PERT) and the *critical path method* (CPM) (Archibald and Villoria 1967; Sage 1992; Wiest and Levy 1977) can be used to structure problems or projects which can be broken down into separate activities or tasks to be accomplished in a particular sequence. The *activity-on-node convention* (as shown in fig. 6.1) is generally used, with the nodes representing activities and the arrows events or time points after activity completion. There are two main stages in the analysis: network planning and scheduling. In the network planning stage, a network planning model is constructed, consisting of one or more paths. Each path consists of activities which follow each other sequentially. Separate paths consist of activities which can occur simultaneously rather than sequentially.

The expected time (t_e) and variance (σ^2) for each activity can be computed (Sage 1992) as

$$t_e = (t_o + 4t_l + t_p)/6 \qquad \sigma^2 = [(t_p - t_o)/6]^2 \tag{6.1}$$

where t_o, t_p and t_l are respectively optimistic, pessimistic and most likely estimates of the activity's duration, based on there being no problems, everything going wrong and past experiences respectively. Time estimates can then be obtained for each activity.

In the scheduling phase the *critical path* (indicated by the heavy line in fig. 6.1) or sequence of activities which takes the longest time to complete is identified. This critical path determines the project duration. The only way to reduce the time required to carry out the whole programme is to remove or speed up activities on the critical path. *Forward pass computation* can be used to determine the earliest completion time for a given starting time, by summing the shortest times required to reach each node. *Backward pass computation* can be used to determine the latest starting time to ensure a given completion time (Sage 1992). There is *slack time* on all paths other than the critical path, as activities on these paths

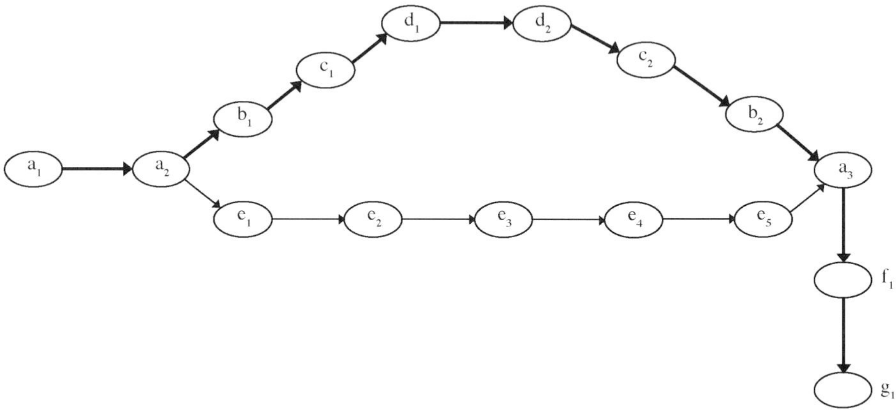

Fig. 6.1 Network of events: activity on node representation

take longer than the minimum possible time. This slack time is equal to the difference between the earliest expected and latest allowable times for completion of all activities on the path.

Network planning and management approaches are generally used by a firm or other organisation, which is undertaking projects or manufacturing activities, to forecast project duration and resource requirements and draw up an activity schedule that is consistent with time and resource constraints. This type of analysis could easily be extended to sustainable project management which takes into account environmental, social and developmental issues with resource constraints defined to minimise negative environmental and social impacts. Scheduling could also be used to improve working conditions and avoid stress through the identification of excessive workloads and the need to reroute activities to reduce pressure. Network planning and management approaches can also be applied to policy management in a sustainable development context. For instance they could be applied to the scheduling of political decision making and policy implementation. This could include the staged implementation of targets for the reduction of emissions of greenhouse or ozone depleting gases. Backward pass computations could be used to determine the latest time policy decisions should be made and implemented to achieve a certain overall reduction by a particular date. Forward pass computations could be used to identify the effects of delays in the implementation of particular targets for emissions reductions on the success of the overall reduction policy.

6.2.3 Interaction Matrices

Interaction matrices can be used to provide a framework for the identification and description of the relationship patterns between sets of elements, including needs, objectives and constraints (Sage 1992). *Self-interaction matrices* can be used to describe and investigate the relations between elements of the same set, whereas *cross-interaction matrices* can be used to investigate the relations between elements of two different sets. There are four different types of self-interaction matrices, which represent nondirected graphs, directed graphs (or digraphs), signed digraphs and weighted graphs. A *graph* consists of a set of variables called *nodes* and a set of links between the variables, called *edges*. In a *digraph* (Harary et al 1965) or *directed graph* some of the edges have associated directions i.e. the relationship occurs in one direction only, whereas in an *undirected graph* none of the edges have associated directions i.e. the relationship is symmetrical.

Undirected relations are often appropriate for cross interactions and directed relationships for self-interactions (Sage 1992). However any type of relationship can be used, as long as its meaning is made clear. To draw up an interaction matrix the (two sets of) elements are listed as headers for the rows and columns, giving a space for each possible combination of elements. A '1' or 'x' is used to indicate that two elements are directly linked by the specified relationship and a '0' or a blank space to indicate that the specified relationship does not hold between the two elements. In the *weighted* case, the strength of the relationship is indicated by a number and in the *signed* case '+1' and '-1' are used.

Self-interaction matrices frequently have a '1' or 'x' on the forward diagonal. Self-interaction matrices are square, whereas cross-interaction matrices are rectangular, since there may be different numbers of elements in the two sets. Self- and cross- interaction matrices are a useful tool for investigating the relationships between the different elements in a problem, such as needs, stakeholders and constraints.

Table 6.1 illustrates the weighted cross-interaction matrix between a number of environmental impacts and the six life cycle stages in the computer industry. In this case the weighted cross interactions (taking values up to 3) give a measure of the significance of a particular environmental impact at the given life cycle stage.

Thus, for instance, the empty space in the place for the impact of landfill exhaustion at the material acquisition life cycle stage indicates that this life cycle stage has no or minimal contribution to landfill exhaustion. The value 3 for local air impacts and waste at the material acquisition life cycle stage indicates that both these impacts are a very significant problem at

Table 6.1 Cross-interaction matrix with weighting for life cycle impacts

LC Stage	EC	RC	LE	LAI	WI	AI	W	FD
1	2	1		3	2	2	3	3
2	2	2						
3	2	3		2	2	2	2	
4	3	2		3		3	2	3
5	3	2	3	1		3	3	3
6	1	2	3	2	2	1	3	

EC = energy consumption
RC = resource consumption
LE = landfill exhaustion
LAI = local air impacts
WI = water impacts
AI = atmospheric impacts
W = waste
FD = forest destruction

LC stage 1 = material acquisition
LC stage 2 = material processing
LC stage 3 = manufacture
LC stage 4 = transport and distribution
LC stage 5 = use and reuse
LC stage 6 = disposal and recycling

this life cycle stage.

6.2.4 Tree Representations

The term *tree* is used to indicate a particular type of graphical representation. There are several different types of trees, including objective, activity, event, intent, decision and attribute trees. The vertices of the tree represent its elements, such as objectives or events and the edges or 'branches' the relations between these elements. A tree can have either a single source or single sink structure. In the single source case the tree starts from a single source and each node has one incoming branch, but may have several outgoing branches.

Trees are particularly useful for structuring objectives, events and activities from a particular starting state. For instance an *activity tree* allows problem activities to be broken down into the simpler activities required to achieve them. In general an objectives tree should be constructed before an activity tree, to allow the objectives to be considered first and then the activities required to achieve them. The first stages of an activity tree for implementing a sustainable development policy is illustrated in fig. 6.2.

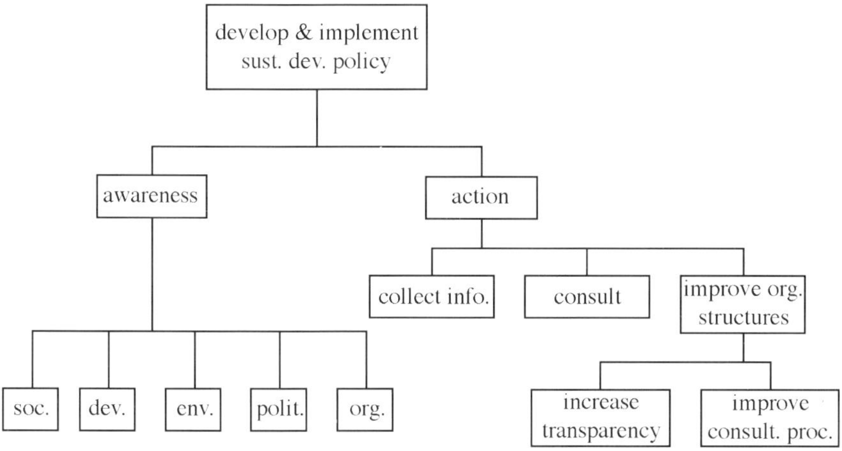

soc. = social
env. = environment
org. = organisation
dev. = development
politic. = political
consult. proc. = consultation procedures

Fig. 6.2 Activity tree for the first stages of a sustainable development policy

6.2.5 Causal Diagrams

A *causal diagram* (as illustrated in fig. 6.3) is a graphical representation of the cause and effect mechanisms that lead to change, particularly over time and sometimes over space or other variables. An arrow between two elements A and B is used to indicate that change in a variable A causes change in another variable B. A '+' sign is sometimes used to indicate that an increase in A leads to an increase in B and a '-' sign that an increase in A leads to a decrease in B. A causal diagram is particularly useful for portraying cycles and feedback and can be used to represent systems with multiple feedback loops (Sage 1992). In the context of sustainable development, causal diagrams can be used to determine tentative conclusions about the major influences on behaviour, such as leverage points, and the reasons why organisations resist change. These conclusions may also indicate which policies are likely to be effective, ineffective or harmful and the reasons why.

A causal diagram can be constructed in the following way (Sage 1992):

- Identify the problem or issues
- Identify the main groupings of causes. Obtain more detailed causes through appropriate techniques such as brainstorming. Eliminate inapplicable causes, while taking care not to eliminate relevant ones.

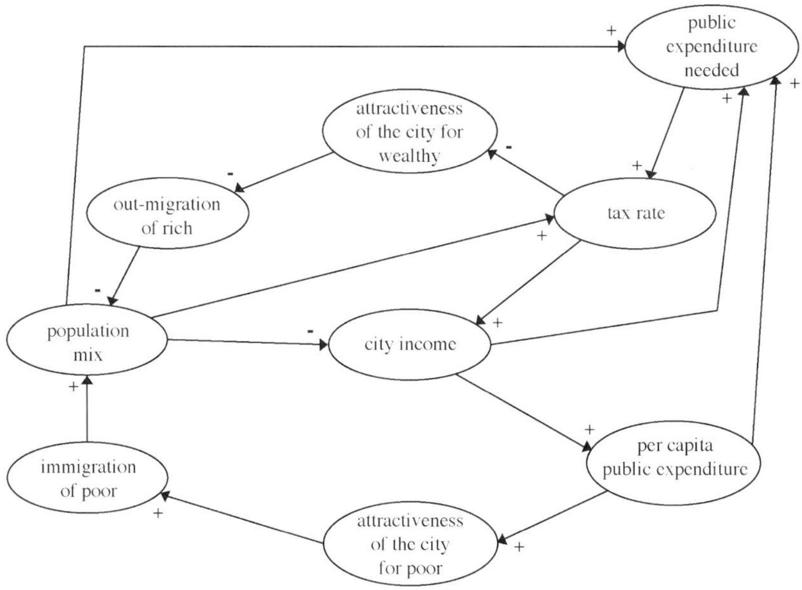

Fig. 6.3 Causal diagram of a model of urban dynamics

- Connect causes and effects by determining predecessor and successor relationships and refine the causal diagram.

6.3 Hard or Engineering Systems Approaches

6.3.1 Introduction

In *hard systems* approaches the systems objectives should be clearly defined and understood to allow desired outputs to be obtained from given inputs and measures of the effectiveness of system performance to be established. There are a number of different systems engineering approaches to structuring problems, many of which have similarities to each other. Related methodologies are also being used in other fields, such as computer science, the social and behavioural sciences and the natural sciences. However it should be noted that, despite considerable similarities between the different methodologies, there are often significant differences of philosophy and detail. An illustrative example of a systems engineering method, the SIMILAR method, is given in the next subsection. Although there are superficial similarities, this methodology is significantly different from Checkland's Soft Systems Methodology presented in section 6.4, particularly in its approach to problem formulation. In Checkland's methodology it is not assumed that the problem is known and determining the problem is an important aspect of the approach. This is a very important philosophical and methodological difference which has a significant impact on how the method is implemented. In the SIMILAR Method the problem can be stated in the first step, so there is a de facto assumption that the problem is known and, at most, requires clarification.

6.3.2 SIMILAR Method

This involves a seven step methodology, consisting of the following steps:

1. *State the problem*: The problem statement should include all the requirements the system must satisfy and should be in terms of what should be done, rather than how to do it. It may involve a description of the top level function to be performed, the deficiency to be addressed or a reason for change plus a vision and mission statement.
2. *Investigate alternatives*: This involves evaluating alternative designs or other options in terms of all the relevant criteria, including environmental, social and developmental criteria. In evaluating the alternatives, tradeoffs will be required and the use of multi-criteria optimisation techniques (see Part II) may be appropriate. It may be necessary to combine features of different alternatives or to modify a particular alternative. If new data becomes available, re-evaluation will be required. Generation of a wide range of alternatives, including unusual ones, should ensure that good options are not omitted.
3. *Model the system*: Models will generally be developed for each alternative, particularly in an engineering environment, and the model for the preferred alternative can then be further developed and used in system management throughout the life cycle. There are a range of different types of system models, including analytical equations, physical analogues, block diagrams, flow diagrams, mental models and computer simulations. These models can be used to clarify requirements and reveal bottlenecks, potential problems and duplication of efforts.
4. *Integrate*: The system should be integrated into the environment and interfaces between the subsystems defined. It is generally preferable to have feedback systems around individual subsystems rather than interconnected subsystems. This allows each individual subsystems to be modified and updated on its own.

5. *Launch the system*: This involves running the system and producing outputs or carrying out whatever activities the system is intended to achieve.
6. *Assess performance*: There are a number of different quantitative and qualitative approaches to evaluating performance to determine whether the system is achieving its goals.
7. *Re-evaluate*: This involves observing outputs and using this information to modify the system as appropriate, either through minor modifications, total system redesign or discontinuing the activity. In many cases re-evaluation should be a continual process with many parallel loops.

6.4 Soft Systems Methodologies

6.4.1 Introduction to Soft Systems Approaches

Soft systems developed out of the need for approaches to analysing ill-defined or ill structured problem situations which could not be analysed with systems engineering approaches. Such approaches should consider both the activities to be carried out and the details of implementation i.e. 'what to do' as well as 'how to do it'. Soft systems analysis involves a systemic process of investigation which is also based on systems models. This uses a human activity system, which consists of a set of activities which are connected to form a purposeful whole. In general there will probably be more than one possible model of the human activity system under consideration, due to the different ways in which a given set of connected activities can be perceived, interpreted and understood. It may therefore be necessary to generate several different models of this human activity system (Checkland and Scholes 1999).

6.4.2 Checkland's Soft Systems Methodology

Soft systems analysis is a process based approach to structuring problems in human activity systems such as organisations. The process may sometimes be even more important than the outcomes and will frequently change the organisation. The area of discussion is kept vague and wide ranging for as long as possible to avoid fixing on one problem and thereby not noticing other problems. The approach is intended to be participative and can be carried out by organisation members with the support of a facilitator familiar with the methodology. It works best for people who can think graphically or visually and involves drawing rich pictures (see section 6.5.1), including to expand on the different models and systems.

It is (loosely) based on a seven stage methodology (Checkland 1996; Checkland et al 1990 1999). However it is not necessary to start with stage 1, proceed sequentially or carry out all the stages and backtracking and iteration are essential. In any study it is likely that several stages will be proceeding concurrently at different levels of detail, as the methodology is itself a system and therefore any change in one stage will affect all the others. Thus the seven stage methodology provides a framework rather than being totally prescriptive. This makes it very open ended, which has both advantages and disadvantages. The focus is on the people involved in the situation. However, although the analysis stage considers power dynamics in the situation, power dynamics between different organisation members, such as workers and management and the way they could affect participation in the process are not considered.

The seven stages can be listed as follows:

Stage 1: Starting with the unstructured problem situation.
Stage 2: Obtaining an expression of the problem situation.
Stage 3: Obtaining the root definition (see below) of the relevant systems.
Stage 4: Obtaining conceptual models drawing on the formal system concept and other systems thinking.
Stage 5: Comparison of the conceptual model of stage 4 with the problem situation of stage 2.
Stage 6: Implementation of feasible, desirable changes.
Stage 7: Taking action to improve the problem situation.

Stages 1, 2, 5, 6, and 7 are *real world* activities which involve the problem situation, whereas stages 3 and 4 are *systems thinking* activities which may or may not involve the problem. *Stages 1 and 2* involve building up the richest possible picture of the situation in which there is perceived to be a problem, rather than the problem itself, without imposing a particular structure. This enables selection of one or more viewpoints from which to further study the situation. The situation is displayed to allow a range of possible and hopefully relevant choices to be revealed. In constructing a neutral picture the concepts of structure and process and the relationship between structure and process may be useful. Structure often involves physical layouts, power hierarchies, reporting structures and formal and informal communication structures. Processes can often be investigated in terms of activities, such as deciding to do something, doing it, monitoring how well it is done and any external effects and taking appropriate corrective action. A difficult relationship between structure and process is frequently a core characteristic of situations perceived to have problems. The first two stages of the methodology will often produce a number of notional systems which seem more relevant to a particular representation or model of the problem rather than to the real problem itself.

One approach to developing this full picture of the situation involves role, social system and political system analyses, sometimes called analyses one, two and three. Alternatively analyses one to three could be used in the comparison stage of the method. The *role analysis* includes consideration of the following:

- The *client* i.e. the person(s) responsible for the systems study and their motivation.
- The *would-be problem solver* (who may also be the client) i.e. whoever wishes to take action on the situation and the proposed actions in terms of their perceptions, knowledge and readiness to make resources available.
- The *problem owner*. There are generally several possibilities and choice of the problem owner will generally determine the choice of system and perspective.

As the analysis proceeds the people or organisations in the problem solver and problem owner roles may change.

One possible approach to the *social system analysis* is based on the three elements of roles, norms and values. *Role* is a social position recognised as significant by people in the problem situation. It may be institutionally defined, such as union member or member of an environmental organisation or behaviourally defined, such as recycler or buyer of fair trade products. *Norms* characterise the expected behaviour of the role, such as paying union dues or not crossing picket lines for a union member. *Values* are beliefs about 'good' or 'bad' performance which are used to judge the role holders. For instance union members who engage in strike breaking are bad union members.

The *political analysis* examines the *power dynamics* between the different interests or stakeholders. It looks at the ways in which power is expressed in the situation and how the *commodities of power* are obtained, used, protected, preserved, passed on and relinquished. For instance in a not very democratic, but nameless organisation, there may be a formal decision making structure, such as an elected committee, but the real power and decisions are made by a clique of three people in the pub. Therefore, opening up the real power mechanisms to public scrutiny can itself change power dynamics and should be approached with a degree of caution.

Stage 3 involves determining a number of systems which are likely to be relevant to the putative problem and preparing concise definitions, referred to as *root definitions*, which are intended to encapsulate the fundamental nature of the systems chosen. The aim is to obtain a carefully phrased explicit statement which can be used to help improve the problem situation. The choice of root definition can be changed if it does not lead to useful models and suggestions at later stages. It can also be tested by briefly investigating what types of model and changes it is likely to lead to.

There are several different possible approaches to obtaining a root definition. One approach contains the following elements, giving the mnemonic *CATWOE*:

- *Customers* (C) of the system i.e. people affected by the system's activities.
- *Actors* (A) or agents who carry out or cause the main activities of the system, particularly the main transformation, to be carried out.
- A *transformation process* (T) which transforms defined inputs into defined outputs.
- A *Weltanschauung* (W) i.e. an outlook or framework which defines the context of the root definition.
- *Ownership* (O) of the system i.e. some agency has prime concern for the system and the ultimate power to abolish it.
- *Environmental constraints* (E) on the system's activities i.e. features which are taken as given.

Example 6.1: Glasgow Campaign to Welcome Refugees

The Glasgow Campaign to Welcome Refugees is defined by its Constitution as a broad based non party political movement made up of organisations and individuals who aim to improve the quality of life of asylum seekers and refugees arriving and settling in Glasgow and the West of Scotland. It does this by undertaking lobbying of relevant authorities and organisations, grassroots lobbying, campaigning for the repeal of detrimental legislation and regulations, awareness raising and providing practical support and advice. The transformation process (in so far as it is successful) changes attitudes, legislation and the facilities and support available to asylum seekers and refugees. Ownership of the system is by the members of the Campaign, who are also the actors or agents. The 'customers' of the system are members of the public, Glasgow City Council, the Scottish Parliament and Executive and the media who are influenced by campaigning and lobbying activities, and refugees and asylum seekers. However, as well as 'customers', refugees and asylum seekers are also 'actors' and 'owners' of the system. The environment is the existing political, social and economic context, including incidents of harassment and attacks on asylum seekers and refugees and negative media coverage. The Weltanschauung or framework, which defines the context of the root definition, is the belief that racism and other forms of prejudice are wrong, that there is a moral and humanitarian responsibility to welcome people suffering from persecution, that societies are generally enriched by a mixture of cultures and that these beliefs are shared by a wide range of different individuals and organisations.

Stage 4 involves making *conceptual models* by starting from the root definition, which can be considered to be the description of a set of purposeful human activities in terms of a transformation process. Therefore stage 4 models the activity system required to achieve this transformation. The minimum set of necessary activities for the system is selected and a root definition is produced for each major activity and its inputs and outputs. The transformation from input to output is described using a set of verbs. Each major activity is considered as a sub-system and can be progressively decomposed into further subsystems as required. There should not be more than five to ten major activities in the first level model.

It should be noted that this model is a structured set of activities required to implement the root definition and not a description of any actual human activity system. As far as possible the tendency to describe actual activity systems existing in the real world should be resisted, as this negates the point of the approach, which is to generate new and radical ideas.

Example 6.2: Conceptual Model

The root definition is of a Scottish based organisation which aims to improve the quality of life of refugees and asylum seekers in the West of Scotland and prevent asylum seekers being deported. The conceptual model should not be based on the Glasgow Campaign to Welcome Refugees or any other existing Scottish organisation to support refugees. Major activities of the system include:

- Campaigning.
- Supporting (refugees).
- Fundraising.
- Publicising.
- Recruiting (new members).

Each of these activities could then be decomposed further. For instance campaigning could be divided into:

- Demonstrating (at Dungavel detention centre for refugees and in town centres).
- Lobbying (Scottish Parliament).
- Petitioning.
- Leafletting.
- Marching.
- Protesting.

6.4.3 Model Validation and Further Discussion

Unlike models which represent the real world, the conceptual model cannot be *validated*, for instance by showing that it simulates observed real world behaviour, as it represents concepts rather than a particular real system or situation. However testing for the properties listed below can still be used to ensure that the model is not fundamentally deficient. This *list of properties* gives a formal structure which can facilitate construction of formal conceptual models. Thus S, which represents the system model, is a formal system if it has the following properties:

1. An ongoing purpose or mission. In a hard systems approach this would be replaced by objectives or goals. In a soft system this purpose may never be fully achievable. For instance reduction of emissions of carbon dioxide and other greenhouse gases to zero is unlikely to be achieved, though it may be possible to reduce them to a relatively small percentage of present values.
2. A measure of progress or lack of it in achieving the system purpose or objectives.
3. A decision taking process, which may take regulatory action relative to the mission, and measures of progress.
4. Component sub-systems.
5. Components which interact and/or are connected, so that effects and actions can be transmitted throughout the whole system.
6. Existence in wider system(s) and/or environment(s) with which it interacts.
7. A boundary separating it from the environment. Inside the boundary the decision taking process has power to cause action to be taken rather than just influencing the environment.
8. Physical and abstract resources which are at the disposal of the decision taking process.

9. Some guarantee of continuity and the ability to recover stability after disturbances. This stability may come from outside the system or from the participants' commitment to the system's main purpose.

However, although this test for validity of the conceptual model has been presented, it is relatively infrequently applied in practice and will not be applied in the case study presented in chapter 7. There are a number of reasons for this. In particular the process based nature of the approach reduces the importance of the formal validity of the conceptual model as a model.

The formal system model allows questions to be framed to reveal any inadequacies in either the conceptual model or the root definition underlying it. Typical questions include the following: Is the measure of performance explicit? What are the sub-systems in the model? Although the formal system model cannot ensure 'validity' of the conceptual model, it can ensure the derivation of a well constructed model which is useful in the comparison stage. Another useful approach is to consider the model in terms of other types of systems theory. This may involve restructuring it, for instance in terms of system dynamics, adaptive control theory, Vickers' (1965 1973) concept of an 'appreciative system', Ackoff's (1971) compendium of system concepts, Beer's (1972) five sub-systems or the Tavistock socio-technical system (Emergy and Trist 1960).

In *stage 5* the conceptual models from stage 4 are compared with perceptions of the real world situation. This will often involve a debate with appropriate stakeholders concerned in the problem situation. This debate should lead to the definition of possible changes in stage 6. These changes should be both desirable and feasible in terms of prevailing attitudes, power structures and the history of the situation. Stage 7 then involves action to implement the changes identified in stage 6 to improve the problem situation. This may result in a new problem which can be investigated using the methodology.

In the *comparison stage*, 5, the problem situation analysed in stage 2 is examined alongside the conceptual models in order to generate a debate about possible changes to improve the situation. In this comparison intuitive perceptions of the problem are brought together with system constructs. There are a number of different possible approaches. For instance the models can be used to provide a source of ordered questions about the existing situation. Alternatively the comparison can be made by reconstructing a sequence of events in the past and comparing this with what would have happened if the relevant conceptual models had been implemented. However care has to be taken to ensure this approach is not interpreted by participants as criticism of them and their past performance.

In some studies the conceptual model stage will raise major strategic questions about the nature of present activities rather than procedures i.e. questions of the type: why do this at all? rather than: was it done well? If this is the case, it may be appropriate to use a general comparison aimed at determining which features of the conceptual models are significantly different from the real situation and why.

In the *model overlay* approach a second model of the existing situation is derived in the same form, as far as possible, as the conceptual model. Direct overlay of the two models can be used to clearly reveal any mismatch between them, which then becomes the source of discussion of change. The root definition implied by the real system can also be obtained and compared to the definition on which the conceptual model is based. If a new rather than existing activity system is being investigated, the comparison needs to be with a defined expectation rather than an existing system. A related approach involves identifying each activity in the conceptual model and then considering whether it exists in the real situation, how it is carried out and how it is judged. Comments can then be made and used as the basis for deriving desirable and feasible changes.

The comparison stage should generate debate about possible changes to be made in the perceived problem situation. However, in practice a number of iterations, involving changes in the initial analysis or the root definitions, are generally required first. As in the 'hard' systems case, the changes envisaged could involve the creation and implementation of a system, but changes are likely to be more modest. There are three types of possible changes: changes in structure, changes in procedures and changes in 'attitudes'. *Structural changes* refer to changes in system elements which generally do not change in the short term, whereas *procedural changes* are changes in dynamic elements. Compared to changes in attitudes, these types of changes are relatively easy to specify and implement, even in the short term. *Changes in attitudes* can be considered to include changes in influence and the expectations of what is considered appropriate behaviour in different roles, as well as changes in readiness to rate certain types of behaviour 'good' or 'bad' relative to others or changes in what types of behaviour are rated 'good' or 'bad'. Attitudes are particularly difficult to change and, when changes do occur, they may not be the desired ones or even in the desired direction. The discussion should involve concerned actors and stakeholders in the problem situation and should aim to identify changes which are desirable as a result of the insight gained from the root definitions and conceptual models and culturally feasible in terms of the characteristics of the situation, the people in it, their shared experiences and their attitudes. It may not be easy to find changes which meet all these criteria.

6.5 Approaches to Understanding Systems Failures

Systems approaches have frequently been applied to technological failure and there have been a number of systems studies of aircraft crashes in particular. This approach is directly applicable to environmental disasters of the Chernobyl and Bhopal type. It can also be extended to examining why the behaviour of both many individual firms and industries as a whole is not sustainable in either environmental or social terms. A further application is evaluating the failure of political and economic systems worldwide to deliver sustainable development. The resulting conclusions could be used to help develop new solutions which will contribute to resolving the present environmental crisis and eliminating poverty and inequalities.

There are a number of similarities between the Systems Failures Method and Checkland's Soft Systems Methodology. Both are process based soft systems approaches and involve a comparison of a formal or conceptual system model with the real system.

6.5.1 Systems Failures Method

The *Systems Failures Method* (Fortune and Peters 1995) has the five main stages listed below:

1. Pre-analysis
2. Identification of significant failures
3. System selection and system modelling
4. Comparison with the formal system model
5. Further analysis and synthesis, leading to development of an agenda for change.

A number of different techniques are available to collect and organise information about the failure situation to be used in *pre-analysis*. These techniques include spray, relationship and multiple-cause diagrams, rich pictures, data-bases and charts. At this stage the situation should not be represented in terms of systems. A *spray diagram* (fig. 6.4a) can be used to record ideas about relationships in the very early stages of analysis, often as a prelude to producing other diagrams. It shows the different factors which come together and either examines them in detail or traces them back through chains of events or to the organisations or people involved. *Rich pictures* (fig. 6.4b) are either depictions of a real situation or abstract pictures which contribute to understanding of a problem situation. A diagrammatic rich picture can present all the main features of a situation on one sheet of paper, generally using cartoon type representations of key ideas and information, in a way that is insightful and can be easily assimilated. *Relationship diagrams* (fig. 6.4c) provide snapshots of situations, with lines drawn to connect components that are significantly related in some way, and the line length used to indicate different degrees of closeness. Clusters of components can suggest where system and subsystem boundaries could usefully be drawn. The depiction of the fire at Manchester airport in figures 6.4a, b and c (Fortune and Peters 1995) illustrates the similarities and differences between spray diagrams, rich pictures and relationship diagrams and should help to clarify their different roles.

In the next stage of the analysis all aspects of the pre-analysis are brought together to identify the focus for the analysis and *specify the systems* from which the failure(s) have emerged. Systems are considered to carry out transformation processes which produce outputs from inputs. Judgements based on purpose, viewpoint and perspective are required, but there are no 'objective' criteria. In the context of the method, *failure* is considered to occur when a system has either inappropriate or no outputs. Putting in trial system boundaries and experimenting with different configurations should allow a notional system, which could carry out these transformations, to be defined, with the aim of carrying the analysis forward. It may be necessary to select a number of interacting systems and

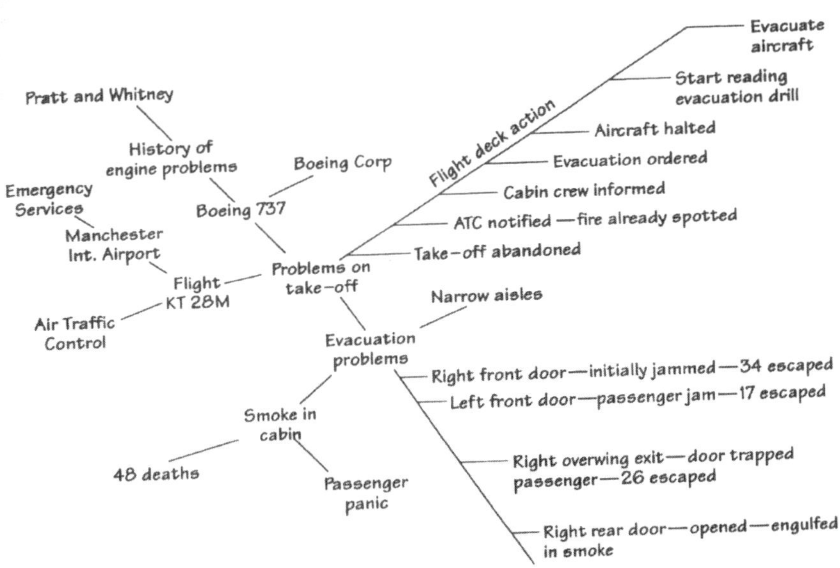

Fig. 6.4a Spray diagram of fire at Manchester Airport

examine them all both individually and as a whole. However care is required to ensure that important inter-relationships are not lost.

The system(s) selected should be modelled in detail to allow switching between levels to be carried out and their structure(s) and process(es) to be represented in appropriate formats for later comparisons. The *modelling process* generally has the following elements:

- Naming and defining appropriate system(s).
- Describing the components and relationships of these systems to the system environment.
- Identifying the wider system and system variables.
- Describing system inputs and outputs.
- Establishing structural relationships between the components.

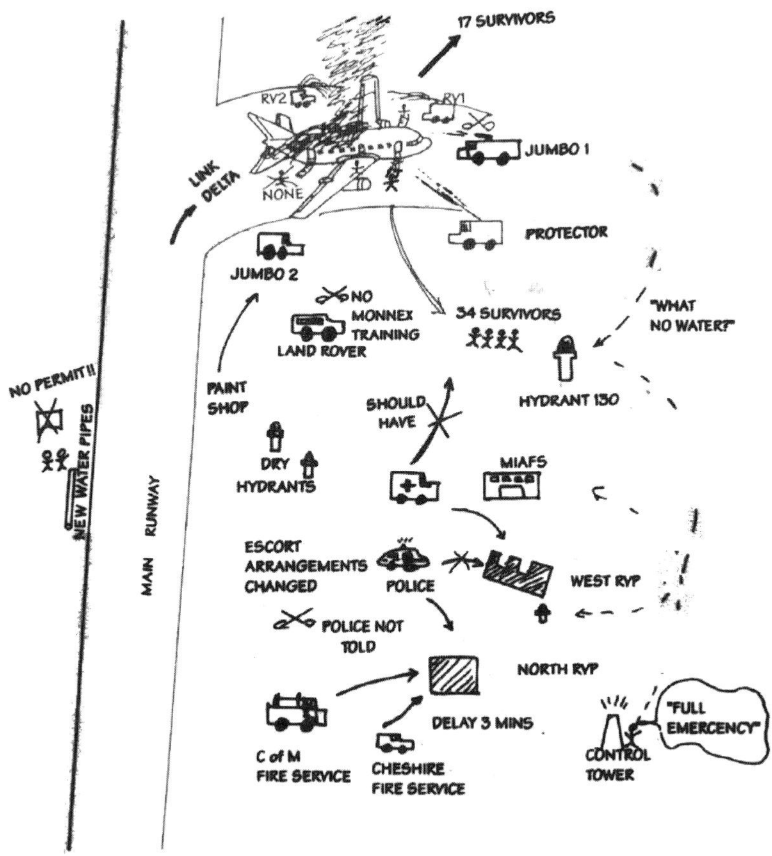

Fig. 6.4b Rich Picture of the Fire at Manchester Airport

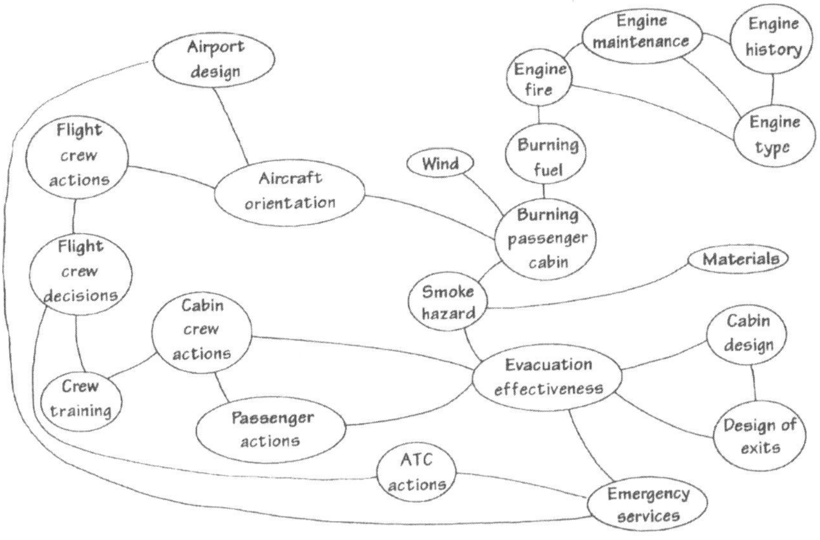

Fig. 6.4c Relationship diagram of fire at Manchester Airport

- Obtaining preliminary indications of the relationships between the variables.

System diagrams, such as input-output diagrams, systems maps and influence diagrams are the main modelling tools. An *input-output diagram* is a block diagram with the system represented as a single box and inputs and outputs shown as labelled arrows. It illustrates the transformation relationships between inputs and outputs. A *systems map* is a snapshot showing the components of a system and its environment at a point in time. It only shows named components and not linking lines and can be used to investigate (sub)system boundaries. The positioning of the components can be used to infer the relative strengths of the relationships between them. *Influence diagrams* are related to the causal diagrams discussed in section 6.2.5 and show named components and arrows, which may be labelled to indicate different types of influence.

Understanding of the reasons for or causes of failure is obtained by *comparing* systems representations of the failure situation with models of how the situation should be structured and managed to allow it to operate without failure. The whole system can be first tested against a system called the *Formal System Model*, which is a robust failure-free system. The formal system consists of:

- A decision making subsystem.
- A performance monitoring subsystem.
- A set of subsystems and elements that carry out the system tasks, including transforming inputs into outputs.

Other significant features include:

- A degree of connectivity between the components.
- An environment with which the system interacts.
- Boundaries separating the system from the wider system and the wider system from the environment.
- Resources.
- A continuous purpose or mission.

Although the formal system will make demands and place restrictions on its subsystems, this does not preclude a certain amount of subsystem autonomy in deciding how these expectations are met. Equally a wider system is able to constrain the system activities by defining goals, exercising control and only providing the resources necessary to meet these goals. However it could also increase the authority of the subsystem(s). The notion of hierarchy is intrinsic to the model and therefore each subsystem could be considered as a formal system with its own decision making, performance monitoring and transformation components.

Comparison of a number of system representations of 'failures' with the formal system model has pinpointed the following differences, which are often *contributory causes to system 'failures'*:

- Deficiencies in the apparent organisational structure, such as a lack of or unsatisfactory performance measures or control and decision making subsystems.
- No clear statement of purpose provided in a comprehensible form by the wider system.
- Deficiencies in performance or inadequate design of one or more subsystems.
- Lack of effective communications between different subsystems.
- Inadequate consideration of environmental influences and the resources required to deal with environmental disturbances.
- Inappropriate resource division between the basic transformation processes and monitoring and control.

Comparison of the system with the formal model requires the system to be put into the same format as the model, so that discrepancies and gaps in situation understanding can be identified. Other comparisons are based on paradigms, including the control strategy used, communication issues and human factors. The concept of cognitive adequacy is useful in investigating the relationship between a system and its environment and relates to the ability of an organisation to respond to observations of hazard. It is shaped by the organisation's culture, including the basic assumptions and beliefs shared by members of the organisation, and factors such as who is responsible for what kinds of actions, what is considered to be appropriate conduct in various positions and how problems are to be resolved (Westrum 1988).

The Systems Failures Method is iterative and may require several passes through the method. When the analysis starts to seem complete, *synthesis* is required to draw together the various components to obtain an understanding of the system failures(s) and their causes. There are a number of possible approaches to further synthesis, depending on the applications of the results. It is often useful to start by drawing up another set of formal system models which include the following three types of factors:

- Factors, changes to which would have prevented the particular failures being studied.
- Factors which would have avoided similar failures.
- (Significant) factors, which did not contribute materially to the set of events.

6.6 Roundup of Chapter 6

This chapter has presented a number of systems techniques and methodologies that can be applied to problems in sustainable development (as well as a range of other applications). Many of the methods use graphical techniques, such as trees, causal diagrams, rich pictures, spray diagrams and relationship diagrams. These approaches all involve a graphical representation of the main components of the problem and the interactions or connections between them.

Other techniques and methodologies include network management, interaction matrices, the SIMILAR Method, Checkland's Soft Systems Methodology and the Systems Failures Method. Network management tools can be used to structure problems or projects consisting of a number of activities to be accomplished in sequence. They involve consideration of the critical path or sequence of activities which takes the longest time to complete. Interaction matrices provide a representations of the relations between elements of the same set (self-interaction) or different sets (cross-interaction).

Both the SIMILAR Method and Checkland's Soft Systems Methodology have seven steps. However the SIMILAR method starts with a pre-specified problem, whereas part of the aim of the Checkland's Soft Systems Methodology is to identify the problem. The process based nature of Checkland's Soft Systems Methodology means that iteration and backtracking frequently occur and that the seven stages provide a framework for the approach rather than being prescriptive.

The SIMILAR method involves investigating alternatives and modelling the system, generally for each alternative. This is followed by integrating the system into the environment, implementing the system, assessing its performance and deciding whether modifications will be required. The aim of Checkland's Soft Systems Methodology is to obtain and implement desirable and feasible changes. This is done by comparing a conceptual model of the situation with perceptions of the real world situation. This generally involves the use of descriptive rather than mathematical modelling techniques. The Systems Failures Method is a five step approach to identifying and analysing the causes of systems failure. It can therefore be applied to analysing why existing political, social and economic systems have failed to deliver sustainable development. There are a number of similarities to Checkland's methodology in that they are both process based soft systems approaches which involve comparison of a formal or conceptual model with the real system. In addition the use of graphical techniques is important in both approaches.

The application of some of the methodologies presented in this chapter will be illustrated in Chapter 7 by a case study of waste management in Glasgow and the Clyde Valley in Scotland.

7 Case Study: Reduction of Domestic Waste

7.0 Learning Objectives

This chapter presents a case study of the application of the SIMILAR Method and Checkland's Soft Systems Methodology to the analysis of waste management in Glasgow and the Clyde Valley in Scotland and a number of tutorial exercises. Specific learning objectives include the following:

- Understanding of a case study of the application of the SIMILAR method and Checkland's Soft Systems Methodology to domestic waste management.
- Practice in applying systems techniques to purely numerical and typical sustainable development problems.
- Increased understanding of the applications of the systems techniques considered in Part I to problems in sustainable development.

7.1 Case Study Reduction of Domestic Waste in Glasgow and the Clyde Valley: Introduction

This case study will apply the SIMILAR Method and Checkland's Soft Systems Methodology to the analysis of the management of municipal solid waste in the Glasgow and Clyde Valley Waste Strategy Area in Scotland, which will be referred to in this case study as Glasgow and CV. It is the largest of the 11 waste strategy areas (SEPA 2003a) into which Scotland is divided, with a population of nearly 1.8 million, about 807,000 households (LAWAS 2003/4) and an area of 3500 square kilometres. Nearly a third of the population and over a third of the households are in Glasgow City. The population is expected to decrease slightly, but the number of households to increase to about 828,000 in 2011 (SEPA 2003b).

In 2003/4 Glasgow & CV produced just over a million (1.09 million) metric tons of municipal waste, of which the bulk (88.9%) was sent to landfill. Only 7.9% of this waste was recycled and 3.2% composted

(LAWAS 2003/4). Landfill is responsible for nearly a quarter of the methane (a global warming gas) generated in Scotland (Salway et al 2001) and only just under half (48.7%) of the waste sent to landfill has methane recovered. Just over three quarters (76.1%) of the waste sent to landfill is treated outside the local authority area, further adding to the impact, due to the transportation energy consumption and emissions. Just over 70% of recycling occurs within the local council areas. However all materials are exported outside Glasgow and CV for reprocessing. In 2003/4 only just over a third (35.8%) of households and 8.1% of commercial premises in the area had a kerbside collection of materials for recycling (LAWAS 2003/4). The percentage of households with kerbside collections for recycling varies in the different local authorities from 16.5% in the largest authority, Glasgow City, to 83.5% in East Dunbartonshire, one of the smaller authorities. Glasgow City intends to provide separate collections to the majority of households, but has not made any significant progress towards this aim over the past few years.

The percentage of waste recycled varies from 3.7% in Glasgow City to 17.5% in East Renfrewshire, the local authority with the smallest population. Examination of the data (LAWAS 2003/4) indicates that increasing the percentage of households with separated household collections contributes to increasing the percentage of waste recycled. However a number of other factors also have an effect and therefore the local authorities in Glasgow and CV do not have the same rank order for recycling waste and provision of separated collections. The percentage of waste composted varies from 0% in two local authorities up to 5.1%.

Materials collected include paper, cans, plastic and green garden waste for composting, though the details vary across the different local authority areas. In 2000-1 household waste in Scotland (LAWAS 2000/1) consisted of:

- 34% paper and card
- 20% putrescibles (biodegradable, mainly kitchen waste)
- 11% plastics
- 9% glass
- 7% metals
- miscellaneous combustibles (8%), miscellaneous non-combustibles (2%), textiles (2%) and fines (7%).

There is currently one dedicated materials recovery facility run by Glasgow City Council and one privately owned partial facility. It is intended to extend this to four clean materials recycling facilities with a capacity of 225,000 metric tons and two mixed waste processing facilities by 2010 and capacity for 546,4000 metric tons of recyclate per annum by 2020 (SEPA 2003b). There are also a number of bring sites (local collection points) for cans, plastic bottles, newspapers and magazines and a relatively smaller number of sites for the collection of textiles.

In addition a certain amount of home composting takes place and just under 6000 home composting bins were distributed by the local authorities in 2003/4, mainly in South Lanarkshire, which provided 5000 bins (LAWAS 2003/4). This is a more than doubling of the 2700 bins distributed by the local authorities in 2002/3 (LAWAS 2002/3), but to date only a very small percentage of the households in the area have been supplied with home composting bins and it is possible that the rate of distribution in South Lanarkshire will not be repeated in future years. Metal is recycled through collection by scrap merchants. Textiles, shoes, household items and some furniture, white goods and electronic items are reused through jumble sales, charity shops and second hand shops. There are also a number of small scale schemes for donating used computers to schools or the 'developing' countries. Unfortunately there are no statistics for the amount of 'waste' reduced or reused in these ways. On the negative side poverty and debt are serious problems in many parts of the area, leading to ill health, which is compounded by a poor diet. Litter is degrading the environment, particularly in urban parts of the area. Many of the features which categorise the waste management situation in Glasgow and CV are illustrated in fig. 7.1.

The Glasgow and Clyde Valley Area Waste Plan (SEPA 2003b) assumes that the proposed waste reduction measures will be sufficient to give a reduction in the rate of growth of waste (to 2% up to 2010 and 1% from 2010 to 2020), but not a reduction in the total amount of municipal waste, leading to 1.244 million metric tons in 2010 and 1.374 million metric tons in 2020, an increase of 26.9%. It is intended to increase the percentage of waste recycled to 28% in 2010 and 40% in 2020 and the percentage composted to 8% in 2010 and 15% in 2020. It is projected that 64% of waste will be landfilled in 2010 and either 25% or 32% in 2020 with the remaining 20% or 13% treated by other means. The difference is due to different assumptions about the other treatments generating ash which is sent to landfill. This means that it is projected that at least 344,000 metric tons of waste will be sent to landfill in 2020. However this will be a significant reduction to only a third of the amount that was sent to landfill in 2002/3.

It has been projected that recyclable and compostable materials will form 91.3% of the waste stream in 2010, with the percentages of the different materials in the waste stream fairly similar to those in 2000-1 and the 7.4% of metals divided between 5.7% steel cans and 1.7% aluminium.

Fig. 7.1 Relationship diagram of waste in Glasgow and the Clyde Valley

Recycling targets vary from 57% for aluminium to 23% for textiles, with targets of 50% for steel cans, 48% for paper, 43% for plastics, 35% for glass and 27% for putrescibles and miscellaneous combustibles (SEPA 2003b). The correlation between ease of recycling from the resident's point of view and the projected collection rate should be noted. Thus the materials which will be collected directly from households (cans, paper and plastics) have the highest projected rates of recycling. They are followed by glass, which will not be collected from households, but for which bring centres (collection points) are fairly frequent. The low rate for textiles is presumably due to the difficulty in finding a collection point, since these are very infrequent. The low rate for putrescibles and

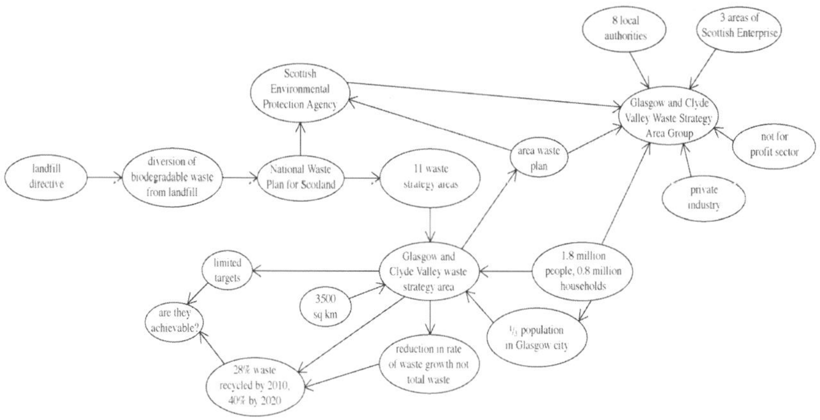

Fig. 7.2 Relationship diagram of the Glasgow & CV Area Waste Plan

combustibles is presumably due to the fact that garden waste, but not the other items in this category will be collected from households. The area waste plan and its stakeholders are illustrated in fig. 7.2.

7.2 Case Study: SIMILAR Method

7.2.1 Problem Statement

The problem situation has been discussed in section 7.1. The vision for the future is a clean, healthy and more prosperous city with an improved quality of life and zero waste achieved by reduction, reuse and recycling.

7.2.2 Investigation of Alternatives

It is probably most useful to approach the choice of alternatives as a two stage decision problem. In the first stage the overall strategy is chosen and in the second stage the various measures that can be used to implement the strategy will be considered. The stage one (choice of strategy) options include the following:

- Zero waste strategy
- Zero waste strategy with several stepped targets for waste reduction, reuse and recycling.

- Minimal compliance with landfill directive targets, mainly achieved through recycling and with some growth in waste.

- Minimal compliance with landfill directive targets, achieved through a mixture of recycling and waste from energy and with some growth in waste.
- More than minimal compliance with landfill directive targets with no growth in waste and a focus on recycling driven by creating demands for recycled products.

These options could be formulated as a decision or optimisation problem and one of the approaches presented in Part II applied to obtain a good solution. However, since decision making and optimisation are considered in detail in Part II, the details of this part of the methodology will not be given here and readers are encouraged to return to this problem after reading Part II. The second option will be chosen for the following reasons:

- Minimal environmental impact.
- Possibility of job creation through the development of recycling, refurbishment, reuse, manufacturing and repair (Renner 2004), particularly if reprocessing facilities for recycling are set up in the area.
- It includes specific targets and is therefore preferable to the first option.

Now that an appropriate option has been chosen, it is necessary to decide on how it will be implemented. This is the second stage of the decision problem. However, rather than choosing from a number of options as in the first stage, the decision approach taken here will be based on constructing the strategy from a number of separate components. This approach is closer to naturalistic than classical decision making (see section 10.3.2), as it is based on deriving a solution that will work rather than the 'best' one in some sense. It is appropriate to this type of problem, as it gives a logical structure for developing a strategy which has the potential to be successful, and this is the aim rather than choosing one option from a list.

The components (building blocks) to be considered include the following:

- *Publicity and information campaign*: The main aims are to increase public awareness of waste and how it can be reduced, reused and recycled, as well as to obtain information on (creative) waste reduction strategies being used elsewhere that could potentially be applied in Glasgow & CV. The campaign would also aim to try and change attitudes to encourage people to have items repaired, upgraded and refurbished rather than automatically replacing them. Existing information gathering on the composition of the waste stream would be continued and enhanced. General principles of the publicity campaign should include developing strategies that have been successful elsewhere, using a variety of approaches to target different segments of the population and accessibility to disabled people and non-native speakers of English e.g. by making materials available in Braille, large print and community languages and in signed as well as text versions on the web. The campaign should aim to be high profile and attention should be paid to the environmental impacts of the materials produced. Therefore, for instance, all brochures and leaflets should be produced on paper consisting of 100% recycled post consumer waste and contain a statement to this effect. (Post consumer waste is waste associated with end-of-life components or products, such as old newspapers, as opposed to waste from the manufacturing process.) Education on waste reduction, reuse and recycling in all schools, including visits to facilities and projects, will be an important part of the campaign. As well as educating children and young people about low waste lifestyles and encouraging them to participate in waste reuse and recycling, it will also lead to them encourage their parents to reduce waste and support waste reuse and recycling.

- *Waste reduction.* Approaches to waste reduction include reduction of packaging and consequently packaging waste, replacement of products by services and trying to reduce consumption. The replacement of products by services has largely been directed at commerce and industry to date and includes leasing rather than purchasing photocopying machines (Renner 2004), but could be extended to domestic consumption. Reduction of packaging waste would include charges for the supply of bags (all of which would be made from 100% post consumer waste) in shops and/or incentives for people to bring their own, as well as a mixture of incentives, charges and legislation to encourage the replacement of disposable bottles by deposit bottles, small items such as buttons and nails to be sold loose and all items to use the minimum packaging necessary for protection and hygiene. Reverse vending schemes would encourage the return of deposit bottles. Junk mail would be reduced by requiring people to opt in rather than opt out and the total amount of advertising would be reduced to try and eliminate the all-pervasive pressures to consume.
- *Repair, refurbishment and exchange centres and schemes*: In order to make it much easier for people to repair and refurbish old items rather than replace them and exchange used items, a number of centres and programmes should be set up. Exchange centres and programme might be similar to the University of Wisconsin-Madison SWAP programme (WWW8 2005) in which surplus property is collected from university departments and agencies of the State of Wisconsin. This property is

either distributed to other university departments, state agencies, schools or non-profit groups that can use these materials or sold to the general public through the SWAP shop and its web site. SWAP collects over 900 tons of surplus material each year, of which 98% is sold for reuse or recycled and earned well over a million dollars in the fiscal year 1997/8 from selling surplus property. Repair and refurbishment would involve the collection, repair and refurbishment or upgrading of goods. Particular centres would probably specialise in items such as shoes and clothes, furniture and rugs, or computers and hi fi systems. After repair and refurbishment to a high standard these items would either be returned to their previous owners for a moderate fee, sold to the general public at a low cost or distributed free of charge to public and community organisations. These centres would also supply parts at a low cost to encourage home repairs. Increasing standardisation of parts and components would make it easier for people to carry out simple repairs and avoid waste. Schemes would also be set up to investigate the use of domestic and commercial waste as inputs to industry.

- *Encouraging recycling and composting*: Materials that can be recycled or reused include glass (either mixed or flint, green and brown), white paper, newspapers and magazines, card, cans (aluminium, steel and mixed), plastics (preferably sorted into the six main types, rather than just plastic bottles), aluminium foil, inkjet and toner cartridges, bulbs and fluorescent tubes, mobile phones, computers, batteries, clothing and other textiles, shoes and oil. Although not all these materials can be collected directly from households, all households should receive recycling collections and collection points for other materials should be increased to make it as easy as possible for residents to participate in recycling. Garden waste for composting should be collected from allotments as well as households and bins for home composting should be supplied to all households with gardens and allotments. The resulting compost should be supplied cheaply or free of charge to all participating households and allotments, sold commercially and used on all local authority and other public properties in the area. This would have the further advantage of reducing the use of peat based compost and the resulting destruction of peat bogs. Food packaging should be replaced by paper or biodegradable cellophane and bioplastics, all of which should be either composted at home or collected with the compostable garden waste, as should kitchen vegetable waste.
- *Encouraging the use of recycled materials* and the purchase of products containing recycled materials and components: This has not yet become a significant issue, due to the low level or recycling in the area. However, as recycling increases, there will be a need for uses of this

recycled material to ensure that separated materials are actually recycled rather than sent to landfill. Agreements could be negotiated or legislation introduced for a minimum recycled content for paper, cardboard, cans, glass bottles and jars and plastic items. There are already some voluntary agreements, such as that reached by the Newspaper Publishers Association with the UK Government in 2000 to increase the amount of recycled content in newsprint to 60% by the end of 2001 and 70% by the end of 2000 (WWW9 2005). Local authorities and other public organisations should commit to only purchasing goods with a high content of recycled materials.

- *Funding strategy*: Some of the costs will come out of the existing budget for waste collection and management. The Scottish Executive has made additional funding of £230 million available for the 2003-6 period (SEPA 2003a), although this will have to be shared between all 11 waste strategy areas. Some aspects of recycling may be profitable, but this will depend on developing markets for recycled materials and possibly also on using taxation or other means to incorporate environmental and social costs (externalities) in the prices of new materials. For instance, it has been calculated that the environmental and social benefits of recycling aluminium as opposed to sending it to landfill and using new materials are about £1769 (3380 dollars) per metric ton (Craighill and Powell 1995). Gate fees per metric ton are £24 (46 dollars) in a materials recovery facility (i.e. for recycling), £10 (19 dollars) for composting and £15 (29 dollars) for landfill (SEPA 2003b). There is therefore a need for the adjustment of gate fees and taxes for different types of processing, to make landfill considerably more expensive than recycling as well as composting. The landfill tax should probably be increased to an extent that anything more than a very minimal use of landfill (for which there would be a lower rate of tax, though higher than at present) would be prohibitively expensive. The additional costs for landfill could contribute to financing schemes for waste reduction, reuse and recycling. An interest free credit scheme should be set up for the purchase of low impact goods with high initial, but reduced total life time costs. This would include long life items, such as long life low energy light bulbs and rechargeable batteries, and items designed to use minimum materials and/or to be easily upgraded and recycled. Local authority halls and rooms should also be made available free of charge to any groups organising events which could reduce waste, such as jumble sales, bring and buy or bring and take events and real i.e. washable nappy groups.
- *Reducing other environmental impacts*: The frequency of waste collection should not be increased to avoid increasing transportation

energy and emissions, and should preferably be reduced. This will require a cycle of collections with materials for recycling, composting and unseparated waste collected in different weeks. This will also have the effect of encouraging participation in the recycling and composting collections. The size of bins for unseparated waste should also be reduced. The vehicles used should also be upgraded to consume less energy and be much quieter, to reduce the resulting noise disturbance. Measures should be taken to investigate and replace or eliminate substances which are either toxic or hazardous in themselves or give off toxic emissions when burnt. This would facilitate combustion with energy recovery of any waste materials that could not be reused, recycled or composted. There is also a need for reprocessing facilities to be set up in the area. In addition to reducing the environmental impacts of transportation, this would have a positive effect on employment opportunities.

7.2.3 Modelling the System

Fig. 7.3 illustrates the main subsystems in the waste reduction system, including reduction, reuse, recycling, financial support and publicity. It shows how financial support and publicity act to drive the reduction, reuse and recycling subsystems. In each of these subsystems the reduction in waste resulting from the subsystem is obtained by comparing the subsystem output with its input and then fed back negatively into the subsystem.

Fig. 7.4 illustrates the component subsystems of the recycling subsystems as well as the presence of both positive and negative feedback. For instance the positive feedback from the output of the *recycled*

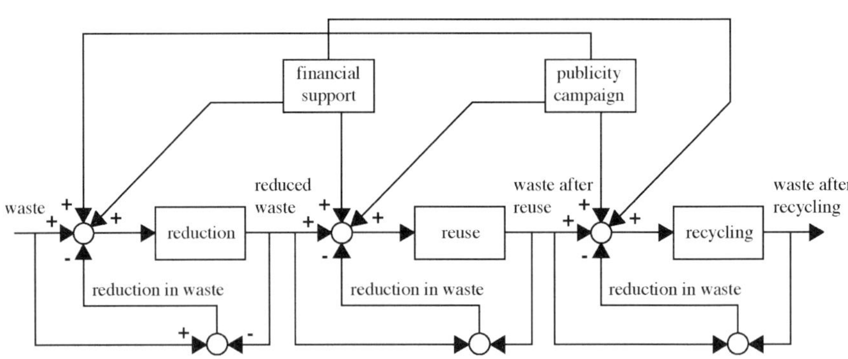

Fig. 7.3 Minimising waste through reduction, reuse and recycling

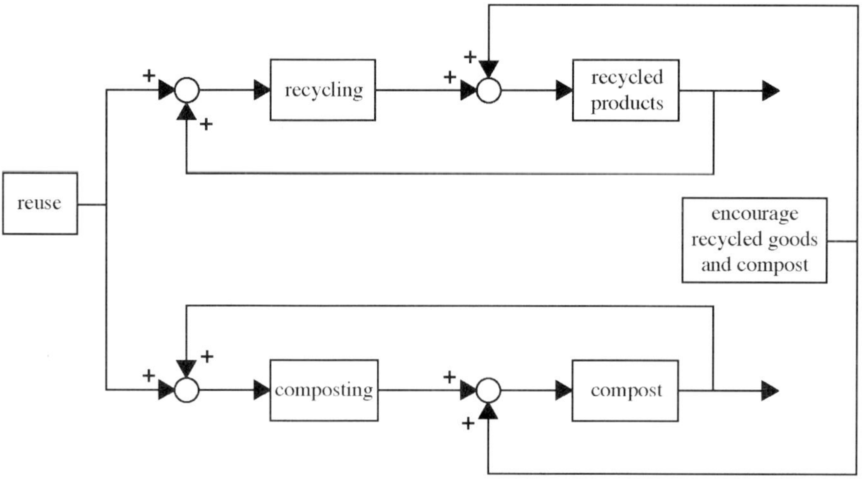

Fig. 7.4 Subsystems of the *recycling* system

products subsystem into the *recycling* subsystem acts as a driver to the latter subsystem. However this positive feedback cannot drive either the overall system or any of the subsystems to instability due to the limits on the availability of materials for recycling. The *reducing other environmental impacts* subsystem has not been included to avoid overcomplicating the diagram. However it could be included, for instance as filters before the *recycling* and *reuse* subsystems and before the *waste* subsystem (to filter out the generation of hazardous waste).

Fig. 7.5 illustrates the different stages involved in recycling, as well as the associated indicators and decisions, including on which materials are recycled and how they are collected. In general participation rates will increase the easier participation is, with separating waste for separated household collections the easiest option to participate in. It is also important that only appropriate items are included in each collection to give a good quality product. This may be influenced by the way waste is collected for recycling, as well as the clarity of the information provided. Other decisions concern the frequency of collection and the size of the containers used.

7.2.4 Integration of the System into its Environment

The environment of the system is the wider context in which waste generation and management takes place. The various measures listed in section 7.2.2 will require action at different levels in this environment. Some types of action will require voluntary agreements, for instance between

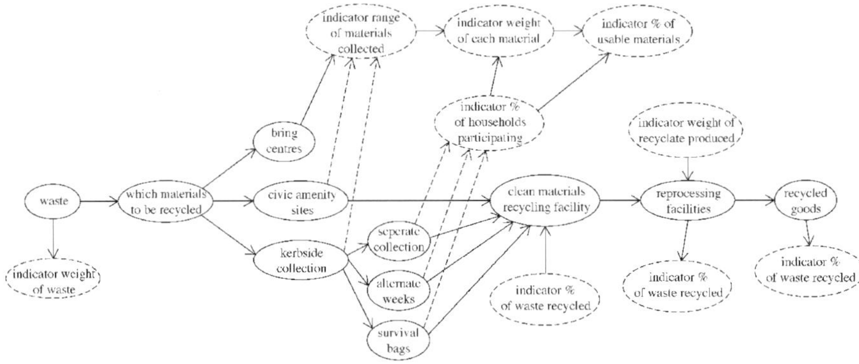

Fig. 7.5 Recycling stages and the associated indicators

the Scottish Executive or Scottish Environmental Protection Agency and manufacturing firms. Others will require legislation, either by the Scottish

Table 7.1 Organisations and individuals required to take action

	Residents	Local Authorities	SEPA	Scottish Executive	Scottish Parliament	UK Parliament	Manufacturing Industry	'Carrots'	'Sticks'
Publicity and Information	√	√	√	√	√	√	√		
Waste reduction	√	√		√	√	√	√	√	√
Repair, refurbish and exchange	√	√		√	√	√	√	√	
(Encourage) recycling and composting		√	√	√	√	√		√	√
Encourage the use of recycled materials		√	√	√	√	√	√	√	√
Reducing other environmental impacts		√	√	√	√	√	√	√	√
Funding Strategy		√		√	√	√		√	

or UK Parliaments and others will require residents in Glasgow & CV to take action. Therefore integration of the system into this environment includes determining who is required to take action to implement the different subsystems. This is indicated in table 7.1, which also indicates where inducements ('carrots') and sanctions ('sticks') will be required.

Integration of the system into its environment also requires determination of subsystems and interfaces between these subsystems and between the system and its environment. There are a number of different ways of dividing the overall systems into subsystems, including by taking each activity listed in section 7.2.2 as a separate subsystem, and this is the approach that will be used here. Initially it is probably sufficient to have one feedback loop round each subsystem. At a later stage it could be useful to determine the effectiveness of, for instance, different components of the *recycling* system, and therefore add additional feedback loops. The interface between the following pairs of systems, *waste reduction* and *reuse*, *reuse* and *recycling* (and *composting*), and *recycling* (and *composting*), and *energy recovery*, is that the output of the first subsystem in each pair is the input to the second subsystem.

7.2.5 Launching the System

Launching the system involves running the system discussed in the previous sections. The outputs obtained will be reused materials, parts and components, materials for recycling, compost, energy (from energy recovery) and a reduced amount of waste sent to landfill.

7.2.6 Assessing Performance

The assessment of system performance requires a number of quantitative and qualitative system goals to be specified as well as the times at which they are to be met. Quantitative goals should include specifications for the following variables for the years 2010, 2015 and 2020:

- The total quantity of waste produced.
- The percentage of waste reused.
- The percentage of waste recycled.
- The percentage of waste composted.
- The percentage and total quantity of waste landfilled.

- The percentage of residents participating in reuse, recycling and composting.
- The costs of waste management.

Qualitative goals to be considered relate to the attitudes and experiences of stakeholders, including residents and workers. They include the following:

- Positive experiences of residents of the area, including ease of obtaining information and ease of participating in material collection for recycling and reuse.
- Positive attitudes of residents to the waste reduction, reuse and recycling strategy and enthusiastic participation in it.
- Improved terms and conditions for workers in waste management.
- An increase in job satisfaction as a result of the new waste management strategy for workers in waste management.
- Some job creation and training in new skills as a result of the new waste management strategy and new opportunities for reuse and recycling of waste, as well as the construction of reprocessing facilities in the area.

7.2.7 Re-Evaluation

It is clearly important to regularly re-evaluate performance, due to a number of factors, including the following:

- To ensure that the system is functioning effectively and to determine whether minor or significant modifications are required.
- The need to monitor participation to ensure that residents do not withdraw or become careless about separating materials.
- The need to update information and publicity materials, including to ensure continued participation by residents.
- Changes in technology which may make it feasible to recycle additional materials and which will probably require changes in collection (and other) arrangements.
- To monitor the functioning of any voluntary agreements and incentives, as well as the degree of compliance with legal requirements, to ensure they are functioning effectively.

This re-evaluation should involve both measurement of outputs, including the use of products with recycled content, and monitoring of satisfaction of the assessment criteria. It may also be necessary to modify or update these criteria, as the situation changes. Re-evaluation will also allow changes, such as the use of smaller wheeled bins and less frequent collections, to be implemented.

7.3 Checkland's Soft Systems Methodology

7.3.1 The Unstructured Problem Situation

An overview of the unstructured problem situation has been given in section 7.1. Further investigation of the unstructured problem situation could include some or all of the following:

- Travelling round Glasgow & CV and obtaining an overview (a 'feel') of the area and its issues and problems, including, but not restricted to waste management.
- Visiting some of the offices of the Scottish Environmental Protection Agency (SEPA).
- Observing some waste collection rounds in Glasgow & CV and unstructured or semi-structured interviews with some of the workers.
- Unstructured or semi-structured interviews with a number of residents in the different local authority areas about their involvement in and attitudes to waste separation, recycling, composting, their knowledge of and attitudes to the existing waste collection and management service and how they would like it to be improved.
- Visiting the materials recovery facilities in Glasgow & CV and unstructured or semi-structured interviews with workers and management.
- Visits to landfill sites and unstructured or semi-structured interviews with workers and management.

7.3.2 An Expression of The Problem Situation

There are generally a number of different ways to consider the problem situation. How the problem situation is formulated will frequently affect the solution obtained. However, if a particular problem formulation is not proving useful, it is always possible to revise it. It may also be useful to consider and compare more than one formulation of the problem situation. Problems and issues to be considered include the following:

- Currently 88.9% of waste goes to landfill. There are no material reprocessing facilities in the area. Other than card being sent to

Aberdeen, materials are transported out of Scotland, either to England or Holland.

- The Scottish Waste Action Plan for Glasgow & CV is largely motivated by legal requirements arising from the landfill directive. It therefore does not go beyond these requirements to, for instance, consider strategies for zero waste and the targets are relatively modest compared to what has been achieved elsewhere. In particular, although the primacy of waste prevention is accepted (in theory), in practice the Plan will allow for some growth of waste and waste prevention is targeted solely at reducing the rate of growth (allowing 27% increase in waste production between 2003 and 2020), rather than overall reduction. It is also not clear that appropriate measures, including a dynamic and wide-ranging education and publicity campaign, are being put in place, to ensure that the targets (limited though they are) will be met.
- Glasgow & CV involves a large number of different types of stakeholders, including residents of the area, eight different local authorities, three enterprise areas, the Scottish Waste Awareness Group and SEPA. Good communications will be vital for achieving existing targets and moving beyond them to try to approach zero waste. There is also a need for communication and co-ordination with the four neighbouring waste strategy areas, as well as coordination with producers of packaging and (potential) manufacturers and users of recycled products. An understanding of human factors and the different cultures of different types of organisations as well as the differences in approaches between large organisations and residents will be crucial for success. The initial consultation process only received 24 responses, though three and a half thousand copies of the document were circulated. All the responses were from organisations, and there seem to have been limited possibilities for individual residents or households to participate, though local authority respondents did consult community councils.
- Many, if not most, residents and households are probably unaware of the targets for diversion of (biodegradable) waste from landfill and the proposals for achieving them. They may also have difficulties finding out where items such as batteries, oil, aluminium foil and textiles, that are not collected from households, can be deposited for recycling and/or find the locations inconvenient or inaccessible. Residents may feel that the system is bureaucratic and a lack of involvement with it and ownership of it.
- Consumerism is becoming a dominant value. The expectation is that people will replace clothes, furniture and other items frequently to keep

up with fashion, new designs and their neighbours. The costs of having items repaired, refurbished and/or upgraded are generally not competitive with the price of new goods, particularly when the added value of having something new is taken into account. There are also items, such as disposable cameras and paper underwear, that are designed to be used once or briefly and then disposed of. Other items, including cheap clothing, seem to be designed to only last for a short period. Spending is sometimes encouraged to boost the economy. However it has not been successful in Glasgow & CV in dealing with serious problems of poverty and providing employment. In addition consumption financed by credit has resulted in serious debt problems.

7.3.3 The Root Definition of the Relevant Systems

Several different systems will be considered. The *root definition of the first system* is: A system for managing waste in Glasgow & CV with the aim of minimising waste sent to landfill particularly by increasing recycling, in accordance with the Landfill Directive (CEC 1999) and the best practice environmental option (SEPA 2003a).

- Customers: People who live and work in Glasgow & CV.
- Actors: Residents of Glasgow & CV, SEPA, Scottish Executive, workers in recycling and other aspects of waste management.
- Transformation: Waste → materials for recycling, compost and energy.
- Weltanschauung: Belief in a best practice environmental option for waste management.
- Owner: SEPA.
- Environmental constraints: The best practice environmental option, the landfill directive, national and European legislation, restrictions on materials that can be composted, financial constraints, the infrastructure of Glasgow & CV.

The *root definition of the second system* is: A system for complying with the requirements of the landfill directive for diverting waste from landfill through increasing recycling and composting and energy recovery.

- Customers: People who live and work in Glasgow & CV.
- Actors: residents of Glasgow & CV, SEPA, the Scottish Executive, workers in recycling and other aspects of waste management.
- Transformation: Waste sent to landfill → waste diverted from landfill (by being turned into materials for recycling, compost and energy).

- Weltanschauung: Complying with legal requirements.
- Owner: SEPA.
- Environmental constraints: The best practice environmental option, the landfill directive, national and European legislation, restrictions on materials that can be composted, financial constraints, the infrastructure of Glasgow & CV.

The *root definition of the third system* is: A system to develop an integrated waste management strategy which takes account of the requirements and potential contributions of all stakeholders.

- Customers: Stakeholders of Glasgow & CV Waste Area Plan, including residents, eight local authorities, SEPA and three enterprise areas.
- Actors: Residents of Glasgow & CV, SEPA, Scottish Executive, workers in recycling and other aspects of waste management.
- Transformation: Existing waste management strategy $\rightarrow$ integrated waste management strategy.
- Weltanschauung: Benefits of an integrated approach to waste management.
- Owner: Stakeholders of Glasgow & CV Waste Area Plan.
- Environmental constraints: The best practice environmental option, the landfill directive, national and European legislation, restrictions on materials that can be composted, financial constraints, the infrastructure of Glasgow & CV.

The *root definition of the fourth system* is: A system to provide an employment creation strategy by taking account of the labour intensive nature of recycling and reuse.

- Customers: potential new employees, particularly residents of Glasgow & CV.
- Actors: residents of Glasgow & CV, SEPA, the Scottish Executive, workers in recycling and other aspects of waste management, potential and actual employers.
- Transformation: Existing number of jobs $\rightarrow$ increased number of jobs.
- Weltanschauung: Recycling and reuse of waste can create jobs.
- Owner: Local authorities and actual and potential employers.
- Environmental constraints: the best practice environmental option, the landfill directive, national and European legislation, restrictions on materials than can be composted, financial constraints, the infrastructure of Glasgow & CV, employment law and regulations.

The *root definition of the fifth system* is: A system to change the attitudes and lifestyles of the people of Glasgow & CV to reduce consumption and waste in ways that improve their quality of life.

- Customers: People who live and work in Glasgow & CV.
- Actors: people who live in work in Glasgow & CV, media and other opinion formers.
- Transformation: existing lifestyles and attitudes $\rightarrow$ low consumption and low waste attitudes and lifestyles.
- Weltanschauung: high consumption and waste do not necessarily lead to a good quality of life and reducing consumption and waste is both desirable and can improve quality of life.
- Owner: people who live and work in Glasgow & CV.
- Environment: Glasgow & CV, the existing political, social and economic system in Scotland and the UK, existing infrastructure and the possibilities of improving it.

The second root definition will be used as a basis for the analysis and therefore this model will be developed further through the use of rich pictures and other diagrams. It is illustrated in fig. 7.6. A similar analysis could be carried out for the other root definitions, though this will not be done here. There would also be benefits in carrying out and combining the analyses for all the different root definitions.

7.3.4 Obtaining Conceptual Models

To facilitate comparison the root definition of the conceptual model will be the same as that for the real system i.e. a system for complying with the requirements of the landfill directive for diverting waste from landfill through increasing recycling and composting and low emissions energy recovery. However the activities to implement the root definition will not be restricted to those in the system and should not relate to any existing system. Major activities in this model include the following:

- Reduce (waste).
- Reuse (waste).
- Recycle (waste).
- Obtain information (about the current waste stream).
- Publicise (measures for waste reduction, reuse and recycling).
- Finance (waste reduction activities).
- Investigate (creative waste reduction and management schemes elsewhere).

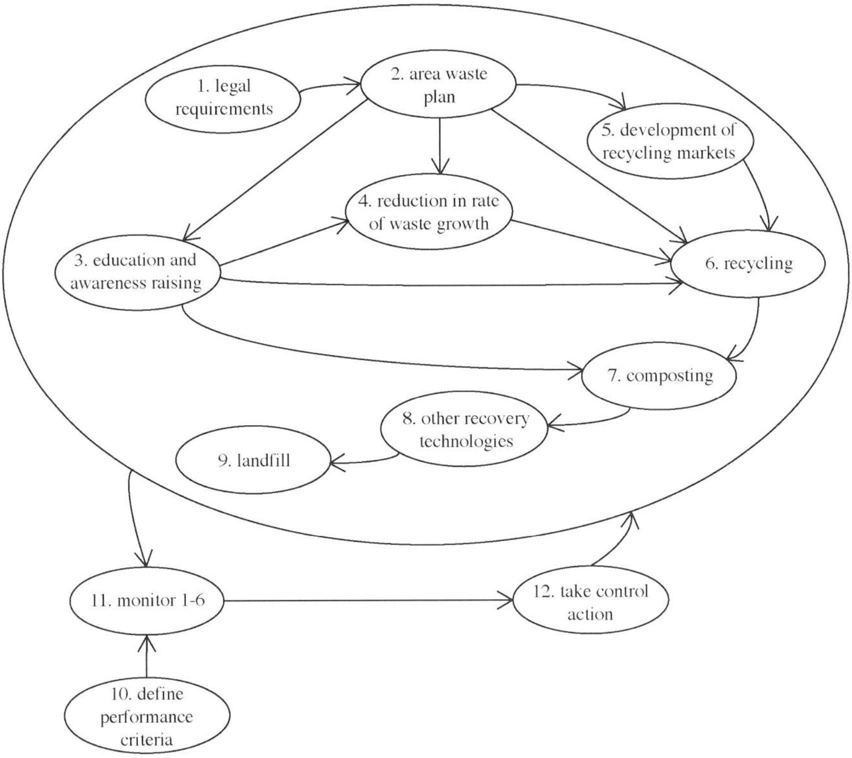

Performance Criteria:

Efficacy: reduction in total waste and landfilled waste, increase in percentage of waste recycled.

Efficiency: the change in costs over the increase in waste recycled or decrease in waste landfilled.

Effectiveness: measures of long term viability of waste reduction and increase in recycling.

Other criteria: user friendliness, environmental costs e.g. of processing or transporting material.

Fig. 7.6 Rich picture of the root definition of the waste management system

- Change (attitudes to recycling, consumption and waste).
- Encourage (recycling).
- Communication (with and between stakeholders).
- Recover (energy from waste with minimal emissions).

The main activities in the root definition are illustrated in fig. 7.7. The various subsystems can then be expanded further. The *Reduce (waste)* subsystem can be further described as follows:

- Reduce (consumption).

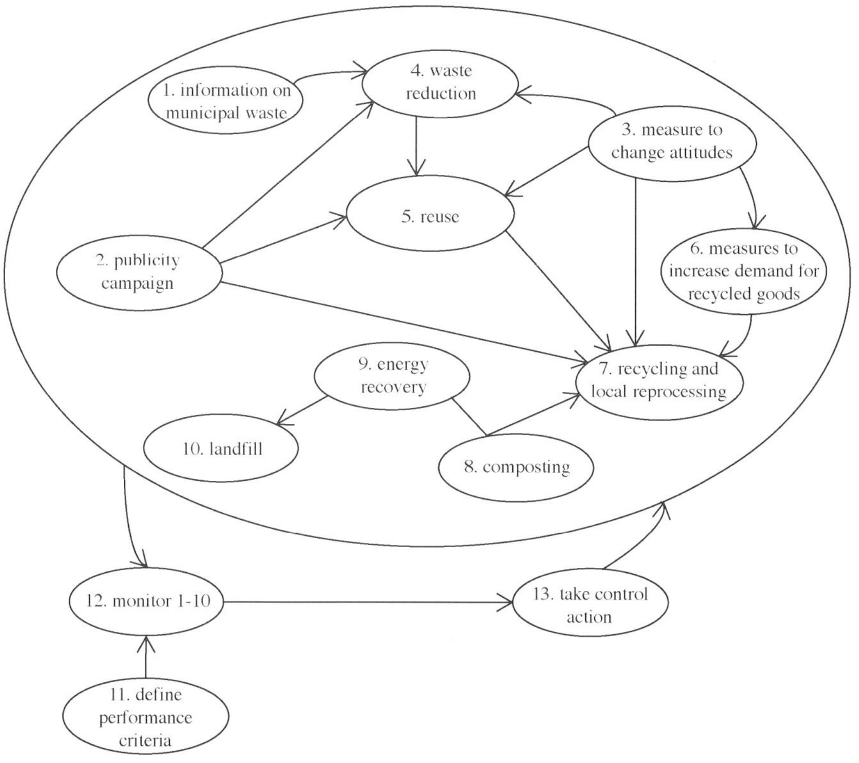

Performance Criteria:

Efficacy: reduction in total waste and landfilled waste, increase in percentage of waste recycled.

Efficiency: the change in costs over the increase in waste recycled or decrease in waste landfilled.

Effectiveness: measures of long term viability of waste reduction and increase in recycling.

Other criteria: user friendliness, environmental costs e.g. of processing or transporting material.

Fig. 7.7 Rich picture of the root definition for the conceptual system

- Choose (products with minimum necessary packaging).
- Use (only deposit and returnable bottles).
- Repair, refurbish and upgrade (old products and devices).
- Replace (goods by services).
- Purchase (second hand and refurbished items).
- Donate (unwanted items to friends, charity shops, jumble sales).
- Exchange (goods with friends and family, at swap shops or bring and take or buy sales).

- Establish (swap programmes and repair, refurbishment and exchange centres).

The *Change (attitudes)* subsystem can be further described as follows:

- Understand (motivation and the social, symbolic and other roles of consumption).
- Find (leverage points).
- Reward (reduced consumption, reuse and recycling in ways that do not involve consumption).
- Reduce (pressures on people and the pace of life).
- Inform and educate (about problems of overconsumption, without making people feel guilty).

The other subsystems can be described similarly. Of these subsystems, the *Change (attitudes)* subsystem will be the most difficult to implement successfully, since changing attitudes is always very difficult even when there is a strategy for doing this. Some of the elements of these subsystems, such as *Understand (motivation)* and *Reduce (pressures)* should be expanded further and the process should continue until no further expansion can usefully be carried out.

7.3.5 Comparison of Conceptual Model with Problem Situation

There are a number of different approaches to carrying out the comparison. Checkland and his colleagues (Checkland and Scholes 1996) suggest a two part approach. The first involves identifying each activity in the conceptual model and then considering whether it exists in the real situation, how it is carried out and how it is judged. Comments on the real system activities can often form the basis for suggested changes. The second involves a cultural analysis, consisting of role, social and political analyses.

In many cases the information from the comparison of the conceptual model and the real situation can usefully be presented in a table. However, the descriptions of how many of the activities are carried out are quite long and therefore the information will be presented under subheadings for each activity. Statements about existing activities and proposals are all drawn from the Glasgow and Clyde Valley Area Waste Plan (SEPA 2003b). The presence of the activities in the real situation will first be listed as follows:

- *Carried out in the real situation*: information on the current waste stream, recycling, composting, promoting recycling, other technologies to obtain energy from waste.

- *Carried out to a limited extent*: publicity, waste reduction, financial support for waste reduction, reuse and refurbishment activities.
- *Not carried out*: changing attitudes, investigating creative approaches used elsewhere.

Publicity

The Area Waste Plan gives the Waste Strategy Action Group the responsibility in conjunction with the Scottish Waste Awareness Group of developing and implementing recycling education and awareness raising programmes to support local recycling and composting initiatives. The Scottish Waste Awareness Group is supposed to be delivering campaigns on domestic waste management and waste reduction throughout Scotland from spring 2002. The promotional materials are supposed to be developed using feedback from focus groups and door to door surveys. There should also be a local network in each waste strategy area, which will provide regular meetings and workshops to stimulate ideas and practices, provide capacity building and other advice and act as a liaison between community groups and local authorities and SEPA. However, to date the campaign in Glasgow & CV has been fairly low profile and there is little evidence of the establishment of a local network. One of the indicators in the Area Waste Plan is the shift in public behaviour and, in particular, the percentage (of households) aware of and participating in recycling and waste prevention. This will be measured using surveys. However it may be difficult to determine how much of any shift in behaviour is due to the publicity campaign and how much to other factors.

Information

Waste arisings i.e. the amount of waste produced each year will be measured in thousands of metric tonnes and categorised according to the requirements of the Waste Data Strategy. In addition the percentages of the total weight of municipal solid waste treated by recycling, composting and to obtain energy and landfill will be measured. Data will also be obtained on the percentage of municipal solid waste collected from segregated kerbside (household) collections, survival bag collections and the number of bring recycling sites per 1000 households will be recorded. The data is published on the web. There do not seem to be any plans to obtain information on (creative) waste management activities in other areas.

Waste Reduction

The Area Waste Plan aims for a reduction in the rate of growth of waste production, rather than a reduction in the amount of waste produced.

Therefore waste is projected to increase by 27% between 2003 and 2020 rather than achieving reduction or even stabilisation. Waste prevention activities will include home composting and reuse of waste materials and be supported by producer responsibility legislation and are intended to be supported by education and awareness initiatives. However activity in this area is likely to be limited, as it is not being promoted very strongly. The main indicator of waste prevention is the measurement of solid waste arisings per year in the Local Authority Waste Arising Surveys, from which increases or reductions in waste production can be calculated.

Reuse
The main components of the Area Waste Plan are recycling, composting and other technologies (largely different approaches to obtaining energy from waste). Reuse and refurbishment are mentioned, but there are no targets or indicators for them. Therefore it is unlikely that reuse and refurbishment activities will increase significantly.

Recycling
Recycling is one of the main activities promoted by the Area Waste Plan. Currently only 11% of waste is recycled and the aim is to increase this to 28% by 2010 and 40% by 2020. There are no large scale reprocessing centres in the area and most materials for recycling have to be transported to England or Holland for reprocessing. In order to encourage participation in recycling, this will need to be made a lot easier. Facilitating participation will require a significant development of separate collection facilities for recycled materials so that all households have segregated collections, as well as an increase in the number of collection points (for materials that are not collected from households). There is also a need to develop reprocessing facilities in the area to reduce the energy and emissions resulting from transportation. This will have the further advantage of creating employment in the area. The Recycling Advisory Group Scotland will set up a community recycling network for Scotland and try to increase expertise and the number of community recycling practitioners and projects. The percentage of municipal waste recycled will be measured.

Promoting Recycled Goods
REMADE Scotland is working to identify potential markets and uses for recovered materials in Scotland. It supports local recycling businesses by encouraging investment, supporting local partnerships and encouraging wider awareness of the uses of recovered materials. The local authorities in the area are encouraged to bargain collectively with industrial reprocessers

to obtain higher and more stable prices for recyclate for reprocessing and to set up an area wide waste supply consortium. This consortium could also be involved in waste exchanges between local authorities. Although the area waste plan suggests that targets should be set for the recycled content of products rather than recovery rates to drive the need to recover for recycling, targets for recycled product content are not included in the list of indicators.

Composting
Composting at present mainly uses open air windrow systems in which biodegradable materials, largely green waste, are placed in long piles which are turned periodically to aid the composting process. There is recognition of the need to use larger scale systems, probably indoors, but no specific plans have been made, as well as the aim to encourage home composting. The percentage of municipal waste composted will be measured.

Energy from Waste
The Waste Strategy Action Group will consider the use of techniques for obtaining energy from waste after 2010. These techniques include incineration (the controlled burning of waste), pyrolosis (in which organic waste is heated in the absence of air to produce a mixture of gas, liquid fuel and a solid inert residue), gasification (in which carbon based wastes are heated in air or steam to produce fuel-rich gases) and refuse derived fuel production (in which the waste is processed to produce a pelletised fuel product). Pyrolosis and gasification have less environmental impacts than incineration. However there is considerable public concern about all these methods, which would have to be addressed in a way that genuinely resolves these concerns before any of these methods could be implemented. If this does not prove possible, then energy from waste technologies should not be introduced. Measures will also be required to, as far as possible, eliminate substances that produce hazardous emissions on pyrolosis or gasification from the municipal waste stream and to ensure that all environmental impacts of any facilities are minimised. The percentage of municipal solid waste used to obtain energy from waste will be measured.

The cultural analysis based on role, social and political analyses will now be considered.

Analysis 1: Role Analysis
- Client: SEPA.

- Client's aspirations: to ensure compliance with the landfill directive and that the Glasgow & CV Area Waste Plan is implemented successfully.
- Resources available: the local authority infrastructure, local residents, Scottish Waste Action Group, Waste Strategy Action Group, financial support from the Scottish Executive (to be divided between all 11 waste strategy areas), experiences of successful waste reduction in other areas.
- Constraints: legal requirements, costs, relative costs of landfill and other treatments (though this may change), target times, existing infrastructure, financial resources.
- Problem owners: SEPA, residents concerned about excessive use of landfill, local authorities, workers in local authorities or waste management organisations.
- Implications of problem owner chosen: SEPA is mainly concerned about meeting legal requirements and has previously expressed the view that landfill is the most appropriate treatment (SEPA 1997), though this view seems to have changed. SEPA as problem owner will not give a very radical approach. Groups of concerned local residents are likely to take a more radical approach and encourage going beyond the targets in the Area Waste Plan. Workers will wish to meet or exceed the targets and support measures that lead to job creation and improve pay and terms conditions. Local authorities will be concerned about budgetary constraints and sources of finance, as well as meeting the Area Waste Plan targets. The budgetary constraints may be a greater concern.
- Reason for regarding the problem as a problem: Currently nearly 89% of waste in the area is sent to landfill (LAWAS 2003/4). Landfill is responsible for about a quarter of the methane (a global warming gas) generated in Scotland, but only just half the waste landfilled has methane recovery. Glasgow & CV is currently very far from meeting the legal requirements for landfill diversion by the target dates and it is therefore possible that the measures in the Area Waste Plan will be insufficient to meet these requirements. In addition these requirements still allow growth in waste and will involve a considerable amount of waste being sent to landfill. There is also considerable public concern about the energy from waste technologies which could be introduced after 2010. The situation involves a large number of different types of stakeholders with different values, aspirations and ways of working.
- Value to the problem owner: SEPA and the local authorities want a system to ensure that the legal requirements are met and that the local authorities remain within budget. The concerned residents also want to ensure that legal requirements are met, but would like to go beyond these requirements to increase recycling and reduce total waste.

Workers want a system that will meet or go beyond the directive and improve their pay and terms and conditions of work.

- Problem content: To obtain procedures which will make it easier for the different stakeholders to work together constructively and to obtain a set of measures that will ensure that the targets for landfill diversion, recycling and reduction of the rate of growth of waste generation are met and exceeded.

Analysis 2: Social Analysis

In the social system analysis important institutionally defined roles include resident, householder, worker for local authority, local authority, SEPA, worker for SEPA, manager, chief executive and worker in a waste management firm. Important behaviourally defined roles include recycler, non-recycler, repairer and reuser (for instance of your own old clothes), home composter, litter lout, bureaucrat.

One of the difficulties in the system is that norms are not clearly defined, other than for the behaviourally defined roles, and that there is not a common set of values. However (possibly slightly idealised) norms for some of the institutionally defined roles can be stated as follows:

- Resident: responsible citizen and member of the local community, active participant in waste separation for recycling and waste reduction
- Local authority: proactive approach to waste reduction and recycling, provides a high quality infrastructure for separate collection and clear information to local residents, consults frequently with local residents and takes their views seriously, responsive to requests from residents.
- Worker: diligent, conscientious, committed to waste reduction and recycling, active in a trade union, engaged in solidarity activity, concerned to improve terms and conditions, knowledgeable about equality, health and safety issues.

Values held by different stakeholders include the following:

- The importance of meeting legal obligations.
- Concern with social justice and equal distribution of resources.
- Sustainable development.
- The importance of maintaining a paper trail.
- The need to make changes now.
- The desirability of a clean environment without litter.
- The value of resources and the importance of using them carefully.
- The importance of financial prudence and remaining within budget.

Analysis 3: Political Analysis

The political analysis is particularly interesting in this problem, since it involves many different types of stakeholders, including local residents, SEPA, eight local authorities, three enterprise areas, the Scottish Waste Awareness Group, the Waste Strategy Action Group, private industry and managers of waste facilities and workers in all the above organisations, amongst others. Resolving the power issues will probably be crucial for success. There are a number of different commodities of power, including:

- Participation: Residents have the power to participate or withhold participation in recycling and composting, for instance by composting their kitchen and garden waste at home, taking glass to recycling centres and separating green garden waste, plastic bottles, paper and cans. Success in meeting targets is going to depend on the participation of a high percentage of households.
- Provision of facilities: Local authorities have the power to decide, for instance, how materials will be collected for recycling. As discussed in section 7.1, the amount of each material collected will depend on the type of collection facilities available and the ease of using them. Residents can lobby their local authority to, for instance, provide kerbside collections to more households and for a wider range of materials. However the costs involved may act as a barrier.
- Financial resources: Much of the additional funding required is coming from the Scottish Executive. Although there may be significant revenues from recycling once a significant amount of recycling takes place, there will be additional capital costs initially and thus the Scottish Executive funding will be crucial for establishing the additional infrastructure. Further funding or investment may be required in order to set up local reprocessing capability to produce recycled materials.
- Legislation: Changes in waste management and the targets for landfill diversion and increased recycling and composting are a direct response to the Landfill Directive (CEC 1999). Any changes in this Directive, for instance stricter targets for landfill diversion and waste reduction, would require a response.
- Landfill tax: This is set by the Westminster Parliament and has a significant impact on the relative costs of landfill and recycling. A significantly higher tax could also act as an incentive to reduce the amount of waste rather than just its rate of growth. The level of this tax can be influenced by lobbying. For instance residents concerned about sustainable waste management might lobby for this tax to be increased significantly and/or the costs of recycling to be subsidised.

- Information: Information is a commodity of power in many systems. In this system the different parties have control over different types of information. For instance local authorities have control over the distribution of information on how local waste collection is carried out.
- Co-operation: Success in meeting (or going beyond) the targets for landfill diversion is dependent on co-operation. Any of the parties can give or withhold co-operation.

Therefore, each of the stakeholders has different types of power and many of the stakeholder have the power to influence the success of the system. The specific types of power are generally related to the particular roles, such as resident or local authority and are therefore generally not transferable between the different types of stakeholders.

7.3.6 Agreement on Feasible, Desirable Changes

Identification of desirable, feasible changes can result from the first two stages of investigating the unstructured problem situation and expressing the problem situation, as well as from comparison of the actual system with the model. To be successful such actions should generally be agreed rather than just identified. This is difficult in both the current problem and other problems which involve both a number of different types of stakeholders, such as members of the general public and statutory organisations such as local authorities. Thus the changes should include measures to improve communication between the different stakeholders, in particular the local authorities and local residents.

Some of the changes will involve strengthening existing measures or making sure that proposed measures are actually implemented. They are therefore desirable and can be considered feasible, since they are based on existing measures. Measures of this type include the following:

- Strengthening the proposed *awareness raising programmes* to support local recycling and composting initiatives. This could be combined with the proposal for setting up a local network in each waste strategy area. At the minimum there should be a high profile poster campaign, combined with advertisements in the media. Designing posters could be set as a competition for local schools, thereby also achieving some educational aims. Schools could also be involved in designing other publicity materials, particularly for use with young people. The awareness programme should also include clear information as to how, when and where different (types of) materials are collected and what exactly can be put in each type of collection, so that the value of the resulting recyclate is not reduced, for instance by adding yellow pages telephone directories to a paper collection, as the yellow dye discolours the paper. All publicity materials should be made available in alternative formats and in minority languages so that they are accessible to all residents of the area, including disabled people and people from ethnic minorities.

- Ensuring that a *local network* is set up. One approach would be to make contact with existing organisations, such as residents and tenants organisations, community councils and green or environmental organisations to identify which groups would be interested to doing some work on waste reduction and recycling and what sort of support they would require. This support could include the provision of publicity materials, free room hire for meetings and events and covering the costs of mailings or the use of local authority facilities to send out mailings to members. Once a (small) number of interested groups and/or individuals in a particular area have been identified they can be invited to a meeting to plan a workshop or other event.
- Providing *bins for home composting* to all households (whether houses or blocks of flats) with gardens and all allotments. This should be done on an opt-out rather than opt-in basis. This should be accompanied by leaflets on home composting as well as meetings in community halls, possibly organised in conjunction with local community councils.
- Obtaining and *publicising up-to-date information* on waste and its treatment on the SEPA website.

Other desirable and feasible measures include the following

- Ensuring that there is *sufficient reprocessing capacity* in the area to process all materials collected for recycling in the area. Although this is both feasible and desirable, reprocessing capacity would probably have to be introduced in a number of stages rather than all at once.
- Obtaining *information on (creative) waste management* activities, including reduction, reuse and recycling, being carried out elsewhere and evaluating them for introduction in Glasgow & CV.
- Negotiating *voluntary agreements for the minimum recycled content* in paper, glass and plastic products. This minimum content should gradually be increased upwards.
- All public bodies in the area to only purchase paper and glass products with a *minimum content of recycled materials.*

- Encouraging the groups in the local network to make *proactive suggestions for change* and the local authorities and SEPA to respond seriously to these suggestions.

These lists of changes are intended to be illustrative of the types of changes that are both feasible and desirable rather than to cover all possibilities.

7.3.7 Action to Improve the Problem Situation

Action to improve the problem situation involves implementing the feasible, desirable changes identified in the previous section and monitoring the effectiveness of these changes. How feasible the changes really are with regards to their acceptability to all stakeholders and whether they do have the intended and desired effects will generally only becomes clear during and following implementation. Therefore some of the suggested changes may be modified during the implementation process. In addition the methodology involves a process that takes place over a period of time and so implementation of desirable, feasible changes may not be the end of the process, particularly as implementing the changes will modify the system and new changes may become desirable and feasible in the new situation.

7.4 Summing Up

This case study has illustrated the application of the SIMILAR method and Checkland's Soft Systems Methodology to waste management in Glasgow and the Clyde Valley and also highlighted some of the important sustainability issues relating to waste management. It was decided not to include the Systems Failures Method in this case study, since it is also a soft systems method and there are some similarities to Checkland's methodology. These include both approaches being process based and using a comparison between the real system and a conceptual model.

Both the SIMILAR Method and Checkland's Soft Systems Methodology have resulted in similar conclusions, though the process by which these conclusions were obtained is quite different in the two cases. In particular current approaches to waste management in the area are totally unsustainable in their excessive reliance on landfill, frequently without even methane recovery. Proposals in the Glasgow & CV Waste

Strategy Area Plan to ameliorate this situation, while a very significant improvement over the current situation, have serious limitations and may not be realisable. Both approaches resulted in a number of positive suggestions to increase the likelihood of realising the targets in the area plan and moving beyond them and there is significant commonality between the two sets of suggestions, despite the different approaches.

The case study has also illustrated some of the differences between hard systems approaches, such as the SIMILAR Method, and soft systems methods, such as Checkland's methodology. In particular the SIMILAR Method started with a particular problem statement in terms of a description of the existing situation plus a vision for the future, whereas deciding which problem to consider was part of Checkland's methodology. In addition Checkland's Soft Systems Methodology is process based and some of the conclusions and suggestions for change arose from the initial investigation of the problem, rather than the comparison of the conceptual model with the real system. It should also be noted that not all the information obtained was actually used in the comparison process and that some of the information obtained in the early stages of the process could be used to make changes at that point, without needing to wait for the formal comparison stage. The description of the case study may give a false impression of a series of discrete steps following one after the other. This is generally not the case in practice and there may not be clear divisions between the different stages of the process. Since Checkland's Soft Systems Methodology is process based, it is frequently carried out over an extended time period, such as three months, whereas results can generally be obtained more quickly with the SIMILAR Method.

The case study involved the application of the two different methods to the same problem to highlight the similarities and differences between the two approaches. However, in practice the two methods are frequently applied to different types of problem.

7.5 Tutorial Exercises

1. Which of the following systems are time invariant or satisfy the superposition principle? Find the impulse response and transfer function for these systems. Why can the impulse response and system transfer function not be obtained for the other systems?

a.
$$\begin{aligned}\dot{x}_1 &= 5x_1 + x_2^2 \\ \dot{x}_2 &= 3x_1x_2 + 4x_2\end{aligned}$$

b.
$$\begin{aligned}\dot{x}_1 &= 2.5x_1 - 3x_2 \\ \dot{x}_2 &= 2x_1 + 5x_2 - 7x_3 \\ \dot{x}_3 &= 3x_2 + 2x_3\end{aligned}$$

c.
$$\begin{aligned}\dot{x}_1 &= 2.5x_1\cos(t) - 3x_2\sin(t) \\ \dot{x}_2 &= 2x_1 + 5x_2 - 7x_3 \\ \dot{x}_3 &= 3x_2 + 2x_3\end{aligned}$$

d.
$$\begin{aligned}\dot{x}_1 &= x_1 - 3x_2 + 4x_4 \\ \dot{x}_2 &= 2x_1 + x_2 - 7x_3 \\ \dot{x}_3 &= 2x_1 + 6x_2 + 2x_3 + x_4 \\ \dot{x}_3 &= 2x_2 + 7x_3 + 3x_4\end{aligned}$$

2. For the time invariant systems in question 1 which satisfy the superposition principle obtain the output when the input is equal to:

   a. the impulse function $\delta(t)$
   b. the unit step function $u(t)$
   c. $e^{-2t}u(t)$
   d. Show that the response of a causal system to an input $x(t)$ can be expressed as

$$y(t) = \int_0^\infty \dot{x}(t-T)a(T)dt$$, with $a(t)$ the response to a unit step at $t=0$.

3. Give two examples of continuous and discrete time systems

4. Consider the following systems:

   a. The arms race model presented in equations (4.39) and (4.40) in section 4.6
   b. The model of a regional system presented in equations (4.7) in example 4.1 in section 4.1.

c. The model of a regional system presented in equations (4.7a) and (4.8) in example 4.1 in section 4.1.
d. The self-sufficient rural community discussed in example 3.6 in section 3.12.
e. The city discussed in example 3.6 in section 3.12.

Are these systems:

i. Discrete or continuous time
ii. Open or closed loop
iii. Deterministic or stochastic
iv. Open or closed

5. Explain the difference between a closed system and a closed loop system.

6. Consider the state space system

$$\dot{\underline{x}} = A\underline{x} + B\underline{u} \qquad y = C\underline{x} + D\underline{u}$$

with the following sets of matrices for A, B, C, D

a. $A = \begin{bmatrix} -2 & 1 \\ 3 & -1 \end{bmatrix}, \quad B = \begin{bmatrix} 1 \\ -3 \end{bmatrix}, \quad C = \begin{bmatrix} -2 & 5 \end{bmatrix}, \quad D = 0$

b. $A = \begin{bmatrix} -4 & 0 & -1 \\ -6 & 5 & 3 \\ 0 & 7 & 2 \end{bmatrix}, \quad B = \begin{bmatrix} 2 \\ -5 \\ 1 \end{bmatrix}, \quad C = \begin{bmatrix} -2 & -1 & 4 \end{bmatrix}, \quad D = 0$

c. $A = \begin{bmatrix} -5 & 3.1 & 0 & 3 \\ -2.3 & -7 & -1.4 & 2 \\ 1 & 4.5 & -3 & 0 \\ 1 & 2 & 7 & -4 \end{bmatrix}, \quad B = \begin{bmatrix} 1.5 \\ 0 \\ 2.3 \\ 1 \end{bmatrix},$

$C = \begin{bmatrix} 1.5 & 0 & 2.3 & 1 \end{bmatrix}, \quad D = 1$

For each system:

a. Obtain the associated transfer function.
b. Determine whether the system is stable.

c. Apply a linear transformation to obtain a state space representation with diagonal state matrix.
d. Obtain the state transition matrix for this system.

7. Repeat question 6 for the discrete time system

$$\underline{x}_{t+1} = A\underline{x}_t + Bu_t \qquad y_t = C\underline{x}_t + Du_t$$

with A, B, C and D as in 6(a), 6(b) and 6(c). Comment on any differences in the answers.

8. Consider the systems with the following transfer functions:

a. $G_1(s) = \dfrac{s^3 + 3s^2 + 2s - 4}{s^3 - 4s^2 + s + 6}$

b. $G_2(s) = \dfrac{s^3 - 2s^2 + 3s + 4}{s^4 + 1.5s^3 + 2s^2 + s + 5}$

c. $G_3(s) = \dfrac{2s^5 + s^4 + s^3 - 3s^2 + 4s + 5}{s^5 + s^4 + 2.5s^3 + 3.5s^2 + 4s + 7}$

Obtain three different state space realisations for system (a) and two different state space realisations for systems (b) and (c). Why is it not possible to obtain a realisation of systems (b) and (c) with the state matrix diagonal? Which of the systems (a), (b) and (c) are stable?

9. Consider the introduction of an environmental policy in a firm, its implementation and monitoring.

a. Derive and analyse the component tasks to obtain a network of events and interconnecting activities.
b. Identify the critical path.
c. Estimate optimistic, pessimist and likely times required for each activity and use these times to calculate the expected time and variance for each activity.
d. Calculate the earliest expected time for successful implementation of the policy.

10. Explain how double loop feedback could be used to change the following mental models based on prejudice or misinformation:

    a. Women are incapable of becoming good scientists or engineers.
    b. Deaf people are less intelligent than hearing people.
    c. It is not necessary to reduce consumption, as technical solutions will be found to current environmental problems.

11. Consider the Bhopal disaster discussed in section 5.6. Apply the following methods to interpreting the reasons for this disaster and how it could have been prevented:

    a. SIMILAR Method
    b. Soft Systems Methodology
    c. Systems Failures Method.

    Comment on the results, including the explanations given by the different methods and the differences in interpretation.

12. Consider the introduction, monitoring and implementation of an environmental policy in a firm. You should either use data based on a real firm or develop a scenario, under the assumption that a number of difficulties are encountered, including opposition from some members of management.

    Apply the following methods to investigate and interpret what happens:

    a. SIMILAR Method
    b. Soft Systems Methodology
    c. Systems Failure Method.

    Comment on the results, including the different explanations given by the different methods.

13. Consider action to make the computer life cycle more sustainable. Using the data in table 2.2 in section 2.4.1 and fig. 2.2 in section 2.4.2, apply the following methods to this problem:

    a. SIMILAR Method
    b. Soft Systems Methodology
    c. Systems Failure Method.

    Comment on the results, including any differences in understanding or different actions resulting from the different methods.

References and Additional Reading for Part I

References for Part I

Ackoff RL (1971) Towards a system of systems concepts. Management Science, 17(11):661-671

Allenby B and Pugh T (1992) What is design for environment? A primer for members of the American electronics association. Amer Electron Associat, p. 3

Alting L and Legarth JB (1995) Life cycle engineering and design. Ann CIRP, 44(2):1-11

Archibald R, Villoria R (1967) Network-based management systems, (PERT/CPM). John Wiley, New York

Argyris C, Putnam R, and Smith DM (1985) Action science. Jossey-Bass, San Fransisco

Bast C, Korpalski T (1996) Hewlett-Packard's life cycle-oriented product stewardship program. In: Fiksel, J (ed) Design for environment, McGraw-Hill, New York, chapter 13

Beer S (1972) Brain of the firm. Allen Lane, London

Bertalanffy L von (1968) General systems theory. George Braziller, New York

Blanchard BS, Fabrycky WJ (1990) System engineering and analysis, Prentice Hall

Bolton W (1994) Laplace and z-transforms. Longman Scientific and Technical

Bourg D and Erkman S (2003) Perspectives on industrial ecology. Greenleaf Publishing

Brinkley A, Kirby JR, Wadehra IL (1996) Life cycle inventory of PVC: manufacturing and fabrication process, IEEE Int Symp Electron Environ, 94-101

Bruner JS (1983) In search of mind, Harper, New York

CEC (1999) Council directive on landfill sites, Official Journal of the European Communities, C 156 010, C 126 011, Brussels

Checkland P (1996) Systems thinking, systems practice. John Wiley and Sons

Checkland P, Forbes P, Martin S (1990) Techniques in soft systems practice: part 3: monitoring and control in conceptual models and in evaluation studies. J of Applied Systems Analysis, 17:29-37

Checkland P, Scholes J (1999) Soft systems methodology in action. John Wiley and Sons

Chestnut H (1967) Systems engineering methods. John Wiley and Sons

Craig J (1964) Laplace and Fourier transforms for electrical engineers. Holt, Rinehart and Winston, New York London

Craighill AL and Powell JC (1995) Lifecycle assessment and economic evaluation of recycling: a case study. CSERGE Working Paper WM 95-05

Derrida J (1981) Limited INC. Evanston, Ill, Northwestern University Press

Emergy FE, Trist EL (1960) Socio-technical systems. In: Churchman CW and Verhulst M (eds) Management sciences, Pergamon, Oxford

ESRC (1999) The politics of GM food, risk science and public trust. Special briefing no. 5

Fiksel J (1996) Design for environment. McGraw-Hill, New York

Forester JW (1969) Urban dynamics. MIT Press

Forester JW (1971) World dynamics. Wright-Allen Press

Fortune J and Peters G (1995) Learning from failure. John Wiley and Sons

Foucault M (1972) Power/Knowledge, Edn Colin Gordon, Pantheon, New York

Friedland B (1987) Control Systems design an introduction to state-space methods. Mc-Graw Hill

Gadamer H-G (1989) Truth and method. Crossroad, New York

GEM (1992) Total quality environment: the primer. Global Management Initiative, Washington DC

Graedel TE, Allenby BR (1995) Industrial ecology. Prentice-Hall, Englewood Cliffs NJ

Harremoës P, Gee D, MacGarvin M, Stirling A, Keys J, Wynne B, Guedes Vaz, S (eds) (2002) The precautionary principle in the 20th century: late lessons from early warnings. Earthscan

Harary F, Norman RZ, Cartwright D (1965) Structural models: an introduction to the theory of directed graphs. John Wiley, New York

Hersh MA (1993) The zeros and poles of delta operator systems. Int J Control, 57(3):357-575

Hersh MA (1998) A survey of systems approaches to green design with illustrations from the computer industry. IEEE Trans on Systems, Man and Cybernetics, 28C(4):528-540

Hersh MA (2001) Eco-Design for all. WACRA, Vienna, Austria

Hersh MA (2002a) Eco-design for all: incorporating sustainable and accessible design into the engineering curriculum. Conference on engineering education in sustainable development, Delft, Netherlands

Hersh MA (2002b) Design issues in assistive technology: applications to alarm and alerting devices for deaf and hearing impaired people. CVHI 2002, Granada, Spain

Hersh, MA (2002c) Whistleblowers – heroes or traitors?: individual and collective responsibility for ethical behaviour. Ann. Reviews in Control, 26:243-262

Hersh, MA. (2004a) Ethical issues in assistive technology. Tutorial Session, CHVI 2004, Granada

Hersh, MA (2004b) Technology development and ethical decision making: gatekeeping, identity and social construction. 16th IFAC World Congress, Prague

Hillary R (2000) ISO 14001, case studies and practical experiences. Greenleaf Publishing

ICFTU (1985) The trade union report on Bhopal. International Confederation of Free Trade Unions (ICFTU), Brussels
Johnson-Laird PN (1983) Mental models. Cambridge University Press
Keoleian GA, Menerey D (1993) Life cycle design guidance manual: environmental requirements and the product system. EPA/600/R-92/226
Kierkegaard S (1944) [1843] Either/or. Swenson DF, Swenson L, Lowrie W (trans), Princeton University Press
Kierkegaard S (1967) [1845] Stages on life's way. Lowrie W (trans), Shocken Press, New York
Klir GJ (1972) Trends in general systems theory. Wiley-Interscience
Korpalski T (1996) The role of the 'product steward' in advancing design for environment in Hewlett-Packard's computer products organization. Proc IEEE Int. Symp Electron Environ, 37-41
Kramer NJTA, Smit J de (1977) System thinking concepts and notions. Martinus Nijhoff Social Sciences Division, Leiden
LAWAS (2001/2) Local authority waste arisings survey 2001/2002. SEPA, (http://www.sepa.org.uk/nws/data/lawas.htm). There are eight separate documents, one for each local authority
LAWAS (2002/3) Local authority waste arisings survey 2002/2003. SEPA, (http://www.sepa.org.uk/nws/data/lawas.htm). There are eight separate documents, one for each local authority
LAWAS (2003/4) Local authority waste arisings survey 2003/2004. SEPA, (http://www.sepa.org.uk/nws/data/lawas.htm). There are eight separate documents, one for each local authority
Lenox M, Ehrenfeld J (1995) Design for environment: a new framework for strategic decision. Total Quality Environment Management, pp. 37-51
Lowe E (1993) Industrial ecology – an organizing framework for environmental management, Total Quality Environ. Management, vol. 3, pp. 73-85
Lowe E (1996) Industrial ecology: a context for design and decision. In: Fiksel J (ed) Design for Environment, McGraw-Hill, New York, chapter 25
Martin MW, Schinzinger R (2004) Ethics in engineering 4th ed. McGraw Hill
Meadows D, Meadows D (1972) The limits of growth. Signet
Meadows DL, Meadows DH (1973) Toward Global Equilibrium Collected Papers. Wright-Allen Press
Meadows DL and Behrens WW III (1974) Dynamics of growth in a finite world. Wright-Allen Press
Meadows DH, Meadows DL, Randers J (1992) Beyond the limits or a sustainable future. Earthscan, London
Mesarovic MD (1972) A mathematical theory of general systems. In: Klir GJ (ed) Trends in General Systems Theory, John Wiley and Sons, pp 251-269
Mesarovic MD, Takahara Y (1975) General Systems theory: mathematical foundations, Academic Press, New York
Meshkati N (1991) Human factors in large-scale technological systems' accidents: Three Mile Island, Bhopal, Chernobyl. Industrial Crisis Quarterly, 5:133-154
Middleton RH, Goodwin GC (1990) Digital control and estimation. Prentice Hall

Miyamoto S, Tamura T, Fujimoto J (1996) Eco-fusion, integrated software for environmentally conscious production. Proc IEEE Int Symp Electron Environ, 179-184

NHS (1996) Guidelines for the use of "soft systems" methodology. NHS Executive Information Management Group

Nielson RP (1996) The politics of ethics. Oxford University Press

O'Connor J, McDermott I (1997) The art of systems thinking: essential skills for creativity and problem solving. Thorsons

O'Riordan T, Cameron J (1994) Interpreting the precautionary principle. Earthscan

Paton B (1994) Design for environment, a management perspective. In: Socolow R, Andrews C, Berkhout F, Thomas C (eds) Industrial ecology and global change, Cambridge University Press

Perrow C (1986) Complex organisations: a critical essay (3rd edn). Random House, New York

Pfleiderer RHH, Dieterich M (1995) New roads generate new traffic. World Transport Policy and Practice, 1(1):29-31

Rapoport A (1972) The uses of mathematical isomorphism in general system theory. In: Klir GJ (ed) Trends in general systems theory, John Wiley & Sons, pp. 42-77

Renner M (2004) Moving toward a less consumptive economy. In: Halweil B, Mastny L, Assadourian E, Starke L State of the world 2004, Starke L (ed), WW Norton and Co, pp 96-119

Rhodes SP (1995) An analysis of blue angel environmental seal-of-approval criteria for workstation computers. Proc IEEE Symp Electron Environ, 295-300

Richardson FL (1960) Arms and insecurity, a mathematical study of the causes and origins of war. Pittsburgh, Boxwood Press

Robinson PD (1968) Fourier and Laplace transforms. Routledge and Paul, London

Rorty R (1982) Consequences of pragmatism. University of Minnesota Press

Rorty R (1991) Essays on Heideger and others: philosophical papers. Vol 2 Cambridge University Press

SACTRA (1994) Trunk roads and the generation of traffic. Standing Advisory Committee on Trunk Road Assessment

Sage AP (1992) Systems Engineering.Wiley.

Salway AG, Murrells TP, Pye S, Watterson J, Milne R (2001). Greenhouse Gas inventories for England, Scotland, Wales and Northern Ireland: 1990, 1995, 1998 and 1999, AET Technology, http://www.aeat.co.uk/netcen/airqual/reports/ghg/regghg99_issue_1.pdf

Sarkis J, Nehman G, Priest J (1996) A systematic evaluation model for environmentally conscious business practices and strategy. Proc IEEE Int Symp Electron Environ, 281-286

SEPA (1997) Scotland, a blueprint for progress 1997-2001 draft national waste strategy. Scottish Environmental Protection Agency, Stirling

SEPA (2003a) National waste strategy Scotland, the national waste plan 2003. http://www.sepa.org.uk/pdf/nws/guidance/national_plan_2003.pdf

SEPA (2003b) National Waste Strategy: Scotland, Glasgow and Clyde Valley Area Waste Plan. http://www.sepa.org.uk/pdf/nws/areas/glasgow_clyde_valley/awp_full.pdf
ShamRao AD (2001) Shaping a Safety culture: core issues. Divaker
Smith D (1992) The Kegworth aircrash – a crisis in three phases?. Disaster Management, 4(2):63-72
Torbert WR (1987) Managing the corporate dream: restructuring for long-term success. Dow-Jones, Irwin
Ulrich H (1968) Die Unternehmung als produktives sociales System. St. Gallen
Vickers G (1965) The art of judgement: a study of Policy Making. Chapman and Hall
Vickers G (1973) Making institutions work. Associated Business Programmes, London
WCED (1987) Our common future, report of the world commission on environment and development. Oxford University Press
Weick KE (1995) Sensemaking in organisations. Sage Publications
Weinberg GM (1975) An Introduction to general systems theory. John Wiley and Sons
Westrum R (1988) Organizational and inter-organizational thought. World Bank
Wiest JD and Levy FK (1977) Management guide to PERT/CPM. Prentice Hall, Englewood Cliffs NJ
Woolman J (1774 [1818] The works of John Woolman in two parts (5th edn). Benjamin and Thomas Kile, Philadelphia
WWW8 (2005) http://www.bussvc.wisc.edu/swap
WWW9 (2005) http://www.defra.gov.uk/environment/waste/topics/paper.htm
Young M (1992) Sustainable investment and resource use. Parthenon, Park Ridge

Additional Reading for Part I

Balakrishnan AV (1983) Elements of state space theory of systems. Optimisation Software, New York
Beer S (1975) Platform for Change. John Wiley
Bertalanffy von L (1975) Perspectives on general systems theory. Braziller
Boulding KE (1985) The world as a total system. Sage Publications
Charter M and Tischner U (2001) Sustainable solutions, developing products and services for the future. Greenleaf Publishing
Checkland P (1981) Systems Theory, Systems Practice. John Wiley
Chen C-T (1993) Analog and digital control system design: transfer-function, state-space, and algebraic methods. Saunders College Publishers
Churchman C (1979) The systems approach and its enemies. Basic Books
CM (1991) Design issues in waste management. Dept Social Decision Sc., Carnegie Mellon University, Pittsburgh, PA.
Coogan CO (1993) Front end analysis. Proc IEEE Int Symp Elect Environ, 132-173

Denbigh P (1998) System analysis and signal processing: with emphasis on the use of MATLAB. Addison-Wesley
Dillon PS (1996) Sustainable production and consumption: a vision for the electronics industry. Proc IEEE Int Symp Elect Environ, 298-301.
Elohim JL (1991) Theory and practice of systems methodology. In: Trappl R, Klir GJ, Pichler FR (eds) Progress in cybernetics and systems research, vol 8, pp 109-118
Eyring G (1993) Policy implication of green product design. Proc IEEE Int Symp Elect Environ,160-163.
Feyerabend P (1982) Science in a free society. Verso/NLB, London
Forester J (1961) Industrial dynamics. Productivity Press
Forester J (1990) Principles of systems. Productivity Press
Garnham A (1985) Psycholinguistics. Methuen.
Hersh MA (1996) A comparison of discrete time system representations based on delta and w operators. IMA J of Mathematical Control and Information, 13:131-149
Hill JD, Warfield JD (1972) Unified program planning. IEEE Trans on Systems, Man and Cybernetics, SMC2:610-622.
Howard RA, Matheson JE (1984) Influence diagrams, readings on the principles and applications of decision analysis, vol II. Strategic Decisions Group, Palo Alto, pp 719-762
Janis IL (1989) Crucial Decisions: leadership in policymaking and crisis management. Free Press, New York
Janis IL, Mann L (1977) Decision making: a psychological analysis of conflict, choice and commitment. Free Press, New York
Jerri AJ (1992) Integral and discrete transforms with applications and error analysis. Marcel Dekker, New York
Johannsen G (1994) Integrated systems engineering: the challenging cross-discipline. IFAC Integrated Systems Engineering Conference, Baden-Baden, Germany
Kahn RL (1964), Organizational stress: studies in role conflict and ambiguity. Wiley
Kaposi AA, Rzevski G (1991) On the aims and scope of a system design methodology. In: Trappl R, Klir GJ, Pichler FR Progress in cybernetics and systems research, vol 8, pp. 123-127
Klir GJ (1972) An approach to general systems theory. Von Nostrand Reinhold
Kucera V (1991) Analysis and design of discrete linear control systems. Prentice Hall
Lewis H and Gertsakis J (2001) Design + environment, a global guide to designing greener goods. Greenleaf Publishing
Lukes S (1991) Moral conflict and politics. Clarendon Press, Oxford.
McGillem CD and Cooper GR (1991) Continuous and discrete signal and system analysis, 3rd. edn Oxford University Press.
March JG, Simon A (1958) Organisations. Wiley, New York.
Mizuki C, Sandborn PA, Pitts G (1996) Design for environment – a survey of current practices and tools. Proc IEEE Int Symp Electron Environ, 1-6.

Orchard RA (1972) On an approach to general systems theory. In: Klir GJ Trends in General Systems Theory. John Wiley and Sons, pp. 205-250.
OTA (1992) Green Products by design: choices for a cleaner environment. OTA-E-541
Perrow C (1984) Normal accidents. Basic Books, New York
Rasmussen J (1983) Skills, rules and knowledge; signals, signs and symbols, and other distinctions in human performance models. IEEE Trans on Systems, Man and Cybernetics, SMC13(3):257-266
Reed M (1991) Organizations and rationality: the odd couple. J. of Management Studies, 28:414-437
Sarkis J, Darnall NM, Nehman GI, Priest JW (1995) The role of supply chain management within the industrial ecosystem. IEEE Int Symp Electron Environ, 229-234
Shachter RD (1986) Evaluating influence diagrams. Operations Research, 34(6):871-882
Schein EH (1985) Organizational culture and leadership. Jossey-Bass, San Francisco
Segerberg T, Hedemalm P (1996) Life cycle assessment of solder and silver-epoxy conductive adhesive. Proc IEEE Int Electron Environ, 64-65
Shrivastava P (1987) Bhopal: anatomy of a crisis. Ballinger, Cambridge MA
Shrivastava P, Mitroff II, Miller D, Miglani A (1988) Understanding industrial crises. J. of Management Studies, 25(4):285-303
Skinner CW (1993) National energy modelling system. Government Information Quarterly, 10(1):41-51
Van Mier GP, C.J.L. M, Sterke M. and Stevels ALN (1996), Life cycle cost calculation and green design options, computer monitors as an example. Proc IEEE Int Symp Elect Environ, 191-196
Van Den Enden AWM and Verhoeck NAM (1989), Discrete-Time signal processing: an introduction. translated by Roberts DAE Prentice Hall.
Weick KE (1988) Enacted sensemaking in crisis situations. J of Management Studies, 25:305-317
Weick KE (1993) The collapse of sensemaking in organizations: the Mann Gulch disaster. Administrative Science Quarterly, 38:628-652
Weston A (1997) A Practical companion to ethics, Oxford University Press.
Workshop on Safety Control and Risk Management, Washington

Part II Decision Making and Multi-Criteria Optimisation

II.1 Introduction

Sustainable development is concerned with a range of complex problems resulting from the need to harmonise the sometimes conflicting requirements of human development and protection and conservation of the natural environment, while maintaining the ability of future generations to meet their own needs (WCED 1987). It also involves consideration of the sometimes conflicting policy goals of environmental integrity, economic efficiency and equity (Young 1992).

Thus decision making at different levels is an important component of sustainable development. This includes:

- Policy formulation in areas such as emissions control, energy and employment at the local, regional and national levels.
- Policy formulation by different organisations on areas such as energy and recycling.
- Practical decisions by organisations on how to, for instance, minimise energy consumption, encourage recycling, encourage employee fitness, reduce waste of all types and make premises and facilities accessible to all disabled people.
- Decisions by individuals on the ethics of accepting a job involving military work or animal experimentation or on holiday locations, activities and means of travel to minimise or, if possible, eliminate any resulting negative social and environmental impacts.

Most of these decisions involve consideration of different factors and tradeoffs, including of the costs involved. In some cases there are tradeoffs between different environmental considerations, for instance whether there are net environmental benefits in recycling household waste e.g. paper, glass jars and plastic bottles if this will involve long trips by car to a collection point or what is the 'environmental break-even point' with regards to weight for a given distance and type of vehicle.

Optimisation provides a range of mathematical techniques for obtaining the 'best' or 'optimal' solution to a problem. Optimisation techniques can be applied to a wide range of problems in science, engineering, economics, sustainable development, decision theory and medicine, amongst other application areas.

Thus there is clearly a range of problems associated with sustainable development where the use of optimisation, decision analysis and decision support methods can clarify the problem and underlying issues and, in many cases, be used to obtain an appropriate solution.

Part II of this book discusses the application of (multi-criteria) optimisation, decision analysis techniques and decision support systems to sustainable development. It is divided into five chapters (chapters 8-12). Chapter 8 introduces some of the basic concepts in optimisation, such as single and multi-criteria optimisation, objective function, decision variable and constraint and illustrates their application to sustainable development through brief examples. The derivation of mathematical expressions for real optimisation problems is discussed and illustrated by a power generation example. The minimisation of functions of one and n variables is considered and a brief overview of the single-criterion optimisation algorithms for functions of n variables is presented.

Chapter 9 introduces the mathematical background to decision making, including binary relations, preference relations and utility functions. It also discusses objectives, criteria and attributes. Binary and preference relations are illustrated by examples relating to energy and emissions and the derivation of utility functions for a decision making problem about reducing the environmental impacts of coal-fired power plants.

Chapter 10 presents an introduction to multi-criteria optimisation problems and discusses related issues, such as problem formulation and types of optimal solution. It also presents a classification of multi-criteria decision methods and gives a brief historical overview of the development of the subject. The discussion is illustrated by social and environmental criteria and objectives, a constructed scale for the attribute *noise* and the multi-stage decision problem of determining the 'best' energy strategy to meet energy needs and environmental objectives in the city of Glasgow in Scotland.

Chapter 11 presents four classes of multi-criteria decision methods:

- Outranking methods, including the ELECTRE and PROMETHEE methods.
- Methods based on aggregation of criteria, including the P/G% method and goal programming.
- The Analytic Hierarchy Process.

- Multi-attribute utility theory.

The four classes of methods presented in Chapter 11 are illustrated in chapter 12 by a case study of waste management options. Each of the methods discussed is applied to resolving a problem in management options for municipal waste to take account of resource consumption and environmental as well as cost considerations. Chapter 12 also contains tutorial exercises, which can be used to give readers practice in applying some of the techniques presented in Part II, as well as an opportunity to gain a deeper understanding of the methods and methodologies and their advantages and drawbacks.

II.2 Learning Objectives

Readers should gain an understanding of optimisation, decision theory and decision support systems and their role in analysing and solving problems in sustainable development. Specific learning objectives for Part II include the following:

- Understanding of the basic principles of optimisation.
- Knowledge of some optimisation methods.
- The ability to obtain mathematical expressions for optimisation problems.
- Knowledge of the ways optimisation techniques can be applied to increase understanding or solve sustainable development problems.
- Understanding of the differences between single and multi-criteria problems.
- Knowledge of the properties of binary and preference functions and their role in decision making.
- Understanding of utility functions and the ability to derive them.
- Understanding of objectives, criteria, alternatives, attributes and different types of optimal solution.
- Understanding of the classification of multi-criteria decision methods.
- Understanding of the differences between naturalistic and classical decision theory.
- Knowledge of and the ability to apply decision support methods, including outranking methods, such as ELECTRE and PROMETHEE, aggregation of criteria methods, such as P/G% and goal programming, the Analytic Hierarchy Process and multi-attribute utility theory.

- The ability to apply decision support methods to increasing understanding of and obtaining solutions to problems in sustainable development.

8 Optimisation

8.0 Learning Objectives

The main aims of this chapter are to introduce the basic concepts of optimisation and give readers an idea of some of their applications in the area of sustainable development. Specific learning objectives include the following:

- Understanding basic concepts, such as cost function, constraints, decision variables and single- and multi-criteria optimisation
- The ability to derive mathematical expressions for (simple) real optimisation problems.
- Knowledge of the conditions for the minimisation of functions of one and n variables.
- The ability to apply the techniques of bracketing and nested multiplication.
- Knowledge of several optimisation methods for functions of n variables.
- Understanding the applications of optimisation techniques to the analysis and solution of problems in sustainable development.

8.1 Introduction to Basic Ideas in Optimisation

Optimisation provides a range of mathematical techniques for obtaining the 'best' or 'optimal' solution to a problem. These techniques can be applied to a wide range of problems in science, engineering, economics, sustainable development, decision theory and medicine, amongst other application areas. This chapter provides an introduction to some of the basic concepts in optimisation. Examples of single and multi-criteria optimisation problems are presented in section 8.1 and the concepts of decision variables and parameters and cost or objective function are introduced in section 8.2. Section 8.3 discusses and presents an example of the derivation of a mathematical expression for a real optimisation problem. Practical issues in single criterion optimisation, including the conditions for the minimisation of a function of one and n variables and mathematical techniques for improving computational efficiency, are discussed in section 8.4. Section 8.5 provides a brief overview of optimisation algorithms for functions of n variables, section 8.6 introduces multi-criteria optimisation problems and section 8.7 sums ups the chapter.

Application of optimisation techniques first requires a mathematical model to be obtained and validated and sometimes also simplified. However real optimisation problems are often complicated and involve a large number of variables. Therefore care has to be taken when obtaining a simplified model to ensure that important variables are not omitted to avoid distorting the problem.

Optimisation problems can be classified as follows:

- Constrained or unconstrained
- Single variable or multivariable
- One criterion or multi-criteria
- Linear or non-linear.

A number of different types of optimisation methods are available for each of these types of problems. Choice of an appropriate method is therefore frequently an important component of problem solution. This requires understanding of a number of different methods, their limitations and the circumstances in which they do and do not work well. Two examples which illustrate the differences between single and multiple criteria optimisation problems are presented in section 8.1.1.

8.1.1 Single and Multi-Criteria Optimisation Problems

Example 8.1: Choice of Route

Consider the problem of choosing the route for a railway line. A number of factors need to be taken into account, including costs, line length, maximum gradient and negative environmental and social impacts. People living in the area close to the route will generally be concerned about noise disturbance and visual intrusion, as well as the possibility of compulsory purchase for demolition of their property. Although there may be a few people who benefit from compulsory purchase of their property, most people will find this very undesirable. Another important factor is choosing the route to avoid areas of scientific interest or natural beauty. It will be assumed that the only significant impact of line length is on costs and therefore line length does not need to be considered separately. It will also be assumed that there is a maximum permitted gradient and a budgetary limit on the total costs. Therefore the gradient and costs should be expressed as constraints rather than in terms of criteria. This gives the following criteria:

1. Minimising noise disturbance to people living close to the route.
2. Minimising visual intrusion to people living close to the route.
3. Minimising the length of route passing through areas of scientific interest or natural beauty.
4. Minimising the number of building compulsory purchase orders.

In this formulation the problem is a multi-criteria problem with two constraints. However the problem could also be expressed as a single criterion problem. For instance criteria 3 and 4 could be converted to constraints, including by requiring that the route does not pass through any areas of scientific interest or natural beauty and that compulsory purchase orders are totally avoided. The remaining two criteria could then be combined to give the single criterion: *Minimising the combination of noise disturbance and visual intrusion to people living close to the route.* Both noise disturbance and visual intrusion are qualitative variables and therefore could, for instance, be expressed on a dimensional scale from one to five. Rules would then be required for combining the values for these two variables. Another approach would be to again treat criteria 3 and 4 as constraints, and to assume that noise disturbance is much more serious than visual intrusion. This would give the single criterion problem of: *Minimising noise disturbance to people living close to the route.*

Example 8.2: Purchasing Strategy

Most organisations are involved in purchasing products, components and/or services and therefore need to make decisions about suppliers. Factors such as cost, reliability and quality are generally important and will initially be treated as constraints. In many cases there are also other important factors, including the following, though not all these issues will be relevant to all purchasing decisions:

1. The supplier's employment practices, including the employment of women, disabled people and people from ethnic minorities.
2. The distance travelled by components and products, with associated transport emissions and energy costs.
3. Any investment or other involvement by the supplier in oppressive regimes (which could either be listed or specified more precisely).
4. The use of animal ingredients or animal testing.

5. Any investment or other involvement in military research, the arms trade, tobacco companies, currency speculation or gambling.

It should be noted that not all these issues will be relevant to all types of suppliers.

This problem could again be expressed as either a single or multi-criterion optimisation problem. For instance factors 1, 3, 4 and 5 could be expressed as constraints. A constraint for factor 1 could specify the minimum acceptable employment practices, such as trade union recognition and more than minimal compliance with all employment legislation. Constraints for factor 3, 4 and 5 could be no investment or other involvement in oppressive regimes, military research and the other activities listed in factor 5 and no use of animal ingredients or animal testing. This would then give the single criterion problem: *Minimising the distance travelled by components and products* (in order to minimise the associated transport energy and emissions). Alternatively factors 3, 4 and 5 could be treated as constraints, giving the following multi-criteria problem:

- Selection of the best possible employment practices, including the employment of women, disabled people and people from ethnic minorities.
- Minimising the distance travelled by components and products, and consequently the associated emissions and energy costs.

A number of organisations operate what are called *contract compliance* conditions. These have the effect of constraints and specify the conditions to be met by suppliers. Decision making about suppliers who meet the conditions is then generally made on the single criterion basis of minimising costs.

Examples 8.1 and 8.2 have illustrated the differences between *single* and *multi-criteria* optimisation problems. In the single criterion case only one factor or criterion is taken into account and can therefore be maximised or minimised as appropriate. In the multi-criteria case there are several different factors or criteria, which have to be considered. The examples also show that many decision problems can be formulated in more than one way and indicate that a multi-criteria approach is often more appropriate and more flexible in sustainable decision making. In some cases the use of a single criterion can artificially limit the problem, whereas in others reducing the number of criteria through the use of constraints is very appropriate.

It is rarely possible to obtain a solution which maximises (or minimises) all the criteria at the same time, as if the variables were independent of each other. Therefore multi-criteria optimisation algorithms obtain solutions which are the best in some pre-defined sense and there may be more than one solution, depending on how 'best' is defined. Different types of optimal solution will be defined in section 10.1.3.

8.2 Some Basic Concepts

This section introduces and defines some important concepts, such as decision variables, parameters, cost functions and constraints.

8.2.1 Decision Variables and Parameters

Decision variables are the numerical variables in an optimisation problem for which values are to be chosen. The number of decision variables required in a particular problem may not always be clear and there may be more than one possible choice of the set of decision variables. It is generally necessary to decide which quantities are to be treated as variables and which quantities will have fixed values. Quantities with fixed values are called *parameters*. Quantities are treated as parameters for a number of reasons, including when a particular value is known to give good results or it is not possible or realistic to change its value for some reason. Any quantity, the values of which can be varied within a range, should be treated as a decision variable.

For instance, in making choices about sustainable uses of particular types of land, the total area is generally treated as a parameter, whereas the sizes of the areas assigned to particular uses should usually be treated as decision variables. However the results of decision choices as well as other factors may affect the total area, making it a time dependent parameter. For instance the area of mangrove forest in the Philippines has dropped from about 5000 squares kilometres historically to 2880 squares kilometres in 1970 and about 1234 square kilometres in 1993, due to the rapidly growing aquaculture industry (Janssen and Padilla 1999). However, although its value has changed over time, the area of mangrove forest in the Philippines can be treated as a parameter in a given decision problem. It is the areas to be assigned to, for instance, total preservation, subsistence forestry, commercial forestry and semi-intensive aquaculture that should be treated as decision variables.

8.2.2 Cost or Objective Function

Optimisation problems are frequently expressed in terms of minimising (or maximising) a particular function, called the *cost* or *objective function.* The cost function is a function of one or (generally in real problems) several variables. Therefore the aim of single criterion optimisation is to find the values of the variables which minimise or maximise the cost function as well as the minimum or maximum value of the cost function itself. Many *single criterion optimisation* algorithms are designed to minimise the cost function, but in some real problems it may be necessary to maximise rather than minimise some quantity h(x). In this case standard minimisation algorithms can be applied to the minimisation of a new function defined as $f(\underline{x}) = -h(\underline{x})$, since the value $\underline{x}^*$ at which the maximum value h^* of $h(\underline{x})$ is obtained is the same as the value at which the minimum value f^* of f(x) is obtained, as shown in fig. 8.1. Therefore minimising $f(\underline{x})$ will obtain the value $\underline{x}^*$ of $\underline{x}$ which also maximises $h(\underline{x})$. In addition the maximum value of $h(\underline{x})$, h^*, is the negative of the minimum value of $f(\underline{x})$, f^*, i.e.

$$h^* = \max_{\underline{x}} h(\underline{x}) = h(\underline{x}^*) = -\min_{\underline{x}} f(\underline{x}) = -f(\underline{x}^*) = -f^* \qquad (8.1)$$

Multi-criteria optimisation problems can also be expressed in terms of cost or objective functions, with the number of functions equal to the number of criteria. For instance, in decision making on different waste management options, criteria could include environmental impact criteria such as minimisation of transportation distance or emissions of sulphur oxides and carbon dioxide, and resource criteria, such as maximising the energy recovered and the percentage of materials recovered (Powell 1996). These objective functions are all quantitative and measured in specific units. In many cases quantitative data is not available or its use may not be appropriate and criteria are based on linguistic variables which are expressed in qualitative or linguistic terms, such as high, low or moderate. In this case it is more likely to be the rank order of the variables than their actual values which are significant. Examples of such qualitative objectives include maximisation of the ease of materials recovery in waste management problems and minimisation of community separation i.e. the extent to which communities will be divided by new roads in transport decision making.

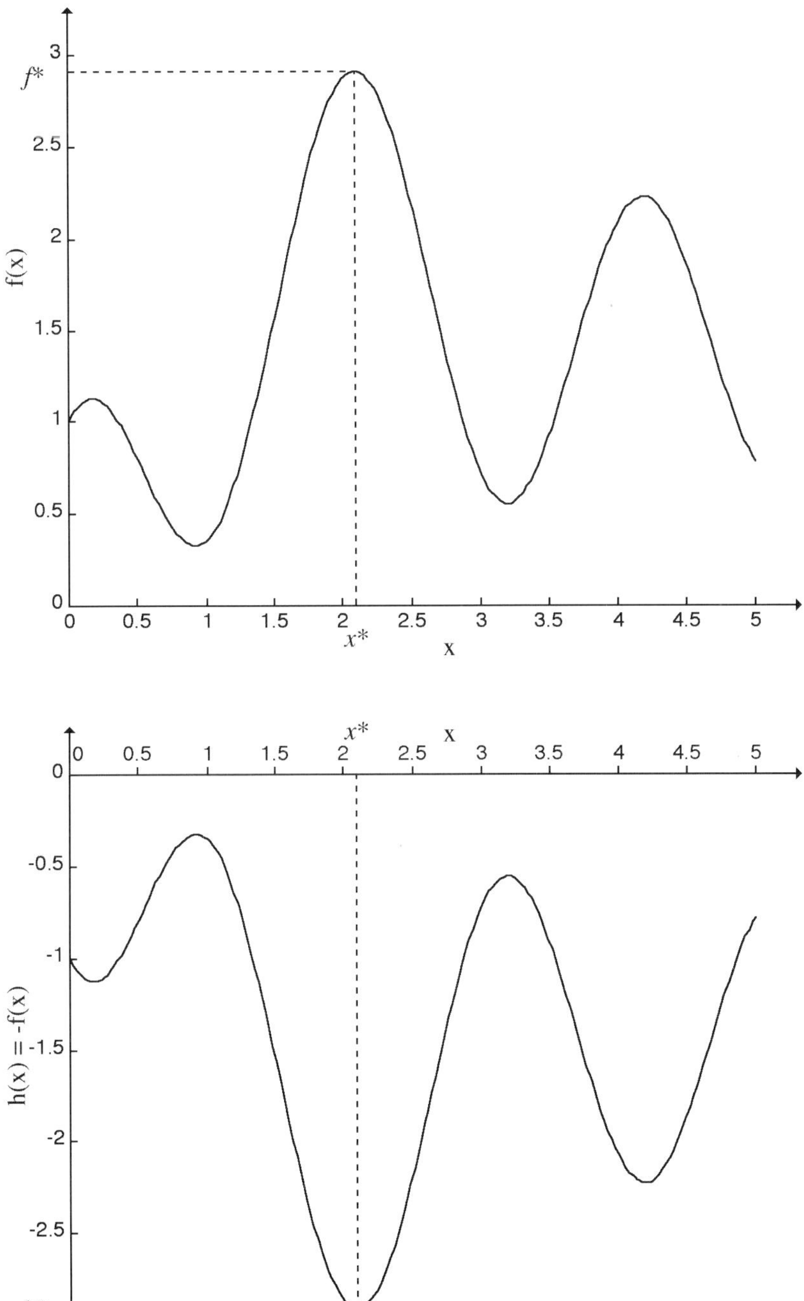

Fig. 8.1 Converting a maximisation to a minimisation problem

8.2.3 Constraints

In most real problems there are conditions or bounds on the values the variables can take. These conditions are called *constraints*. One of the simplest examples of such constraints is the requirement that most real variables, such as energy consumption, the size of forested areas and time duration, must be positive. Other simple constraints result from the limitations of available resources of personnel, raw materials and finances, which must not be exceeded. Constraints of this type are generally expressed in the form of *inequalities*, which give upper bounds on the values the variables are allowed to take. Some variables may have constraints in the form of both *upper and lower bounds*. For instance, the population of a particular species in a given habitat generally has a lower limit for viability as well as an upper limit, so as not to exceed the available food supply. Constraints may involve more than one variable and may also be non-linear. Performance specifications and legislative requirements can both lead to non-linear constraints or constraints involving more than one variable. Examples of this type of constraint include the performance specifications (Melton 1993) for replacement materials for lead solder, such as functional strength, low fatigue in thermal cycling and moisture resistance.

Multi-criteria optimisation problems can sometimes be reduced to single criterion problems with constraints by expressing some of the objectives or criteria as constraints, generally through the use of upper or lower bounds. For instance the objective of maximising efficiency may be replaced by putting a lower bound, for instance 0.6, on the acceptable values of efficiency. However it may not be obvious which factors should be treated as objectives or criteria and which as constraints. In some cases changing the problem in this way will result in less satisfactory or even poor decisions, whereas in others constraints on some variables will arise naturally.

8.3 Derivation of a Mathematical Expression

The application of optimisation algorithms to an optimisation problem often requires the problem first to be expressed in a suitable *mathematical form*, to which the algorithm can then be applied. This includes the determination of decision variables and parameters and the derivation of objective and constraint functions to represent the criteria and constraints on the problem. As in other modelling problems, deriving an appropriate model requires understanding of the problem, including the type of results required, to determine what factors should be included in the model and the type of problem, including whether it is single or multi-criteria, linear or nonlinear and whether or not there are constraints.

In some cases the objective to be met will be very clear and the cost function will therefore result from expressing this objective in mathematical form. In other cases there will be a number of objectives. It is generally possible to include more than one objective in the cost function. Weights on the different terms can be used to express their relative importance. However care must be taken not to make the cost function too complicated by inclusion of too many different terms, as this can result in a function which is not sufficiently sensitive to changes in the variables and may make it difficult to obtain a minimum. In this case it is preferable to use multi-criteria optimisation. An objective can be made into a constraint by putting a *bound* on it, for instance by replacing the objective of minimising costs by the constraint that costs should be less than a particular value, such as 1000 euros, or the objective of maximising efficiency by the constraint that efficiency should be greater than a particular value, for instance 0.6. The choice of objectives will be discussed further in section 9.3.

In some cases mathematical expressions for the objective functions and constraints can be obtained by using *physical laws*. As in other situations where the real world is modelled, it may be necessary to use *approximations* or make *assumptions* in deriving the mathematical expressions.

Example 8.3: Environmental Impacts of Energy Generation

In the problem of the choice of additional power generating capacity, environmental criteria include reduction of the emissions of sulphur dioxide, nitrogen oxides, solid particulate matter and carbon dioxide. This problem can be considered either as a single criterion problem in which all the environmental criteria are combined into a single objective function using a weighted sum of appropriate damage terms or a multi-criteria problem in which the criteria are considered separately. The multi-criteria approach is generally preferable, as it does not reduce the available information and there may be difficulties in choosing appropriate weights, which could also vary over time (Kalika and Frant 1992).

The presentation here is a slight modification of Kalika and Frant's (1992) work. It is assumed that there are a number of alternative strategies for generating a prespecified energy capacity. Each alternative involves specific types of power stations in specific locations e.g. two wind farms in locations A and B and one micro hydro-power scheme in location C or a wind farm in location A and a combined heat and power station in location D. The emissions of a particular pollutant i vary with the type of power station, the location and the year t. These emissions can be expressed in appropriate units, generally metric tons or kilograms, as $X_j(i, k, t)$ where j denotes the particular alternative and minimisation will be carried out over all these alternatives. It is assumed that the jth alternative has n(j) power stations/locations. The parameter k indexes both the location and type of power station and its dependence on the alternative j is not indicated explicitly.

It is further assumed that the damage resulting from the ith pollutant in location/power station k in year t is obtained by multiplying $X_j(i, k, t)$ by a weighting factor K(i, k, t) which depends on the type of pollutant, the location/power station and year. This factor can be expressed as the ratio of the background concentration G(i,k,t) of the pollutant i in the atmosphere in year t and location k to the limiting standard L(i,k,t) for the pollutant i in the atmosphere in year t and location k, with both values expressed in micrograms per cubic metre of atmosphere, i.e.

$$K(i, k, t) = \frac{G(i, k, t)}{L(i, k, t)} \tag{8.2}$$

The total emissions of the ith pollutant for alternative j in year t is obtained by summing over all the power stations/locations, as follows:

$$f_j(t, i) = \sum_{k=1}^{n(j)} K(i, k, t) X_j(i, k, t) \tag{8.3a}$$

Decision making on the appropriate mix of types and locations/power stations is based on minimising the total emissions over a period of years, denoted $[t_o, t_f]$. Therefore the criterion for the ith pollutant is obtained by summing expression (8.3a) between t_o and t_f. A discount factor D_t can be used to allow the emissions in different years to be weighted differently, but it is preferable not to discount future emissions, which would give $D_t = 1$. The criterion $C_j(i)$ for the ith pollutant is therefore given by

$$C_j(i) = \sum_{t=t_o}^{t_f} D_t f_j(t, i) = \sum_{t=t_o}^{t_f} D_t \sum_{k=1}^{n(j)} K(i, k, t) X_j(i, k, t) \tag{8.3b}$$

where K(i, k, t) is obtained from equation (8.2). Expression (8.3b) is fairly general and can therefore be used for a wide range of different alternative generating strategies. The values of K(i, k, t) and X_j(i, k, t) can be obtained from measured data. This gives a multi-criteria optimisation problem based on minimising the criteria C_j(i) for all the pollutants.

8.4 Single Criterion Optimisation

This section discusses some of the practical issues in single criterion optimisation, including the conditions for minimisation of a function of one and n variables, methods for the minimisation of functions of one and n variables and mathematical techniques for improving computational efficiency.

8.4.1 Conditions for the Minimum of a Function of One Variable

Functions of one variable can have both a global minimum and one or more local minima. A *global minimum* is distinguished from a local minimum by taking the overall lowest function value, whereas a *local minimum* just has values which are lower than those at the surrounding points. This can be expressed in more mathematical terms by stating that a function f(x) has a global minimum at the point x_o if it satisfies the following three conditions, where the notation $f^{(m)}(.)$ is used for the mth derivation of the function and g(.) is used in addition to $f^{(1)}(.)$ for the gradient or first derivative:

1. $f(x) \geq f(x_o)$ for all x
2. $f^{(1)}(x_o) = g(x_o) = 0$
3. $f^{(2)}(x_o) > 0$ or $f^{(1)}(x_o) = 0$ (i = 1...m-1) and $f^{(m)}(x_o) > 0$ i.e. the first derivative which is non-zero at x_o is positive.

Both local and global minima satisfy conditions 2 and 3, whereas only global minima also satisfy condition 1. Therefore condition 1 can be used to distinguish between local and global minima by comparing function values for points satisfying conditions 2 and 3.

Fig. 8.2 illustrates a function with a local minimum at xa and a global minimum at xo.

8.4.2 Conditions for the Minimum of Functions of n Variables

In the case of functions of n variables, the gradient is now a *vector of partial derivatives* given by

$$\underline{g}(\underline{x}) = \underline{\nabla} f(\underline{x}) = \left[\frac{\partial f}{\partial x_1}, \frac{\partial f}{\partial x_2}, \ldots, \frac{\partial f}{\partial x_n}\right]^T \tag{8.4}$$

where $\frac{\partial f}{\partial x_j}$ denotes the partial derivative of $f(\underline{x})$ with respect to x_j, and the second derivative is now the *Hessian matrix* $G(\underline{x})$ of second order partial derivatives with ijth term given by

$$G_{ij} = \frac{\partial^2 f}{\partial x_i \partial x_j} \tag{8.5}$$

The function itself now has the form

$$f(\underline{x}) = f(x_1, x_2, \ldots, x_n) \tag{8.6}$$

Therefore the conditions for a global minimum of a function of one variable can be generalised to functions of n variables by taking account of

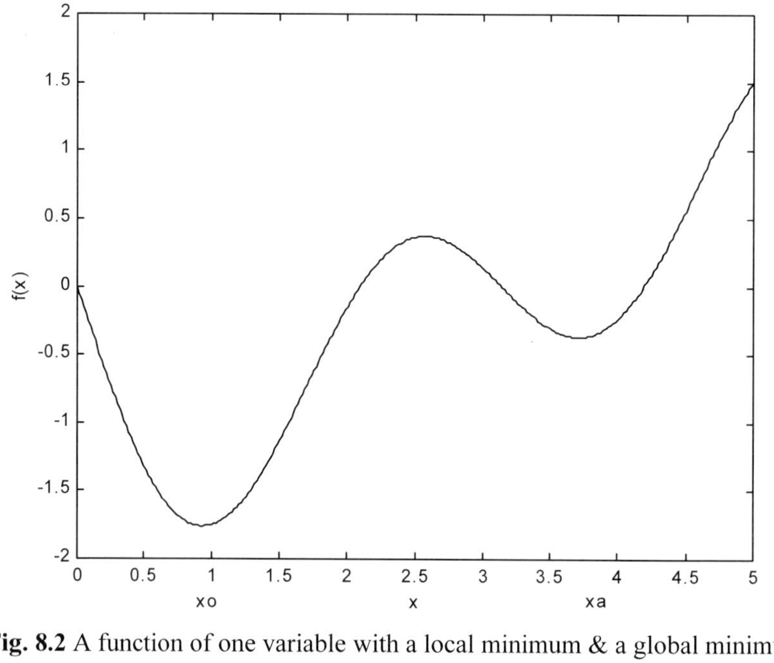

Fig. 8.2 A function of one variable with a local minimum & a global minimum

the form of the function, gradient and second order derivative. This gives the following three conditions for the function of n variables $f(\underline{x})$ to have a global minimum at $\underline{x}_o$:

1. $f(\underline{x}) \geq f(\underline{x}_o)$ for all $\underline{x}$
2. $\nabla g(x_o) = f(\underline{x}_o) = 0$ i.e $\dfrac{\partial f}{\partial x_j} = 0 \ (j=1,...,n)$
3. The Hessian matrix $G(\underline{x})$ of second order derivatives is positive definite at $\underline{x}_o$ i.e. all its eigenvalues are positive.

8.4.3 Bracketing

Many optimisation methods work best when the initial approximation is close to the minimum and it is often advisable to obtain a bracket for the minimum i.e. an interval in which the minimum is situated before applying an optimisation algorithm. The following method can be used to do this.

Consider the points x_1, $x_2 = x_1 + h$, $x_3 = x_2 + 2h$, $x_4 = x_3 + 4h$ ... for a particular step length h, with $f(x_i) = f_i$.

The points x_{i-1} and x_{i+1} bracket a minimum, as shown in fig. 8.3, i.e. the minimum is between x_{i-1} and x_{i+1} if

$$f(x_i) < f(x_{i-1}) \text{ and } f(x_i) < f(x_{i+1}) \qquad (8.7)$$

Example 8.4: Bracketing the Minimum of a Function

Bracket the minimum of $f(x) = -\tanh(x)/(1 + x^2)$. Take $x_1 = 0.1$ and $h = 0.1$

$x_1 = 0.1 \quad f_1 = -0.09868;$ $\qquad x_2 = 0.2 \quad f_2 = -0.18978;$
$x_3 = 0.4 \quad f_3 = -0.32754;$ $\qquad x_4 = 0.8 \quad f_4 = -0.49005;$
$x_5 = 1.6 \quad f_5 = -0.25890$

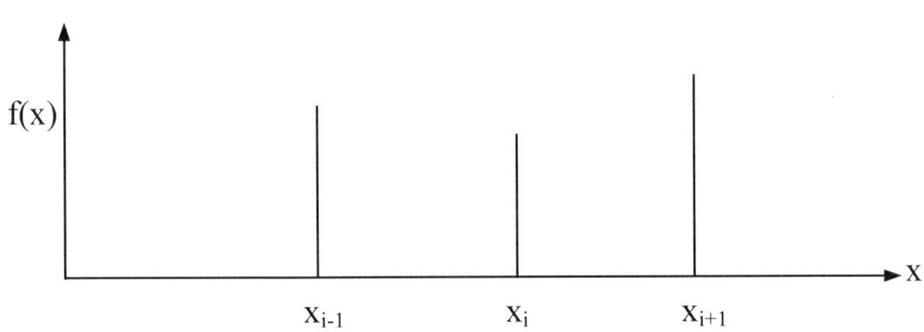

Fig. 8.3 Bracketing a minimum

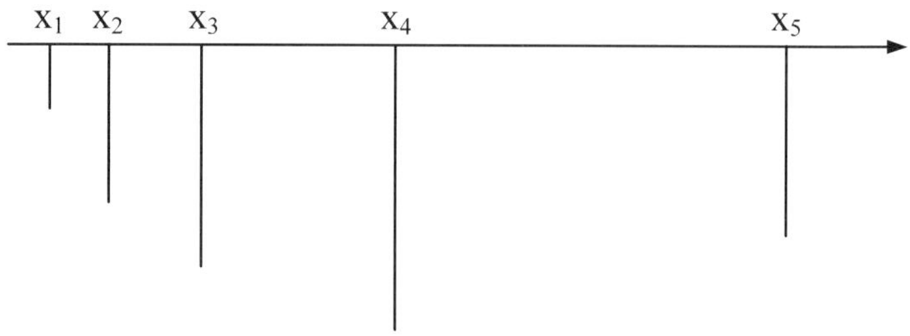

Fig. 8.4 Bracketing the minimum of $f(x) = -\tanh(x)/(1 + x^2)$

The minimum is bracketed by x_3 and x_5 i.e. by 0.4 and 1.6, as shown in fig. 8.4.

8.4.4 Performance of Single Criterion Optimisation Algorithms

Single criterion optimisation problems almost always have some *error*. This is due to roundoff and other numerical errors in computation, approximations in the model of the system and formulation of the problem and inaccuracies in the data, amongst other factors. Therefore, there are rarely exact answers to optimisation problems, and a good optimisation algorithm should give a meaningful solution to a wide range of problems with numerical results which are 'close' to the true minimum. The concept of 'error' is less clearly defined for multi-criteria problems, though numerical errors and errors in data and problem formulation are still relevant in this case.

In general it is the *relative error* i.e. the ratio of the error to the true solution which is important and not the *absolute value* of the error. To avoid dividing small numbers by each other, this relative error is often replaced by $|x_e - x_o|/(1+|x_o|)$, where x_o and x_e are respectively the true solution and an estimate of it. This expression is approximately equal to the relative error when the true solution is large and the absolute value of the error when the true solution is small.

Fig. 8.5 can be used to demonstrate the concepts of relative and absolute error. The measured values f(1) and f(3) are the true values, whereas f(2) and f(4) are estimates of them. It can be seen from the figure that the absolute error is about the same in the two cases, whereas the relative error is quite significant in the first case and fairly small in the second case. The absolute error in both cases is equal to 0.5. However in the first case the relative error is equal to 100% (0.5/0.5 x 100%), whereas in the second case it is only 5% (0.5/10 x 100%).

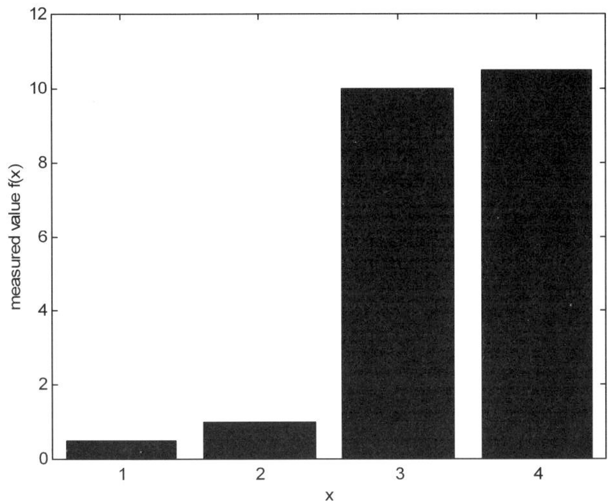

Fig. 8.5 Relative and Absolute Error

Since optimisation algorithms are iterative, a *termination criterion* is required in order to obtain a solution within a reasonable length of time. This is normally based on the closeness together of succeeding estimates or the smallness of the gradient of the function being minimised. The algorithm terminates when the termination criterion is less than some predetermined value ε. This value should be chosen appropriately to avoid indefinite looping if it is too small and insufficient accuracy if it is too large.

It is often advisable to check that approximately the same minimum is obtained when the algorithm is started from different initial points, particularly in the case of constrained minimisation problems, to ensure that the global rather than a local minimum has been found. Other checks on *accuracy* include applying different optimisation algorithms to the problem and varying the algorithm parameters, such as the desired accuracy, at each step.

8.4.5 Nested Multiplication

Numerical optimisation algorithms often require the repeated evaluation of polynomials. Therefore there is a need for numerical methods for evaluating polynomials efficiently to reduce computational time and memory requirement. One such algorithm is *nested multiplication.* It evaluates the polynomial of order n

$$p(x) = a_0 + a_1 x + a_2 x^2 + \dots + a_n x^n$$

at $x = z$ using an n-step iteration as

$$p(z) = b_n \quad \text{for any real } z \tag{8.8}$$

where the sequence $\{b_k\}$ is obtained from

$$b_0 = a_n, \quad b_1 = a_{n-1} + zb_0, \quad b_2 = a_{n-2} + zb_1, \dots, \quad b_n = a_0 + zb_{n-1} \tag{8.9}$$

i.e. start with the coefficient of the highest power, multiply by z, add the coefficient of the next highest power, multiply by z and continue until all the coefficients have been used.

8.4.6 Test Functions

Before applying an optimisation algorithm to a real problem, it is advisable to test its performance on a function with a *known minimum.* The *De Jong* (1975) *five function test suite* consists of five functions with different properties that can be used to investigate independently of each other some of the main difficulties that are found in optimisation problems. These functions are:

Sphere: $f_1(x) = \sum_{i=1}^{n} x_i^2 \quad -5.12 \le x_i \le 5.12$

$\min(f_1) = f_1(0, \dots, 0) = 0$

Sphere is smooth, unimodal and symmetric and does not have any specific problems. Therefore it can be used to test the general efficiency of an algorithm.

Generalised Rosenbrock's function:

$$f_2(x) = \sum_{i=1}^{n-1} \left[100(x_{i+1} - x_i^2)^2 - (x_i - 1)^2\right] \quad -5.12 \le x_i \le 5.12$$

$\min(f_2) = f_2(1, \dots, 1) = 0$

This function has a very narrow ridge with a sharp tip and runs round a parabola. Most optimisation algorithms converge very slowly and require a large number of iterations when trying to minimise this function, particularly if they are unable to discover good search directions. It is a generalisation of Rosenbrock's function $f_2(\underline{x}) = 100(x_2 - x_1^2)^2 + (1 - x_1)^2$, which is illustrated in fig. 8.6. Rosenbrock's function is a banana shaped function of two variables with a long valley and a minimum of 0 at (1, 1).

Step function: $f_3(x) = 6n + \sum_{i=1}^{n} [x_i] \quad -5.12 \le x_i \le 5.12$

$\min(f_3) = f_3[(-5.12,-5), ..., (-5.12,-5)] = 0$

The step function can be used to test the performance of optimisation algorithm on flat surfaces. Flat surfaces cause difficulties for optimisation functions, as they do not give any information on the best search direction. An algorithm can get stuck on a flat plateaux unless it has variable step sizes.

Quartic function with noise: $f_4(x) = \sum_{i=1}^{n} ix_i^4 + 1 - e^{-x^2/2\sigma^2}$

$-1.28 \le x_i \le 1.28$

$\min(f_4) = f_4(0, ..., 0) = 0$

This is a simple unimodal function with gaussian noise. Due to the noise, algorithms generally obtain different values when evaluating the same point repeatedly. Therefore algorithms which have trouble with this function are unsuitable for noisy data.

Shekel's Foxholes: $\frac{1}{f_5(x)} = \frac{1}{K} + \sum_{j=1}^{25} \frac{1}{c_j + \sum_{i=1}^{2} (x_i - a_{ij})^2}$

$-65.536 \le x_i \le 65.536$

$$(a_{ij}) = \begin{pmatrix} -32 & -16 & 0 & -16 & 32 & \cdots & 0 & 16 & 32 \\ -32 & -32 & -32 & -32 & -32 & \cdots & 32 & 32 & 32 \end{pmatrix}$$

$K = 500 \quad f(a_{1j}, a_{2j}) \approx c_j = j$
$\min(f_5) = f_5(-32,-32) \approx 1$

This is an example of a function with many (in this case 25) local optima. Many standard optimisation algorithms get stuck in the first peak they encounter rather than looking for the global optimum.

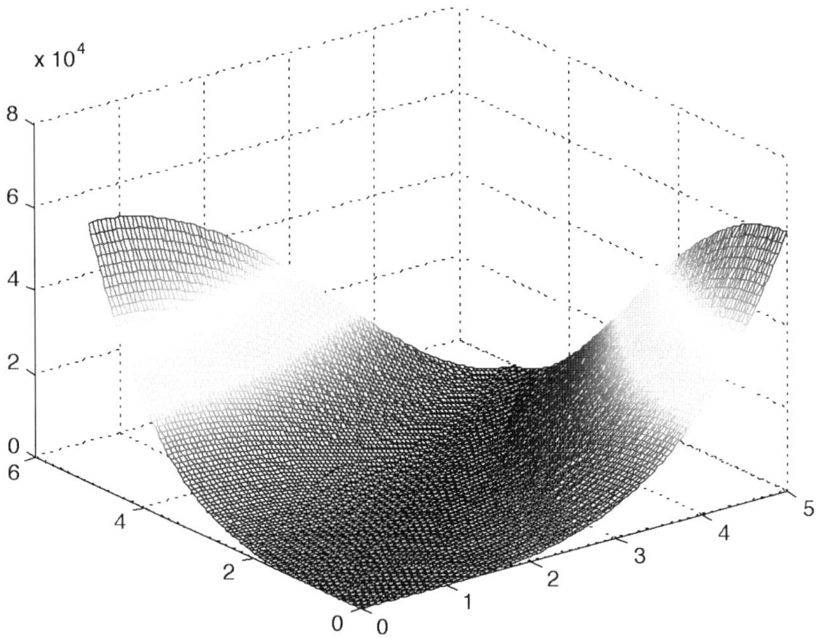

Fig. 8.6 Rosenbrock's function

Test functions have been developed only for single criterion, but not for multi-criteria optimisation algorithms, which do not have a single well-defined global optimum for a given problem. However, even in the multi-criteria case, algorithm performance should be tested before the algorithm is applied to real problems, possibly by applying it to artificially constructed problems with known solutions.

8.4.7 Minimisation of Functions of One Variable

Methods for minimising functions of one variable can be divided into the following three categories:

1. Algorithms for solving $f'(x) = 0$. These include Newton's method.
2. Search algorithms which find an interval containing the minimum by comparison of function values. These include the Fibonacci and golden search methods.
3. Curve fitting algorithms which fit a polynomial to the function and approximate the minimum by the minimum of this function.

These methods are fairly straightforward and will therefore not be discussed further here. More information can be found in (Beale 1988; Bunday and Garside 1986).

8.5 Functions of n Variables

This section presents a brief overview of optimisation algorithms for functions of n variables. Further information can be obtained in (Beale 1988; Bunday and Garside 1986). Optimisation algorithms for functions of several variables are generally based on minimisation in a number of different directions sequentially and differ from each other largely on the basis of how these directions are chosen. On this basis optimisation methods for functions of several variables can be divided into two categories:

1. Direct methods
2. Gradient methods.

Direct methods require only function values and do not need any information about the derivatives, whereas *gradient methods* use information about the derivatives to determine a series of directions to minimise the function in. Direct methods include Hooke and Jeeves Method and simplex methods. *Hooke and Jeeves Method* has two main parts: a sequence of *exploration steps* about a *base point* to determine the direction the minimum is likely to be in followed by a *pattern move* in this direction.

Simplex methods are based on comparing values at the vertices of a simplex. A regular simplex in n-dimensional space is a figure formed by n+1 equidistant points e.g. an equilateral triangle in two dimensions. The values of the function at the n+1 vertices are compared and the simplex is moved in the direction of the vertex with the lowest function value. In *Nelder and Mead's method* the simplex is allowed to become irregular and is moved towards the lowest function value using the three operations of *reflection*, *expansion* (with expansion coefficient $e > 1$) and *contraction* (with contraction coefficient $c < 1$). Fig. 8.7 illustrates the points of contraction ($\underline{x}_c$), reflection ($\underline{x}_r$) and expansion ($\underline{x}_e$) relative to the centroid of the vertices ($\underline{x}_o$) for an initial point $\underline{x}$. The point of reflection is an equal distance the other side of the centroid from the point $\underline{x}$, whereas the points of expansion and contraction are on the same side of the centroid as the point $\underline{x}$, with the point of contraction closer to the centroid and the point of

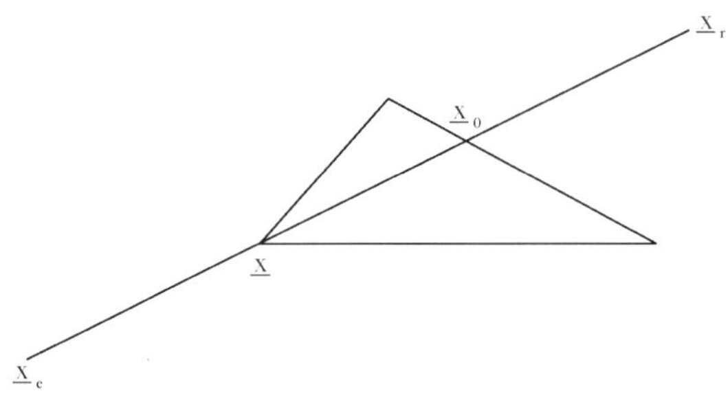

Fig. 8.7 Contraction, reflection and expansion

expansion further from the centroid than $\underline{x}$.

In general, functions of several variables can be minimised by performing a sequence of single variable searches. It is possible to obtain the minimum by searching along each axis in turn (method of alternating variables), but this is not an efficient procedure and is generally very slow. Therefore knowledge of the derivatives of the function and, in particular, the first derivative or gradient is used to determine the search direction, $\underline{h}_k$, as indicated by the name *gradient methods*. This gives an iterative procedure, with the following steps carried out at the kth iteration:

1. Determine the direction of search $\underline{h}_k$.
2. Find λ_k to minimise $f(\underline{x}_k + \lambda h_k)$ with respect to λ.
3. Set $\underline{x}_{k+1} = \underline{x}_k + \lambda_k \underline{h}_k$.

Different choices of the search direction $\underline{h}_k$ give different methods. For instance taking the direction $\underline{h}_k$ as the direction of the negative gradient gives the method of *steepest descent*, which is slow. The *Davidon Fletcher Powell* method is much faster and includes information on the second derivatives (Bunday and Garside 1986).

8.6 Introduction to Multi-Criteria Problems

Multi-criteria problems differ from single criterion ones in involving more than one criterion or objective. The terms maximum and minimum are clearly defined for single criterion problems and mathematical expressions for a point to be a local or global maximum or minimum can be obtained relatively easily, including in the case of many variables and/or many constraints. However *problem definition* is more complicated in the case of multi-criteria problems. In general there will not be a set of points or alternatives which minimise or maximise all the criteria, but rather some alternatives will minimise or maximise some criteria and other alternatives will minimise or maximise other criteria. Therefore some sort of tradeoff between the different criteria will generally be required. In a real application this will generally be based on the *preferences* of the decision maker(s) (and stakeholders). However many individuals do not have a clear preference structure.

There are a number of analogies between multi-criteria optimisation and multi-variable optimisation problems. However in the *multi-variable* case there is a global minimum, which is generally unique, though there may be several local minima as well, whereas in the *multi-criteria* case the optimum value obtained will generally depend on the relative importance given to the different criteria. In the multi-variable case, as in the single variable case, most optimisation algorithms are for continuous problems, whereas most multi-criteria algorithms are for discrete problems.

Decision support methods, including multi-criteria algorithms can help in structuring problems, identifying the issues on which decisions are required and the factors to be taken into consideration. However, a number of algorithms also allow investigation of the effects of changing the preference structure on the resulting decision, for instance by changing the weights assigned to different factors or variables. There are also a number of interactive algorithms which use a series of questions to elucidate the preference structure of the decision maker (Keeney 1980; Linden et al 1997). Many of these interactive algorithms assume consistency in the decision maker's responses and do not allow the preference structure to be updated, for instance in response to any inconsistencies in replies. There is therefore a role for decision support systems which give more support to all stages of decision making, as will be discussed further in section 10.1.1. Decision support systems, including multi-criteria algorithms, can be used to restructure the problem to make the underlying issues on which the decision is based clearer, identify good alternatives and replace more complex decisions by simpler choices. At their present state of development they cannot actually make decisions. Although decision making is difficult, there are advantages in decisions being made by people with the support of decision algorithms rather than the decisions algorithms being used to actually make decisions. These advantages include being able to take account of factors which cannot (easily) be incorporated into the decision problem considered by the algorithm and the importance of human involvement in the decision making process.

As will be discussed further in section 10.1.3, there are also a number of different types of 'best' or 'optimal' solution, including satisficing solutions, non-dominated solutions and Pareto optimal solutions. Each of these has its own advantages and disadvantages. In some cases the different types of solutions will give similar results, but in many cases the outcomes will be significantly different, so that the type of solution aimed for should depend on the particular problem.

8.7 Roundup of Chapter 8

Chapter 8 has introduced the basic ideas in optimisation and given a brief overview of the methods that can be used in single variable optimisation problems. The field of optimisation provides mathematical and computing techniques that can be used to obtain a solution to sustainable development and other problems, that is optimal or the best in some sense. Mathematical and other modelling techniques can frequently be used to obtain an appropriate form of the problem for applying optimisation techniques. The problem is generally stated in terms of the criteria to be satisfied. There can either be one criterion or more than one criterion, giving rise to single-criterion and multi-criteria optimisation problems respectively. Optimisation problems are constrained if there are restrictions or constraints on permissible solutions.

Single-criterion optimisation problems are generally expressed in terms of minimising a cost or objective function. Redefining the cost function $f(\underline{x})$ as the negative of another function $h(\underline{x})$ i.e. $f(\underline{x}) = -h(\underline{x})$ allows minimisation algorithms to be used when the aim is maximisation of a particular variable. In the single-criterion case there is generally a well-defined optimal solution. Conditions for the minimum of functions of one and n variables in terms of their (partial) derivatives can be used to check whether the minimum has been obtained. Single-criterion optimisation algorithms are generally applied first to test functions with a known minimum to check their performance.

There are three different types of minimisation algorithms for functions of one variable, based on finding zeros of the first derivative, obtaining an interval which contains the minimum (search algorithms) and curve fitting. Optimisation algorithms for functions of n variables are based on minimisation in a number of different directions sequentially and differ from each other largely on the basis of how these directions are chosen.

The introduction to optimisation techniques in this chapter is followed by a presentation of the mathematical background to decision making in chapter 9.

9 Mathematical Background to Decision Making

9.0 Learning Objectives

The main aim of this chapter is to introduce some of the mathematical techniques which form the background to decision support methods. Specific learning objectives include the following:

- Knowledge of the main properties of binary relations.
- Knowledge of the main properties of preference relations and an understanding of their role in decision support systems.
- The ability to define appropriate objectives, criteria and attributes for sustainable development problems.
- Knowledge of some of the properties of utility functions, including the forms they take for risk prone, risk averse and risk neutral decision makers, and an understanding of the role of utility functions in decision making.

9.1 Introduction

This chapter introduces the mathematical techniques which are required to understand decision support methods. Decision problems in sustainable development generally involve more than one criterion. Therefore there are frequently a number of possible solutions and the decision process will involve the subjective preferences of the decision maker in deciding between the different options. It is therefore useful to have a mathematical theory to represent preferences. This involves preference relations, which are a particular type of binary relation and presented in section 9.2. Defining a decision problem involves the use of objectives, criteria and attributes which are discussed in section 9.3. Understanding of the mathematical theory of utility and the necessary (and sufficient) conditions for a multi-attribute utility function to be represented in multi-linear form is required for the application of multi-attribute utility theory, one of the methods considered in chapter 12. Utility theory is discussed in section 9.4 and section 9.5 sums up the chapter.

9.2 Binary and Preference Relations

9.2.1 Binary Relations

Since the preference structure of decision makers plays an important role in multi-criteria optimisation, the mathematical representation of preference structures will now be considered. It is generally only the pairwise preference i.e. the preference between two rather than multiple items that is modelled. Therefore preference is generally modelled in terms of binary relations. A *binary relation* is a relation between two elements of a set. It can therefore be defined (Yu 1985) as a subset R of the cross product of the set with itself, Y x Y. The notation $(y^1, y^2) \in R$ means that two elements y^1 and y^2 of Y satisfy the binary relation R. The main properties of binary relations will now be presented. All these properties are used in discussing the preference structure of decision makers and the conditions under which all the options can be ordered in increasing or decreasing rank order or the conditions for a function to be a utility function. This statement of the properties of binary relations is also useful for reference when reading the literature on decision making.

A binary relation R on Y is (Klir and Yuan 1995):

1. *reflexive* if $(y, y) \in R$ for every $y \in Y$, *irreflexive* if $(y, y) \notin R$ for some $y \in Y$, and *antireflexive* if $(y, y) \notin R$ for all $y \in Y$.

2. *symmetric* if $(y^1, y^2) \in R \Rightarrow (y^2, y^1) \in R$ for all $y^1,y^2 \in Y$, *asymmetric* if $(y^1, y^2) \in R \Rightarrow (y^2, y^1) \notin R$ for all $y^1,y^2 \in Y$, and *antisymmetric* if $(y^1, y^2) \in R$ and $(y^2, y^1) \in R \Rightarrow y^1 = y^2$ for all $y^1,y^2 \in Y$.

3. *transitive* if $(y^1, y^2) \in R$ and $(y^2, y^3) \in R \Rightarrow (y^1, y^3) \in R$ for all $y^1,y^2,y^3 \in Y$, *nontransitive* if $(y^1, y^2) \in R$ and $(y^2, y^3) \in R \Rightarrow (y^1, y^3) \notin R$ for some $y^1,y^2,y^3 \in Y$ and *antitransitive* if $(y^1, y^2) \in R$ and $(y^2, y^3) \in R \Rightarrow (y^1, y^3) \notin R$ for all $y^1,y^2,y^3 \in Y$.

4. *complete* or *connected* if $(y^1, y^2) \in R$ or $(y^2, y^1) \in R$ for all $y^1,y^2 \in Y$ for which $y^1 \neq y^2$; otherwise it is incomplete or unconnected.

5. an *equivalence relation* if it is reflective, symmetric and transitive.

6. a *preorder* if it is reflexive and transitive.

7. an *order* if it is irreflexive and transitive.

Therefore a binary relation is *reflexive* if the relation holds between each element and itself, *irreflexive* if some, but not all elements have the relation with themselves and *anti-reflexive* if no element has the relation with itself. It is *symmetric* if the binary relation holds independently of the order of the two elements in the pair of elements and *asymmetric* if the order matters i.e. the relation holds for one ordering, but not the other. It is *antisymmetric* if, whenever the relation holds between a pair of elements independently of the order, the two elements are equal. The binary relation is *transitive* if, whenever the relation holds between two pairs of elements (y^1 and y^2) and (y^2 and y^3), the relation also holds between the pair of elements y^1 and y^3. It is *nontransitive* if the relation does not hold for at least one pair of elements y^1 and y^3 when it holds for both the pairs (y^1 and y^2) and (y^2 and y^3) and *antitransitive* when it does not hold for any pair of elements y^1 and y^3 when the relation holds for the pairs of relations (y^1 and y^2) and (y^2 and y^3). A binary relation is *complete* or *connected* if it can be used to connect all pairs of distinct elements (with the order of the elements in the relation unimportant).

Example 9.1: Processes in the Chemical Industry

Consider the following two simple binary relations which are defined on the set Y of processes in the chemical industry

- R_1: generation of less emissions of nitrogen oxides
- R_2: consumption of less energy

Neither relation R_1 nor R_2 is reflexive, as a process cannot generate less emissions or consume less energy than itself. Both relations R_1 and R_2 are asymmetric: if $(y^1, y^2) \in R_1$, then process 1 generates less emissions than process 2, so the reverse cannot hold. Similarly if $(y^1, y^2) \in R_2$, then process 1 consumes less energy that process 2, so the reverse cannot hold.

Both relations are transitive: $(y^1, y^2) \in R_1$ and $(y^2, y^3) \in R_1$ implies that process 1 generates less emissions than process 2 and process 2 generates less emissions than process 3. It therefore follows that process 1 generates less emissions than process 3. Similarly $(y^1, y^2) \in R_2$ and $(y^2, y^3) \in R_2$ implies that process 1 consumes less energy than process 2 and process 2 consumes less energy than process 3. It therefore follows that process 1 consumes less energy than process 3.

Therefore neither relation is an equivalence relation or a preorder, as both relations are transitive, but not reflexive or symmetric.

When R is an equivalence relation, for any given $y \in Y$, the subset of elements $x \in Y$, for which $(x, y) \in R$ is an *equivalence class*. Equivalence classes are disjoint and their union is equal to the whole set, Y, so that they give a partition of Y. Therefore equivalence classes could be used to divide up the set of options into groups of options which are equally preferred by the decision maker.

9.2.2 Preference Relations

Preference relations are a particular type of binary relation that can be used to describe the preferences of a decision maker for a pair of options or alternatives. Therefore preference relations only allow two rather than multiple options to be compared at one time. If the possible outcomes of a decision are considered one pair at a time, for any pair y^1, y^2, then the decision maker's preference for the two options is described by one and only one of the following:

1. y^1 is better than or preferred to y^2, denoted by $y^1 \succ y^2$.
2. y^1 is worse than or less preferred to y^2, denoted by $y^1 \prec y^2$.
3. y^1 is equivalent, equally preferred or indifferent to y^2, denoted by $y^1 \sim y^2$ or $y^1 \approx y^2$.
4. there is insufficient information to determine the preference structure, denoted by $y^1 ? y^2$, so that the preference between y^1 and y^2 is indefinite or has not yet been clarified.

Therefore the four operations $\succ$ (preferred to), $\prec$ (less preferred to), $\sim$ (equally preferred) and ? (insufficient information) can be used to describe the preferences of a decision maker. Since each of these operations is a binary relation it describes a subset of Y x Y. The four subsets are denoted by $\{\succ\}$, $\{\prec\}$, $\{\sim\}$ and $\{?\}$. Since, for any pair of options either one is preferred to the other, they are equally preferred or there is insufficient information to determine the preference structure, $\{\succ\}$, $\{\prec\}$, $\{\sim\}$ and $\{?\}$ form a partition of Y x Y. This means that for each element y_i in Y all the elements can be divided into four classes of elements that are equally preferred, more preferred and less preferred than y_i and where there is insufficient information to decide. In addition, the further preferences $\{\succeq\} = \{\succ\} + \{\sim\}$ and $\{\preceq\} = \{\prec\} + \{\sim\}$ can be defined, where $y^1 \succeq y^2$ implies that y^1 is either preferred to or indifferent to y^2 and $y^1 \preceq y^2$ implies that y^1 is either less preferred to or indifferent to y^2.

Example 9.2: Preference Relations on Industrial Processes

Consider the set Y = {y: y = (x, z) of processes that consume x GJ of energy per annum and produce z metric tons of waste per annum}, with $y_1 = (250, 5800)$, $y_2 = (300, 6000)$, $y_3 = (300, 5500)$ $y_4 = (255, 5750)$.

Then $y_1 \succ y_2$ and $y_2 \prec y_1$ as process y_1 consumes less energy and produces less waste than process y_2.

y_1 ? y_3, as process y_1 consumes less energy, but produces more waste than process y_3, so additional information would be required to determine which of these processes is preferred.

y_1 ? y_4, as process y_1 consumes less energy, and produces more waste than process y_4.

Similarly $y_2 \prec y_3$, $y_2 \prec y_4$ and y_3 ? y_4.

These preference relations are illustrated in fig. 9.1, where variables to the left and lower down are preferred to those to the right and higher up.

9.2.3 Properties of Preference Relations

The preference $\{\succ\}$ is in general irreflexive, since an option cannot be preferred to itself, and asymmetric, as the preferences $\{\succ\}$, $\{\prec\}$, $\{\sim\}$ and {?} are mutually exclusive, so that either y^1 is preferred to y^2 or y^2 is preferred to y^1, but not both. Similarly the preference $\{\prec\}$ is in general irreflexive and asymmetric. If $(y^1,y^2) \in \{\succ\}$, then $(y^2,y^1) \in \{\prec\}$ i.e. y^1 is preferred to y^2 implies that y^2 is less preferred than y^1. A preference $\{\prec\}$ or $\{\succ\}$ is a *partial order* if it is transitive and a *weak order* if its induced preference $\{\preceq\}$ or $\{\succeq\}$ is also transitive. A weak order that is also complete or connected is called a *strict order*. Despite its rather misleading name, weak ordering is in fact a very strong property (Yu, 1985). When it holds, all the different options can be ranked in increasing or decreasing order of preference. However, although transitivity of preferences seems to be necessary for logical consistency, in many cases it does not hold. Therefore preference of option 1 to option 2 and option 2 to option 3 does not necessarily imply that option 1 is preferred to option 3.

It is often assumed that 'rational' decision makers have a *preference structure* for which assumptions 1 and 2 and either 3a or 3b hold:

1. $\sim$ and $\succ$ are disjoint.
2. $\sim$ is reflexive and symmetric.

3a. $\succeq$ is transitive.
3b. $\succ$ is asymmetric and transitive.

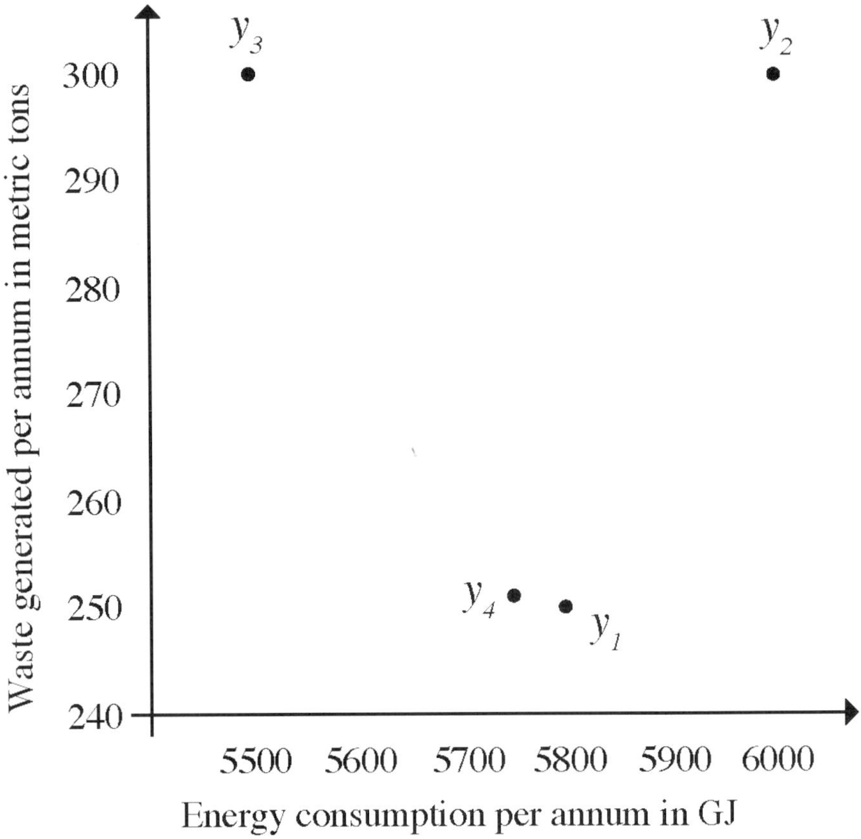

Fig. 9.1 Preference relations

Assumption 1 means that the decision maker is clear about whether two options are equally preferred or one is preferred to the other. Assumption 2 means that each option is equally preferred to itself and that, if option 2 is equally preferred to option 1, then option 1 is equally preferred to option 2. Assumption 3a means that, if option 1 is more or equally preferred to option 2 and option 2 is more or equally preferred to option 3, then option 1 is more or equally preferred to option 3. Assumption 3b means that, if option 1 is preferred to option 2 and option 2 is preferred to option 3, then option 1 is preferred to option 3, and if option 1 is preferred to option 2, then option 2 is less preferred than option 1. Assumption 2 seems almost tautologous, whereas assumption 1 may not always hold, as decision makers may be unsure whether two options are equally preferred or one is slightly preferred to the other. In assumption 3a both preference and indifference are transitive, whereas in assumption 3b only preference is transitive.

Preference relations do not distinguish between *slight* and *strong preference* i.e. whether one option is very slightly or strongly preferred to another. This is clearly a disadvantage. In decision problems where it is important to distinguish between strong and weak preference this could be done either by defining an additional operator to represent strong preference or using $\succ$ to denote strong preference and $\succeq$ to denote slight preference. However the mathematical structure would probably have to be reworked before using either of these options.

When assumptions 1, 2 and 3a hold, $\succ$ is an order, $\sim$ is an equivalence relation and $\succeq$ is a preorder. When assumptions 1, 2, and 3b hold the preference structure is a partial order. It has been assumed that the preference $\succ$ or $\succeq$ is transitive. However in some cases the assumption of transitivity is either false or too strong. This can occur for two main types of reasons:

- Comparison of options when one or more variables have very small increments or preference is slight rather than strong.
- Comparison of very different types of quantity, so that it is not meaningful to express (relative) preferences for all the options.

Inclusion of degrees of preference would probably remove problems of non-transitivity of preference structures, but would have the problem of making preference relations more complicated.

Example 9.3: Nontransitive Relation

As an example of the case of variables with very small increments, consider the number of lead particles in the air, as a result of, for instance, car exhausts. Clearly the general preference is for totally lead-free air. However most people would be indifferent between levels of 0.31 and 0.32 micrograms per cubic metre, both of which are well below the World Health Organisation's (WHO) recommended maximum of between 0.5 and 1.0 micrograms per cubic metre. Similarly indifference between levels of 0.32 and 0.33 micrograms per cubic metre can be assumed. However continually increasing the level by 0.01 microgram per cubic metre will lead to the deduction of indifference between levels of 0.31 and 1.2 micrograms per cubic metre, one of which is below and the other above the WHO recommended maximum, and this is clearly false. The problem here is the fact that a large multiple of a very small increment is no longer very small, but can become significant.

The difference between assumptions 3a and 3b relates to transitivity of indifference. As can be seen from the example, this is the case where transitivity most frequently breaks down. This non-transitivity can be taken into account by defining a *threshold of indifference*. Two options a and b can be considered indifferent, written a I b or a $\simeq$ b if and only if $|g(a) - g(b)| \leq s$, where s is the threshold of indifference and the function g(.) is an appropriate measure of preference. In example 9.2 it would be more logical to take $y_1 \simeq y_4$ rather than y_1 ? y_4, as, in relative terms, process y_1 consumes only slightly less energy, and produces only slightly more waste than process y_4, making their performance approximately equivalent. This can be achieved by choosing the thresholds of 5 GJ and 50 tonnes of waste. This example also illustrates the fact that threshold values are not necessarily small in absolute terms, though they should generally be small relative to other variables in the problem, particularly those being compared using the threshold.

9.3 Objectives, Criteria and Attributes

This section defines and discusses a number of concepts, such as objectives, criteria, attributes and alternatives (Keeney 1980; Keeney and Raiffa 1993), which are important in multi-criteria optimisation.

9.3.1 Objectives

The terms objectives and criteria are often used interchangeably. However here the term *objectives* will be used to represent what decision makers (and other stakeholders) wish to achieve by carrying out a particular decision and *criteria* to represent the more formalised, possibly mathematical, expression of one or more of the objectives used in decision analysis. Although achievement of objectives should be the main purpose of any decision, in many cases objectives are not clearly expressed. Objectives are generally characterised by three features:

- a decision context.
- an object.
- a direction or some other measure of relative preferences.

Objectives can be divided into fundamental objectives and means objectives. A *fundamental objective* describes one of the essential aims of the decision situation, whereas a *means objective* is important as a means of achieving a fundamental objective. The decision context is most easily specified by the activity or activities under consideration. The decision context and the fundamental objectives should be compatible. Individuals and organisations generally have strategic objectives, which should guide all decision making activities. In the context of sustainable decision making these strategic objectives will include choosing projects and processes and making other decisions which promote sustainable development.

Objectives generally express the underlying values of decision makers (and stakeholders) in the decision context. Therefore sustainable decision making will generally include objectives in one or more of the following categories:

- Preserving and enhancing the natural environment.
- Improving health and safety.
- Equity concerns.
- Social justice.
- Human and other rights, frequently of particular groups such as disabled people or in specific contexts, such as at work.
- Employment creation and the terms and conditions of employment.
- Positive rather than negative social impacts.
- Positive developmental impacts.
- Acceptability to all or particular groups of stakeholders.

In order to cover all the issues and to avoid missing or excluding any issues or narrowing down the decision context, the initial list of objectives should be as broad and all inclusive as possible. After the initial list has been drawn up it can be made more manageable by deleting redundant or repeated objectives, combining objectives where appropriate and structuring objectives. A full list of objectives can be obtained by using a number of different approaches, including the following:

- Drawing up a full list of categories of objectives, including those in the list above, and including all the objectives in each of the categories.
- Involvement of all stakeholders to obtain their objectives.
- Brainstorming i.e. listing all possible objectives, regardless of relevance.
- Consideration of consequences, goals, constraints and different perspectives.

The objectives can be structured by identifying fundamental objectives and linking objectives through means-ends relationships. This can be done by considering why each objective is important in the decision context. This will generally lead to the identification of fundamental objectives as important in their own right as one of the essential reasons for interest in the decision and lead to a chain of means-end relations for the other objectives.

Example 9.4: Fundamental and Means Objectives

Examples of *fundamental objectives* include the following:

- Reduction of pollution.
- Maintenance of biodiversity.
- Reduction of waste.
- Improvement of education.

Examples of *means objectives* for achieving the fundamental objective of reduction of pollution include the following:

- Reduction of carbon dioxide emissions from a particular industrial process.
- Building new cycle routes.
- A modal shift from car use to public transport.

Examples of *means objectives* for achieving the fundamental objective of improvement of education include the following:

- Building a new school in a particular area.
- Providing adequate financial support for all students.
- Improving the pay of teachers and lecturers.

Examples of *means objectives* for achieving the fundamental objective of maintenance of biodiversity include the following:

- Protection of a particular endangered species, such as the panda or snow leopard (WWW10 2005), illustrated in figures 9.2 and 9.3.
- Preserving habitats.
- Avoiding monoculture or the cultivation of only one variety of one species of plant. Monoculture reduces species resistance to disease. Crop diversity, including the maintenance of traditional crop varieties, is important for global food security and wild relatives of crops continue to be used to increase disease resistance and vigour (Tuxill 1999). Monoculture was largely responsible for the scale of the deaths in the Irish potato famine in 1846-50 where up to one million people out of a population of eight million died from famine and disease, and emigration and other factors reduced the population after the famine to five million.
- Not introducing exotic species or species which are not native to the area, as they could have an unexpected and possibly devastating impact

Fig. 9.2 Example of an endangered species: panda

on local ecosystems. The introduction of the rabbit to Australia illustrates some of the problems that can occur in terms of habitat destruction and extinction of native fauna and flora.

In defining objectives an appropriate level of generality or specificity is required, as this will largely determine the types of options that are appropriate. For instance, with regards to education, if the objective is to improve education in general the options could include policy and financial measures such as doubling the education budget, increasing the pay of all education workers and a range of measures to make it easier for people to return to college or university in later life. However if the aim is improving the education of children between 5 and 12 in a particular area the options would be more specific and could include recruiting more teachers and improving the facilities at schools (and pre-schools) serving the 5-12 year age group in that area and might include building one or more new schools. Options which improved education in general might have negligible impact on the education of 5-12 year olds in the area of interest.

9.3.2 Criteria

A *criterion* can generally be represented as a real valued function g(.) on the set A of *alternatives*. It allows two (or more) alternatives a and b $\in$ A to be compared on the basis of the two numbers g(a) and g(b). In a single criterion problem one criterion is used to describe the problem. This criterion may be obtained by focussing on the most important aspects of the problem or by combining different factors or subsuming them into the single criterion. A criterion can be modelled as a preorder i.e. a reflexive and transitive relation which expresses the preferences of the decision maker on the set of options and only considers one dimension of the problem, called an *attribute* or *evaluation axis*. A criterion can often be used as a measure of the extent to which an *objective is satisfied*. Criteria often involve consideration of particular attributes. Criteria can be defined for both means and fundamental objectives. Since implementation is generally in terms of means rather than fundamental objectives, it may be appropriate to first consider criteria for means objectives and then the extent to which satisfaction of these criteria leads to satisfaction of (criteria for) the fundamental objectives. Two detailed examples of single and multi-criteria problems are given in examples 8.1 and 8.2 in section 8.1.1.

Other examples of a *single criterion* include minimisation of the following linear functions:

- Carbon dioxide emissions from power sources in Japan (Amagai and Leung 1991).
- The weighted sum of sulphur dioxide and nitrogen oxide emissions, with the weights based on the emission rates for oil and coal fired power plants in Japan (Amagai and Leung 1989).
- A weighted sum of the environmental damage due to sulphur dioxide, nitrogen oxides and solid particulate matter (Chattopadhyay 1994).

Other examples of *multiple criteria* include the following:

- Minimisation of the environmental damage due to each of sulphur dioxide, nitrogen oxides and solid particulate matter separately in power generation choices (Kalika and Frant 1992).
- Minimisation of the change in oxygen content, increase in humidity, disturbance of slope stability, flooding of forest tracks and flooding of bogs in decisions on hydro power development (Khrisanov et al 1990).
- Maximisation of air quality and water quality, minimisation of clean-up costs, minimisation of the effects on terrestrial ecology, aquatic ecology and aesthetic and scenic value and minimisation of increases in unemployment in acid rain policy choices (Anandalingam 1987).

In some cases composite multiple criteria are used, for instance composite economic, environmental and socio-economic criteria. As the above examples illustrate, criteria are generally more specific and precisely formulated than objectives and should be measurable in either quantitative or qualitative terms. However in some cases objectives or combinations of objectives can be translated directly into criteria.

Example 9.5: Criteria for the Objective of Species Preservation

Criteria for the means objective of preserving a particular species include the following:

- Increasing the overall number of members of the species.
- Increasing the size and numbers of the regions or areas with viable populations of the species.

Criteria for the fundamental objective of maintaining biodiversity include the following:

- Decreasing the rate at which species extinction occurs.
- Maintaining current numbers and types of different species.

In setting up a decision problem it is generally preferable to first obtain and structure the objectives and then obtain the criteria from these objectives. The set of criteria chosen should cover all relevant objectives without duplication and give a formulation of them which is appropriate for the particular decision problem. The criteria should be understood and interpreted in the same way by all stakeholders, though there may be disagreements on their relative importance. As already indicated, the choice of objectives will be influenced by decision makers' (and stakeholders') values. It should also be recognised that decisions about sustainable development are frequently taken in a political context, which is unlikely to be value neutral. It is therefore particularly important that decision makers and analysts should be aware of their own underlying values and how these values may affect their decision processes.
Criterion definition should take into account the quality of the available data and it may be necessary to re-express certain criteria in terms of variables for which qualitative or quantitative data is available. However it is important that significant factors are not excluded from the problem due to a lack of appropriate data. Sufficient criteria are required to cover all significant aspects of the problem, but an excessive number of criteria should be avoided, as decision makers will find it more difficult to have an overview of the problem when the number of criteria becomes large. In addition computational time will generally increase with the number of criteria. It has been suggested that a reasonable maximum is about twelve, though these criteria can be decomposed into subcriteria if a more detailed representation is required. Analogously to the structuring of objectives, criteria can be *structured*, to give a hierarchy with increasing detail going down the levels. This is likely to make the data more accessible and facilitate understanding of the results by decision makers. It should be

Fig. 9.3 Example of an endangered species: snow leopard

recognised that the choice of criteria is in general *non-unique* and that the particular set of criteria chosen is likely to have an effect on the outcomes.

9.3.3 Attributes

Attributes are required to indicate the extent to which criteria or objectives are achieved. Other terms such as *measure of effectiveness*, *measure of performance* or even criterion are also used. Attributes can be direct or proxy. *Direct attributes* directly measure the achievement of an objective or criterion, whereas *proxy attributes* reflect the degree to which an associated objective or criterion is met, but do not directly measure its achievement. However care has to be taken in the choice of proxy attributes to ensure that they really do reflect the desired objectives or criteria sufficiently. For instance for the objective of decreasing the number of cases of asthma due to air pollution a direct attribute would be the number of cases of asthma due to air pollution and a proxy indirect one the total emissions of air pollutants. The first attribute is a direct one, since reducing the numerical value of the attribute would directly achieve the objective. The second objective is a proxy one since reducing total emissions of air pollutants generally leads to a reduction in the number of cases of asthma due to air pollution. However the proxy attribute may be easier to measure, since there are likely to be difficulties in deciding how many cases of asthma are due to air pollution and how many to other causes or to a mixture of causes, including air pollution.

Attributes can either be quantitative or qualitative. Qualitative attributes can be treated as linguistic variables and measured in qualitative terms or words, such as high, low or moderately high. Alternatively a constructed scale of numerical values can be used. In the latter case the attribute is sometimes called a *constructed attribute*. For example a constructed scale taking values between 1 and 6 could be drawn up as shown in table 9.1 for the attribute *noise*. This would be useful in decision making in a number of different contexts, including transport problems and industry.

An appropriate set of attributes should

- Cover all aspects of the particular problem.
- Be capable of incorporation into the problem through the use of qualitative or quantitative measures.
- Aid in reducing the complexity of the problem.
- Avoid double counting, redundancy or other duplication.
- Be relevant to the particular decision problem.
- Be familiar and meaningful to decision makers and stakeholders.
- Avoid omitting significant factors due to difficulties in expressing them in terms of appropriate attribute(s).

Example 9.6: A Constructed Scale for the Attribute Noise

Table 9.1 A constructed scale for the attribute *noise*

Attribute Level	Description of Attribute Level
1	Quiet ($<$ 20 dB).
2	Low levels of background or external noise (20-40dB).
3	Moderate levels of background or external noise, which are, for instance, likely to cause difficulties in understanding speech to people with moderate hearing impairments (40-90 dB).
4	High levels of continuous background noise, which interfere with speech and concentration (90-120 dB).
5	High levels of continuous background or external noise (90-120 dB) plus moderate levels of intermittent or rhythmic noise, for instance from an electric drill or electric lawn mower (70-110 dB).
6	Very high levels of continuous background or external noise ($\geq$ 120 dB) plus high levels of intermittent or rythmic noise, for instance from an ambulance siren or an aeroplane taking off ($\geq$ 120 dB).

9.4 Utility Functions

9.4.1 Basic Concepts

Utility functions can be used to represent the preferences of decision makers and therefore decision problems can be expressed in terms of maximising the appropriate utility function. A certain amount of mathematical formalism is required to state the conditions under which a function is a utility function (also called a *value* or *preference function*) i.e. to determine what is particular about the class of utility functions. This is given in terms of a preference relation $\succeq$ (more or equally preferred to) which is a complete preorder. This means that the preference relation is reflexive and transitive (see section 9.2.1) and that all pairs of elements are connected by the preference relation. A utility function is a function which maintains the *preference order* i.e. U(x) is equally or more preferred to U(y) when x is equally or more preferred to y.

This can now be expressed more formally as follows. A utility function is a real function U on Y which represents the preference structure given by the complete preorder $\succeq$ on Y if and only if

$$x \succeq y \Leftrightarrow U(x) \succeq U(y) \tag{9.1}$$

Utility functions can be cardinal or ordinal, by analogy with the ordinal numbers first, second, third and the cardinal numbers one, two, three. Therefore *cardinal utility functions* can be used to give a rank order to the utility function evaluated at different points, whereas *ordinal utility functions* also give the magnitude of the values at these points. In general obtaining a cardinal utility function for a given decision problem requires considerably more information than obtaining an ordinal one. In the multi-criteria case the utility function represents the preferences of the decision maker(s) aggregated in some way over all the different criteria. This allows the multi-criteria optimisation problem to be reduced to the much easier single objective problem of maximising the utility function.

There are a number of different ways of combining several single criterion utility functions to give one multi-criteria utility function. When appropriate utility independence conditions, which will be discussed in section 9.4.3, are satisfied the *multi-attribute utility function* can be expressed as either the sum or product of single criterion (or attribute) utility functions. Slightly different conditions are required in the *additive* and *multiplicative* cases. These forms have the advantage of relative simplicity, which increases computational speed, and separating out contributions from the different utility functions. The additive form has the additional advantage of linearity. It also forms the basis of multi-attribute utility theory, which is one of the decision support methods discussed in chapter 11.

When appropriate conditions are satisfied the additive and multiplicative forms can be obtained from the single criterion utility functions as (Keeney 1980)

$$U(x) = k_1U_1(x) + k_2U_2(x) + ... + k_nU_n(x) \quad (9.2)$$

$$\text{and } 1 + kU(x) = [1 + k\,k_1\,U(x)][1 + k\,k_2\,U(x)] \ldots [1 + k\,k_n\,U(x)] \quad (9.3)$$

respectively. Thus the additive form requires the determination of the weights $k_1, ..., k_n$ and the multiplicative form the parameters $k, k_1, ..., k_n$.

Example 9.7: Coal-Fired Power Plants: Environmental Impacts

In a decision making problem about reducing the environmental impacts of coal-fired power plants (Lincoln and Rubin 1979), the utility function for the ith of six pollutants can be represented by either the straight line

$$U_i(x) = 1 - \frac{x_i - x_{min}}{x_{max} - x_{min}}$$

or the curve

$$U_i(x) = 1 + b_i\left\{1 - \exp\left(\frac{c_i(x_i - x_{min})}{x_{max} - x_{min}}\right)\right\}$$

obtained on the assumption that lower emission rates have greater utilities.

The specific utility function for an individual pollutant for a particular decision maker can be obtained by interviewing the decision maker to obtain five discrete data points which are then fitted to the curve using the least squares method. Lincoln and Rubin (1979) obtained complete data for six of the nine decision makers they interviewed. They also found that appropriate conditions (see section 9.4.3) hold for the utility function for all six pollutants for a given decision maker to have a multiplicative form. Therefore the overall utility function for each of the six decision makers can be obtained using expression (9.3) as

$$1 + kU(x) = \prod_{i=1}^{6}[1 + kk_i U_i(x_i)]$$

where U(x) is the total utility of the pollutant emission mix and the constants k and k_i are derived from the test data. If it is assumed that the given decision maker has a linear utility function for the first three pollutants, particulates, sulphur dioxide and heat emitted to air, and a non-linear utility function for the other three, heat to water, dry ash to land and wet sludge to land, then the utility function is given by

$$U(x) = -1/k + \frac{1}{k}\prod_{i=1}^{3}\left[1 + kk_i\left(1 - \frac{x_i - x_{min}}{x_{max} - x_{min}}\right)\right]\prod_{i=4}^{6}\left[1 + kk_i\left(1 + b_i\left\{1 - \exp\left(\frac{c_i(x_i - x_{min})}{x_{max} - x_{min}}\right)\right\}\right)\right]$$

Suppose further data collection is carried out for some additional decision makers and that it is found that the attributes i.e. the levels of the different pollutants are mutually utility independent (see section 9.4.3) for these additional decision makers. Then an additive utility function can be obtained for these decision makers, as follows:

$$U(x) = \sum_{i=1}^{3}k_i\left(1 - \frac{x_i - x_{min}}{x_{max} - x_{min}}\right) + \sum_{i=4}^{6}k_i\left[1 + b_i\left\{1 - \exp\left(\frac{c_i(x_i - x_{min})}{x_{max} - x_{min}}\right)\right\}\right]$$

The use of utility functions in decision making has the advantage of relative simplicity, but the disadvantage of potential loss or degradation of information, through the replacement of different types of information, such as energy consumption, waste generation and resource consumption, represented in different units, by one function. This often requires conversion of all the information to comparable or common units, which are frequently monetary. The use of monetary units is generally inappropriate in sustainable decision problems, as market values are not available for many environmental and social variables. There is also considerable ethical, philosophical and political opposition to valuing environmental and social goods, such as clean air or civil rights, in market terms. The exact numerical values of utility functions for particular decision makers are to a certain extent arbitrary and rarely robust to changes of scale and context. Consequently care should be taken in their use to avoid giving the impression that the utility functions for decision makers have exact numerical values, which is in general not the case.

The general shape of the utility function is determined by the decision maker's attitude to risk. Attitudes to risk can be defined in terms of preferences for an event with a known outcome and an event with an uncertain outcome, but the same expected value as the event with the known outcome. *Risk averse* decision makers prefer the known outcome, *risk prone* decision makers the uncertain event and *risk neutral* decision makers are indifferent to the two options.

Example 9.8: Reduction of Carbon Dioxide Emissions I

Consider a hypothetical situation in which two sets of control actions are possible. The first control action would reduce carbon dioxide emissions by 10% relative to present values over a five year period and the second action would reduce them by either 0% or 20% with equal probabilities. A risk averse decision maker would prefer the certain 10% reduction, a risk prone decision maker the possibility of a 20% reduction and a risk neutral decision maker would have no preference.

9.4.2 Risk Averse, Risk Neutral & Risk Prone Decision Makers

The utility functions for risk averse, risk neutral and risk prone decision makers are respectively *concave* (no linear segment above the graph or a negative second derivative if it exists), *linear* and *convex* (no line segment below the graph or a positive second derivative if it exists). They can therefore be represented respectively by the general forms (Keeney 1980)

$$U_1(x) = a + b(-x^{-cx}) \tag{9.4a}$$
$$U_2(x) = a + b(cx) \tag{9.4b}$$
$$U_3(x) = a + b(x^{cx}) \tag{9.4c}$$

where a, b > 0 are constants which can be chosen to give appropriate scaling of U(x), frequently from 0 to 1, and c indicates the degree of risk aversion or proneness. The parameter c is positive or negative in the linear case according to whether the function is increasing or decreasing and positive for risk averse and risk prone decision makers. Equations (9.4a), (9.4b) and (9.4c) are examples of general concave, linear and convex functions and this is why they are appropriate for the representation of the utility functions of risk averse, risk neutral and risk prone decision makers.

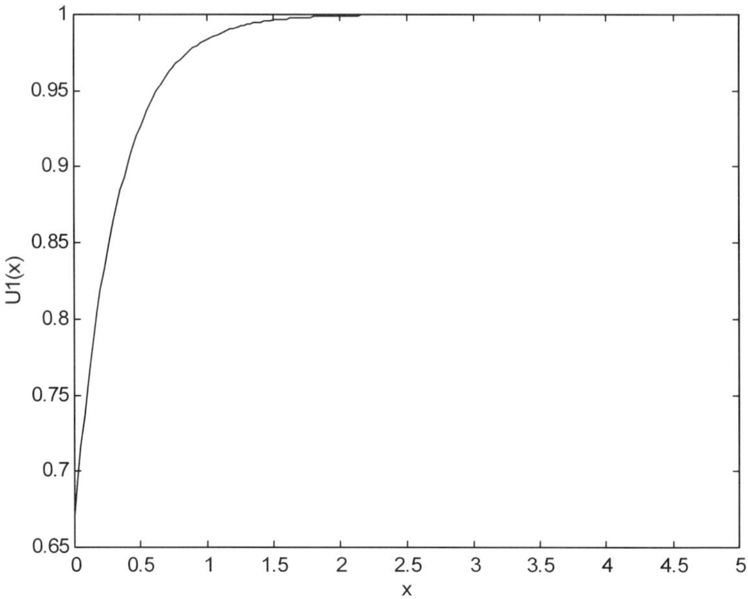

Fig. 9.4a Utility function for a risk averse decision maker

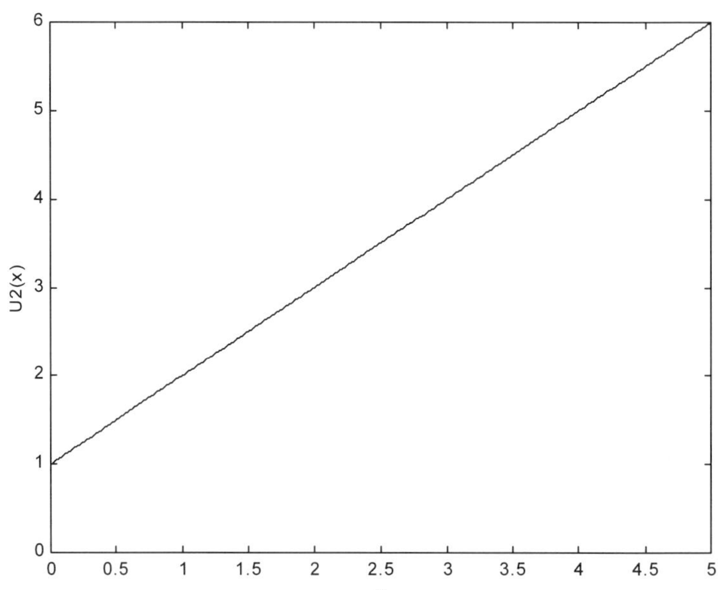

Fig. 9.4b Utility Function for a Risk Neutral Decision Maker

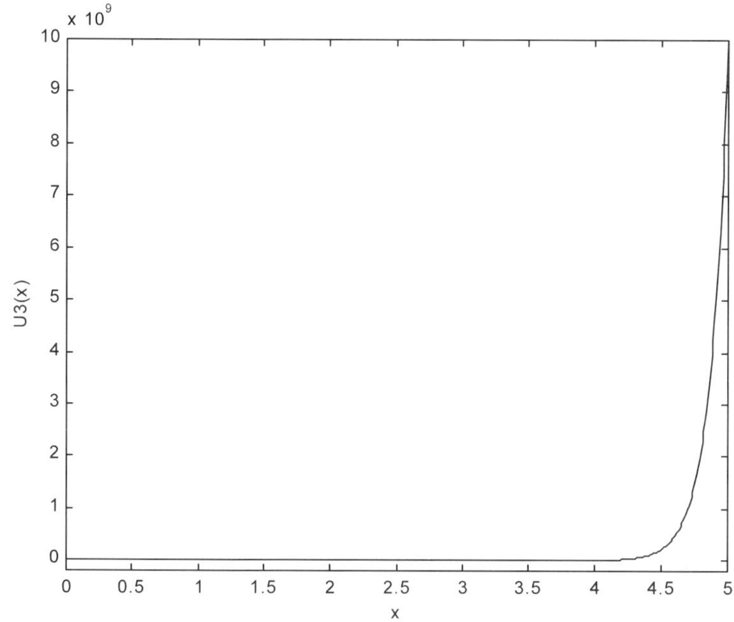

Fig. 9.4c Utility function for a risk prone decision maker

The three utility functions are illustrated in figures 9.4a, 9.4b and 9.4c for the values 1, 0.33 and 3 of the constants a, b and c. The values of a, b and c for all three decision makers have been chosen to give the utility function for the risk averse decision maker a maximum value of 1. The figures show that the utility function $U_1(x)$ is concave, as it is only possible to draw a straight line segment below the graph in fig. 9.4a, but not above it, and that the utility function $U_3(x)$ is convex, as it is only possible to draw a straight line segment above the graph in fig. 9.4c, but not below it. The utility function $U_2(x)$ in equation (9.4b) and fig. 9.4b is clearly linear. It can also be seen from the figures that the utility function for a risk adverse decision maker reaches a steady state value, whereas those for risk neutral and risk prone decision makers increase linearly and exponentially respectively.

9.4.3 Independence Conditions and Additive Utility Functions

Utility functions are generally dependent on all the criteria or all the attributes of these criteria. However there are clearly advantages in being able to express utility functions in terms of functions of only one criterion or one attribute. If appropriate conditions are satisfied a utility function can be expressed as the product or sum of terms which are functions of a single criterion or a single attribute. *Multi-attribute utility theory* is concerned with these independence conditions. There are four main independence conditions, preferential independence, weak-difference independence, utility independence and additive independence (Keeney 1980).

Consider a set F of criteria $\{C_1, C_2, ..., C_m\}$ and the set X of attributes $\{X_1, X_2, ..., X_n\}$ on these criteria. *Indifference curves* can be considered to be contours of values of the attributes with the same value of preference or the same value of the utility function. A *lottery* is defined by specifying a set of possible consequences and the probability that each consequence will occur.

Example 9.9: Reduction of Carbon Dioxide Emissions II

Consider two sets of (hypothetical) control actions to reduce carbon dioxide emissions. This gives two possible lotteries. The first lottery has one possible consequence: reduction of carbon dioxide emissions by 10% relative to present values over a five year period. This consequence has a probability of one. The second lottery has two possible consequences: reduction of carbon dioxide emissions by 0% or 20% relative to present values over a five year period. Each of these consequences has a probability of 0.5.

If the decision maker is risk neutral, then they will have no preference for either lottery and both possible sets of outcomes will be on the same indifference curve.

The definitions will be stated here in terms of attributes, but could also be given in terms of criteria. The pair of attributes $\{X_1, X_2\}$ is *preferentially independent* of the other attributes X_3 to X_n if the preference order for consequences which only involve changes in the values of X_1 and X_2 is independent of the particular fixed values taken by the other attributes X_3 to X_n. Therefore the indifference curves over X_1 and X_2 do not depend on the fixed values taken by the other attributes i.e. the preferences for particular values of X_1 and X_2 remain the same whatever the values taken by the attributes X_3 to X_n.

The attribute X_1 is *weak-difference independent* of the other attributes X_1 to X_n if the rank order between preference differences for different pairs of values of X_1 does not depend on the fixed values taken by the other attributes. Thus weak-difference independence differs from preferential independence in concerning the independence of differences between pairs of values taken by one attribute rather than the values taken by a pair of attributes.

Preferential and weak-difference independence are defined for situations where there is no uncertainty, whereas utility and additive independence involve uncertainty and are therefore defined in terms of lotteries rather than outcomes. The attribute X_1 is *utility independent* of the attributes X_2 to X_n if the preference order for lotteries which only involve changes in the value of X_1 does not depend on the fixed values of the other attributes i.e. if there are two lotteries L_1 and L_2 giving probabilities for different values of X_1 then if lottery L_1 is preferred to lottery L_2 for one set of values of the attributes X_2 to X_n then lottery L_1 will still be preferred for another set of values of X_2 to X_n.

Unlike utility independence, additive independence concerns lotteries over more than one attribute. The n attributes X_1 to X_n are *additive independent* if the preference order for lotteries is only dependent on the marginal probability distributions of these lotteries and not on their joint probability distributions as well. The marginal probability of an event is the probability of this event, ignoring any information about the other event(s). In the two attribute case additive independence is a necessary and sufficient condition for the utility function being additive. Utility independence implies preference independence, but the converse does not necessarily hold.

Example 9.10: Impact Reduction for an Industrial Process I

Consider the criteria of reducing the environmental and social impacts of a particular industrial process. Quite a large number of different attributes would be required to fully represent all aspects of this problem. However, for illustrative purposes only the following three attributes will be considered here:

- *Noise levels*
- *Emissions of carbon dioxide*
- *Jobs created.*

Again, in the interests of simplification, only a small number of values for each attribute will be considered. The values to be considered for each attribute are set out in table 9.2. It should be noted that the values of the attribute *noise levels* are different from those given in table 9.1 for *noise* due to the different context in which the difference between background and foreground noise is less relevant. The total number of values has also been reduced in the interest of simplification.

Table 9.2 Values of the attributes *noise levels*, *emissions* and *jobs created*

Noise Levels	Emissions of CO_2	Jobs Created
Very low	Very low	Negative (jobs lost)
Low	Low	Low
Moderate	Moderate	Moderate
High	High	High

In general the values taken by the attributes *noise levels*, and *emissions of carbon dioxide* will be independent of the value taken by *jobs created.* Therefore there will be a preference for very low *noise levels* and very low (or low) emissions whether *jobs created* takes the value negative or high. Clearly there will generally also be a preference for high *jobs created*, but this preference does not affect the preferences for *noise levels* and *emissions of carbon dioxide.* Thus the pair of attributes *noise levels* and *emissions of carbon dioxide* is preferentially independent of the attribute*jobs created.*

The attribute *noise levels* is weak-difference independent of the attributes *emissions* and *jobs created* if preference differences between the pairs of values taken by the attribute *noise levels* are ranked the same, regardless of the values taken by the other attributes. Ranking of the pairs of differences will generally be subjective and depend on the particular decision maker's preferences. Thus all decision makers will rank the difference 'high - very low' above the difference 'low - very low', but some decision makers may rank 'high – moderate' above 'low - very low', whereas others may rank 'low - very low above 'high – moderate'.

However decisions makers who rank the difference 'high – moderate' above 'low - very low' when *emissions* are high and *jobs created* is negative will still rank the difference 'high – moderate' above 'low - very low' when *emissions* are moderate and *jobs created* is low. Similarly decisions makers who rank the difference 'low - very low' above 'high – moderate' when *emissions* are high and *jobs created* is negative will still rank the difference 'low - very low' above 'high – moderate' when *emissions* are moderate and *jobs created* is low. Since analogous arguments can be made for the rank of other preference differences on *noise levels* remaining fixed when the values of *emissions* and *jobs created* change, *noise levels* is weak-difference independent of the attributes *emissions* and *jobs created.* The preferences for low *emissions* and high *jobs created* do not affect this.

Consider sets of control actions to regulate *noise levels* in the area of the industrial process. Since it is not certain that the control actions will have the desired effects, the outcomes can be expressed in terms of lotteries. Consider three possible control actions, which result in the following lotteries:

1. Very low *noise levels* with a probability of 0.6 and low *noise levels* with a probability of 0.4.
2. Low *noise levels* with a probability of 0.6 and moderate *noise levels* with a probability of 0.4.
3. Very low *noise levels* with a probability of 0.5 and moderate *noise levels* with a probability of 0.5.

In general the first preference will be for lottery 1, independently of the fixed values taken by the attributes *emissions* and *jobs created.* Some people will prefer lottery 2 to lottery 3 and others will prefer lottery 3 to lottery 2. However these preferences will generally be independent of whether *emissions* and *jobs created* have the values low and moderate or high and negative and this will remain the case for other values takes by *emissions* and *jobs created.* Therefore the attribute *noise levels* is utility independent of the attributes *emissions* and *jobs created.*

Consider the two attributes *noise* and *emissions of carbon dioxide.* There is likely to be some uncertainty associated with the outcomes of regulatory actions and therefore they can be expressed in terms of lotteries. Consider two possible control actions, which result in the following lotteries for *noise* and *emissions*:

1. Very low *noise levels* and low *emissions* with a probability of 0.5 and low *noise levels* and moderate *emissions* with a probability of 0.5.
2. Low *noise levels* and low *emissions* with a probability of 0.5 and very low *noise levels* and moderate *emissions* with a probability of 0.5.

Thus both lotteries have marginal probabilities of 0.5 for the values low and moderate for *emissions of carbon dioxide* and marginal probabilities of 0.5 for the values very low and low for *noise levels.* Therefore if the attributes *noise levels* and *emissions* are additive independent of each other, both lotteries will be equally preferred. However it is possible that some decision makers will prefer the first lottery, since it gives the option for a nearly best case scenario and that other decision makers will prefer the second lottery, as when the value of one attribute gets worse, that of the other attribute improves. If this is the case, then the two attributes are not additive independent of each other.

The criteria $\{C_1, C_2, ..., C_n\}$ or attributes $\{X_1, X_2, ..., X_n\}$ are *mutually utility independent* if every subset of the criteria or attributes is utility independent of its complement. Checking for mutual utility independence over a set of n attributes $\{X_1, X_2, ..., X_n\}$ involves checking the utility independence of 2^n - 2 subsets, which grows very fast with n. However the work involved can be reduced to checking n assumptions i.e. from checking 1022 to 10 conditions for n=10, as follows:

The set X_1 is *utility independent* and the pair of sets $\{X_1, X_i\}$ is *preferentially independent* for i =2,3,...,n if there exist n increasing functions U_1 to U_n, such that

$$U = U_1(C_1) + U_2(C_2) + \dots + U_n(C_n) \tag{9.5}$$

Therefore condition (9.5) implies that utility independence holds if the utility function can be expressed as the sum of the utility functions for the component single criterion (or attribute) problems. Unfortunately this result is not particularly useful, since the aim of checking for mutual utility independence is generally to obtain conditions for the existence of an additive form of the utility function. Expressing the utility functions in a multi-criteria (or -attribute) problem as the sum of the utility functions for the component single criterion (or attribute) problems generally requires a number of other conditions in addition to mutual utility independence.

Example 9.11: Impact Reduction for an Industrial Process II

Consider example 9.10 of reducing the adverse environmental and social impacts of an industrial process with the three attributes of *noise levels* (X_1), *emissions of carbon dioxide* (X_2) and *jobs created* (X_3). In this case n=3 and therefore checking for mutual utility independence requires checking the utility independence of $2^n - 2 = 8 - 2 = 6$ subsets. This requires utility independence conditions to be checked for the following:

1. X_1 of $\{X_2, X_3\}$
2. X_2 of $\{X_1, X_3\}$
3. X_3 of $\{X_1, X_2\}$
4. $\{X_2, X_3\}$ of X_1
5. $\{X_1, X_3\}$of X_2
6. $\{X_1, X_2\}$of X_3

Case 1: Consider two lotteries: L_1 has low *noise levels* with probability 0.6 and very low *noise levels* with probability 0.4 and very *low emissions* and high *jobs created*; L_2 has moderate noise levels with probability 0.5 and high *noise levels* with probability 0.5 and very low *emissions* and high *jobs created.* Decision makers will generally prefer L_1 to L_2, as it has lower expected *noise levels*, but the same levels for *emissions* and *jobs created.* Now consider the following lotteries: L'_1 has low *noise levels* with probability 0.6 and very low *noise levels* with probability 0.4 and moderate *emissions* and moderate *jobs created*; L'_2 has moderate *noise levels* with probability 0.5 and high *noise levels* with probability 0.5 and moderate *emissions* and moderate *jobs created*. Decision makers will generally still prefer L'_1 to L'_2, as it has lower expected *noise levels* and the same levels for *emissions* and *jobs created*. (They will also generally prefer L_1 to L'_1 and L_2 to L'_2, but this will not affect the preferences of L_1 to L_2 and L'_1 to L'_2.) Thus preferences will only depend on the expected values of *noise levels*, but not on the fixed values taken by *emissions* and *jobs created*.

Case 2: Consider two lotteries: L_3 has very low *emissions* and high *jobs created* with probability 0.6 and low emissions and moderate *jobs created* with probability 0.4 and low *noise levels*; L_4 has moderate *emissions* and moderate *jobs created* with probability 0.6 and high *emissions* and low *jobs created* with probability 0.4 and low *noise levels*; Decision makers will generally prefer L_3 to L_4, as it has lower *expected emissions* and higher expected *jobs created*, but the same *noise levels*. Now consider the lotteries: L'_3 has very low *emissions* and high *jobs created* with probability 0.6 and low *emissions* and moderate *jobs created* with probability 0.4 and moderate *noise levels*; L'_4 has moderate *emissions* and moderate *jobs created* with probability 0.6 and high *emissions* and low *jobs created* with probability 0.4 and moderate *noise levels*; Decision makers will generally prefer L'_3 to L'_4, as it has lower expected *emissions* and higher expected *jobs created*, but the same *noise levels*. (They will also generally prefer L_3 to L'_3 and L_4 to L'_4, but this will not affect the preferences of L_3 to L_4 and L'_3 to L'_4.) Thus preferences will only depend on the expected values of *emissions* and *jobs created*, but not on the fixed values taken by *noise levels*.

Similar arguments can be used to show that the other utility independence conditions hold and that, consequently, the set of three attributes *noise levels*, *emissions* and *jobs created* is mutually utility independent.

When there are at least three attributes, an additive value function v(.) can be used to describe preferences over the attributes. If one of the attributes X_1 is utility independent of the other attributes X_1 to X_{i-1} and X_{i+1} to X_n, and $\{X_1, X_i\}$ is preferentially independent of the other attributes for all i, then the utility function U(x) must have one of the following forms (Keeney 1980).

$$U(x) = -e^{-cv(x)} \qquad U(x) = v(x) \qquad U(x) = e^{cv(x)}, \qquad c>0 \tag{9.6}$$

where v(x) is an additive value function

9.4.4 Tradeoffs

The existence of a utility function which allows some degree of aggregation of the criteria implies that a certain amount of satisfaction of one criterion can be traded off against satisfaction of one or more of the other criteria. More formally the *tradeoff* w_{ij} between the ith and jth criteria is the amount that the decision maker is willing to concede on the jth criterion to obtain one unit more on the ith criterion. This amount may vary according to the values of the ith and jth criteria and is often difficult to evaluate. However when the criteria are sufficiently regular and the utility function is strictly concave, the tradeoff can be obtained as the ratio of the partial derivatives of the two utility functions (Rosenthal 1985) i.e.

$$w_{ij} = \frac{\partial U}{\partial g_i} \Big/ \frac{\partial U}{\partial g_j} \tag{9.7}$$

In economics the tradeoff w_{ij} is known as the *marginal rate of substitution*. Concave functions of n variables can only be defined on convex sets. A set S of vectors is *convex* if

$$(1-\lambda)\underline{x}_1 + \lambda\underline{x}_2 \in S \text{ whenever } \underline{x}_1, \underline{x}_2 \in S, \text{ and } \lambda \in [0,1] \tag{9.8}$$

The function $f(x_1, ..., x_n)$ on the convex set S is

- *concave* on the set S if
  $f([1-\lambda]x_1 + \lambda x_2) \geq (1-\lambda)f(\lambda) + \lambda f(x_2)$ for all $x_1, x_2 \in S$, $\lambda \in (0,1)$
- *convex* on the set S if
  $f([1-\lambda]x_1 + \lambda x_2) \leq (1-\lambda)f(\lambda) + \lambda f(x_2)$ for all $x_1, x_2 \in S$, $\lambda \in (0,1)$

The function f(., ..., .) is strictly concave or convex if the inequalities $\geq$ and $\leq$ are replaced by $>$ and $<$ respectively. It can be shown that the Hessian matrix of partial derivatives of f(., ..., .) is

- Negative definite when the function is strictly concave.
- Negative semi-definite when the function is concave.
- Positive definite when the function is strictly convex.
- Positive semi-definite when the function is convex.

These conditions are useful, as they are fairly easy to check.

It has been shown (Ting 1971) that a subset K of the set F of criteria is preferentially independent in F if and only if

$$\frac{\partial w_{ij}}{\partial g_k} = 0 \quad \forall i, j \in K, \forall k \in F \setminus K \tag{9.9}$$

Example 9.12: Tradeoffs Between Criteria

The substitution condition (9.7) can only be applied when the utility function is a concave function of the criteria. To obtain a concave utility function, consider an acid rain problem with the criteria $g_1(.)$ of maximising sulphur dioxide emitted to air, $g_2(.)$ of maximising negative impacts on aquatic ecology and $g_3(.)$ of maximising increases in unemployment. It is assumed that all these criteria are defined on dimensionless scales. Utility is a decreasing function of the criteria i.e. utility will decrease as the values of the criteria increase. It will be assumed that appropriate conditions (see section 11.5.2) are satisfied for utility to be an additive function of the single criterion utility functions and that the utility function is negative quadratic in each of the criteria. This gives the following utility function

$$U(g_1, g_3, g_3) = 1 - k_1 g_1^2 - k_2 g_2^2 - k_3 g_3^2$$

where k_1, k_2 and k_3 are positive constants whose values are determined by the preferences of the decision maker, with (see section 11.5.2)

$$k_1 + k_2 + k_3 = 1$$

$U(g_1, g_2, g_3)$ can be shown to be a strictly concave function by calculating the Hessian matrix of second derivatives, which is given by

$$G(g_1, g_2, g_3) = \begin{bmatrix} -2k_1 & 0 & 0 \\ 0 & -2k_2 & 0 \\ 0 & 0 & -2k_3 \end{bmatrix}$$

Therefore the three eigenvalues of $G(g_1, g_2, g_3)$ are equal to $-2k_1$, $-2k_2$ and $-2k_3$ and clearly strictly negative, as k_1, k_2 and k_3 are positive. Therefore $G(g_1, g_2, g_3)$ is strictly negative definite.

The reduction that the decision maker is willing to allow on the jth criterion to gain one more unit on the ith criterion is given by

$$w_{ij} = \frac{\partial U}{\partial g_i} \bigg/ \frac{\partial U}{\partial g_j} = \frac{-2k_i g_i}{-2k_j g_j} = \frac{k_i g_i}{k_j g_j} \tag{9.10}$$

Consider the specific case $w_{12} = k_1g_1/k_2g_2$. If low emissions of sulphur dioxide to air ($g_1(.)$) are more important to the decision maker than low negative impacts on aquatic ecology ($g_2(.)$), then the decision maker will be willing to trade more than one unit on criterion 2 for a reduction in one unit on criterion 1 and therefore from (9.10)

$$w_{12} = \frac{k_1 g_1}{k_2 g_2} > 1 \Rightarrow k_1 g_1 > k_2 g_2$$

It can then be shown that the following conditions hold:

- $k_1g_1 > k_2g_2$ when it is more important to the decision maker to reduce sulphur dioxide emissions to air than negative impacts on aquatic ecology.
- $k_1g_1 < k_2g_2$ when it is more important to the decision maker to reduce negative impacts on aquatic ecology than sulphur dioxide emissions to air.
- $k_1g_1 = k_2g_2$ when it is equally important to the decision maker to reduce sulphur dioxide emissions to air and negative impacts on aquatic ecology.

Preferential independence of the attributes *sulphur dioxide emitted to air* and *impacts on aquatic ecology criteria* of the attribute *increases in unemployment* will now be considered. This is obtained by considering the derivative (9.9), giving

$$\frac{\partial w_{12}}{\partial g_3} = \frac{\partial}{\partial g_3}\left(\frac{k_1 g_1}{k_2 g_2}\right) = 0$$

$$\frac{\partial w_{21}}{\partial g_3} = \frac{\partial}{\partial g_3}\left(\frac{k_2 g_2}{k_1 g_1}\right) = 0$$

and therefore the attributes *sulphur dioxide emitted to air* and *impacts on aquatic ecology criteria* are preferentially independent of the attribute *increases in unemployment*

The utility function is this example has deliberately been chosen to be very simple. In practice, even when decision makers are willing to trade between satisfaction of different criteria, the tradeoff values will frequently depend on the current values of the different criteria. In particular there may be values above which the decision maker does not consider it worth increasing satisfaction of a particular criterion and values below which they will not allow the value of even a relatively unimportant criterion to drop, regardless of the increases on the other criteria.

In addition a number of decision makers will not be willing on ethical grounds to make certain types of tradeoffs, for instance between loss of life and a number of other variables, such as the loss of significant numbers of jobs or loss of species. In this case it is clearly most appropriate to look for alternative approaches to the problem which lead to different types of solutions that do not involve unacceptable tradeoffs.

9.5 Roundup of Chapter 9

Chapter 9 has presented the mathematical background required to understand and apply decision support methods (to sustainable development decision problems). Three main topics were covered: binary and preference relations, objectives, criteria and attributes, and utility theory. Preference relations are a particular type of binary relation which can be used to model the preferences of decision makers. Since most decision problems in sustainable development involve more than one criterion and multi-criteria problems generally have multiple solutions, the preferences of the decision maker have to be taken into account in choosing the appropriate solution in the particular context. Two elements of a set satisfy a binary relation if there is a particular type of connection between them, such as being equal to or greater than. Binary relations can be classified according to their properties, such as being transitive and reflexive. Any two outcomes of a decision are related by one of the following preference relations: the first outcome is preferred to the second; the first outcome is less preferred than the second; the outcomes are equally preferred; there is insufficient information to make a decision.

Objectives, criteria and attributes are three of the main components of decision problems. Objectives are what decision makers and other stakeholders hope to achieve by carrying out a particular decision and criteria are a more formalised, possibly more mathematical, expression of one or more objectives. Attributes measure the extent to which objectives or criteria have been achieved. They can be direct or proxy, with proxy criteria generally easier to measure.

Multi-linear utility functions form the basis of multi-attribute utility theory, one of the methods considered in chapter 11. Utility functions can be used to represent the preference structure of decision makers. Multi-linear functions are functions of n variables which are linear in each variable on its own and therefore consist of terms which are the sums of products of up to n terms. A number of utility independence conditions need to be satisfied for a utility function to be multi-linear. Additive representations, in which the utility function is the sum of the separate utility functions of one variable, are the easiest to use. One of the prerequisite conditions for the existence of an additive utility function is that the criteria or attributes are mutually utility independent i.e. each subset of the criteria or attributes is utility independent of its complement. Utility functions have particular general forms based on the decision maker's attitude to risk i.e. whether they are risk prone, neutral or averse.

Chapter 10 draws on the mathematical background to decision making presented in this chapter to discuss the main issues in multi-criteria decision making.

10 Multi-Criteria Problems

10.0 Learning Objectives

The main aims of this chapter are to present the basic principles of multi-criteria decision and optimisation problems and give an understanding of their role in sustainable development. Specific learning objectives include the following:

- Understanding of the basic principles of multi-criteria optimisation and decision making.
- The ability to formulate multi-criteria problems.
- Knowledge of different types of optimal solution.
- Understanding of the role of multi-stage decision making and the ability to formulate multi-stage decision problems.
- Knowledge of the classification of multi-criteria decision methods.
- Knowledge of the main differences between classical and naturalistic decision theory and understanding of the applications of the two approaches to sustainable development.
- Understanding of some of the applications of multi-criteria decision support methods to sustainable development.

10.1 Introduction

This chapter presents an introduction to the basic principles of multi-criteria decision and optimisation problems and gives an understanding of their role in sustainable development. There are three main sections followed by a chapter summary in section 10.4. Section 10.1 discusses the main stages in decision problems and the different types of optimal solutions. This is of particular importance in multi-criteria decision and optimisation problems, since there is generally more than one possible solution. This is followed in section 10.2 by consideration of problem formulation and the concepts of alternatives and multi-stage decision problems. Section 10.3 presents an overview and categorisation of the different decision support methods and discusses their application to sustainable decision problems.

10.1.1 Basic Concepts

A *multi-criteria* problem can often be formulated in terms of a set of *objectives*, a set of *criteria*, a set of *alternatives* and a set of *attributes* which are used to determine the extent to which the criteria or objectives are satisfied by a particular alternative. The alternatives and criteria can often be expressed in terms of a decision matrix, though this is not always the most appropriate approach. The values and *preferences* of the *decision maker(s)* (and *stakeholders*) generally play an important role in decision problems, with regards to both choice of objectives and criteria and prioritisation between them.

Decision problems have been divided into the four stages (Simon 1977) of intelligence, design, choice and review, as illustrated in table 10.1. *Intelligence* refers to the information gathering process and may result in the setting up of one or more data bases to support the decision maker(s). *Design* refers to problem formulation and generally includes determining families of alternatives, criteria and attributes. It may also involve clarification or more precise specification of the particular decision problem to be resolved. *Choice* generally involves the selection of an alternative from the set of alternatives. However in multi-stage decision problems, for instance relating to transport and energy planning, it could involve the choice of a particular type of approach or type of solution.

The *review* stage involves assessment of both the decision process and the final decision reached. Assessment of the decision process allows appropriate lessons to be learnt to improve decision making, so that successes can be repeated and failures avoided in the future. Assessment of the decision reached can lead to its acceptance or modification, possibly through repetition of one or more of the previous decision stages. This assessment can also involve reconsideration of some of the alternatives that have been rejected and investigation of whether there are better alternatives, particularly in the vicinity of alternatives that perform well on all except one of the criteria.

For instance, in energy planning problems the first choice stage could involve determination of the particular energy strategy required to meet projected energy needs in terms of the balance between saving energy through energy conservation and efficiency, reducing energy requirements and generating additional energy through building new power plants and expanding the capacity of existing power plants. The strategy will also

Table 10.1 Four stages of decision making

Stage	Activities
Intelligence	Collecting relevant data to support decision maker. Setting up data bases.
Design	Problem formulation. Determination of alternatives, criteria and attributes.
Choice	Selection of alternative or type of approach.
Review	Assessment of decision process. Acceptance or rejection of the decision reached.

involve decisions about promoting renewable sources of energy, for instance that all additional generation capacity should be from renewable sources and that there should be a (gradual) replacement of non-renewable by renewable generation capacity. In many cases the energy strategy chosen at this stage may involve no additional energy generation or even a reduction in generation capacity.

The next decision stage would involve choices about how the strategy is implemented, including decisions about specific conservation measures, as well as the details of additional generation capacity and a timetable for the conversion of non-renewable to renewable generation capacity.

The majority of decision support systems and, in particular, multi-criteria algorithms only support the third choice stage of decision making. However the other stages are equally important, since good decision making is critically dependent on the availability of good information and appropriate problem formulation, as well as review to ensure that a good or at least a satisfactory solution has been obtained.

Since there is not a unique absolute maximum in the multi-criteria case, multi-criteria decision algorithms can be used to obtain one of the following:

- An overall optimum solution.
- A solution set which contains all the good or satisfactory alternatives.
- A rank order for all (or a subset) of the alternatives.

10.1.2 Problem Formulation

As discussed above, *decision support methods* have generally focused on either the choice of the 'best' alternative or ranking the alternatives from a given set using a pre-defined set of criteria and objectives. If an appropriate algorithm is chosen and implemented correctly this will result in a good solution to the problem as formulated. Whether this solution is useful in real terms depends on how well the important issues have been represented in the decision problem, as well as whether the solution can be implemented in practice. Applying a very good algorithm to the wrong problem will not lead to a useful result. In this context important considerations include the following:

- Acting proactively as well as retroactively.
- Problem analysis to determine and understand the real issues rather than assumptions being made about what the most important issues are.

Acting *proactively* rather than retroactively has been described in terms of identifying or creating decision opportunities rather than waiting for decision problems (Keeney 1992). This is particularly relevant in the area of sustainable decision making. For instance many firms consume unnecessary resources, including energy, and generate (excessive) waste. In addition most industrial processes generate emissions and/or solid waste and effluent. Thus the actions of many and probably most companies have negative environmental and social consequences, many of which could be reduced by appropriate action. A retroactive approach involves dealing with decisions as they arise, for instance through the tightening of environmental legislation, in response to a public complaint or in the context of discussions about introducing a new process or product. A more proactive approach could involve regular evaluation of the sustainability of all operations, processes and products over their whole life cycle and consideration of how sustainability could be increased. This should include the implementation, monitoring and updating of equal opportunities policies and improvement of terms and conditions of employment, as well as environmental factors. This could also include the development and regular updating of the sustainability criteria to be met by new products and processes.

A proactive approach is likely to have a number of advantages over a retroactive one, including the following:

- Control over how and when changes are introduced rather than having to act in response to legislation or other external pressures.
- Cost savings, for instance through reduction in waste and energy consumption.
- Competitive advantage, through better environmental, social or labour relations performance than competitors.
- The possibility of discovering new products.

There are a number of schemes for linking up companies to allow the 'waste' products of one firm to become the raw materials of another. For instance a pulp and paper mill on the Androscogin River in the USA is making 'waste' materials available to local firms and a number of companies have located near it to take advantage of this 'waste'. The community of Kalundborg in Denmark is becoming well known for industrial ecology, including the creative use of each other's waste by different firms. For instance the natural gas which was previously flared off by Denmark's largest refinery is being used in a plasterboard factory (Renner 2004). Some firms have also found creative ways of using their own raw materials. For instance a California based manufacturer has used the coloured scrap cloth from making outdoor clothing and tents to produce multicoloured parkas, which sell for more than single coloured parkas in a number of markets (Keeney 1992). There are also a number of examples where small changes in processes or minor reorganisation can lead to great savings of materials or energy and consequently great reductions in costs. However many firms are not using these opportunities.

Few countries or regions have a cohesive short and long term infrastructure, energy or transport policy or a cohesive strategy for meeting commitments under the Climate Convention (WWW11 2005) to reduce emissions of carbon dioxide and other greenhouse gases. Although a number of countries have introduced individual or packages of measures to address climate change, emissions of greenhouse gases are increasing in many countries and only relatively few countries are meeting their commitments (Dunn and Flavin 2002). Decisions about the need for new power stations, roads or railways are generally made on an ad hoc basis, using the projected growth in demand for energy and car or rail journeys, leading to high environmental and frequently also high social costs. However a more sustainable approach would involve consideration of the demand for energy or car journeys as a variable to be regulated. The decision problem would then include determination of the levels at which these variables should be regulated and different strategies for attaining and maintaining these regulated levels.

Example 10.1: Analysis of a Transport Decision Problem

The need for problem analysis to identify the real issues is particularly apparent in the transport area. Here the problem is often seen as congestion and the solution as building one or more new roads, leading to the decision problem of the best choice of route to maximise benefits and reduce costs. However both historical and theoretical evidence show that road construction has led to traffic generation (Pfleiderer and Dieterich 1995) and therefore increased rather than solved the problem of congestion. In addition there are a number of other problems associated with current approaches to transport, (in industrialised countries), including:

- Climate change and other environmental problems resulting from emissions.
- Respiratory and other health problems resulting from emissions, which generally affect poorer people more than wealthier ones.
- Inequalities between those with and without private cars, particularly when facilities are sited away from population centres, so they can only be accessed easily by car owners.
- The possibility of insufficient energy to meet the development needs of the 'developing' countries.
- Land use problems due to roads and parking facilities taking away land from other potential uses.
- Health problems resulting from a sedentary life style to which excessive car use contributes.
- Noise pollution.
- Inadequate public transport and facilities for cyclists and pedestrians.

The priority to be given to the different problems is partly subjective and likely to depend on political and ideological viewpoints, life style choices and other factors. A sustainable approach would involve investigation of the various needs served by transportation and the ways in which these needs can be met at minimal environmental and social cost. A number of different techniques, including brainstorming, group discussions, lateral thinking, focusing on specific objectives and best and worst case scenarios, could be used to generate a wide range of alternatives. The various options could then be combined to give a range of packages of measures from which the best option could be chosen.

10.1.3 Different Types of Optimal Solution

Unlike in the single criterion case, there is not a unique set of conditions to be satisfied by the optimal value or a unique best value for a multi-criteria problem. Instead there are a number of different ways of defining the optimal value (Pomerol and Sergio 1993), including Pareto (Yu 1985) and satisficing (Simon 1955) solutions.

Consider a multi-criteria problem with n criteria C_1 to C_n and associated criterion functions g_1 to g_n and two options a and b, which take values $g_1(a)$ to $g_n(a)$ and $g_1(b)$ to $g_n(b)$ respectively under the different criteria C_1 to C_n. If a performs better than b on criterion j, then $g_j(a) > g_j(b)$, which can be written a $\succ_j$ b. Similarly if a performs better than or equally well to b on criterion j, then $g_j(a) \geq g_j(b)$ which can be written a $\succeq_j$ b. The relation $\succeq$ is a preorder as it is reflexive and transitive.

Then a *dominates* b if a performs better or as well as b on all the criteria and better than b on at least one criterion, written a $\succeq$ b or a $\succeq_j$ b for j=1...n and a strictly dominates b if a performs better than b on all the criteria, written a $\succ$ b or a $\succ_j$ b for j=1...n. An option a in Y is said to be *admissible*, *efficient* or a *Pareto optimum* for the preorder $\succeq$ if there is no element b in A which dominates a i.e. which performs at least equally well on all the criteria and strictly better on one of them. Thus a Pareto optimum performs as least as well as all the other options on all the criteria and better than the other options on at least one criterion.

The *Pareto set* is the set of Pareto optima or admissible points. If an alternative is a Pareto optimum it is not possible to find another alternative which performs better on one criterion, unless it also performs worse on another. This makes Pareto optima an obvious choice for multi-criteria optimisation problems. However Pareto optima do not take equity considerations or the distribution of benefits into account. Therefore in sustainable decision making it may be desirable to give greater weight to environmental, social and developmental criteria, even if this is at the expense of economic ones.

A point x is *properly efficient* (Geoffrion 1968) if there exists a positive constant M, such that for any y, i and j

$$\frac{g_i(x) - g_i(y)}{g_j(y) - g_j(x)} \leq M \tag{10.1}$$

This ratio is the improvement in the ith criterion divided by the decrement in the jth criterion resulting from a change of solution from x to y. Putting an upper bound on this ratio ensures that a small or moderate improvement on the ith criterion is not obtained at the expense of a very large or infinite reduction in performance on the jth criterion. However, in some cases changes to the status quo may require significant decrements on some criteria. For instance the criterion of increases in equity would require reduced satisfaction of the criterion that the income received by individuals should not be reduced, as it would require a redistribution of wealth and resources from richer to poorer countries and richer to poorer individuals within countries. The constant M should be chosen to take account of the extent to which improvement in one criterion can be traded against a decrease in satisfaction of another criterion.

Example 10.2: Acid Rain Policy Choices I

Consider the three criteria of maximising air quality (C_1), maximising water quality (C_2) and minimising increases in unemployment (C_3) in acid rain policy choices. It will be assumed that the associated criterion functions can be defined as follows:

- $g_1(.)$: *air quality* with values bad, poor, reasonable, good.
- $g_2(.)$: *water quality* with values bad, poor, reasonable, good.
- $g_3(.)$: *increases in unemployment* with values negative (decrease), none, low, high.

Consider four different policy choices, P_a, P_b P_c and P_d with

- $(g_1(P_a), g_2(P_a), g_3(P_a))$ = (reasonable, poor, none)
- $(g_1(P_b), g_2(P_b), g_3(P_b))$ = (reasonable, good, none)
- $(g_1(P_c), g_2(P_c), g_3(P_c))$ = (good, good, low)
- $(g_1(P_d), g_2(P_d), g_3(P_d))$ = (good, good, negative)

Then the policy choice P_b dominates P_a as it performs equally well on the criteria maximising *air quality* and minimising *increases in unemployment* and better on the criterion maximising *water quality*. P_b neither dominates P_c nor is dominated by P_c, as P_b performs better on the *unemployment* criterion and worse on the *air quality* criterion and they perform equally well on the *water quality* criterion. P_d strictly dominates P_a, as it performs better on all the criteria. P_d is also an admissible, efficient or Pareto optimum, as it performs better than the three other policy choices on the *unemployment* criterion and equally well or better on the *air quality* and *water quality* criteria. It is the only Pareto optimum in this set of policy choices.

In order to determine which points are properly efficient the criterion functions will now be given numerical rather than linguistic values i.e. numbers rather than verbal descriptions. The *air quality* and *water quality* criteria will be expressed on a scale of 1 to 4, with 1 denoting bad *air* or *water quality* and 4 good *air* or *water quality*. To avoid treating *unemployment* disproportionately to the other criteria, it will also be measured on a scale of 1 to 4 rather than in terms of a particular number of jobs lost (or created). For compatibility with the *air quality* and *water quality* criteria, increasing values will also denote increasing satisfaction of the unemployment criterion, so that 1 denotes high increases and 4 negative increases (decreases). Then the criteria have the following numerical values for the four points:

- $(g_1(P_a), g_2(P_a), g_3(P_a))$ = (reasonable, poor, none) = (3, 2, 3)
- $(g_1(P_b), g_2(P_b), g_3(P_b))$ = (reasonable, good, none) = (3, 4, 3)
- $(g_1(P_c), g_2(P_c), g_3(P_c))$ = (good, good, low) = (4, 4, 2)
- $(g_1(P_d), g_2(P_d), g_3(P_d))$ = (good, good, negative) = (4, 4, 4)

However it should be noted that in some cases it will be desired to give greater weight to some criteria than to others. In addition not all approaches are able to take into account the extent to which an option performs better than another.

If it is assumed that all the criterion functions are monotonically increasing (after redefinition if necessary, for instance by replacing C_i by $-C_i$), then a 'rational' decision maker will only choose non-dominated alternatives. However dominance relations have a number of limitations, including the fact that they do not allow for tradeoffs, even where considerable improvement on one or more criteria can be obtained at the expense of a small reduction on another criterion. The limitations of dominance relations have led to the development of outranking relations, which will be discussed in section 11.2.

The assumption of *monotonicity* is realistic for many types of criteria. However there are examples such as population levels of, for instance, deer in forest management problems for which there is (a range) of optimal values, with both excessive and very low populations undesirable. If the population becomes too high, it will exceed the capacity of the habitat to support it. This gives an upper bound on the population level. On the other hand if the population becomes too low it will no longer be able to reproduce itself and will therefore die out. This gives a lower bound on the population level. Therefore constraints will be required to ensure that population levels remain within the upper and lower bounds. It should be noted that the value of the upper bound for a particular decision maker may depend on whether the species is valued in its own right or solely for its value to humanity. The value of the upper (and possibly also the lower) bound may also vary over time due to factors such as changes in climate and decisions taken on forest management, which may effect the forest as a habitat for deer and other animals.

The *ideal point* is the point in the criterion space with coordinates:

$$\left(\max_{a\in A} C_1(a), \max_{a\in A} C_2(a), \dots, \max_{a\in A} C_n(a)\right) \tag{10.2}$$

where A is the set of alternatives i.e. the point with coordinates which maximise the different criteria in turn. If an alternative a could be found which maximised each criterion simultaneously it would clearly be the overall best point. However in general there are tradeoffs between the different criteria and therefore there is only very rarely an alternative which maximises all the criteria at once.

Satisfaction levels are one way of obtaining an appropriate performance compromise on all the criteria. A *satisfaction level* for a particular criterion can be defined as a level of performance which is acceptable to the decision maker, although it is not the best possible. When the criterion C_j is defined by a function $g_j(.)$, the *satisfaction threshold* can be defined by a number u_j, such that the option a is satisfactory if $g_j(a) \geq u_j$ and unsatisfactory if $g_j(a) < u_j$. Defining a satisfaction level for each criterion gives an overall vector of satisfaction $\underline{u} = (u_1, u_2, ..., u_n)$ for all the criteria. A *satisficing solution* is a solution which performs acceptably on all the criteria, but which is in general not optimal. For instance a satisficing solution meets or exceeds the satisfaction threshold on all the criteria, but may have lower values than other option(s) on some of the criteria.

Example 10.3: Acid Rain Policy Choices II

Consider example 10.2. The ideal point has the coordinates (good, good, negative) or (4, 4, 4) in numerical terms.

The satisfaction level will be defined as $\underline{u} = (3, 3, 3)$. In numerical terms the coordinates of the four options can be written as:

- $(g_1(P_a), g_2(P_a), g_3(P_a))$ = (reasonable, poor, none) = (3, 2, 3)
- $(g_1(P_b), g_2(P_b), g_3(P_b))$ = (reasonable, good, none) = (3, 4, 3)
- $(g_1(P_c), g_2(P_c), g_3(P_c))$ = (good, good, low) = (4, 4, 2)
- $(g_1(P_d), g_2(P_d), g_3(P_d))$ = (good, good, negative) = (4, 4, 4)

Then the options P_a and P_c are unsatisfactory, as $g_2(P_a) = 2 < 3$ and $g_3(P_c) = 2 < 3$. The options P_b and P_d are satisficing solutions, as $g_1(P_b) = 3 \geq 3$, $g_2(P_b) = 4 \geq 3$ and $g_2(P_b) = 3 \geq 3$; $g_1(P_d) = 4 \geq 3$, $g_2(P_d) = 4 \geq 3$ and $g_2(P_d) = 4 \geq 3$. Thus it has proved possible to distinguish between the options P_b and P_c by classifying them as satisfactory (satisficing solutions) or unsatisfactory, though neither option dominates the other. P_d is also the idea point, as all the criteria take their maximum values.

10.2 Alternatives and Multi-Stage Decision Problems

10.2.1 Alternatives

As discussed in section 10.1.1, one stage of decision making frequently involves the choice of one or more *alternatives* or options from a set of alternatives or options, though these alternatives may be implicitly defined. A number of multi-criteria algorithms and other decision support systems have been developed to facilitate making this choice, but considerably less attention has been given to drawing up the set of alternatives. It is frequently assumed that the set of alternatives should be well defined and fixed and that alternatives are mutually exclusive. However this would prevent modification or combination of alternatives to obtain new and possibly preferable alternatives or trying to improve on the alternative selected by considering additional alternatives. A better solution can sometimes be obtained through consideration of additional alternatives if some of the alternatives perform well on all criteria except one, so that there may be related alternatives which will perform well on that particular criterion as well.

The initial set of alternatives should be as large and varied as possible to ensure that no potentially good, but less obvious options are missed. This large set can then be reduced by eliminating options which are not feasible or do not satisfy certain conditions before detailed data gathering and analysis of the remainder. Analogous techniques to those described for generating objectives can also be used to generate alternatives. Other approaches to *generating alternatives* include:

- Consideration of how individual objectives, pairs of objectives and groups of objectives can be met.
- Lateral thinking.
- Modification and combination of the previously generated alternatives.

After the full set of alternatives has been generated and any duplicated alternatives removed, it may be useful to carry out a preliminary analysis of the problem to eliminate all the alternatives which do not meet certain conditions. Such conditions could include:

- Not being dominated.
- Not meeting appropriate aspiration or satisfaction levels on some or all the criteria.
- Not dominating a vector of satisfaction levels.
- Having unsatisfactory performance on particular criteria.

The problem can also be simplified by the removal of criteria which either do not have a significant effect or are considered less important than the others. The aim of the removal of some alternatives and/or criteria is to simplify the problem and make it more tractable. This is particularly relevant when the costs, resource implications and time requirements of collecting the detailed data required for full problem analysis are high. However care has to be taken not to eliminate criteria, the importance of which is not fully realised, or options which do give reasonable performance. In particular, it should make little difference whether options or criteria are eliminated first. Therefore, if preliminary analysis is carried out to eliminate alternatives and/or criteria, it can be useful to compare the performance of the alternative obtained with that of some of the previously rejected alternatives under all the criteria to ensure that an appropriate decision has been made.

10.2.2 Multi-Stage Decision Problems

As already indicated, a number of decisions are in practice *multi-stage* i.e. they involve a consecutive sequence of decisions, rather than a single one-off decision. When this is the case a number of different approaches can be taken to obtaining and structuring the alternatives, including the following:

- Identification of all possible alternatives and then structuring them in a hierarchy with two or more levels to correspond to the decision levels of the problem.
- Identification of the highest level alternatives first, followed by identification of the next level alternatives corresponding to the 'best' alternative at the level above, after this has been chosen. The process of identifying alternatives at subsequent levels corresponding to the 'best' alternative at the previous level can then be repeated as often as required.
- Identification of all possible alternatives and treating them all equally i.e. treating the problem as a one-stage decision problem.

These approaches all have their advantages and disadvantages. Structuring the problem can increase efficiency and reduce the costs of data collection. However care must be taken, whichever approach is used, to ensure that good alternatives are not missed due to the structuring or lack of structuring of the problem.

Example 10.4: Multi-Stage Decision Problem

It will be assumed that studies indicate that there will be a ten per cent growth in energy demand in the city of Glasgow in Scotland over the next twenty years if measures are not taken to reduce this demand. The problem of determining the 'best' energy strategy to simultaneously meet energy needs and environmental objectives can be treated as a multi-stage decision problem. Otherwise there would be a very large number of alternatives of the following types:

- Provide y% of subsidy for insulation to x% of homes in Glasgow.
- Build combined heat and power station(s) in location(s) z.
- Build hydro-power stations in location(s) w.

where w, x, y and z are to be determined.

In this case a hierarchical approach is appropriate. As shown in fig. 10.1, the alternatives can be arranged in a decision tree. Only some of the possible alternatives are given at each level to make the diagram easier to read. At Level 1 there is one alternative A0, which is the original decision problem of determining the 'best' energy strategy to simultaneously meet energy needs and environmental objectives.

The alternatives at Level 2 include the following:

1. To meet all energy requirements by energy conservation measures (A1).
2. To meet all energy requirements by building combined heat and power stations and small scale hydro-power stations (A2).
3. To meet all energy requirements by a combination of energy conservation measures and building both combined heat and power and small scale hydro-power stations (A3).

It should be noted that this list includes only a small subset of all the possible options. Level 3 from alternative 1 consists of a number of packages of measures for reducing energy consumption and using energy efficiently. This would include measures to encourage appliances to be turned off when not in use, the use of low energy light bulbs and the installation of effective insulation in buildings. Level 3 from alternative 2 consists of different combinations of locations and energy generation capacities for specific types of power stations. Level 3 from alternative 3 consists of different combinations of packages of energy efficiency measures and locations and energy generation capacities for power stations. This could include giving away low energy bulbs, subsidies for building insulation, social or other sanctions for people who do not turn off appliances when they have finished using them, measures to increase the efficiency of existing power stations and not building new power stations.

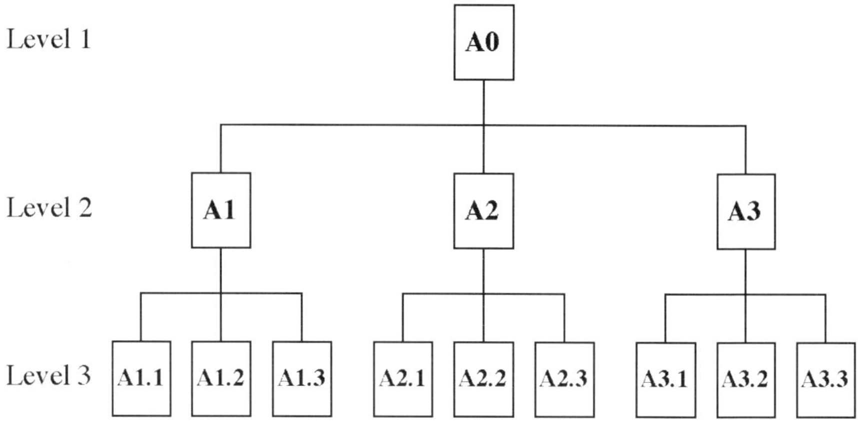

Fig. 10.1 Decision tree for a multi-stage decision problem on energy choices

One approach to solving the problem would be to find the best option at the first level for, for instance, alternative 1 of using only energy conservation measures and then work down the alternative tree, making choices at each level. The other approach is to go to the lowest level, which consists of a number of branches and find the best choice for each branch, for instance the best combination of energy conservation measures or the best set of locations and generation capacities for small scale hydropower and combined heat and power stations and then compare the best alternatives for all the branches. The first approach is probably simpler and less time consuming, but the alternatives may not be specific enough to identify and obtain the data required to allow adequate discrimination between the alternatives. In addition the difficulties of defining and obtaining appropriate data may increase the likelihood of decision makers choosing alternatives in line with their own existing views, rather than to meet criteria and fulfil objectives. The more detailed comparisons in the second approach may be more likely to result in the 'best' overall solution being chosen, but have considerably greater data requirements and therefore costs. Another possibility would therefore be to start at a middle level and find the best alternative at this level and then work down the alternative tree, making choices at each level.

10.3 Decision Making and Decision Support Systems

This section considers the use of decision support methods to support decision making, including the classification of multi-criteria decision methods, different approaches to decision making and the general principles of sustainable decision making.

10.3.1 Classification of Multi-criteria Decision Methods

Both choice of an appropriate problem structure and use of an appropriate method are important in obtaining a satisfactory decision (Ozernoy 1992) and the use of different *multi-criteria decision methods* (MCDM) can lead to very different decisions. For instance the application of two different methods to rank 542 hydro-electric projects in Norway resulted in totally different rank orders (Wenstop and Carlsen 1988). Since different types of problems generally require different approaches and the limitations on available information may also affect the choice of method, there is no universally best MCDM which is appropriate for all problems. Thus there is a need for techniques to evaluate and classify the different MCDM so the best choice can be made for a given problem situation.

A number of criteria for evaluating MCDM have been suggested (Evans 1984; Hobbs 1986; Ozernoy 1987) and experimental comparisons of discrete alternative MCDM have been carried out, but have not yet produced a universally accepted set of guidelines for selection of the best method for a given application. Many authors consider accuracy in representing the decision maker's preferences and quality of prescriptive insight to be the most important criteria for MCDM, but these conditions are difficult to define precisely and evaluate (Ozernoy 1992). There may be a role for the development of expert systems to facilitate choice of the most appropriate MCDM (Ozernoy 1992).

Multi-criteria decision problems can be categorised in a number of different ways to facilitate choice of MCDM for a given problem. One important distinction is between discrete and continuous decision alternatives. *Discrete* problems can be classified according to whether:

- They have small or large numbers of alternatives and criteria.
- The alternatives and values of the criteria are a priori known or unknown.
- The criteria are explicitly or implicitly specified.

Multiplying the number of options for the different types of problems together, noting that small/large numbers and known/unknown options hold for both alternatives and criteria, gives 32 (4 x 4 x 2) different types of problems.

Continuous problems can be classified according to whether:

- The underlying model is linear or nonlinear.
- The variables are continuous or integer.
- There is a small or large number of criteria.
- The problem is large or small scale (in terms of the number of constraints and variables).
- Relationships between the variables are quantitative or qualitative.
- The decision variables are a priori known or unknown.

Multiplying the numbers of options for different types of problems together gives 64 (2^6) different types of problems (Korhonen et al 1992).

Discrete MCDM methods can be divided into compensatory, non-compensatory and partially compensatory approaches, depending on whether high performance on some criteria can balance low performance on some or all of the others (Colson and de Bruyn 1989). Most decision support systems developed to date are aggregation methods of the criteria in the sense (Roy 1985; Vincke 1989) that a (finite) set of feasible alternatives and criteria is used to give an evaluation matrix in table or data file form. The main design features of *aggregation methods* are common, but the different methods can be distinguished from each other (Colson and de Bruyn 1989) by:

- The type of information and the step at which it is required.
- The extent of interaction between the procedure and the decision maker.
- The measurement scales of the data and results.
- The types of preference structures, criteria and aggregation structure.

Another important division is into holistic, heuristic and wholistic techniques (Sage 1981). *Holistic methods* evaluate the alternatives independently and choose the most highly rated one. The different methods are defined by the scoring functions used. The best known method is based on multi-attribute utility theory (Keeney and Raiffa 1976). This approach is the most popular, but is not always valid and is demanding of information.

Heuristic elimination is a process of sequential comparison of alternatives to determine the preferred choice. Methods include the dominance decision rule and outranking techniques in which the most 'satisficing' alternative is obtained by pairwise preference ordering of a finite set of alternatives. Holistic methods are generally compensatory, as a high value on one attribute can offset a low score on another, whereas most elimination techniques are non-compensatory (Silver 1991). *Wholistic judgement*, unlike both holistic and heuristic rules, is based on previous experience and includes intuition, standard operating procedures and reasoning by analogy. Mixed techniques have elements of two or more of these types of methods. Wholistic techniques are based on naturalistic decision models, whereas holistic and heuristic ones are based on classical decision theory. These concepts are discussed in more detail in the next section. Most existing MCDM are holistic or heuristic and very few wholistic methods have been developed. A number of different types of multi-criteria decision methods are presented in chapter 11 and their use is illustrated by a case study of waste management options in chapter 12.

10.3.2 Decision Making and Decision Support Systems

There are a number of different models of decision making, but still considerable gaps in understanding. Little attention has been given to defining the term decision and there has been a tendency to focus on choice from a set of alternatives which is not always appropriate. Some decision models are prescriptive and normative, while others are intended to describe the behaviour of real decision makers. However there has been little discussion in the literature of what constitutes a good decision or the relationship between good decision processes and good decisions (Hersh 1999).

There are two main approaches to modelling decision making, based on classical (Beach and Mitchell 1978) and naturalistic decision theory (Lipschitz 1993a) respectively. The main features of the two approaches are summarised in table 10.2 and discussed in more detail in the text.

Classical decision models focus on the decision event or choice of an alternative from a fixed set. However *naturalistic* decision models consider decision making to be a dynamic and evolving process. Naturalistic decision making is based on intuitive strategies, whereas classical approaches require more analytical ones. Classical theory is normative and prescriptive, in that it describes the choices of an ideal hypothetical decision maker, and is often used as a standard against which actual decision making is evaluated, whereas naturalistic decision theory is intended to model the behaviour of real decision makers.

Classical theory assumes that decision models are based on the idea of rational choice and frequently also on multi-attribute utility theory. Decision processes are assumed to be consequential and preference based in the sense that a course of action is chosen from the set of alternatives on the basis of personal preferences for the (expected) future consequences of current actions. Although rational decision makers are frequently assumed to maximise or choose the best available alternative, in practice they often satisfice or choose an alternative which exceeds predefined criteria or targets. *Maximising* requires comparison of all possible alternatives and

Table 10.2 Natural and classical decision making

Classical Decision Making	Naturalistic Decision Making
Choice of an alternative from a fixed set.	Decision making is a dynamic and evolving process.
Analytical strategies.	Intuitive strategies
Normative and prescriptive.	Models the behaviour of real decision makers.
Decision models based on rational choice.	Decision models based on situation assessment and mental imagery.
Maximising decisions.	Satisficing decisions.
Chooses the best of the available alternatives.	Looks for new alternatives by changing the problem constraints.
Calculation.	Reasoning about available information.

choice of the best one and *satisficing* sequential comparison of alternatives until a satisfactory one is found (March 1994). A maximising decision maker is likely to select the best of a number of poor alternatives, whereas a satisficing decision maker may try to improve the set of possible options by changing the problem constraints, making the latter approach more appropriate to sustainable decision making, since existing alternatives may not be sustainable.

Naturalistic decision models are generally based on situation assessment, understanding of the surrounding context and frequently also on mental imagery or mental simulation. This allows decision makers to forecast the adequacy of a course of action by imagining it in a specific situation, anticipating important steps and considering the most likely reaction to it. (Klein 1989). Reasoning about the available information is more important in naturalistic decision models than the performance of calculations on it. There are a number of different approaches (Lipshitz 1993b).

There has been a certain amount of *polarisation* between the two main approaches to decision making with, for instance, naturalistic decision theory generally concentrating on the performance of expert decision makers and classical theory on the biases which prevent all decision makers from making optimal decisions. However there has been little consideration of what makes a decision unsatisfactory in real terms and what types of decision processes to avoid. In terms of classical theory (Janis and Mann 1977) good decision making requires thorough investigation of the full range of objectives and alternatives and all their positive and negatives consequences. However, according to naturalistic decision theory there is no benefit in generating a large number of options, particularly for experienced decision makers under time pressures (Klein 1993).

It should be noted that there is not a unique universal decision strategy which is appropriate for all types of decisions, but that a number of factors, such as the importance and complexity of the decision, the possible consequences of making a 'wrong' decision, the experience and training of the decision maker(s), the extent to which available information meets information needs and the degree of time pressure, determine the most appropriate approach. It has been found that intuitive naturalistic approaches are most likely to be used by decision makers who are experienced, when there is time pressure or the situation is changing. Analytical classical strategies are most likely to be used when data is alphanumeric rather than graphical, there is an organisational need to justify decision choices, the problem is computationally complex or different populations representing varying interests have to be reconciled (Hammond et al 1987).

This raises the question of which type(s) of approaches to decision making are most appropriate for *sustainable decision making* and under what circumstances. The *classical decision* approach suffers from requiring objective, quantitative and precise data and criteria and the lack of a satisfactory framework for incorporating environmental issues and abstract and subjective concepts such as equity, poverty, social justice and quality of life into the analysis. However there is ongoing work to derive measurable indicators and criteria for sustainable development (UNESCO 1997; Simonovic et al 1997), though progress has been moderate to date. There has also been considerable work on obtaining measures of quality of life, welfare and associated concepts, frequently as part of an attempt to obtain alternative economic indicators to gross domestic product (GDP) (Anderson 1991; Gallopin and Öberg 1992). A number of measures of intragenerational (within a generation) or intergenerational (between generations) equity which could be used in decision making have been derived (Matheson et al 1997). However this type of approach tends to transform the original problem into one which is frequently just as difficult to resolve and has similar advantages and disadvantages to the use of classical decision methods. On the positive side problem transformation by the use of indicators and criteria of sustainable development can help to structure the problem and increase understanding by clarifying the conditions to be satisfied. On the negative side it may lead to degradation of information and quantitative evaluation of the indicators may be difficult, inexact or even inaccurate.

To a lesser extent classical decision theory also suffers from the limitation of having a philosophical basis in economic utility theory, which tends to give this approach a rather narrow and even selfish perspective. On the other hand it has the advantage of providing an organised framework for structuring all the available information, analysing a large volume of data and justifying or clarifying the various tradeoffs made. This is particularly relevant in the case of the different interests and stakeholders involved in sustainable decision making. However the benefits of this last factor are frequently reduced by the highly mathematical nature of this approach, which will generally reduce the comprehensibility and transparency of the analysis.

However *naturalistic decision* strategies also have their disadvantages. Despite the benefits of making full use of the expertise of experienced decision makers, it is less clear how suited these strategies are to relatively inexperienced decision makers. Making use of expertise may not even be advantageous in the case of sustainable decision making, since this expertise may be based on old decision making patterns which do not consider sustainability issues. Discussion of naturalistic decision making has also largely focused on the types of decisions that are a prelude to immediate action, for example in the case of fire fighting, rather than on policy formulation and decisions relating to project choice or industrial design, which may be more relevant in the context of sustainable decision making.

Approaches in some areas of sustainable policy formulation and decision making can be seen to follow *explanation based models*, in which the goodness of fit of different explanations or accounts of the available evidence is evaluated and the account with the fewest questionable assumptions is accepted (Pennington and Hastie 1993).

Example 10.5: Accounts of Global Warming

Responses to global warming seem to be based on explanations of the evidence which can be divided into two main accounts. One of them explains observed changes in climate in terms of anthropogenic (human generated) emissions, leading to calls for the immediate formulation of policy and the investigation of measures for a very significant staged reduction of emissions. The other explains the same observations in terms of natural climate variations and further postulates the ability of present and future technological developments to resolve any problems. However improvements in modelling the relationship between anthropogenic emissions and climate changes and the availability of additional data have reduced the scope for divergent interpretations of the evidence and led to general, though not universal, acceptance, including by the overwhelming majority of scientific opinion, that anthropogenic emissions are largely responsible for observed climate changes (Houghton et al 2001).

As this example illustrates, explanation based decision making does not always lead to sustainable decision making. It also has the basic drawback of lacking independent mechanisms for validating conclusions and ensuring that the importance attached to items of evidence is not determined by preconceived ideas and evidence which does not fit the account is not ignored.

10.3.3 Sustainable Decision Making: Some General Principles

Sustainable decision making generally involves a range of environmental, economic, political, social, ethical and other factors, in addition to considerations of costs and the quality and quantity of service provision, and requires a mixture of quantitative and qualitative, precise and imprecise, subjective and objective data. It requires a focus on long as well as short terms effects and both local and global impacts. This may be best formulated as a multi-scale approach to allow consideration of impacts and consequences over a range of different time scales and regions.

Sustainable decision problems may be unstructured and characterised by uncertain dynamic environments, shifting, ill-defined or competing goals, action feedback loops, time stress, high stakes, multiple players and multiple organisational goals and norms. *Uncertainty* and *risk* are also important. In addition to the uncertainty from measurement error and poor quality data, incomplete understanding of some of the underlying issues may lead to controversy about what is and is not sustainable. For instance the causal relationship between anthropogenic emissions and global climate change has only fairly recently gained general acceptance. Although considerable progress has been made towards understanding the mechanisms involved, there are still many open questions in this area.

Thus the *precautionary principle* (Harremoës et al 2002; O'Riordan and Cameron 1994) of avoiding action which might have unforeseen effects on parts of the complex interacting environmental system which are poorly understood should be an important part of sustainable decision making. According to this principle, for instance, nuclear power stations should not have been built until the effects of radiation on the environment were better understood and the problem of disposal of radioactive waste had been resolved. Wider issues relating to the use of nuclear power include the fact that in some countries workers in the nuclear industry have been forbidden to join trade unions. The precautionary principle should probably be extended to cases where the introduction of new technology could have effects which may be socially or developmentally damaging, for instance through being used to restrict trade union organisation.

The problem of *qualitative* data has been mentioned. If a rank order can be obtained for qualitative data of a particular type, it is generally possible to convert this rank order to a numerical scale. This data can then be integrated with other types of data and data on different scales can be put on an equal footing by using normalisation. There are four main types of *normalisation*:

$$v_{i.1} = \frac{a_i}{\max a_i} \qquad v_{i.2} = \frac{a_i - \min a_i}{\max a_i - \min a_i}$$

$$v_{i.3} = \frac{a_i}{\sum_i a_i} \qquad v_{i.4} = \frac{a_i}{\left(\sum_i a_i^2\right)^{1/2}} \qquad (10.3)$$

where the a_i are the original coefficients and the $v_{i.m}$ the normalised ones, and the second suffix m indicates the type of normalisation. The first method is the simplest, can be easily interpreted and maintains proportionality. The second one is designed so that the values cover the whole interval [0, 1] and maintain the respective magnitudes of the coefficients, but not proportionality, so that the ratio of the normalised coefficients $v_{ij.m}/v_{kj.m}$ is not necessarily equal to the ratio of the original coefficients a_{ij}/a_{kj}. The third procedure is frequently used, for instance in Saaty's Analytic Hierarchy Process (1980 1990), Nagel's P/G% method (Nagel 1988; Nagel and Bievenue 1989 1990ab 1992) and for the normalisation of the weights associated with different criteria. However it has the disadvantage of giving values which are small and bunched together. The fourth method allows comparison of unimodular vectors i.e. vectors which can be extended to a square matrix with determinant 1. However this method has the same disadvantages as the previous method and is rarely used (Pomerol and Sergio 1993).

Despite the considerable advantages of being able to quantise qualitative data, it should be noted that this may involve a loss or sometimes even a distortion of the information. It should also be recognised that any numerical values obtained will not be precise. It may therefore be more appropriate to apply fuzzy methods, which are discussed in Part III, to systems with qualitative as well as quantitative data, as the use of fuzzy logic allows the representation of linguistic variables and uncertain data (Fortin and Bobee 1995; Zadeh 1994).

Sustainable decision making frequently involves uncertainty and inadequate information. In some cases full understanding of the situation would require data on environmental effects over an extended period of possibly several hundred years, but decisions have to be made within the limitations of existing data and time constraints. However decision making based on such imperfect or uncertain information is preferable to the exclusion of ecological considerations, as has occurred in the past. Since much of the available information is uncertain, sensitivity analysis should be used to investigate the dependence of decisions on particular parameters, weights and models.

Most sustainable decisions involve a number of groups and individuals, often with very different or even conflicting interests. This gives rise to the question of how all these different interests can best be taken into account in decision making. There are two approaches, which can be considered satisfactory:

- *Enlarging* the decision problem to allow equal or weighted consideration to be given to the values of all the decision makers.
- *Prioritising* particular decision makers or points of view, for instance based on equity considerations.

However it is still common for one individual or organisation to take decisions for all stakeholders, with differing degrees of consultation. The approach to be discussed now is that of enlarging the decision problem when multi-criteria decision methods are used, to take into account all the decision makers by increasing the size of the problem and allowing (possibly weighted) values for each decision maker on each criterion to be considered.

Thus it is assumed that there are n decision makers (or stakeholders whose interests should be considered) $D_1, D_2, ..., D_n$ and m criteria $C_1, C_2, ..., C_m$. The dimensions of the problem can be increased to give nm criteria $C_{11}, ..., C_{1m}, C_{21}, ..., C_{2m}, ..., C_{n1}, ..., C_{nm}$, with the criteria $C_{i1}, C_{i2}, ..., C_{im}$ associated with the ith decision maker D_i. The criteria associated with each decision maker can be given weights w_{ij} $(i = 1...n, j = 1...m)$, with $\sum_{j=1}^{m} w_{ij} = 1$ for all i to allow each decision maker to give different relative weights to the different criteria. Similarly all the criteria C_{i1} to C_{im} associated with a decision maker D_i could be given a (normalised) weight w_i for $i = 1...n$ to allow different degrees of importance to be attached to the different decision makers, as shown in fig. 10.2. The decision problem with the weighted criteria $w_i w_{ij} C_{ij}$ $(i = 1...n, j = 1...m)$ can then be analysed by any multi-criteria decision method.

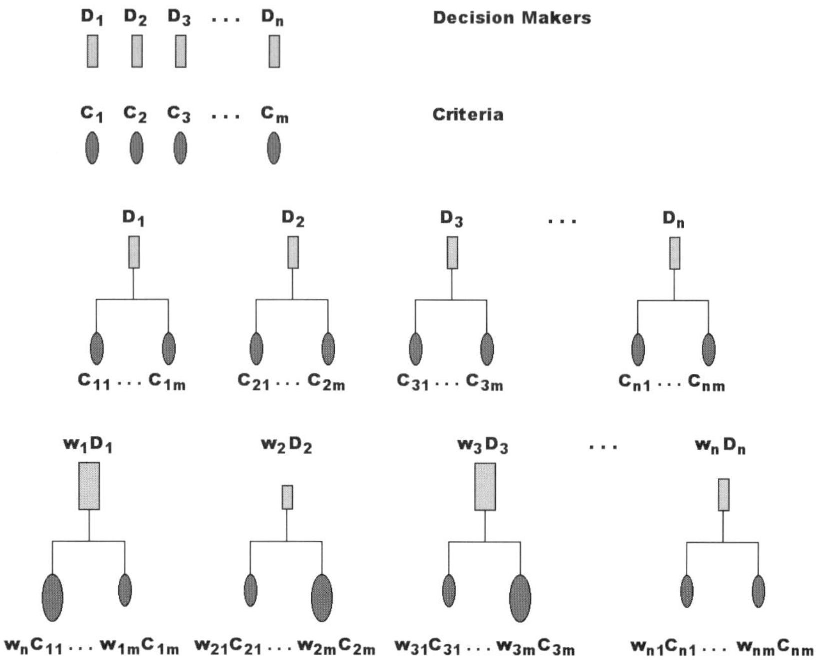

Fig. 10.2 Expansion of criteria to take account of multiple decision makers

Most multi-criteria decision methods target the case of explicitly stated courses of action, whereas many sustainable decision problems should only be defined implicitly in terms of the achievement of a set of goals or satisfaction of a number of criteria. Use of an explicit decision making process when the underlying problem is implicit generally assumes the prior decision stage required to arrive at an explicit statement of alternatives. For instance the decision process in the case of relieving road congestion is frequently restricted to choice of route, but this approach assumes an a priori decision to resolve the problem by building a new road, though this is rarely the most sustainable approach. Therefore a more sustainable approach to the decision problem would be in terms of the goal of minimising traffic congestion, which might lead to the solution of providing new bus routes and reducing bus fares to attract traffic from cars to buses. Thus, in some cases implicit decision making can best be divided into two stages, in the first of which a set of explicit alternatives is derived by determining the approach which best meets the criteria and in the second the best alternative from this set is chosen. There is also often the question of what the appropriate *decision question* is. Thus, in the transport case, though one of the most frequently posed questions is about reducing congestion, a more sustainable approach might involve obtaining transport and urban planning strategies to make facilities and resources equally accessible to all sections of the population while minimising the need to travel to reach these facilities.

An important aspect of sustainable decision making is policy decision making on the appropriate responses to changes in the levels of important system variables, whether by regulation or by changes in service provision. There are a number of examples of inappropriate responses in the area of public decision making. For instance in the past policy decisions about the requirements for power generation capacity have been based on (unfounded) assumptions about the rate of growth of electricity demand, rather than measures being taken to regulate the level of demand and generation capacity provided to meet this regulated level of demand. An analogous situation holds for car traffic where (in this case underestimated) levels of growth have been assumed and new road capacity built to deal with the expected increases in the volume of car traffic. However this has resulted in traffic generation and a more sustainable approach would regulate traffic volumes rather than increase road capacity to deal with expected increases in demand. Increasing road capacity also gives rise to equity issues, as it has generally been at the expense of the provision of public transport and facilities for cyclist and pedestrians. It has therefore disadvantaged poorer sections of the community, women, children and older people, who are the main groups of pedestrians and users of public transport. The use of feedback to regulate demand was discussed in section 3.5.

10.4 Roundup of Chapter 10

Chapter 10 presented an introduction to the basic principles of multi-criteria decision and optimisation problems and discussed their role in sustainable development. There were two main topics: the structure of multi-criteria decision problems, and the categorisation of decision support methods and their application to sustainable decision problems.

Decision making has been considered to have four main stages, involving collecting information, problem formulation in terms of alternatives, criteria and attributes, choice of alternative or type of approach, and assessment of the decision process. Most decision support algorithms only support the third stage of choice of alternative. In multi-stage decision problems there is more than one choice stage.

There are a number of different types of optimal solutions. Admissible, efficient or Pareto optima dominate i.e. perform better or at least as well as the other solutions on all the criteria and strictly better on one criterion. Satisficing solutions meet a satisfaction level or level of performance on all the criteria, but are not necessarily optimal on any of them. The ideal point is the point in criterion space which maximises each criterion separately. Unfortunately the ideal point frequently does not correspond to a realisable option.

One of the main categorisations of decision making is into classical and naturalistic approaches. Classical decision making approaches use analytical techniques to choose the best alternative (in terms of the criteria) from the available alternatives. Naturalistic approaches use intuitive strategies and decision models based on situation assessment and mental imagery to obtain a solution which performs well rather than necessarily the optimal solution.

Sustainable decision making often involves qualitative and unstructured data, a number of groups and individuals with different or even competing interests and unstructured problems in uncertain environments. The views of different stakeholders can be taken into account by increasing the dimension of the problem to include the different criteria from the perspective of the different stakeholders.

This introduction to multi-criteria decision making in chapter 10 is followed by discussion of a number of multi-criteria decision support methods in chapter 11. It should be noted that these methods are only able to support the third choice stage of the decision problem.

11 Multi-Criteria Decision Support Methods

11.0 Learning Objectives

This chapter presents four different types of multi-criteria decision support methods. The specific learning objectives are knowledge of and ability to apply the following types of decision support methods:

- Outranking methods, including ELECTRE and PROMETHEE.
- Aggregation of criteria methods, including P/G% and goal programming.
- The Analytic Hierarchy Process.
- Multi-attribute utility theory.

11.1 Introduction

The previous chapters in this section have considered the mathematical and conceptual framework in which decision making takes place. This chapter draws on this material in the presentation of methods which can be used to support practical decision making. It should be noted that decision support methods can only be used in the third choice stage of decision making and do not support the other three stages of obtaining information, problem formulation and evaluation of the decision process and the decision reached.

The decision methods presented here are divided into four main categories and at least one example of a method in each category is given. Section 11.2 defines outranking relations, on which outranking methods are based and then presents the ELECTRE IS, ELECTRE III, PROMETHEE I and PROMETHEE II methods. Section 11.3 presents the P/G% and goal programming methods, which use different approaches to aggregate the criteria to give one criterion. Section 11.4 presents the Analytic Hierarchy Process and section 11.5 multi-attribute utility theory.

11.2 Outranking Methods

Outranking methods are based on outranking relations, which will be defined in section 11.2.1. They generally can be applied to multi-criteria problems which are defined explicitly and involve choosing an alternative from a finite set A of n possible alternatives or potential actions to maximise a consistent family F of k evaluation criteria $g_j(x)$, $j = 1...k$ i.e.

$$\text{Max}\{g_1(x), g_2(x), ..., g_k(x) | x \in A\} \tag{11.1}$$

11.2.1 Outranking Relations

An *outranking relation* (Pomerol and Sergio 1993; Vetschera 1988) is a binary relation between two alternatives (in a multi-criteria problem). Two options a and b satisfy an outranking relation if, in an appropriate sense, the option a is at least as good as (or not worse than) the option b relative to the n criteria of interest. This can be written a S b. It should be noted that, unlike a dominance relation, an outranking relation is reflexive, as clearly a S a for all possible options a i.e. each option is at least as good as itself. However S is not necessarily transitive. (The properties of binary relations are defined in section 9.2.1.)

This then raises the issue of how the concept 'at least as good as' should be interpreted. In terms of a single criterion with evaluation function $g_j(.)$ this can be defined in terms of indifference thresholds as

$$a \text{ S } b \text{ if and only if } g_j(a) \geq g_j(b) - q_j, \quad g_j(.) \geq 0 \tag{11.2}$$

where the real positive number q_j is the *indifference threshold* for the jth criterion and the expression can easily be generalised to allow for variable thresholds. Therefore outcomes a and b are considered equally acceptable in terms of the criterion with evaluation function $g_j(.)$ if their values for this criterion do not differ by more than q_j. Then, a outranks b if the above inequality is satisfied for each of the k criteria i.e.

$$a \text{ S } b \text{ if and only if } g_j(a) \geq g_j(b) - q_j, \quad q_j \geq 0, \quad j = 1...k \tag{11.3}$$

A dominance relation can be obtained from an outranking relation by putting the threshold q_j equal to zero for all j. Therefore dominance is a stricter condition than outranking and satisfaction of a dominance relation D by two outcomes a and b, written a D b, implies satisfaction of an outranking relation, but the converse does not generally hold i.e.

$$a\,D\,b \Rightarrow a\,S\,b \tag{11.4a}$$
$$a\,S\,b \not\Rightarrow a\,D\,b \tag{11.4b}$$

Outranking relations have a wider definition of indifference between two options than dominance relations. Consequently outranking relations allow *tradeoffs* between performance on the different criteria and this makes them more appropriate for use in decision algorithms. The different outranking algorithms, which include the ELECTRE and PROMETHEE methods discussed below, use the outranking relation in different ways to find the overall best outcome or the set of acceptable outcomes.

Example 11.1: Acid Rain Policy Choices III

Consider examples10.2 and 10.3 on acid rain policy choices in section 10.1.3, with the three criterion functions, $g_1(.)$ for maximising air quality, $g_2(.)$ for maximising water quality, and $g_3(.)$ for minimising increases in unemployment, all taking values 1-4, with increasing values satisfying the criterion to an increasing extent. The four policies P_a, P_b, P_c and P_d have

- $(g_1(P_a), g_2(P_a), g_3(P_a)) = (3, 2, 3)$
- $(g_1(P_b), g_2(P_b), g_3(P_b)) = (3, 4, 3)$
- $(g_1(P_c), g_2(P_c), g_3(P_c)) = (4, 4, 2)$
- $(g_1(P_d), g_2(P_d), g_3(P_d)) = (4, 4, 4)$

Example 10.2 showed that P_d dominates all the other criteria and strictly dominates P_a, P_b dominates P_a and P_c neither dominates or is dominated by P_a and P_b. Thus P_d outranks all the other criteria and P_b outranks P_a, whatever values are chosen for the indifference thresholds.

If the indifference threshold on the unemployment criterion is taken equal to 1 and the indifference thresholds on the air and water quality criteria are taken equal to 0, then

$$g_1(P_c) \geq g_1(P_b) \qquad g_1(P_c) \geq g_1(P_a)$$
$$g_2(P_c) \geq g_2(P_b) \qquad g_2(P_c) \geq g_2(P_a)$$
$$g_3(P_c) \geq g_3(P_b) - 1 \qquad g_3(P_c) \geq g_3(P_a) - 1$$

and P_c outranks both of P_b and P_a. though it does not dominate either of them.

11.2.2 ELECTRE Methods

This class of methods (Crama and Hansen 1983; Rogers and Bruen 1998; Roy 1968, 1971 1978 1990; Roy and Bouyssou 1993; Vetschera 1986;

Vincke 1989) uses concordance and discordance indices and also takes into account the (relative) importance of the different criteria. The jth criterion agrees (is *concordant*) with the outranking relation a S b if and only if

$$g_j(a) \geq g_j(b) - q_j \tag{11.5a}$$

disagrees (is *discordant*) with the outranking relation a S b if and only if

$$g_j(b) \geq g_j(a) + p_j \tag{11.5b}$$

and neither agrees nor disagrees with the outranking relation a S b if and only p and q are such that both inequalities (11.5a) and (11.5b) hold.

$$\Rightarrow g_j(a) + p_j \leq g_j(b) \leq g_j(a) + q_j \tag{11.5c}$$

where the *indifference* and *preference thresholds* q_i and p_i can be chosen to take account of imprecision, uncertainty and inaccuracies. The three inequalities (11.5) can be used to partition the family F of criteria into three subsets, denoted by C(a S b), C(a P b) and C(a Q b) respectively, of criteria in agreement (concordant), in disagreement (discordant) and indifferent to the outranking relation a S b. These families of criteria are not disjoint. In particular the criteria which are indifferent to the outranking relationship i.e. those in C(a P b) are those which satisfy both inequalities (11.5a) and (11.5b) and consequently are in both the concordance subset C(a S b) and the discordance subset C(a Q b). Fig. 11.1 shows the three subsets for the case q > p. When this condition is not satisfied there is no indifference region as it is not possible for both inequalities (11.5a) and (11.5b) to hold simultaneously.

The *concordance index* c(a, b) of two options a and b is a measure of the extent to which a outranks b. It is defined as follows (Roy 1990)

$$c(a,b) = \frac{1}{k}\sum_{j \in C(aSb)} k_j + \frac{1}{k}\sum_{j \in C(aQb)} k_j \frac{p_j + g_j(a) - g_j(b)}{p_j - q_j}, \quad \text{with } k = \sum_{j \in F} k_j \tag{11.6}$$

where the coefficient $k_j \geq 0$ measures the relative importance of the jth criterion, so that the ratio k_j/k gives a measure of the relative strength of each of the criteria which agrees with the concordance relation in the family F of criteria. Therefore

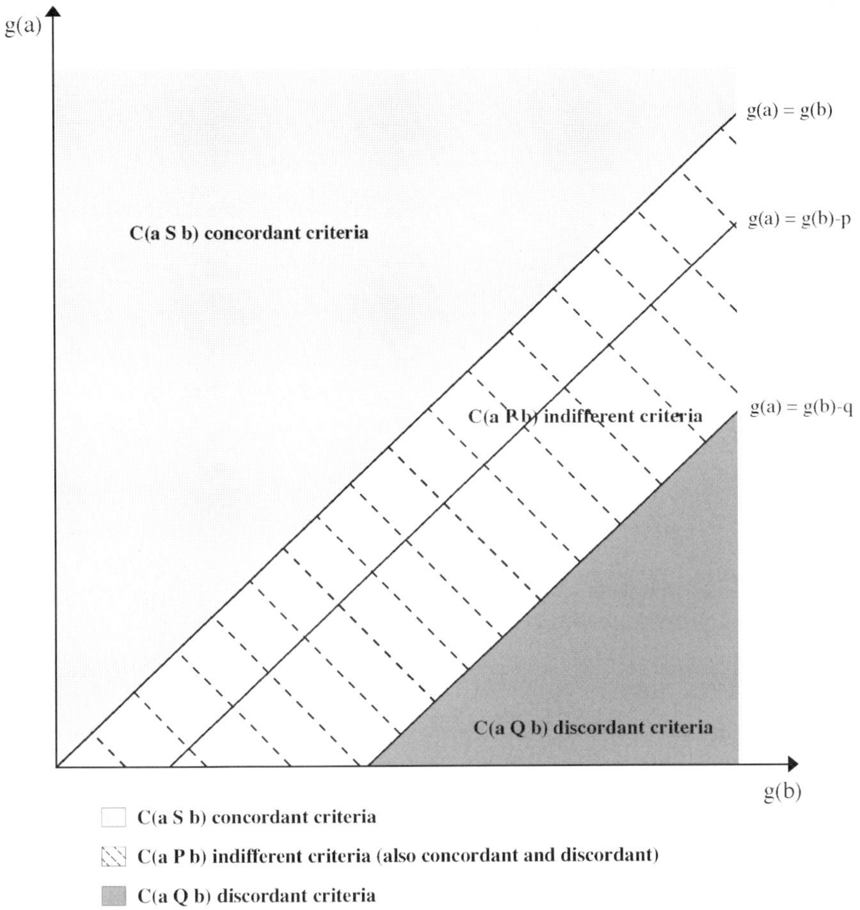

Fig. 11.1 Concordance, discordance and indifference

$$0 \leq c(a, b) \leq 1 \tag{11.7}$$

with

- c(a, b) = 0   if all the criteria disagree (are discordant) with the outranking relation, so that the subset of discordant criteria C(a P b) is equal to the whole set i.e. c(a P b) = F.
- c(a, b) = 1   if all the criteria agree (are concordant) with the outranking relation, so that the subset of concordant criteria C(a S b) is equal to the whole set F of criteria i.e. c(a S b) = F.

Therefore the concordance index takes values between 0 and 1.

The *discordance index* of the jth criterion is defined as

$$d_j(a,b) = \begin{cases} 0 & \text{if } g_j(b) - g_j(a) \leq p_j \\ \dfrac{g_j(b) - g_j(a) - p_j}{v_j - p_j} & \text{if } p_j < g_j(b) - g_j(a) \leq v_j \\ 1 & \text{if } g_j(b) - g_j(a) > v_j \end{cases} \tag{11.8}$$

where v_j is the *veto threshold* of the jth criterion, so that the outranking relation a S b does not hold if

$$g_j(b) - g_j(a) > v_j \tag{11.9}$$

regardless of the values of a and b for the other criteria. It is often useful to investigate the effect of changing the threshold values on the concordance and discordance coefficients.

The different ELECTRE methods include ELECTRE I, ELECTRE IS, ELECTRE Iv (veto), ELECTRE II, ELECTRE III, ELECTRE TRI AND ELECTRE IV (Figueira et al 2005). The methods ELECTRE IS and ELECTRE III will be discussed further here. Software has been developed (WWW12 2005) to run some of the methods, including ELECTRE IS, ELECTRE III-IV and ELECTRE TRI.

In the method *ELECTRE IS*, the outranking relation a S b is considered valid if the following two conditions hold:

$$c(a,b) \geq s \quad \text{for } 0.5 < s \leq 1 - \frac{1}{k} \min_{j \in F} k_j \tag{11.10a}$$

where s is a parameter called the *concordance level*

$$g_j(a) + v_j \geq g_j(b) + \min(q_j, v_j - p_j) w(s, c) \tag{11.10b}$$

$$\text{with } \; w(s,c) = \frac{1 - c(a,b) - k_j/k}{1 - s - k_j/k} \tag{11.11}$$

Condition (11.10a) implies that a sufficiently high percentage of the criteria must support outranking, whereas condition (11.10b) implies that no criterion so strongly opposes outranking that there is a veto effect. The coefficient w(s, c) models the reinforcement of the veto effect, with no reinforcement for $w(s, 1 - k_j/k) = 0$ and maximum reinforcement (magnitude q_j) for $w(s, s) = 1$.

The outranking relation in *ELECTRE III* is a fuzzy binary relation, with *credibility index*

$$\sigma(a,b) = c(a,b) \,.\, \prod_{j \in D_c(a,b)} \frac{1 - d_j(a,b)}{1 - c(a,b)} \tag{11.12}$$

where $D_c(a,b) = \{j : j \in F,\ d_j(a,b) > c(a,b)\}$

i.e. $D_c(a, b)$ is the set of indices for which there is greater disagreement than agreement with the outranking relation. The credibility index is thus the product of the concordance index with a factor which depends on the terms for which the discordance index is greater than the concordance index. The ELECTRE outranking methods can be used to obtain a comprehensive model of preference structure to be used in decision making.

11.2.3 PROMETHEE Methods

This class of methods (Brans et al 1984 1985 1986; Briggs et al 1990; Mladlineo and Margeta 1987) is based on a preference index $\pi(a, b)$ and a valued outranking graph, with nodes which are the outcomes in K and arcs with the value $\pi(a, b)$ for all a, and b in K. If a dominates b, $\pi(b, a) = 0$, but $\pi(a, b)$ is not necessarily equal to 1, as a can perform better than b on each criterion without the preference being strict. The *preference index* for an outcome a relative to b can be defined as the weighted average of the preference functions for six different types of criteria i.e.

$$\pi(a,b) = \sum_{i=1}^{6} w_i P_i(a,b) \quad \text{with} \sum_{i=1}^{6} w_i = 1 \tag{11.13}$$

where $P_i(a, b)$ is the preference function of a relative to b for the ith type of criterion and w_i the relative weight or importance of the ith type of criterion.

A *preference function*, taking values between 0 (indifference) and 1 (strict preference) can be defined for each criterion to give the decision maker's preference for a particular alternative a with respect to another alternative b under that criterion. The preference function P(a,b) associated with a criterion g(.) can be defined in terms of the difference x between g(a) and g(b) as follows

$$P(a,b) = \begin{cases} 0 & \text{if } g(a) \le g(b) \\ p(x) & \text{if } g(a) > g(b) \text{ for } x = g(a) - g(b) \end{cases} \tag{11.14}$$

Six classes of preference function p(x) are used in the PROMETHEE outranking method. They are defined as follows and illustrated in fig. 11.2:

Type 1 Usual Criterion

$$p_1(x) = \begin{cases} 0 & x = 0 \\ 1 & |x| > 0 \end{cases} \tag{11.15a}$$

Type 2 Quasi-Criterion

$$p_2(x) = \begin{cases} 0 & |x| \le l \\ 1 & |x| > l \end{cases} \tag{11.15b}$$

Type 3 Criterion with Linear Preference

$$p_3(x) = \begin{cases} x/m & |x| \le m \\ 1 & |x| > m \end{cases} \tag{11.15c}$$

Type 4 Level-Criterion

$$p_4(x) = \begin{cases} 0 & x \le q \\ 0.5 & q < x \le q+p \\ 1 & x > q+p \end{cases} \tag{11.15d}$$

Type 5 Criterion with Linear Preference and Indifference Area

$$p_5(x) = \begin{cases} 0 & |x| \le s \\ (x-s)/r & s < |x| \le s+r \\ 1 & |x| > s+r \end{cases} \tag{11.15e}$$

Type 6 Gaussian Criterion

$$p_6(x) = \begin{cases} 1 - e^{-x^2/2\sigma^2} & |x| \le 1, \\ 1 & |x| > 1 \end{cases} \tag{11.15f}$$

For the type 1 criterion there is indifference only when g(a) and g(b) are equal and otherwise there is strict preference. In type 2 the indifference region is given by $|g(a) - g(b)| \le l$. In type 3 there is indifference when g(a) and g(b) are equal and strict preference when their difference is greater than m and the preference function increases linearly between indifference and strict preference. Types 4 and 5 both have indifference when the modulus of the difference $|g(a) - g(b)|$ is below a threshold, denoted by q and s for the type 4 and type 5 criteria respectively, and strict preference for this difference above another threshold, denoted by q+p and s+r respectively.

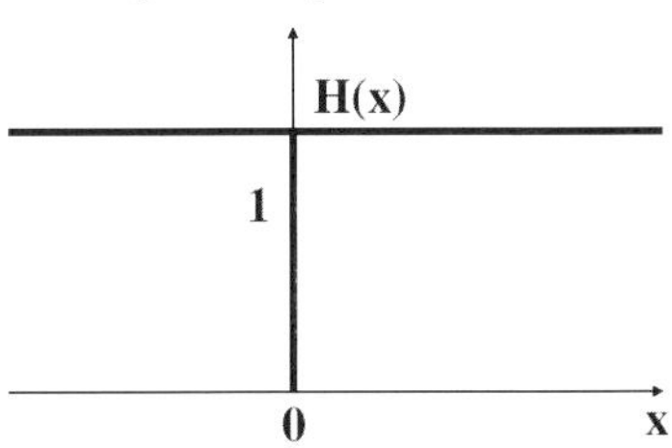

Fig. 11.2a Type 1 criterion

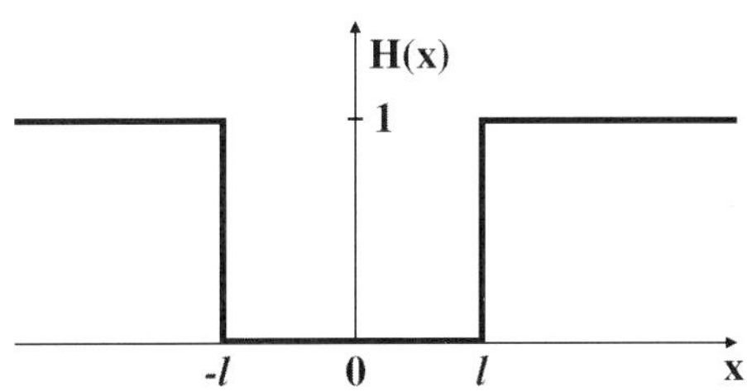

Fig. 11.2b Type 2 criterion

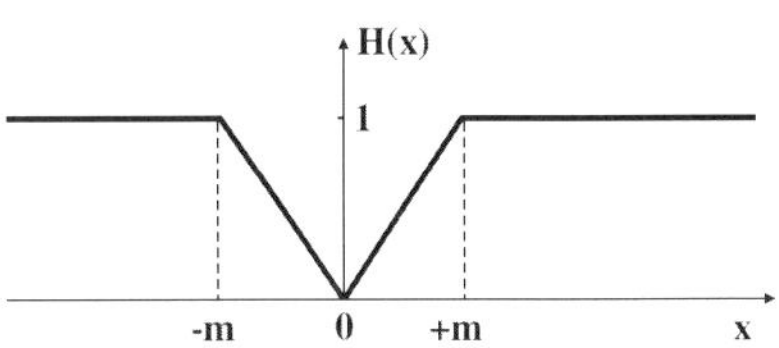

Fig. 11.2c Type 3 criterion

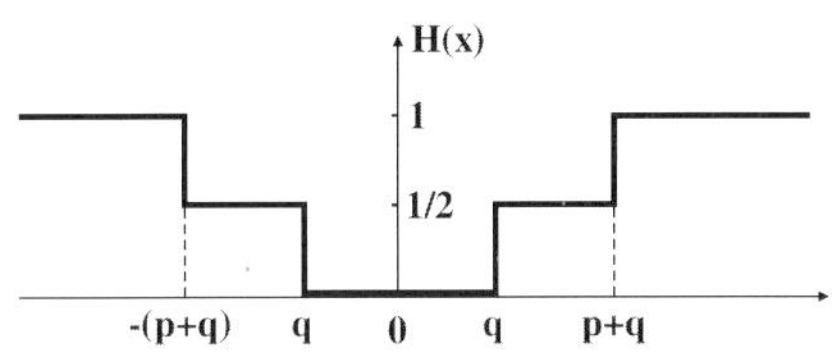

Fig. 11.2d Type 4 criterion

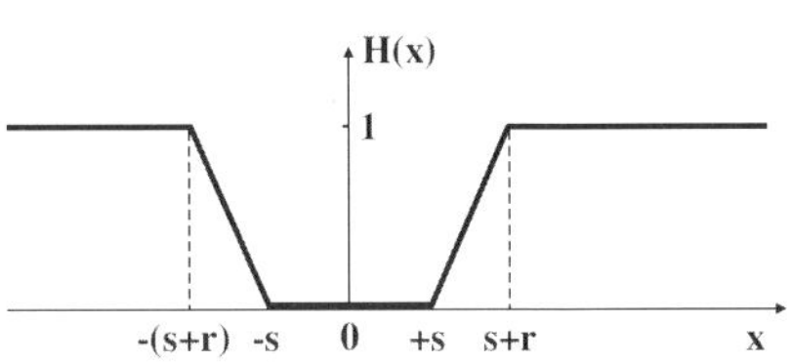

Fig. 11.2e Type 5 Criterion

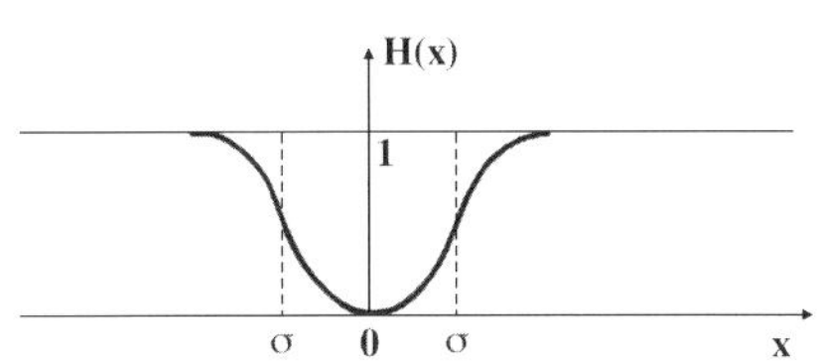

Fig. 11.2f Type 6 Criterion

They differ in the intermediate region, with the type 4 function equal to 0.5 and the type 5 function increasing linearly from 0 to 1 over this region. The type 6 function has indifference for g(a) and g(b) equal, strict preference for the difference greater than σ and increases as a Gaussian curve between these points.

In *PROMETHEE I* the outcomes are ranked by a partial preorder (transitive and symmetric relation) and for each node a the *outgoing flow* is defined as

$$\phi^+(a) = \sum_{b \in A} \pi(a, b) \tag{11.16a}$$

and the *incoming flow* as

$$\phi^-(a) = \sum_{b \in A} \pi(b, a) \tag{11.16b}$$

The extent to which an alternative a dominates the other (n-1) alternatives in the set of alternatives A increases with $\phi^+(a)$ and the extent to which it is dominated increases with $\phi^-(a)$. The PROMETHEE I partial relation is defined as

$$\begin{cases} \text{a outranks b (a } P^{(1)} \text{ b):} & \text{if } \begin{cases} a\,P^+\,b \text{ and } a\,P^-\,b \\ \text{or } a\,P^+\,b \text{ and } a\,I^-b \\ \text{or } a\,I^+\,b \text{ and } a\,P^-b \end{cases} \\ \text{a is indifferent to b (a } I^{(1)} \text{ b)} & \text{if } a\,I^+\,b \text{ and } a\,I^-b \\ \text{a and b are incomparable (a R b)} & \text{otherwise} \end{cases} \tag{11.17}$$

where

$$a\,P^+\,b \quad \text{if and only if} \quad \phi^+(a) > \phi^+(b) \tag{11.18a}$$

$$a\,P^-\,b \quad \text{if and only if} \quad \phi^-(a) < \phi^-(b) \tag{11.18b}$$

$$a\,I^+\,b \quad \text{if and only if} \quad \phi^+(a) = \phi^+(b) \tag{11.18c}$$

$$a\,I^-\,b \quad \text{if and only if} \quad \phi^-(a) = \phi^-(b) \tag{11.18d}$$

Thus an action a *outranks* another action b if one of the following holds:

- The outgoing flow of a is greater than that of b and the incoming flow is less.
- The outgoing flow of a is greater than that of b and the incoming flows of a and b are equal.
- The outgoing flows of a and b are equal and the incoming flow of a is less than that of b.

This results in a *valued outranking graph* in which some actions are comparable, whereas others are not, and this information can be used in decision making.

Example 11.2: Ranking Hydroelectric Power Station Projects

Fig. 11.3 shows the graph of the partial preorder obtained by ranking six hydroelectric power station projects on the criteria of person power required, power output, construction costs, maintenance costs, number of villages to be evacuated and security level (Brans et al 1986). However it should be noted that projects which require villages to be evacuated should be considered incompatible with sustainable development, due to the high social costs to the villagers involved. Therefore it would be more appropriate to treat this particular problem as a constrained optimisation problem with the constraint that projects should not require the evacuation of any villages. This would allow the elimination of projects which do not meet this constraint. In addition, in the large part of the world where there are job shortages, the criteria of person power should be modified to allow a certain minimum level of job creation. Fig. 11.3 shows that option x_5 outranks all the other options and that option x_4 outranks options x_6 and x_1 and is incomparable with options x_3 and x_2.

The *PROMETHEE II* method is based on a total preorder (a reflexive and

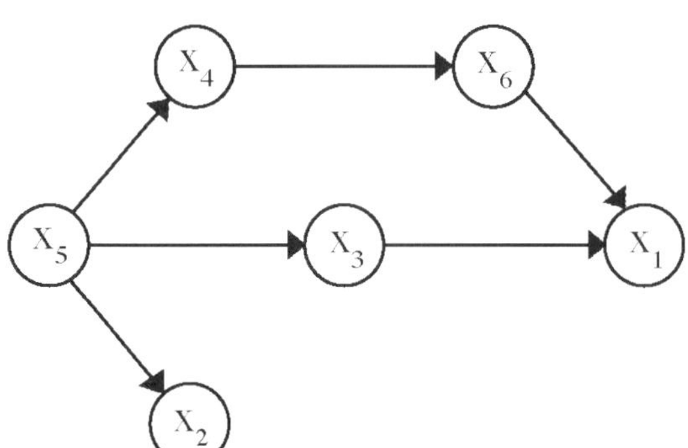

Fig. 11.3 Graph of PROMETHEE I preferences

transitive relation which gives a complete ranking without incomparabilities). This preorder is obtained by subtracting the inflow from the outflow to give the *net flow*:

$$\phi(a) = \phi^+(a) - \phi^-(a) = \sum_{j=1}^{k} \sum_{b \in A} [P_j(a,b) - P_j(b,a)] w_j \qquad (11.19)$$

where A is the set of alternatives, k is the number of criteria and $P_j(.,.)$ the preference function associated with the jth criterion. $P_j(a,b)$ (j=1...k) can be calculated for each pair of actions for each criterion and gives the preference of an action a relative to an action b for the jth criterion.

Then a outranks b (a $P^{(2)}$ b) if and only if $\phi(a) > \phi(b)$ (11.20a)

a is indifferent to b (a $I^{(2)}$ b) if and only if $\phi(a) = \phi(b)$ (11.20b)

Thus the PROMETHEE II method has the advantage of giving a complete ranking, but the disadvantages of a reduction in information and realism due to the balancing effects between outgoing and incoming flows. In addition there can also be advantages in allowing inconsistencies or lack of comparability, as in PROMETHEE I.

The *GAIA visual modelling technique* gives a geometrical interpretation of the PROMETHEE approach. It indicates which alternatives are particularly good (or bad) on particular criteria, allows clusters of similar criteria and alternatives to be obtained and gives a decision axis along which the best alternatives should be selected according to the weights allocated to each criterion. For each of the n alternatives the *normed flow* can be defined as

$$\gamma(a) = \frac{\pi(a)}{n-1} = \sum_{j=1}^{k} \gamma_j(a) w_j \quad \text{where } \gamma_j(a) = \frac{1}{n-1} \sum_{b \in A} [P_j(a,b) - P_j(b,a)] \qquad (11.21)$$

Since the terms $\gamma_j(a)$ are only dependent on j for each a, each $\gamma_j(a)$ is a function of only one unicriterion flow. Therefore each of the n alternatives is characterised by k unicriterion flows and can be represented as a point in $\mathcal{R}^k$ space, the axes of which correspond to the different criteria. The unicriterion flows are centred on the origin of the $\mathcal{R}^k$ space. Since there are generally more than 2 criteria, the k-dimensional space is projected orthogonally onto a plane. This plane, called the *GAIA plane*, is chosen to minimise the information lost through projection and is given by

$$u^T C u + v^T C v = n(\lambda_1 + \lambda_2) \tag{11.22}$$

where C is the covariance matrix of the γ_j, j=1...k and λ_1 and λ_2 are respectively the largest and second largest eigenvalues of C. A measure of the quantity of information preserved is given by (Brans et al 1994).

$$\delta = \frac{\lambda_1 + \lambda_2}{\sum_{j=1}^{k} \lambda_j} \tag{11.23}$$

i.e. the ratio of the sum of the two largest eigenvalues to the sum of all the eigenvalues.

Similar criteria have large positive covariances, whereas conflicting ones have negative covariances. Alternatives with a projection located in the direction of a particular set of criteria give good performance on those criteria and the projections of similar alternatives are close together. The weights associated with the different criteria can be expressed as a vector $\underline{w} = (w_1, w_2, ..., w_k)$. The following expression holds for the projection α_I of any criterion a_i (i=1...k) onto $\underline{w}$:

$$(\alpha_i, w) = \sum_{j=1}^{k} w_j \gamma_j(a_i) = \gamma(a_i) \tag{11.24}$$

The projection of a given alternative onto the weight vector $\underline{w}$ gives the net flow of this alternative and consequently the complete PROMETHEE II ranking. Therefore the vector $\underline{w}$ can be considered to be on the *decision axis*. However it is generally not in the GAIA plane, but an image π of it can be obtained by projecting the unit vector $\underline{e}$ in the $\underline{w}$ direction onto the GAIA plane. When the projection is short, the weight vector $\underline{w}$ is nearly orthogonal to the GAIA plane, the criteria are strongly conflicting and a good compromise should be selected in the neighbourhood of the origin.

When the decision axis is long, the decision maker should select alternatives which are as far as possible in this direction. When the scalar product of π and γ_j is large and positive, the jth criterion will be in agreement with the ranking, whereas when it is strongly negative it will be in conflict with it.

Example 11.3: GAIA Plane for Nuclear Waste Management

Fig. 11.4 (Briggs et al 1990) illustrates the Gaia plane for a nuclear waste management problems with 27 alternatives, based on the combination of three different time scenarios with three different methods of financing and three different disposal sites (Briggs et al 1990). Fig. 11.4 is drawn for one of the sites. The nine options from the combination of three time scenarios and three methods of financing are marked as points on this diagram and then clustered into groups of points which are in the same sector of the diagram and therefore belong to similar alternatives. The bold lines in the figure are the criterion axes which point to the direction where the best alternatives on a particular criterion are located. In this case the four criteria are total financial costs (C_1), present costs (C_2), future costs (C_3) and risk (C_4). As shown in the figure, none of the groups of alternatives performs well on all the criteria for this choice of site, illustrating the problems associated with the disposal of nuclear waste.

11.3 Aggregation of Criteria

This section presents two methods based on the aggregation or combination of criteria to give one criterion. They are the P/G% method and goal programming.

11.3.1 Dimensionless Scaling: P/G% Method

The *P/G% method* (Nagel 1988; Nagel and Bievenue 1989 1990ab 1992)

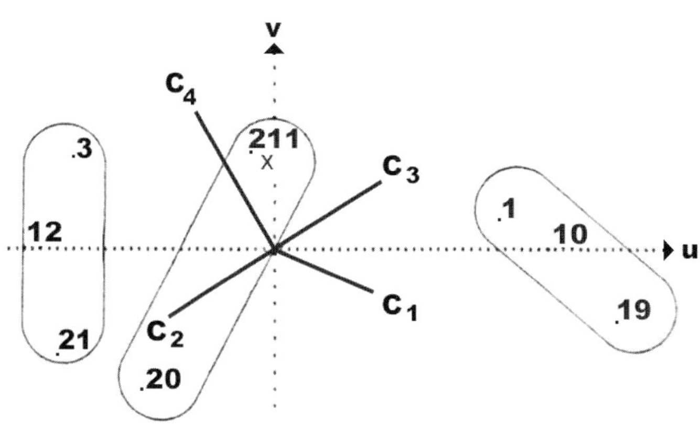

Fig. 11.4 The GAIA plane

allows different criteria to be combined using percentaging techniques to give *dimensionless scaling*. Despite the inevitable and sometimes very serious loss of information in moving from multiple to a single criterion, this approach allows comparison and combination of criteria measured in different units and avoids the problems of combining data in monetary units. This is particularly problematic in the context of sustainable development, where many important environmental and social goods do not have market prices.

A multi-criteria decision matrix is set up in spreadsheet format, with alternatives listed down the rows and criteria across the columns and values for each alternative on each criterion in the appropriate cell. The total score for all the alternatives for each goal is calculated by summation. The percentage (part/whole) score for a particular alternative on a given criterion is then obtained as a percentage score with values between 0 and 100. This percentage score is obtained by division of the score for the particular alternative on the given criterion by the total score for all alternatives on that criterion. Non-linear relations and negative goals can be taken into account by algebraic manipulation. The different criteria can be given weights to reflect their relative importance and each percentage score multiplied by the associated weight. The total score for each alternative is calculated by adding the weighted scores for the different criteria and then dividing by the sum of the total weights to give a dimensionless number between 0 and 100. Logs or antilogs can be used rather than the raw data in the case of diminishing returns.

The effect of constraints is considered after the optimum solution has been obtained from the part/whole percentages. When this solution does not satisfy the constraints, it is modified slightly in an iterative process to meet the constraints in order of importance.

11.3.2 Goal Programming

Goal programming involves identification of goals and their achievement levels, prioritising goals and derivation of an objective function using these *goals*, *achievement levels* and *priorities*. The decision problem can then be solved using a modified simplex approach (McKenna 1980; Stanton 1992). Formulation of a goal program has three main steps:

- Identification of goals and listing the desired or required achievement level for each goal as a constraint for program solution.
- Prioritisation of goal achievement by listing the goals in a hierarchy.

OK, I'll let you make the decisions in future!

- Construction of an objective function using the goals, achievement levels and priority levels.

The aim of the goal program is to obtain a solution which minimises deviations, denoted (D_1, D_2, ..., D_n) from the goals (G_1, G_2, ..., G_n), taking the priorities (P_1, P_2, ..., P_n). This gives the following *objective* or *cost function* to be minimised:

$$S = P_1(D_1G_1) + P_2(D_2G_2) + ... + P_n(D_nG_n) \quad (11.25)$$

11.4 The Analytic Hierarchy Process

The *Analytic Hierarchy Process* (AHP) (Saaty 1980 1990) is both descriptive and prescriptive. It is related to *multi-attribute utility theory*, but allows decision makers to make their own decisions on whether inconsistency in preferences and rank reversals should be permitted and, if so, the amount of permissible inconsistency. As its name implies, the AHP uses a hierarchical approach to structure complex information. The first step in applying the method is development of a *hierarchy*, the three main levels of which are in order from the top down: the *overall goal*, the *criteria* and the *alternatives*. The model can be refined by adding subcriteria to give further details of the problem objectives. Scenario models can be developed to enable decision alternatives to be considered under a range of different circumstances. The views of different decision makers can also be incorporated, as an additional layer in the hierarchy between the goal and the criteria. Intensities of achievement or preference for each of the criteria can be defined and used instead of alternatives in the first stage of evaluation.

After the hierarchical problem structure has been obtained, the next step is *evaluation* of each element of the problem by pairwise comparisons of all nodes at a given level in relation to the parent node at the previous level. Therefore the different criteria are compared in terms of their satisfaction of the overall goal and the different alternatives are compared in terms of their satisfaction of each criterion in turn. To facilitate these pairwise comparisons the elements at each level, such as the criteria and alternatives, are used to label the rows and columns of a matrix, so that the results of the comparison become the elements of the matrix.

This *comparison* is carried out using the scale from 0 to 9.0 given in table 11.1. If finer gradations are required, increments of 0.1 between these values could be used. If the ith element contributes equally or more strongly than the jth element to the level above it in the hierarchy, then the ijth element takes the value in the table and the jith element its reciprocal. Otherwise the jith element has the value in the table and the ijth element its reciprocal. Therefore the matrix has ones on the diagonal and the product of the ijth and jith elements is always one.

This matrix can then be used to calculate a priority vector by normalising the *principal eigenvector*, so that its elements sum to one. This priority vector gives the relative importance of the different criteria in achieving the overall goal. In general the matrix elements representing the relative importance or preference will not be totally consistent, due to factors such as lack of information, real world inconsistencies, an inadequate

Table 11.1 Comparison scale for Analytic Hierarchy Process

Scale	Interpretation
1.0	Two elements contribute equally to the objective.
3.0	Experience and judgement favour one element over another.
5.0	An element is strongly favoured.
7.0	An element is strongly dominant.
9.0	An element is favoured by a least an order of magnitude.
2.0, 4.0, 6.0, 8.0	Intermediate values between two judgements.

model structure, lack of concentration or fatigue and errors in entering data. To give a measure of the degree of consistency of the data, the *consistency index* has been defined as

$$\text{C.I.} = (\lambda_{max} - n)/(n-1) \tag{11.26}$$

where λ_{max} is the maximum eigenvalue of the n-square matrix of the relative weights w_i (i=1...n) of the different alternatives. This expression is zero for a set of totally consistent judgements and 1 for totally inconsistent judgements, for instance made on a random basis. The *consistency ratio* is defined as the ratio of the consistency index for the data to the average consistency index over a large number of reciprocal matrices of the same order with random entries. If this ratio is sufficiently small, generally taken to be 10% or less, the data is accepted.

There are a number of different causes of *inconsistency*, including the following:

- Clerical error, which is one of the commonest sources of inconsistency.
- Lack of information which can lead to judgments seeming to have been made on a random basis.
- Lack of concentration on the judgment process due to, for instance, fatigue or lack of interest in the problem.
- Inconsistency in the real data.
- The model structure.

Low inconsistency is necessary, but not sufficient for a good decision and should not become the main goal of the decision making process. There may be specific reasons for accepting data with a higher consistency ratio and it is possible to be perfectly consistent, but consistently wrong (Forman 1990).

The last step is to obtain the composite or *global priorities* for each alternative. The local priorities for each alternative with respect to each criterion are arranged in a matrix, with the rows labelled by the alternatives and the columns by the criteria. The global priorities for each alternative are then obtained by multiplying the elements in each column of vectors by the priority of the corresponding criterion and then adding the elements in each row. *Sensitivity analysis* can be carried out to determine the sensitivity of the alternatives to changes in the relative importance of the criteria and the results displayed graphically. Subsidiary models can be used to evaluate the importance of the different decision makers or the likelihood of particular scenarios.

11.5 Multi-attribute Utility Theory

In methods based on utility theory, such as *multi-attribute utility theory*, the decision makers' preferences with regards to the criteria are expressed in terms of utility functions. When appropriate conditions are satisfied an n-attribute utility function can be expressed as the sum or product of n one-attribute utility functions and an n-attribute utility function is not required. The main advantage of the *additive utility function* is simplicity and its main disadvantage the restrictiveness of the assumptions. However multi-linear utility functions will be discussed first. Multi-attribute decision analysis can be applied when the number of alternatives is small and there is uncertainty.

11.5.1 Multi-linear Utility Functions

Multi-linear functions are functions of n variables which are linear in each variable on its own. The simplest example is a *bilinear function* which has the form

$$f(x_1, x_2) = k_1 f_1(x_1) + k_2 f_2(x_2) + k_{12} f_1(x_1) f_2(x_2) \tag{11.27}$$

where $f_1(.)$ and $f_2(.)$ are linear functions of one variable and k_1, k_2 and k_{12} are constants. This can be generalised to a multi-linear function of n variables which contains the sum of products of up to n terms.

When the attributes are *mutually utility independent* (see section 9.4.3), the n-attribute utility function can be expressed as a *multi-linear function* of the n single-attribute utility functions as follows (Keeney and Raiffa 1976 1993):

$$U(x_1, x_2, ..., x_n) = \sum_{i=1}^{n} k_i U_i(x_i) + \sum_{i=1}^{n} \sum_{j>i} k_{ij} U_i(x_i) U_j(x_j) + ...$$
$$+ \sum_{i=1}^{n} \sum_{j>i} \sum_{m>j} k_{ijm} U_i(x_i) U_j(x_j) U_m(x_m) + k_{12...n} U_1(x_1) U_1(x_1) \dots U_n(x_n) \tag{11.28}$$

where

1. $U(x_1, x_2, ..., x_n)$ is normalised so that $U(x^o_1, x^o_2, ..., x^o_n) = 0$ and $U(x^*_1, x^*_2, ..., x^*_n) = 1$ for given values x^o_i and x^*_i, i=1...n.
2. $U(x_i)$ is a conditional utility function on the attribute X_i normalised by $U(x^o_i) = 0$ and $U(x^*_i) = 1$, i =1...n.
3. The scaling constants are obtained as follows

$$k_i = U(x^*_i, \overline{x}^o_i), \ i=1...n \tag{11.29a}$$

$$k_{ij} = U(x^*_i, x^*_j, \overline{x}^o_{ij}) - U(x^*_i, \overline{x}^o_i) - U(x^*_j, \overline{x}^o_j) \tag{11.29b}$$

$$k_{12...n} = 1 - \sum_i U(x^o_i, \overline{x}^*_i) + ... + (-1)^{n-2} \sum_{i,j>i} U(x^*_i, x^*_j, \overline{x}^o_{ij})$$

$$+ (-1)^{n-1} \sum_i U(x^*_i, \overline{x}^o_i) \tag{11.29c}$$

where $\overline{x}_i$ denotes the compliment of x_i i.e. all the attributes other than x_i and $\overline{x}^o_{ij}$ denotes the complement of x_i and x_j i.e. all the attributes other than x_i and x_j.

It should be noted that $U(x_1, x_2, ..., x_n)$ is multi-linear in the n single attribute utility functions $U_1(x_1)$ to $U_n(x_n)$ and not in the variables x_1 to x_n. Therefore the multi-linearity of $U(x_1, x_2, ..., x_n)$ in (11.28) does not require the functions $U_1(x_1)$ to $U_n(x_n)$ to be linear in the variables x_1 to x_n.

Example 11.4: Reduction of Carbon Dioxide Emissions III

Consider examples 9.10 and 9.11 in section 9.4.3 of minimising the negative social and environmental impacts of an industrial process with the three attributes *noise levels* (X_1), emissions of *carbon dioxide* (X_2) and *jobs created* (X_3). It has already been shown that these three attributes are mutually utility independent. The multi-linear utility function for this problem is

$$U(X_1, X_2, X_3) = \sum_{i=1}^{3} k_i U_i(X_i) + k_{12} U_1(X_1) U_2(X_2) + k_{13} U_1(X_1) U_3(X_3)$$

$$+ k_{23} U_2(X_3) U_3(X_3) + k_{123} U_1(X_1) U_2(X_2) U_3(X_3)$$

It should be noted that maximum utility is obtained when the *attributes noise levels* and *emissions of carbon dioxide* are minimised and *jobs created* is maximised. Therefore the minimum and maximum values of utility are obtained for the following values of the three attributes:

- (x^o_1, x^o_2, x^o_3) = (high, high, negative)
- (x^*_1, x^*_2, x^*_3) = (very low, very low, high)

The multi-linear utility function is then normalised so that:

- U(high, high, negative) = 0
- U(very low, very low, high) = 1

In deriving the scaling constants the following assumptions will be made

1. U(very low *noise levels*) = 0.2
2. U(very low *emissions*) = U(high *jobs created*) = 0.3
3. $U(x^*_i, x^*_j, x^o_k) = U(x^*_i) + U(x^*_j) + U(x^*_i)\,U(x^*_j)$

This gives the following values

k_1 = U(very low, high, negative) = 0.2

k_2 = U(high, very low, negative) = 0.3

k_3 = U(high, high, high) = 0.3

k_{12} = U(very low, very low, negative) - U(very low, high, negative) - U(high, very low, negative)

= 0.56 – 0.2 – 0.3 = 0.06

k_{13} = U(very low, high, high) - U(very low, high, negative) - U(high, high, high)

= 0.56 – 0.2 – 0.3 = 0.06

k_{23} = U(high, high, high) - U(high, very low, negative) - U(high, high, high)

= 0.69 – 0.3 – 0.3 = 0.09

$$k_{123} = 1 - U(x^*_1, x^*_2, x^o_3) - U(x^*_1, x^*_3, x^o_2) - U(x^*_2, x^*_3, x^o_1) + U(x^*_1, x^o_2, x^o_3) + U(x^*_2, x^o_1, x^o_3) + U(x^*_3, x^o_1, x^o_2)$$

= 1 – (very low, very low, negative) - (very low, high, high) - (high, very low, high) + (very low, high, negative) + (high, very low, negative) + (high, high, high)

= 1 – 0.56 – 0.56 – 0.69 + 0.2 + 0.3 + 0.3 = -0.01

The values of U(very low, very low, negative), U(very low, high, high) and U(high, high, high) were obtained using assumption 3 as:

- U(very low, very low, negative) = 0.2 + 0.3 + 0.2x0.3 = 0.56
- U(very low, high, high) = 0.2 + 0.3 + 0.2x0.3 = 0.56
- U(high, high, high) = 0.3 + 0.3 + 0.3x0.3 = 0.69

This gives the multi-linear representation

$$\begin{aligned} U(X_1, X_2, X_3) = {} & 0.2U_1(X_1) + 0.3U_2(X_2) + 0.3U_3(X_3) \\ & + 0.06U_1(X_1)U_2(X_2) + 0.06U_1(X_1)U_3(X_3) \\ & + 0.09U_2(X_2)U_3(X_3) - 0.01U_1(X_1)U_2(X_2)U_3(X_3) \end{aligned}$$

11.5.2 Additive Utility Functions

The n-attribute utility function can be expressed as the sum of n one attribute utility functions under *either* of the following necessary and sufficient conditions (Fishburn 1964 1972; Pollak 1967; Keeney and Raiffa 1993)

1. The *additive independence* condition holds among the attributes $\{X_1, X_2, ..., X_n\}$ i.e. preferences over lotteries on $\{X_1, X_2, ..., X_n\}$ depend only on their marginal probability distributions and not on their joint probability distributions. (See section 9.4.3 for the definitions.)
2. The preference between any two *lotteries*
   $L_1 \equiv \left\langle (x_i, \bar{x}_i'), (x_i^a, \bar{x}_i^a) \right\rangle$ and $L_2 \equiv \left\langle (x_i, \bar{x}_i''), (x_i^b, \bar{x}_i^b) \right\rangle$ is the same for all x_i for any $\bar{x}_i', \bar{x}_i'', x_i^a, x_i^b, \bar{x}_i^a$ and $\bar{x}_i^b$
   where $\bar{x}$ denotes all the attributes other than x.

When either condition 1 or condition 2 holds the utility function can be represented as follows

$$U(x_1, x_2, ..., x_n) = \sum_{i=1}^{n} k_i U_i(x_i) \tag{11.30}$$

where

1. U is normalised so that $U(x^o_1, x^o_2, ..., x^o_n) = 0$ & $U(x^*_1, x^*_2, ..., x^*_n) = 1$.
2. $U(x_i)$ is a conditional utility function of X_i which is normalised such that $U(x^o_i) = 0$ and $U(x^*_i) = 1$.
3. $k_i = U(x^*_1, \bar{x}_i^o)$ for i = 1, 2, ..., n.

For *additive utility functions* the weighting factors k_1 satisfy the consistency requirement $\sum_{i=1}^{n} k_i = 1$. Therefore, only n-1 of these factors need to be calculated. The nth factor can then be obtained from $\sum_{i=1}^{n} k_i = 1$.

Evaluation of the scaling constants k_i can be simplified by the choice of appropriate values of the attributes, with some attributes taking their most desirable values and the others their least desirable values. For any subset T of the indices, $\underline{x}_T$ is defined as the point with its ith attribute having its most desirable value x^*_i if i is in T and its least desirable value x^o_i otherwise. Therefore for n = 5 and T ={1, 2, 4}, $\underline{x}_T=\{x^*_1, x^*_2, x^o_3, x^*_4, x^o_5\}$ i.e. the attributes take their most desirable values for i=1,2 and 4 and least desirable values for i=3,5. The utility of $\underline{x}_T$ will be represented by $U(\underline{x}_T) = k_T$.

Due to the normalisation condition, there is no contribution to utility from attributes which take their least desirable value. Since the utility is additive the total utility is the sum of the utilities from the attributes which take their maximum values. The scaling constant k_i is equal to the expected utility of the ith attribute when it takes its most desirable value. Therefore the expected utility of the point $\underline{x}_T$ is given by the sum of the utilities from each of the attributes which takes its maximum value or the sum of the scaling constants k_i for $i \in T$ i.e. $k_T = \sum_{i \in T} k_i$.

Consider the lottery $L_T \equiv < \underline{x}^*, p_T, \underline{x}^o>$, where p_T is the probability of indifference between this lottery and the consequence $\underline{x}_T$. Therefore the expected utility of this lottery is p_T. Since the utility of the consequence $\underline{x}_T$ is k_T, $k_T = p_T$ for all T. This can be used to simplify the evaluation of the scaling constants k_i through appropriate choices of T.

Example 11.5: Impact Reduction for an Industrial Process III

Consider examples 9.10 and 9.11 in section 9.4.3 of minimising the negative social and environmental impacts of an industrial process with the three attributes *noise levels* (X_1), *emissions of carbon dioxide* (X_2) and *jobs created* (X_3). The additive independence condition for this example requires that preferences among the following lotteries depend only on marginal and not joint probability distributions:

1. Lotteries over the attributes *noise levels* and *emissions*.
2. Lotteries over the attributes *noise levels* and *jobs created*.
3. Lotteries over the attributes *emissions* and *jobs created*.
4. Lotteries over the attributes *noise levels*, *emissions* and *jobs created*.

Consider the following lotteries for case 1:

a) Very low *noise levels* and low *emissions* with a probability of 0.5 and low *noise levels* and moderate *emissions* with a probability of 0.5.
b) Low *noise levels* and moderate *emissions* with a probability of 0.5 and very low *noise levels* and low *emissions* with a probability of 0.5.
c) Very low *noise levels* and high *emissions* with a probability of 0.25; low *noise levels* and low *emissions* with a probability of 0.25; moderate *noise levels* and very low *emissions* with a probability of 0.25; high *noise levels* and moderate *emissions* with a probability of 0.25.
d) Very low *noise levels* and moderate *emissions* with a probability of 0.25; low *noise levels* and very low *emissions* with a probability of 0.25; moderate *noise levels* and low *emissions* with a probability of 0.25; high *noise levels* and high *emissions* with a probability of 0.25.
e) Very low *noise levels* and very low *emissions* with a probability of 0.25; low *noise levels* and high *emissions* with a probability of 0.25; moderate *noise levels* and moderate *emissions* with a probability of 0.25; high *noise levels* and low *emissions* with a probability of 0.25.

Thus lotteries a and b have the same marginal probabilities (0.5) for the values low and moderate for *emissions* and the values very low and low for *noise levels*, but different joint probabilities for the combinations of values. Similarly lotteries c, d and e have the same marginal probabilities (0.25) for all possible options, but different joint probabilities for the different combinations of options. Checking for additive independence requires checking that the decision maker(s) prefer lotteries a and b equally and that they also prefer lotteries c, d and e equally, as well as any other combinations of lotteries that have the same total probabilities for each value of the attributes *noise levels* and *emissions*. Similarly the preference for lotteries which have the same total probabilities for each value of the attributes *noise levels* and *jobs created* for case 2, *emissions* and *jobs created* for case 3 and all three attributes for case 4 should be compared. If the preferences are the same regardless of how the values on the different attributes are combined then the additive independence condition is satisfied and the utility function can be expressed in additive form as

$$U(X_1, X_2, X_3) = k_1U_1(X_1) + k_2U_2(X_2) + k_2U_3(X_3) \quad \text{with } k_1 + k_2 + k_3 = 1$$

For instance, taking

U(*emissions*) = U(*jobs created*) = 2U(*noise levels*)

gives

$$U(X_1, X_2, X_3) = 0.2U_1(X_1) + 0.4U_2(X_2) + 0.4U_3(X_3)$$

11.5.3 Checking the Satisfaction of Independence Conditions

Before additive or multiplicative forms of the utility function can be derived, it is necessary to check whether the prerequisite independence conditions are satisfied. This requires asking the decision maker(s) questions about their preferences and carrying out various checks to determine whether the preferences under different criteria are independent of each other. There are a number of different approaches, most of which involve partitioning the set X of criteria. One approach involves partitioning X into the subsets Y_1 and Y_2 of attributes, where the elements of Y_1 take variable values and the elements of Y_2 have fixed values. The evaluations of the elements of Y_1 and Y_2 are compared. It is then investigated whether the decision maker's preferences change when the fixed values for the elements of Y_2 change.

In another approach pairs of elements (x_{1a}, x_{2c}) and (x_{1b}, x_{2c}), with x_{1a}, x_{1b} in Y_1, x_{1a} and x_{1b} indifferent to each other, and x_{2c} in Y_2, are obtained. The value of x_{2c} is varied over desirable and undesirable levels in X_2 to check whether indifference is maintained for all levels of x_{2c} and is consequently independent of the value of x_{2c}. Similarly independence of preference is checked by taking pairs of elements (x_{1d}, x_{2c}) and (x_{1e}, x_{2c}) with x_{1d} and x_{1e}, in X_1, x_{1d} preferred to x_{1e}, and x_{2c} in X, and varying over X_2. If indifference and preference are both independent of the value in X_2, then X_1 is preferentially independent of X_2.

However it is often the case that the preferences of decision maker(s) under different criteria are not independent of each other. When the requisite assumptions do not hold it may be possible to use additive or multiplicative utility functions as an approximation or to use different additive and multiplicative utility functions over different regions of the attributes.

Example 11.6: Impact Reduction for an Industrial Process IV

Consider example 11.5 of minimising the negative social and environmental impacts of an industrial process with the three attributes *noise levels* (X_1), *emissions of carbon dioxide* (X_2) and *jobs created* (X_3). This set of attributes is partitioned into the set Y_1, consisting of the attributes *noise levels* (X_1) and *emissions of carbon dioxide* (X_2) and the set Y_2 consisting of the attribute *jobs created* (X_3). If the decision maker's preferences for the attributes *noise levels* and *emissions* remain the same when the fixed values of the attribute *jobs created* changes, for instance from negative to high and this holds for all possible values of the attribute *jobs created*, then the decision maker's preferences for Y_1 are preferentially independent of those for Y_2.

Alternatively consider the pairs of elements (x_{1a}, x_{2c}) and (x_{1b}, x_{2c}), with x_{1a} = (very low, low) for the attributes *noise levels* and *emissions* in Y_1 and x_{1b} = (low, very low) for the attributes *noise levels* and *emissions*. It will be assumed that x_{1a} and x_{1b} are indifferent i.e. equally valued or of equal utility when x_{1c} = moderate for the attribute *jobs created* in Y_2. Then it needs to be checked that x_{1a} and x_{1b} remain indifferent to each other when the attribute *jobs created* takes the other three values. Similarly independence of preference is checked by considering the pairs of elements (x_{1d}, x_{2c}) and (x_{1e}, x_{2c}), with x_{1d} = (very low, very low) and x_{1e} = (moderate, low) for the attributes *noise levels* and *emissions* in Y_1. Then x_{1d} is preferred to x_{1e} when x_{1c} is the value moderate of the attribute *jobs created* in Y_2. It then needs to be checked that this preference is maintained when the attribute *jobs created* takes the other three values. If these conditions, hold then Y_1 is preferentially independent of Y_2 i.e. *noise levels* and *emissions of carbon dioxide* are preferentially independent of *jobs created.*

11.6 Roundup of Chapter 11

Chapter 11 has presented four different types of multi-criteria decision support methods, namely outranking methods, aggregation of criteria methods, the Analytic Hierarchy Process and multi-attribute utility theory. Outranking methods use outranking relations to classify the alternatives. An alternative outranks another alternative if it performs at least as well relative to a threshold on all the criteria. Outranking relations are therefore weaker than dominance relations which do not have thresholds. The use of outranking methods allows trade-offs to be made between performance on the different criteria. Outranking methods include the ELECTRE and PROMETHEE methods. The ELECTRE methods use concordance and discordance indices and also use veto thresholds on particular criteria. The PROMETHEE methods use a preference index and a valued outranking graph and can give either a complete or partial ranking.

Aggregation of criteria methods involve the aggregation or combination of criteria to obtain one criterion. They include the P/G% method and goal programming. The P/G% combines the different criteria using percentaging techniques to give dimensionless scaling. Goal programming involves a structured approach based on identification of goals, prioritisation of goal achievement and construction of an objective function using the goals, achievement levels and priority levels.

The Analytic Hierarchy Process (AHP) uses a hierarchical approach to structure the decision problem. The three main levels of the hierarchy are the overall goal, the criteria and the alternatives. The hierarchy can also include subcriteria. The method involves evaluation of each element of the problem by pairwise comparison of nodes with relation to the parent nodes at the previous level. The results are presented in matrix form. Priority vectors are also obtained at each stage. Finally the global priorities for each alternative are obtained.

Multi-attribute utility theory expresses the decision maker's preferences in terms of a utility function. When appropriate conditions are satisfied the utility function can be expressed as the sum of n single-attribute utility functions. This simplifies the analysis and the identification of the optimal solution.

These methods will be illustrated in chapter 12 by application to decision making on waste management options for municipal waste.

12 Case Study: Waste Management Options

12.0 Learning Objectives

This chapter presents a case study of the application of the four different types of decision support methods introduced in chapter 11. The case study applies the methods to the analysis of decision making on waste management options for municipal waste, to take into account resource consumption and environmental as well as cost considerations. The chapter concludes with tutorial exercises.

Specific learning objectives include the following:

- Understanding of a case study of the application of the four different types of decision support systems presented in chapter 11 to sustainable development problems.
- Practice in applying optimisation and decision support techniques to both purely numerical and typical sustainable development problems.
- Increased understanding of the applications of the optimisation and decision support techniques considered in Part II to problems in sustainable development.

12.1 Introduction to the Case Study

In this case study waste management options for municipal waste are considered in the context of resource consumption and environmental as well as cost factors. The aim is to obtain the waste management option which gives the overall 'best' performance on fifteen criteria, which cover costs, resource use and environmental impacts. Six different waste management options are considered, namely incineration, refuse derived fuel and landfill, both on their own and in combination with 25% source separation, as well as ferrous metal and energy (electricity) recovery. Data for the performance of the different waste management options under the criteria (as measured by specific attributes, one per criterion) is given in table 12.1. The data and choice of alternatives and attributes have been obtained from Powell (1996). The abbreviations Incin, RDF and Land are used to denote incineration, refuse derived fuel and landfill respectively. These abbreviations are combined with '+' and '-' to indicate 25% source separation, as well as ferrous metal and energy (electricity) recovery and no source separation and metal or energy recovery respectively. Criteria 7, 8 and 13-15 take values between 0 and 3, with 3 high and the best value.

Table 12.1 Data for different waste management options

	Incin+	Incin-	RDF+	RDF-	Land+	Land-
1. Internal cost (£/tonne)	26.25	25.00	26.26	25.00	18.75	15.00
2. Savings in transport costs (£/tonne)	2.88	2.88	2.88	2.88	0	0
3. Land used (ha/0.01 sq km)	3	3	1	1	13	13
4. % waste eliminated	80	74	98	97	25	0
5. Energy recovered (MJ/metric ton)	3611	1754	5416	2394	3360	96
6. % materials recovered	27	4	29	6	25	0
7. Waste categories Handled	3	3	1	1	3	3
8. Ease of materials Recovery	2	2	3	3	1	1
9. Transport distance (km)	22.28	7.69	22.22	7.63	21.86	6.93
10. % waste incinerated	75	100	17	27	0	0
11. Local air pollution (g SO_x equiv)	-3428	-3385	-4692	-5504	-1626	-300
12. Global air pollution (kg CO_2 equiv./t)	-486	390	-1214	-584	-721	156
13. Water/soil pollution	2	2	2	2	1	1
14. Relative concentrations of toxics	1	1	1	1	3	3
15. Disamenity	2	2	3	3	1	1

12.2 ELECTRE Methods

For use in the ELECTRE methods the following monotonic criterion functions are defined to take values between 0 and 10:

$$g_1(x_i) = \frac{30 - x_i}{1.5}, \quad g_2(x_i) = \frac{10x_i}{3}, \quad g_3(x_i) = \frac{13 - x_i}{1.2}$$

$$g_4(x_i) = \frac{x_i}{10}, \quad g_5(x_i) = \frac{x_i}{550}, \quad g_6(x_i) = \frac{x_i}{3},$$

$$g_7(x_i) = \frac{10x_i}{3}, \quad g_8(x_i) = \frac{10x_i}{3}, \quad g_9(x_i) = \frac{25 - x_i}{2},$$

$$g_{10}(x_i) = \frac{100 - x_i}{10}, \quad g_{11}(x_i) = \frac{-x_i}{560}, \quad g_{12}(x_i) = \frac{-(x_i - 400)}{170}$$

$$g_{13}(x_i) = \frac{10x_i}{3}, \quad g_{14}(x_i) = \frac{10x_i}{3}, \quad g_{15}(x_i) = \frac{10x_i}{3}$$

The values of the criterion functions for the different alternatives are stated in table 12.2.

The following weights are chosen to give the relative importance of the different criteria:

$k_1 = 0.4$ $k_2 = 0.4$ $k_3 = 0.6$ $k_4 = 1.0$ $k_5 = 0.7$
$k_6 = 0.7$ $k_7 = 0.5$ $k_8 = 0.5$ $k_9 = 0.6$ $k_{10} = 0.6$
$k_{11} = 0.8$ $k_{12} = 1.0$ $k_{13} = 0.8$ $k_{14} = 0.8$ $k_{15} = 0.6$
$\Rightarrow k = 10$

The indifference (q_j), preference (p_j) and veto (v_j) thresholds are chosen as follows (j=1...15):

$q_i = 0.2$, i=1...8, 10-15 $q_9 = 0.25$ $p_i = 0.5$, i=1...15

$v_1 = 8$ $v_2 = 9.6$ $v_3 = 10$ $v_4 = 5$ $v_5 = 6$
$v_6 = 6$ $v_7 = 7$ $v_8 = 7$ $v_9 = 9$ $v_{10} = 10$
$v_{11} = 6$ $v_{12} = 5.5$ $v_{13} = 5$ $v_{14} = 8$ $v_{15} = 9$

The chosen values for the indifference and preference thresholds allow each criterion to discriminate between alternatives with moderately or significantly different values and give indifference between alternatives with close values. The veto threshold values have been chosen to allow the criteria with higher priorities, specifically of at least 0.7, to exercise a veto and exclude alternatives solely on their performance on those criteria.

Table 12.2 Values of the criterion functions for the different alternatives

	Incin.+	Incin.-	RDF+	RDF-	Land+	Land-
1. Internal cost (/metric ton)	2.50	3.33	2.49	3.33	7.5	10
2. Savings in transport costs (/metric ton)	9.6	9.6	9.6	9.6	0	0
3. Land used (ha)	8.33	8.33	10.0	10.0	0	0
4. % waste eliminated	8.0	7.4	9.8	9.7	2.5	0
5. Energy recovered (MJ/tonne)	6.57	3.19	9.85	4.35	6.11	0.17
6. % materials recovered	9.00	1.33	9.67	2.00	8.33	0
7. Waste categories handled	10	10	3.33	3.33	10	10
8. Ease of materials recovery	6.67	6.67	10	10	3.33	3.33
9. Transport distance (km)	1.36	8.66	1.39	8.69	1.57	9.04
10.% waste incinerated	2.5	0	8.3	7.3	10	10
11.Local air pollution (g SO_x equiv.)	6.12	6.04	8.38	9.83	2.90	0.54
12.Global air pollution (kg CO_2 equiv./metric ton)	5.21	0.06	9.49	5.79	6.59	1.44
13.Water/soil pollution	6.67	6.67	6.67	6.67	3.33	3.33
14. Relative concentrations of toxics	3.33	3.33	3.33	3.33	10	10
15. Disamenity	6.67	6.67	10	10	3.33	3.33

12.2.1 ELECTRE IS

The values of the criterion functions are first checked for satisfaction of the 'veto' condition $g_j(a) < g_j(b) - v_j \Rightarrow g_j(b) - g_j(a) - v_j > 0$. If this condition holds satisfaction of the outranking relation a S b if not possible, regardless of all other values of the criterion functions.

$g_4(1) - g_4(5) - v_4 = 8 - 2.5 - 5 = 0.5 > 0 \Rightarrow g_4(5) < g_4(1) - v_4$ so the outranking relation 5 S 1 does not hold

$g_4(2) - g_4(6) - v_4 = 7.4 - 0 - 5 = 2.4 > 0 \Rightarrow g_4(6) < g_4(2) - v_4$ so the outranking relation 6 S 2 does not hold

Similarly the 'veto' on criterion 4 excludes satisfaction of the outranking relations:
5 S 3, 5 S 4, 6 S 1, 6 S 3 and 6 S 4.

The 'veto' on criterion 5 excludes satisfaction of the outranking relations:
2 S 3 and 6 S 3.

The 'veto' on criterion 6 excludes satisfaction of the outranking relations:
2 S 1, 2 S 3, 2 S 5, 4 S 1, 4 S 3, 4 S 5, 6 S 1, 6 S 3 and 6 S 5.

The 'veto' on criterion 11 excludes satisfaction of the outranking relations:
5 S 4, 6 S 3 and 6 S 4.

The 'veto' on criterion 12 excludes satisfaction of the outranking relations:
2 S 3, 2 S 4, 2 S 5 and 6 S 3.

Therefore it remains to check satisfaction of the following outranking relations:
1 S 2, 1 S 3, 1 S 4, 1 S 5, 1 S 6, 2 S 6, 3 S 1, 3 S 2, 3 S 4, 3 S 5, 3 S 6, 4 S 2, 4 S 6, 5 S 2 and 5 S 6.

The concordance level s is chosen to satisfy $0.5 \le s \le 1 - \frac{1}{k}\min_{j \in F} k_j$

$$1 - \frac{1}{k}\min_{j \in F} k_j = 1 - 0.04 = 0.96 \quad \Rightarrow 0.5 \le s \le 0.96$$

Take the concordance level s = 0.75. The details of the calculations of the concordance coefficients for the outranking relations 1 S 2 and 3 S 1 are given below. The calculations of the coefficients for the other outranking relations are analogous. For the outranking relation a S b to hold the following two conditions are required:

$$c(a, b) > s$$

$$g_j(a) + v_j \geq g_j(b) + \min(q_j, v_j - p_j)w(s, c) = q_j(b) + q_j\, w(s, c)$$

(as $q_j < v_j - p_j$ for the threshold values chosen here)

$$w(s, c) = \frac{1 - c(a, b) - k_j/k}{1 - s - k_j/k} \leq 1 \text{ for } c(a,b) > s$$

$$\Rightarrow g_j(a) - g_j(b) + v_j - q_j \geq 0$$

Using the values of the criterion functions in table 12.2 and the given values for the indifference and preference thresholds, criteria 2-8 and 10-15 are in agreement with the outranking relation 1 S 2 i.e. in C(1 S 2); criteria 1 and 9 are in disagreement with it i.e. in C(1 P 2); and no criteria are indifferent to it i.e. in C(1 Q 2).

$$c(1,2) = \sum_{j=2\text{-}8,10\text{-}15} \frac{k_j}{10}$$

$$= 0.1[0.4 + 0.6 + 1.0 + 0.7 + 0.7 + 0.5 + 0.5 + 0.6 + 0.8 + 1.0 + 0.8 + 0.8 + 0.6]$$

$$= 0.9 > s = 0.75$$

$g_j(1) + v_j \geq g_j(2) + q_j$, j = 2-8, 10-15, as $q_j(1) \geq q_j(2)$, j = 2-8, 10-15,
and $v_j > q_j$, j=1...15

$$g_1(1) - g_1(2) + v_1 - q_1 = 2.5 - 3.33 + 8 - 0.2 = 6.97 > 0$$
$$g_9(1) - g_9(2) + v_9 - q_9 = 1.36 - 8.65 + 9 - 0.25 = 1.46 > 0$$

∴ All the conditions are satisfied for alternative 1 to outrank 2 i.e. 1 S 2.

Using the values of the criterion functions in table 12.2 and the given values for the indifference and preference thresholds, criteria 1-6 and 8-15 are in agreement with the outranking relation 3 S 1 i.e. in C(3 S 1);

criterion 7 is in disagreement with it i.e. in C(3 P 1); and no criteria are indifferent to it i.e. in C(3 Q 1).

$$c(3,1) = \sum_{j=1\text{-}6,8\text{-}15} \frac{k_j}{10}$$
$$=0.1[0.4+0.4+0.6+1+0.7+0.7+0.5+0.6+0.6+0.8+1+0.8+0.8+0.6]$$
$$= 0.95 > s = 0.75$$

$q_j(3) + v_j \geq q_j(1) + \min(q_j, v_j - p_j),\ j = 1...6, 8...15,$
as $q_j(3) \geq q_j(2), j = 2...6, 8...15$

$q_1(3) \approx q_1(3)$ and $v_j > \min(q_j, v_j - p_j), j=1...15$

$q_7(3) + v_7 = 3.33 + 7 \geq 10 + 0.2 = q_7(1) + \min(q_7, v_7 - p_7)$

∴ All the conditions are satisfied for alternative 3 to outrank 1 i.e. 3 S 1.

It can be shown analogously that:

c(1, 3) = 0.35	c(1, 4) = 0.39	c(1, 5) = 0.72
c(1, 6) = 0.76	c(2, 6) = 0.69	c(3, 2) = 0.85
c(3, 4) = 0.82	c(3, 5) = 0.77	c(3, 6) = 0.71
c(4, 2) = 0.95	c(4, 6) = 0.75	c(5, 2) = 0.47
c(5, 6) = 0.90		

∴ the following concordance indices are greater than s:
c(1, 2), c(1, 6), c(3, 1), c(3, 2), c(3, 4),
c(3, 5), c(4, 2), c(4, 6), c(5, 6).

It can then be shown that the second condition is satisfied

i.e. $g_j(a) + v_j \geq g_j(b) + q_j$ (j=1…15) for the following pairs:
(1, 2), (1, 6), (3, 1), (3, 2), (3, 4), (3, 5), (4, 2), (4, 6), (5, 6)

⇒ alternative 1 outranks alternatives 2 and 6;
alternative 3 outranks alternatives 1, 2, 4 and 5;
alternative 4 outranks alternatives 2 and 6;
alternative 5 outranks alternative 6.
alternative 2 and 6 do not outrank any of the other alternatives.

The outranking relations are illustrated in fig. 12.1, where directed arrows indicate outranking, the relations are not necessarily transitive and

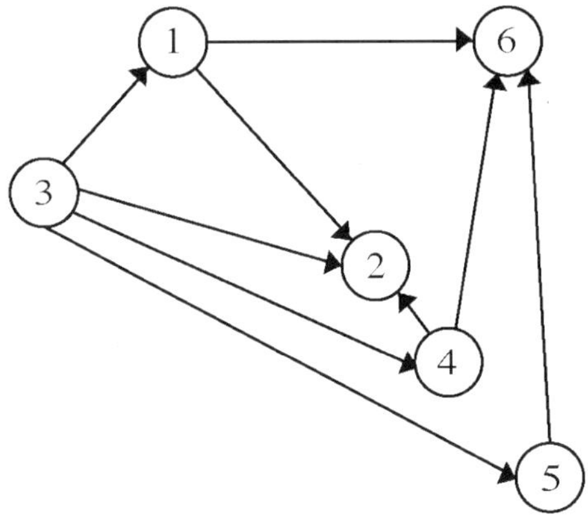

Fig. 12.1 Outranking graph for ELECTRE IS method

the numbers 1-6 indicate the alternatives Incin+, Incin-, RDF+, RDF-, Land+ and Land- .

This implies that alternative 3 (RDF+) is the best overall alternative, particularly since reducing the concordance coefficient to 0.7 would allow it to additionally outrank alternative 6 (Land-).

12.2.2 ELECTRE III

The following outranking relations have been shown to be excluded by the veto condition:

2 S 1, 2 S 3, 2 S 4, 2 S 5, 4 S 1, 4 S 3, 4 S 5, 5 S 1, 5 S 3, 5 S 4, 6 S 1, 6 S 2, 6 S 3, 6 S 4 and 6 S 5.

It is therefore only necessary to calculate the credibility indices for the following outranking relations:

1 S 2, 1 S 3, 1 S 4, 1 S 5, 1 S 6, 2 S 6, 3 S 1, 3 S 2, 3 S 4, 3 S 5, 3 S 6, 4 S 2, 4 S 6, 5 S 2 and 5 S 6.

The details of the calculations of the credibility indices for the outranking relations 1 S 3 and 1 S 4 are given below.

$g_j(3) - g_j(1) \leq p_j$ for $j = 1, 2, 7, 9, 13, 14$

$d_j(1, 3) = 0, \quad j = 1, 2, 7, 9, 13, 14$

$$d_3(1,3) = \frac{g_3(3) - g_3(1) - p_3}{v_3 - p_3} = \frac{10 - 8.33 - 0.5}{8.5 - 0.5} = 0.12$$

$$d_4(1,3) = \frac{g_4(3) - g_4(1) - p_4}{v_4 - p_4} = \frac{9.8 - 8 - 0.5}{5 - 0.5} = 0.29$$

$d_5(1,3) = 0.51$ $d_6(1,3) = 0.03$ $d_8(1,3) = 0.44$
$d_{10}(1,3) = 0.56$ $d_{11}(1,3) = 0.32$ $d_{12}(1,3) = 0.76$
$d_{15}(1,3) = 0.33$

$\therefore\ d_j(1,3) > c(1,3) = 0.35$ for $j = 5,8,10,12,$

$$\sigma(1,3) = \frac{0.35(1 - 0.51)(1 - 0.44)(1 - 0.56)(1 - 0.76}{(1 - 0.35)^4} = 0.06$$

$d_j(1,4) = 0,\ j = 2, 5\text{-}7, 13, 14$

$$d_1(1,4) = \frac{g_1(4) - g_1(1) - p_1}{v_1 - p_1} = \frac{3.33 - 2.5 - 0.5}{8 - 0.5} = 0.044$$

$$d_3(1,4) = \frac{g_3(4) - g_3(1) - p_4}{v_3 - p_3} = \frac{10 - 8.33 - 0.5}{10 - 0.5} = 0.12$$

$d_4(1,4) = 0.27$ $d_8(1,4) = 0.44$ $d_9(1,4) = 0.80$
$d_{10}(1,4) = 0.45$ $d_{11}(1,4) = 0.58$ $d_{12}(1,4) = 0.02$
$d_{15}(1,4) = 0.33$

$d_j(1,4) > c(1,4) = 0.39$ for $j = 8,9,10,11$

$$\sigma(1,4) = \frac{0.39(1 - 0.44)(1 - 0.80)(1 - 0.45)(1 - 0.58)}{(1 - 0.39)^4} = 0.07$$

The remaining credibility indices can be calculated similarly to give the following values:

$\sigma(1,2) = 0.90$ $\sigma(1,5) = 0.43$ $\sigma(1,6) = 0.10$
$\sigma(2,6) = 0$ $\sigma(3,1) = 0.95$ $\sigma(3,2) = 0.29$
$\sigma(3,4) = 0.82$ $\sigma(3,5) = 0.13$ $\sigma(3,6) = 0.02$
$\sigma(4,2) = 0.95$ $\sigma(4,6) = 0.08$ $\sigma(5,2) = 0$
$\sigma(5,6) = 0.90$

It should be noted that the credibility indices $\sigma(1, 2)$, $\sigma(3, 1)$, $\sigma(3, 4)$, $\sigma(4, 2)$, and $\sigma(5, 6)$ are equal to the associated concordance coefficients c(1, 2), c(3, 1), c(3, 4), c(4, 2) and c(5, 6), as there are no criteria for these pairs of alternatives for which the discordance index takes a higher value than the concordance index. This is the case in which the outranking relation holds with the highest credibility.

From the credibility indices,

- The following outranking relations hold with high to very high credibility:
  1 S 2, 3 S 1, 4 S 2 and 5 S 6.
- The following outranking relations have moderate credibility:
  1 S 5 and 3 S 2.
- The following outranking relations have some credibility:
  1 S 6 and 3 S 5.

All the other outranking relations either have very low credibility or are excluded by the veto condition. Thus, in this case, the ELECTRE III method does not give any clear indications as to the best alternative(s), though the results indicate that neither alternative 2 (Incin-) nor 6 (Land-) is in the running. The effects of varying the preference, indifference and veto thresholds could be investigated to determine how this affects the credibility indices and whether some values of the thresholds give a clearer indication of the overall preference(s).

12.3 PROMETHEE Methods

In the PROMETHEE methods the alternatives are evaluated on the 15 criteria using the six preference functions (11.15). This requires one of the preference functions to be associated with each criterion and the associated parameter values to be chosen. The choice of preference function determines how the alternatives are compared on the criterion. Table 12.3 gives the type of preference function associated with each criterion, its parameter values, the associated weights (one tenth of those used previously to give a sum of one) and whether values on the criterion are to be minimised or maximised.

The preference indices can now be calculated. Details are given of the calculations for $\phi(1)$.

$$g'_1(1) - g'_1(2) = 26.25 - 25.00 = 1.25 \text{ Minimisation} \Rightarrow p_1(1, 2) = 0$$

$$g'_2(1) - g'_2(2) = 2.88 - 2.88 = 0 < l = 0.5 \Rightarrow p_2(1, 2) = 0$$

Table 12.3 Preference functions for the PROMETHEE Method

Criterion	Min/Max	p(x)	Parameters	Weight
1. Internal cost (£/tonne)	min	$p_5(x)$	s = 0.2, r = 0.8	0.04
2. Savings in transport costs (£/tonne)	max	$p_2(x)$	l = 0.5	0.04
3. Land used (ha)	min	$p_4(x)$	q = 0.3, p = 0.7	0.06
4. % waste eliminated	max	$p_5(x)$	s = 1, r = 5	0.1
5. Energy recovered (MJ/tonne)	max	$p_5(x)$	s = 20, r = 100	0.07
6. % materials recovered	max	$p_5(x)$	s = 1, r = 2	0.07
7. Waste categories handled	max	$p_1(x)$		0.05
8. Ease of materials recovery	max	$p_1(x)$		0.05
9. Transport distance (km)	min	$p_5(x)$	s = 0.3, r = 0.7	0.06
10.% waste incinerated	min	$p_5(x)$	s = 1, r = 2	0.06
11.Local air pollution (g SO_x equiv.)	min	$p_5(x)$	s = 20, r = 100	0.08
12.Global air pollution (kg CO_2 equiv./t)	min	$p_3(x)$	q = 20, p = 80	0.1
13.Water/soil pollution	max	$p_1(x)$		0.08
14.Relative concentrations of toxics	max	$p_1(x)$		0.08
15.Disamenity	max	$p_1(x)$		0.06

$$g'_3(1) - g'_3(2) = 3 - 3 = 0 < q = 0.5 \Rightarrow p_3(1, 2) = 0$$

$$g'_4(1) - g'_4(2) = 80 - 74 = 6 > r = 5 \Rightarrow p_4(1, 2) = 1$$

$$g'_5(1) - g'_5(2) = 3611 - 1754 = 1857 > r = 500 \Rightarrow p_5(1, 2) = 1$$

$$g'_6(1) - g'_6(2) = 27 - 4 = 23 > r = 3 \Rightarrow p_6(1, 2) = 1$$

$$g'_7(1) - g'_7(2) = 3 - 3 = 0 \Rightarrow p_7(1, 2) = 0$$

$$g'_8(1) - g'_8(2) = 2 - 2 = 0 \Rightarrow p_8(1, 2) = 0$$

$$g'_9(1) - g'_9(2) = 22.79 - 7.69 = 14.59 \text{ Minimisation} \Rightarrow p_9(1, 2) = 0$$

$$g'_{10}(1) - g'_{10}(2) = 75 - 100 = -25$$

Minimisation and $|-25| > r \Rightarrow p_{10}(1, 2) = 1$

$g'_{11}(1) - g'_{11}(2) = -3428 - -3385 = -43,\ s < |-43| < s+r$

$\Rightarrow p_{11}(1, 2) = (43 - 20)/100 = 0.13$

$g'_{12}(1) - g'_{12}(2) = -486 - 390 = -876$

Minimisation and $|-876| > r \Rightarrow p_{12}(1, 2) = 1$

$g'_{13}(1) - g'_{13}(2) = 2 - 2 = 0 \Rightarrow p_{13}(1, 2) = 0$

$g'_{14}(1) - g'_{14}(2) = 1 - 1 = 0 \Rightarrow p_{14}(1, 2) = 0$

$g'_{15}(1) - g'_{15}(2) = 2 - 2 - 0 \Rightarrow p_{15}(1, 2) = 0$

$$\begin{aligned}\phi(1,2) &= \sum_{j=1}^{15} w_j p_j(1, 2)\\ &= 1 \text{ x } (0.1 + 0.07 + 0.07 + 0.06) + 0.13 \text{ x } 0.08 + 1 \text{ x } 0.1\\ &= 0.410\end{aligned}$$

The other values can be calculated analogously to give:

$\phi(1, 3) = 0.050$	$\phi(1, 4) = 0.190$	$\phi(1, 5) = 0.575$
$\phi(1, 6) = 0.710$	$\phi(2, 1) = 0.010$	$\phi(3, 1) = 0.615$
$\phi(4, 1) = 0.610$	$\phi(5, 1) = 0.290$	$\phi(6, 1) = 0.240$

$$\phi^+(1) = \sum_{j=2}^{6} \pi(1, j) = 0.410 + 0.050 + 0.190 + 0.575 + 0.710 = 1.935$$

$$\phi^-(1) = \sum_{j=2}^{6} \pi(j, 1) = 0.054 + 0.511 + 0.323 + 0.291 + 0.239 = 1.855$$

so that

$$\phi(1) = \phi^+(1) - \phi^-(1) = 1.935 - 1.855 = 0.80$$

Analogously it can be shown that

$\phi^+(2) = 1.44$	$\phi^-(2) = 2.280$	$\phi(2) = -0.84$
$\phi^+(3) = 2.915$	$\phi^-(3) = 0.905$	$\phi(3) = 2.10$
$\phi^+(4) = 2.610$	$\phi^-(4) = 1.274$	$\phi(4) = 1.336$
$\phi^+(5) = 1.835$	$\phi^-(5) = 2.505$	$\phi(5) = -0.70$
$\phi^+(6) = 1.174$	$\phi^-(6) = 3.090$	$\phi(6) = -1.92$

Therefore in PROMETHEE II the net flows can be ordered as follows:

$\phi(3) > \phi(4) > \phi(1) > \phi(5) > \phi(2) > \phi(6)$

and consequently the rank order of the alternatives is as follows:

RDF+ RDF- Incin+ Incin- Land+ Land-,

implying that alternative 3, RDF+, is preferred overall.

In PROMETHEE I the following inequalities hold:

$\phi^{+}(1) > \phi^{+}(2)$ and $\phi^{-}(1) > \phi^{-}(2) \Rightarrow 1\ P^{+}\ 2$ and $1\ P^{-}\ 2$
alternative 1 outranks alternative 2 i.e. Incin+ is preferred to Incin-

$\phi^{+}(2) < \phi^{+}(5)$ and $\phi^{-}(2) < \phi^{-}(5) \Rightarrow 5\ P^{+}\ 2$ and $2\ P^{-}\ 5$
alternatives 2 and 5 are not comparable.

Similarly it can be shown that:

- Incin+ is preferred to Incin-, Land+ and Land-
- RDF+ is preferred to Incin+, Incin-, RDF-, Land+ and Land-
- RDF- is preferred to Incin+, Incin-, Land+ and Land-
- Land+ is preferred to Land-
- Incin- and Land- are not comparable

This again implies that alternative 3 (RDF+) is preferred overall. The outranking graph is given in fig. 12.2. Outranking is indicated by a directed arrow and the alternatives Incin+, Incin-, RDF+, RDF-, Land+ and Land- by the numbers 1-6 respectively.

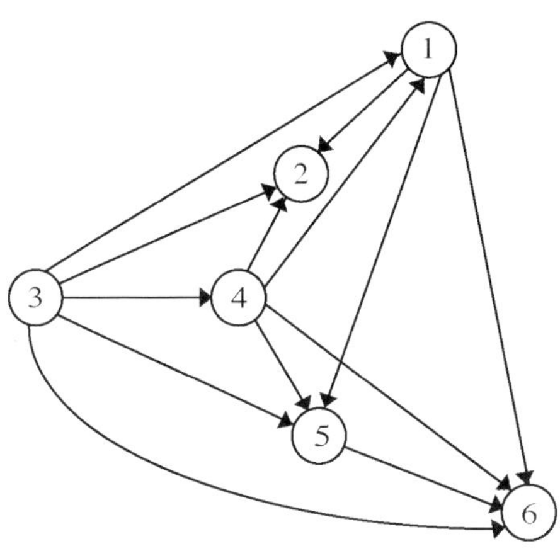

Fig. 12.2 Outranking Graph for PROMETHEE Method

12.4 P/G% Method

The data on alternatives and criteria given in table 12.1 has been rewritten in spreadsheet form and is presented in this format in table 12.4.

Table 12.4 Data on alternatives and criteria in spreadsheet format

	Crit 1	Crit 2	Crit 3	Crit 4	Crit 5	Crit 6	Crit 7	Crit 8
Incin+	26.25	2.88	3	80	3611	27	3	2
Incin-	25.00	2.88	3	74	1754	4	3	2
RDF+	26.26	2.88	1	98	5416	29	1	3
RDF-	25.00	2.88	1	97	2394	6	1	3
Land+	18.75	0	13	25	3360	25	3	1
Land-	15.00	0	13	0	96	0	3	1
Total	136.25	11.52	34	374	16,631	91	14	12

	Crit. 9	Crit. 10	Crit. 11	Crit. 12	Crit. 13	Crit. 14	Crit. 15
Incin+	22.28	75	-3428	-876	2	1	2
Incin-	7.69	100	-3385	0	2	1	2
RDF+	22.22	17	-4692	-1604	2	1	3
RDF-	7.63	27	-5504	-974	2	1	3
Land+	21.86	0	-1626	-1111	1	3	1
Land-	6.93	0	-300	-234	1	3	1
Total	86.61	219	-18,935	-4799	10	10	12

Data manipulation, consisting of the subtraction of 390 from all values, has been used in criterion 12 to take account of the fact that some options have positive values and others negative values under this criterion. The same weights are used as in the ELECTRE method and are repeated below for ease of reference:

$k_1 = 0.4$ $k_2 = 0.4$ $k_3 = 0.6$ $k_4 = 1.0$ $k_5 = 0.7$
$k_6 = 0.7$ $k_7 = 0.5$ $k_8 = 0.5$ $k_9 = 0.6$ $k_{10} = 0.6$
$k_{11} = 0.8$ $k_{12} = 1.0$ $k_{13} = 0.8$ $k_{14} = 0.8$ $k_{15} = 0.6$

sum of weights = 10

The percentage score for each alternative for each criterion is calculated by dividing the value in the table above by the total score on that criterion.

These values are then multiplied by the weights for the criterion and the results are given in table 12.5.

Table 12.5 Percentage scores

	Crit. 1	Crit. 2	Crit. 3	Crit. 4	Crit. 5	Crit. 6	Crit. 7	Crit. 8
Incin+	7.71	10.0	5.29	21.39	15.20	20.80	10.71	8.33
Incin-	7.34	10.0	5.29	19.79	7.38	3.08	10.71	8.33
RDF+	7.71	10.0	1.76	26.20	22.80	22.31	3.57	12.5
RDF-	7.34	10.0	1.76	25.94	10.08	4.62	3.57	12.5
Land+	5.50	0	22.94	6.68	14.14	19.23	10.71	4.17
Land-	4.40	0	22.94	0	0.40	0	10.71	4.17

	Crit. 9	Crit. 10	Crit. 11	Crit. 12	Crit. 13	Crit. 14	Crit. 15
Incin+	15.09	20.55	14.48	18.25	16.0	8.0	10.0
Incin-	5.21	27.40	14.30	0	16.0	8.0	10.0
RDF+	15.05	4.66	19.82	33.42	16.0	8.0	15.0
RDF-	5.17	7.40	23.25	20.30	16.0	8.0	15.0
Land+	14.80	0	6.87	23.15	8.0	24.0	5.0
Land-	4.69	0	1.27	4.88	8.0	24.0	5.0

The total score for each alternative is then obtained by adding the weighted scores for the different criteria and dividing by the sum of the weights, which is equal to 10. This gives the following calculation for the total score for the Incin+ alternative:

$$\begin{aligned}\text{Score} &= 0.1[7.71 + 10 + 5.29 + 21.39 + 15.20 + 20.80 + 10.71 + 8.33] \\ &\quad + 0.1[15.09 + 20.55 + 14.48 + 18.25 + 16 + 8 + 10] \\ &= 20.18\end{aligned}$$

The total scores for the other alternatives can be obtained analogously:

Incin+: 20.18 Incin-: 15.28
RDF+: 21.88 RDF-: 17.09
Land+: 16.52 Land-: 9.05

The highest score is obtained for RDF+, showing that it is the preferred option.

12.5 Goal Programming

The desired levels of goal achievement for the fifteen goals are given in table 12.6.

Table 12.6 Desired goal levels

	Criterion	Desired Goal Level
1	Internal cost	£10/tonne
2	Savings in transport cost	£4/tonne
3	Land used	1 ha (10000 m^2)
4	Waste eliminated	99%
5	Energy recovered	6000 MJ/tonne
6	Materials recovered	40%
7	Waste categories handled	3
8	Ease of materials recovery	3
9	Transport distance	5km
10	Waste incinerated	0%
11	Local air pollution	-6000 g SO_x equivalent
12	Global air pollution	-1500 kg CO_2
13	Water/soil pollution	3
14	Relative concentration of toxics	3
15	Disamenity	3

The same relative priorities will be used as previously, which gives $P_j = k_j$, $j = 1...15$. This gives the cost function

$$\begin{aligned} S_1 &= P_1(D_1G_1) + P_2(D_2G_2) + \ldots + P_{15}(D_{15}G_{15}) \\ &= 0.4|x_1 - 10| + 0.4|x_2 - 4| + 0.6|x_3 - 1| + 1.0|x_4 - 99| \\ &\quad + 0.7|x_5 - 6000| + 0.7|x_6 - 40| + 0.5|x_7 - 3| + 0.5|x_8 - 3| \\ &\quad + 0.6|x_9 - 5| + 0.6|x_{10} - 0| + 0.8|x_{11} - -6000| + 1.0|x_{12} - -1500| \\ &\quad + 0.8|x_{13} - 3| + 0.8|x_{14} - 3| + 0.6|x_{15} - 3| \end{aligned}$$

where the variables x_1 to x_{15} represent the values of an alternative under the criteria 1 to 15 respectively. Since the deviations in the original formula have been replaced by the differences between the values of an alternative and the desired values modulus signs are required, to ensure that all these differences are positive.

$$\begin{aligned} S_1 = {} & 0.4|26.25 - 10| + 0.4|2.88 - 4| + 0.6|3 - 1| + 1.0|80 - 99| \\ & + 0.7|3611 - 6000| + 0.7|27 - 40| + 0.5|3 - 3| + 0.5|2 - 3| \\ & + 0.6|22.28 - 5| + 0.6|75 - 0| + 0.8|-3428 - -6000| \\ & + 1.0|-486 - -1500| + 0.8|2 - 3 + 0.8|1 - 3| + 0.6|2 - 3| \\ = {} & 4839.02 \end{aligned}$$

$$\begin{aligned} S_2 = {} & 0.4|25.00 - 10| + 0.4|2.88 - 4| + 0.6|3 - 1| + 1.0|74 - 99| \\ & + 0.7|1754 - 6000| + 0.7|4 - 40| + 0.5|3 - 3| + 0.5|2 - 3| \\ & + 0.6|7.69 - 5| + 0.6|100 - 0| + 0.8|-3385 - -6000| \\ & + 1.0|390 - -1500| + 0.8|2 - 3| + 0.8|1 - 3| + 0.6|2 - 3| \\ = {} & 7077.16 \end{aligned}$$

$$\begin{aligned} S_3 = {} & 0.4|26.26 - 10| + 0.4|2.88 - 4| + 0.6|1 - 1| + 1.0|98 - 99| \\ & + 0.7|5416 - 6000| + 0.7|29 - 40| + 0.5|1 - 3| + 0.5|3 - 3| \\ & + 0.6|22.22 - 5| + 0.6|17 - 0| + 0.8|-4692 - -6000| \\ & + 1.0|-1214 - -1500| + 0.8|2 - 3| + 0.8|1 - 3| + 0.6|3 - 3| \\ = {} & 1780.78 \end{aligned}$$

$$\begin{aligned} S_4 = {} & 0.4|25.00 - 10| + 0.4|2.88 - 4| + 0.6|1 - 1| + 1.0|97 - 99| \\ & + 0.7|2394 - 6000| + 0.7|6 - 40| + 0.5|1 - 3| + 0.5|3 - 3| \\ & + 0.6|7.63 - 5| + 0.6|27 - 0| + 0.8|-5504 - -6000| \\ & + 1.0|-584 - -1500| + 0.8|2 - 3| + 0.8|1 - 3| + 0.6|3 - 3| \\ = {} & 3890.43 \end{aligned}$$

$$\begin{aligned} S_5 = {} & 0.4|18.75 - 10| + 0.4|0 - 4| + 0.6|13 - 1| + 1.0|25 - 99| \\ & + 0.7|3360 - 6000| + 0.7|25 - 40| + 0.5|3 - 3| + 0.5|1 - 3| \\ & + 0.6|21.86 - 5| + 0.6|0 - 0| + 0.8|-1626 - -6000| \\ & + 1.0|-721 - -1500| + 0.8|1 - 3| + 0.8|3 - 3| + 0.6|1 - 3| \\ = {} & 6236.92 \end{aligned}$$

$$
\begin{aligned}
S_6 = {} & 0.4|15.00 - 10| + 0.4|0 - 4| + 0.6|13 - 1| + 1.0|0 - 99| \\
& + 0.7|96 - 6000| + 0.7|0 - 40| + 0.5|3 - 3| + 0.5|1 - 3| \\
& + 0.6|6.93 - 5| + 0.6|0 - 0| + 0.8|-300 - -6000| \\
& + 1.0|156 - -1500| + 0.8|1 - 3| + 0.8|3 - 3| + 0.6|1 - 3| \\
= {} & 10491.56
\end{aligned}
$$

According to this method, the third option, refuse derived fuel with source separation and ferrous recovery at plant is clearly the best. However the fact that the different variables and consequently also their differences have very different orders of magnitude gives a much greater weight to some of the criteria than others. This problem could be reduced by scaling. The scaling factors used here will be one third of the greater of the desired value and the maximum value obtained by all alternatives for each criterion, with the values appropriately rounded. Therefore the different terms in the cost function will be divided respectively by:

G_1: 9	G_2: 1	G_3: 4	G_4: 33	G_5: 2000
G_6: 13	G_7: 1	G_8: 1	G_9: 7	G_{10}: 33
G_{11}: 2000	G_{12}: 500	G_{13}: 1	G_{13}: 1	G_{13}: 1

This gives the new values:

$S'_1 = 12.98$ $\quad S'_2 = 15.97$ $\quad S'_3 = 8.28$

$S'_4 = 10.41$ $\quad S'_5 = 16.32$ $\quad S'_6 = 20.40$

so that the third option, RDF+, is still the preferred one.

12.6 Analytic Hierarchy Process

The first step is to draw up the hierarchy, consisting of the goal, criteria and alternatives. The criteria and alternatives have already been obtained. There are a number of different possibilities for the goal and the one chosen is likely to have a significant effect on the resulting decision. The following goal will be considered here:

Managing municipal waste so as to minimise environmental impacts, maximise resource recovery and, as far as possible, reduce costs.

The weights given to the different criteria give a measure of their relative importance and therefore the extent to which they meet this goal. Therefore these weights will be used to calculate the matrix of pairwise comparisons of the extent to which the different criteria meet this goal.

This matrix is presented in table 12.7. The priority vector for the different criteria is the normalised eigenvector corresponding to the largest eigenvalue and is given in the last column of the table.

Table 12.7 Pairwise Comparison of the extent to which the criteria meet the goal

	1	2	3	4	5	6	7	8
1	1.0	1.0	0.25	0.111	0.167	0.167	0.333	0.333
2	1.0	1.0	0.25	0.111	0.167	0.167	0.333	0.333
3	4.0	4.0	1.0	0.167	0.333	0.333	3.0	3.0
4	9.0	9.0	6.0	1.0	4.0	4.0	7.0	7.0
5	6.0	6.0	3.0	0.25	1.0	1.0	4.0	4.0
6	6.0	6.0	3.0	0.25	1.0	1.0	4.0	4.0
7	3.0	3.0	0.333	0.143	0.25	0.25	1.0	1.0
8	3.0	3.0	0.333	0.143	0.25	0.25	1.0	1.0
9	4.0	4.0	1.0	0.167	0.333	0.333	3.0	3.0
10	4.0	4.0	1.0	0.167	0.333	0.333	3.0	3.0
11	7.0	7.0	4.0	0.333	3.0	3.0	6.0	6.0
12	9.0	9.0	6.0	1.0	4.0	4.0	7.0	7.0
13	7.0	7.0	4.0	0.333	3.0	3.0	6.0	6.0
14	7.0	7.0	4.0	0.333	3.0	3.0	6.0	6.0
15	4.0	4.0	1.0	0.167	0.333	0.333	3.0	3.0

	9	10	11	12	13	14	15	Priority vector
1	0.25	0.25	0.143	0.111	0.143	0.143	0.25	0.032
2	0.25	0.25	0.143	0.111	0.143	0.143	0.25	0.032
3	1.0	1.0	0.25	0.167	0.25	0.25	1.0	0.092
4	6.0	6.0	3.0	1.0	3.0	3.0	6.0	0.553
5	3.0	3.0	0.333	0.25	0.333	0.333	3.0	0.178
6	3.0	3.0	0.333	0.25	0.333	0.333	3.0	0.178
7	0.333	0.333	0.167	0.143	0.167	0.167	0.333	0.051
8	0.333	0.333	0.167	0.143	0.167	0.167	0.333	0.051
9	1.0	1.0	0.25	0.167	0.25	0.25	1.0	0.092
10	1.0	1.0	0.25	0.167	0.25	0.25	1.0	0.092
11	4.0	4.0	1.0	0.333	1.0	1.0	4.0	0.308
12	6.0	6.0	3.0	1.0	3.0	3.0	6.0	0.553
13	4.0	4.0	1.0	0.333	1.0	1.0	4.0	0.308
14	4.0	4.0	1.0	0.333	1.0	1.0	4.0	0.308
15	1.0	1.0	0.25	0.167	0.25	0.25	1.0	0.092

The next stage is obtaining the 15 matrices of pairwise comparisons of the extent to which the different alternatives meet each of the criteria in turn. These matrices are stated in table 12.8 for criteria 1- 4. The priority vector for each of these four criteria can be obtained as the normalised eigenvector corresponding to the largest eigenvalue of the associated matrix. The coefficients of the priority vectors are presented in the last columns of the matrices.

Table 12.8a The Extent to which the alternatives meet the criterion 1

	Incin+	Incin-	RDF+	RDF-	Land+	Land-	Priority vector
Incin+	1.0	0.5	1.0	0.5	0.2	0.167	0.091
Incin-	2.0	1.0	2.0	1.0	0.2	0.167	0.143
RDF+	1.0	0.5	1.0	0.5	0.2	0.167	0.091
RDF-	2.0	1.0	2.0	1.0	0.2	0.167	0.143
Land+	5.0	5.0	5.0	5.0	1.0	0.333	0.503
Land-	6.0	6.0	6.0	6.0	3.0	1.0	0.830

Table 12.8b The extent to which the alternatives meet the criterion 2

	Incin+	Incin-	RDF+	RDF-	Land+	Land-	Priority vector
Incin+	1.0	1.0	1.0	1.0	9.0	9.0	-0.499
Incin-	1.0	1.0	1.0	1.0	9.0	9.0	-0.499
RDF+	1.0	1.0	1.0	1.0	9.0	9.0	-0.499
RDF-	1.0	1.0	1.0	1.0	9.0	9.0	-0.499
Land+	0.111	0.111	0.111	0.111	1.0	1.0	-0.055
Land-	0.111	0.111	0.111	0.111	1.0	1.0	-0.055

Table 12.8c The Extent to which the alternatives meet the criterion 3

	Incin+	Incin-	RDF+	RDF-	Land+	Land-	Priority vector
Incin+	1.0	1.0	0.2	0.2	7.0	7.0	-0.212
Incin-	1.0	1.0	0.2	0.2	7.0	7.0	-0.212
RDF+	5.0	5.0	1.0	1.0	9.0	9.0	-0.673
RDF-	5.0	5.0	1.0	1.0	9.0	9.0	-0.673
Land+	0.143	0.143	0.111	0.111	1.0	1.0	-0.048
Land-	0.143	0.143	0.111	0.111	1.0	1.0	-0.048

Table 12.8d The extent to which the alternatives meet the criterion 4

	Incin+	Incin-	RDF+	RDF-	Land+	Land-	Priority vector
Incin+	1.0	3.0	0.25	0.25	7.0	7.0	0.306
Incin-	0.33	1.0	0.25	0.25	7.0	7.0	0.209
RDF+	4.0	4.0	1.0	1.0	8.0	9.0	0.655
RDF-	4.0	4.0	1.0	1.0	8.0	9.0	0.655
Land+	0.143	0.143	0.125	0.125	1.0	1.0	0.053
Land-	0.143	0.143	0.111	0.111	1.0	1.0	0.050

The priority vectors for the other criteria can be obtained in a similar way and are tabulated in table 12.9. The weight given to each criterion is stated in brackets after the criterion number. These weights are the coefficients of the priority vector of the extent to which the criteria meet the goal i.e. of the priority vector in table 12.7.

Table 12.9 The extent to which the alternatives meet the criteria

	1 (0.032)	2 (0.032)	3 (0.092)	4 (0.553)	5 (0.178)	6 (0.178)	7 (0.051)	8 (0.051)
Incin+	0.091	-0.499	-0.212	0.306	-0.384	0.524	-0.497	-0.230
Incin-	0.143	-0.499	-0.212	0.209	-0.126	0.132	-0.497	-0.230
RDF+	0.091	-0.499	-0.673	0.655	-0.840	0.717	-0.083	-0.664
RDF-	0.143	-0.499	-0.673	0.655	-0.186	0.169	-0.083	-0.664
Land+	0.503	-0.055	-0.048	0.053	-0.310	0.407	-0.497	-0.080
Land-	0.830	-0.055	-0.048	0.050	-0.306	0.038	-0.497	-0.080

	9 (0.092)	10 (0.092)	11 (0.308)	12 (0.553)	13 (0.308)	14 (0.308)	15 (0.092)
Incin+	0.066	0.080	-0.252	0.192	-0.943	-0.115	-0.230
Incin-	0.429	0.043	-0.202	0.082	0.189	-0.115	-0.230
RDF+	0.066	0.358	-0.802	0.795	0.189	-0.115	-0.664
RDF-	0.429	0.220	-0.498	0.292	0.189	-0.115	-0.664
Land+	0.094	0.638	-0.052	0.487	0.047	-0.688	-0.080
Land-	0.784	0.638	-0.046	0.049	0.047	-0.688	-0.080

The overall priority of each of the alternatives can be obtained from table 12.9 by multiplying the extent to which each alternative meets each criterion (obtained from the priority vector) by the weight given to that criterion and summing over all the criteria for each alternative. Thus the total score or overall priority for Incin+ can be calculated as follows:

Incin+: 0.032 x 0.091 - 0.032 x 0.499 - 0.092 x 0.212 + 0.553 x 0.306
- 0.178 x 0.384 + 0.178 x 0.524 - 0.051 x 0.497 - 0.051 x 0.230
+ 0.092 x 0.066 + 0.092 x 0.080 - 0.308 x 0.252 + 0.553 x 0.192
- 0.308 x 0.943 - 0.308 x 0.115 - 0.092 x 0.230
= -0.32

The overall priorities for the other alternatives can be calculated similarly, to give the following overall scores:

Incin+: -0.32	RDF+: 0.42	Land+: 0.05
Incin-: -0.02	RDF-: 0.28	Land-: 0.09

From these priorities the RDF+ option clearly has the highest overall score and therefore the best performance. The Land+, Land- and Incin- options only meet the goal to a minimal extent. It is not clear why this method gives such a negative result for the performance of Incin+, which is shown not to meet the goal.

12.7 Multi-Attribute Utility Theory

Examination of the different alternatives indicates that preferences on criteria 1, 3, 9, 11 and 12 will be monotonically decreasing and preferences on criteria 2, 4-8, 10 and 13-15 will be monotonically increasing for most 'rational' decision makers. Therefore it can be assumed that most decision makers will prefer alternatives with lower values on criteria 1, 3, 9, 11 and 12 and higher values on criteria 2, 4-8, 10 and 13-15 independently of the values taken by the other criteria. Consequently the 'lottery' condition is satisfied (see section 11.5.2) and the utility function can be expressed as the weighted sum of 15 single attribute utility functions $U_i(x_i)$ i=1,...15, with the weights k_i used to indicate the relative importance of the different criteria i.e.

$$U(x_1, x_2, ..., x_{15}) = \sum_{i=1}^{15} k_i U_i(x_i) \quad \text{with} \sum_{i=1}^{15} k_i = 1$$

The fifteen utility functions now have to be chosen to give a measure of the preferences of the decision maker(s) relative to the different criteria. This will generally involve questioning the decision maker(s) to try and determine their preferences. However, in this case, it will be assumed that preferences are linear, as well as monotonically decreasing on criteria 1, 3, 9, 11 and 12 and monotonically increasing on criteria 2, 4-8, 10 and 13-15. It will also be assumed that the relative importance of the different criteria is given solely by the weights and therefore the utility functions should be scaled to take the same range of values. One way of doing this is to take the utility functions equal to the criterion functions used with the ELECTRE methods. Since the weights are required to sum to one, one possibility is to obtain them from the values previously used by dividing by the sum of these weights. This sum is equal to 10. This then gives the utility function.

$$
\begin{aligned}
U(x_1, x_2, \dots, x_{15}) = {} & \frac{0.4}{10}\frac{30 - x_1}{1.5} + \frac{0.4}{10}\frac{10x_2}{3} + \frac{0.6}{10}\frac{13 - x_3}{1.2} + \frac{1.0}{10}\frac{x_4}{10} + \frac{0.7}{10}\frac{x_5}{550} \\
& + \frac{0.7}{10}\frac{x_6}{3} + \frac{0.5}{10}\frac{10x_7}{3} + \frac{0.5}{10}\frac{10x_8}{3} + \frac{0.6}{10}\frac{25 - x_9}{2} + \frac{0.6}{10}\frac{100 - x_{10}}{10} \\
& + \frac{0.8}{10}\frac{(-x_{11})}{560} + \frac{1.0}{10}\frac{-(x_{12} - 400)}{170} + \frac{0.8}{10}\frac{10x_{13}}{3} + \frac{0.8}{10}\frac{10x_{14}}{3} \\
& + \frac{0.6}{110}\frac{10\,x_{15}}{3}
\end{aligned}
$$

The utility function for the Incin+ alternative can be calculated as follows

$$
\begin{aligned}
U_{Incin+}(x_1, x_2, \dots, x_{15}) = {} \\
& 0.1\left[0.4\frac{30 - 26.25}{1.5} + 0.4\frac{10 \times 2.88}{3} + 0.6\frac{13 - 3}{1.2} + 1.0\frac{80}{10} + 0.7\frac{3611}{550}\right] \\
& + 0.1\left[0.7\frac{27}{3} + 0.5\frac{10 \times 3}{3} + 0.5\frac{10 \times 2}{3} + 0.6\frac{25 - 22.28}{2} + 0.6\frac{100 - 75}{10}\right] \\
& + 0.1\left[0.8\frac{--3428}{560} + 1.0\frac{-(-486 - 400)}{170} + 0.8\frac{10 \times 2}{3} + 0.8\frac{10 \times 1}{3} + 0.6\frac{10 \times 2}{3}\right] \\
= {} & 6.15
\end{aligned}
$$

The values of the utility functions for the other alternatives can be calculated analogously to give:

$$
\begin{array}{ll}
U_{Incin+}(x_1, x_2, \dots, x_{15}) = 6.15 & U_{Incin-}(x_1, x_2, \dots, x_{15}) = 5.12 \\
U_{RDF+}(x_1, x_2, \dots, x_{15}) = 7.02 & U_{RDF-}(x_1, x_2, \dots, x_{15}) = 6.92 \\
U_{Land+}(x_1, x_2, \dots, x_{15}) = 5.08 & U_{land-}(x_1, x_2, \dots, x_{15}) = 3.67
\end{array}
$$

Therefore the RDF+ alternative has the highest value for its utility function, indicating that it is the preferred option.

12.8 Discussion

The above case study has illustrated the application of a number of different methods. In this case all the methods considered, except ELECTRE III, came to the same conclusion that refuse derived fuel with recycling is the best overall waste management option. This is the same conclusion as was reached in the original paper. No real conclusions could be drawn from application of ELECTRE III. However, in all cases, it would be interesting to investigate the effects of changing the relative weighting and other parameters of the algorithm on the decision obtained.

To illustrate the multi-criteria decision support methods, the six alternatives and 15 criteria given in the paper (Powell 1996) from which the data was obtained, were used. However they are clearly not the only possibilities and further consideration of the problem, as discussed in section 9.3, would have led to other alternatives and criteria and possibly a better decision. For instance, as discussed in chapter 7, there is considerable public concern about both incineration and some of the treatments associated with refuse derived fuel. Therefore waste management options based on waste reduction and recycling would have advantages and probably give better performance.

The six different waste management options presented here were considered purely as alternatives, using discrete multi-criteria methods. However in some cases there might be advantages in considering a combination of the options and continuous versions of some of the methods could be used to determine the percentage composition that best satisfies the criteria. Of the methods given, PG%, goal programming and multi-attribute utility theory could be most easily applied in the continuous case.

12.9 Tutorial Exercises

1. Bracket the minimum of the following functions:

   a. $f_1(x) = \dfrac{\cos(x) + \sin(x)}{1 + x^3}$

b. $f_2(x) = \dfrac{\cos^2(x) + \tan(x)}{1 + x^4}$

c. $f_3(x) = \dfrac{x^5 + 2x^4 - x^3 + 1.5x^2 + 5.4x - 7}{6x^5 + x^4\cos(x) + x^3\sin(x) + 2.3x^2 - 4.7x + 9}$

2. Evaluate the following polynomials using nested multiplication:

a. $f_1(x) = x^7 + 23x^6 - 16x^5 + 4x^4 - 2x^3 - 1.2 + x^2 + 5x - 16$
for $x = 0.2$

b. $f_2(x) = x^7 + 23x^6 - 16x^5 + 4x^4 - 2x^3 - 1.2x^2 + 5x - 16$ for $x = 1.3$

c. $f_3(x) = x^9 + 15x^8 + 3x^7 + 4x^6 - 2x^5 - 1.5x^4 + 2.5x^3 + 20x^2 + 7x + 3$
for $x = 2$

3. Investigate the choice of different search directions $\underline{h}_k$ in the application of gradient algorithms to the following functions

a. $f_1(x_1, x_2) = 1 - 3x_1 - 3x_2 - 6x_1x_2 + 11x_1^2 + 5x_1^2$

b. $f_3(x_1, x_2, x_3) = x_1^4 + x_2^4 + x_3^4 - 2x_2^2 - 2x_3^2 - 3x_1$

c. The de Jong five function test suite in section 8.4.6.

4. Are the following binary relations:

i. Consuming more energy than
ii. Having a better human rights policy than

a. Reflexive?
b. Symmetric?
c. Transitive?

5. Consider the following situations:

a. Minimising the negative environmental impacts of coal fired power plants, including emissions of sulphur dioxides and particulates, waste heat generation, sludge and ash and negative impacts on human health.

b. Ranking hydroelectric power plants according to their impacts on the environment, including flooding and erosion, recreational activities, agriculture, drinking water supplies, transportation facilities and the regional economy.

c. Traffic regulation measures, which take account of equity considerations, reduce energy consumption and emissions and contribute to local and regional economic development.

d. Choice of a solid waste management system which supports sustainable development criteria.

Derive:

a. Objectives
b. Criteria
c. Attributes
d. Alternatives

for these situations.

6. Formulate the situations described in question 5 as optimisation problems.

7. Consider the planning, design and evaluation of geothermal energy projects on technical, economic, social and environmental criteria. The performance of five different projects under the three criteria of energy use, return on investment and creation of new jobs is given in table 12.10 (Goumas et al 1999).

Table 12.10 Performance of different geothermal energy projects

Scenario	Criterion A Energy Use (metric tons of oil equivalent per year)	Criterion B Return on Investment (per year)	Criterion C Jobs (person days per year)
1	1335	1.06	15 400
2	1972	0.88	16 900
3	3052	0.64	21 400
4	3010	0.60	17 400
5	3640	0.51	21 700

Apply the following methods to the choice of project and compare the results:

a. ELECTRE IS
b. ELECTRE III
c. PROMETHEE I
d. PROMETHEE II
e. P/G%
f. Goal Programming
g. Analytical Hierarchy Process
h. Multi-attribute Utility Theory.

8. Consider the problem of choosing an appropriate solid waste management strategy using a mixture of landfill, open composting and refuse derived fuel (RDF) combustion, using the eight criterion of:

net cost per metric ton (g_1), technical reliability (g_2), global effects (g_3), local and regional effects (g_4), acidificative releases (g_5), surface water dispersed releases (g_6), number of employees (g_7), amount of waste recovered (g_8).

Data for the seven options below under the criteria is given in table 12.11 (Hokkanen and Salminen 1997):

Table 12.11 Performance of different solid waste management options

Option	g_1 (min) cost	g_2 (min) technical reliability	g_3 (min) global effects	g_4 (min) health effects	g_5 (min) acidicative releases	g_6 (min) surface water dispersed releases	g_7 (max) employees	g_8 (max) resource recovery
1	656	5	552,678,100	609	1190	670	14	13,900
2	912	4	408,565,400	670	1222	594	24	39,767
3	589	9	559,780,715	411	1191	443	10	13,900
4	706	7	532,286,214	325	1191	404	14	23,600
5	838	6.5	465,356,158	499	1230	361	17	42,467
6	579	9	568,674,539	495	1193	558	7	13,900
7	827	7	457,184,239	651	1237	513	16	45,167

1. decentralised, 17 landfill sites.
2. decentralised, 17 landfill sites, 17 composting plants and one RDF-combustion site.
3. intermediate, four landfill sites.
4. intermediate, four landfill sites and four composting sites.
5. intermediate, three landfill sites, three composting plants and one RDF-combustion site.
6. centralised, one landfill site.
7. centralised, one landfill site, one composting site and one RDF-combustion site.

where decentralised, intermediate and centralised refer to each municipal area processing its own waste, the region being divided into a number of areas each of which has responsibility for its own waste processing, and waste being treated in one centralised plant respectively.

Apply the following methods to the choice of a solid waste management system and compare the results:

a. ELECTRE IS
b. ELECTRE III
c. PROMETHEE I
d. PROMETHEE II
e. P/G%
f. Goal Programming
g. Analytical Hierarchy Process
h. Multi-attribute Utility Theory.

9. Consider the problem of choosing an appropriate management strategy for a mangrove forest in the Philippines. Data is given in table 12.12 (Janssen and Padilla 1999) for the performance of the following options:

a. preservation
b. subsistence forestry
c. commercial forestry
d. intensive aquaculture
e. commercial forestry and intensive aquaculture

under the following three criteria:

- The value of products and services generated (economic efficiency).
- The equal distribution of this value across the various stakeholders (equity).
- Sustainability of the type of use (environmental quality).

Table 12.12 Management strategies for a mangrove forest in the Philippines

Option	a	b	c	d	e
Subsistence forestry (1000 pesos)	0	349	0	0	0
Commercial forestry (1000 pesos)	0	0	416	0	229
Fishponds (1000 pesos)	0	0	0	9294	3417
Fisheries (1000 pesos)	165	161	161	8	40
Emissions (tons/year)	0	0	0	100	50
Soil accretion (cm/year)	1.0	0.34	0.42	0.05	0.13
Biodiversity (index)	1.0	0.61	0.39	0.06	0.15
Shore protection (index)	1.0	0.36	0.14	0.06	0.14
Ecotourism (index)	0.8	1.00	0.38	0.08	0.21

Apply the following methods to the choice of management strategy and compare the results:

a. ELECTRE IS
b. ELECTRE III
c. PROMETHEE I
d. PROMETHEE II
e. P/G%
f. Goal Programming
g. Analytical Hierarchy Process
h. Multi-attribute Utility Theory.

References and Additional Reading for Part II

References for Part II

Amagai H, Leung P (1989) Multiple criteria analysis for Japan's electric power generation mix. Energy Systems and Policy, 13(3):219-236

Amagai H, Leung P (1991) The trade-off between economic and environmental objectives in Japan's power sector. Energy Journal, 12(4):95-104

Anandalingam G (1987) A multiple criteria decision analytic approach for evaluating acid rain policy choices. European J. of Operational Research, 29:336-352

Anderson V (1991) Alternative Economic Indicators. Evanston, IL, Routledge

Beach LR, Mitchell TR (1978) A contingency model for the selection of decision strategies. Academy Management Review, 3:434-449, July

Beale EML (1988) Introduction to Optimization. John Wiley and Sons

Brans JP, Mareschal B, Vincke Ph (1984) PROMETHEE: a new family of outranking methods in multicriteria analysis. In: Brans JP (ed) Operational Research '84, Elsevier Science Publishers, pp 477-490

Brans JP, Vincke Ph (1985), A preference ranking organisation method. Management Science, 31(6):647-656

Brans JP, Vincke Ph, Mareschal B (1986) How to select and how to rank projects: the PROMETHEE method. European J of Operational Research, 24:228-238

Brans J-P, Mareschal B (1994) The PROMCALC & GAIA decision support system for multicriteria decision aid. Decision Support Systems 12:297-310

Briggs Th, Kunsch PL, Mareschal B (1990) Nuclear waste management: an application of the multicriteria PROMETHEE methods. European J of Operational Research, 44:1-10

Bunday BD, Garside GR (1986) Optimisation Methods in Pascal. Edward Arnold

Chattopadhyay D (1994) Systems approach to emissions reduction from a power system in India. Energy Sources, 16 (3):425-438

Colson G, de Bruyn C (1989) Models and methods in multiple objectives decision making. Mathematical and Computer Modeling, 12(10-11):1201-1211

Crama Y, Hansen P (1983) An introduction to the ELECTRE research programme. In: Hansen P (ed) Essays and surveys on multiple criteria decision making, Springer, pp 31-42

De Jong K (1975) An analysis of the behaviour of a class of genetic adaptive systems. PhD thesis, University of Michigan.

Dunn S, Flavin C (2002) Moving the climate change agenda forward. In: Flavin C, French H, Gardner State of the world 2002, progress towards a sustainable society, Starke L (ed), Earthscan, London, pp 24-50

Evans GW (1984) An overview of techniques for solving multiobjective mathematical problems. Management Science, 30(11):1268-1282.

Fishburn PC (1964) Decision and value theory. Wiley, New York

Fishburn PC (1972) Interdependent preferences on finite sets. J of Mathematical Psychology, 9:225-236

Forman EH (1990) Multi criteria decision making and the analytic hierarchy process. In: Bana e Costa CA (ed) Readings in multiple criteria decision aid, Springer Verlag

Fortin V, Bobee B (1995) Détermination floue des zones hydrologiques homogènes. In: Simonovic S, Kundzevicz Z, Rosbjerg D, Takeuchi K (eds) Modeling and management of sustainable basin-scale water resources, Wallingford UK, IAHS, no 231, pp 367-376

Figueira, J, Mousseau V, Roy B (2005) Chapter 1 ELECTRE methods. In: Figueira J, Greco S, Ehrgott M (eds) Multiple criteria decision analysis: state of the art surveys, Springer Verlag, pp 133-162.

Gallopin GC, Öberg W (1992) Quality of life. In: Dooge JCI, Goodman G, Rivière JWM, Martin-Lefèvre J, O'Riordan T An agenda of science for environment and development in the 21st century, Cambridge MA, Cambridge University Press

Geoffrion AM (1968) Proper efficiency and the theory of vector maximization, J Math Anal App, 22:618-630

Goumas MG, Lygerou VA, Papayannakis LE (1999) Computational methods for planning and evaluating geothermal energy projects. Energy Policy, 27:147-154

Hammond KR, Hamm RM, Grassia J, Pearson T (1987) Direct comparison of the relative efficiency on intuitive and analytical cognition. IEEE Trans on Systems, Man and Cybernetics, 17(5):753-770

Harremoës P, Gee D, MacGarvin M, Stirling A, Keys J, Wynne B, Guedes Vaz, S (eds) (2002) The precautionary principle in the 20th century: Late lessons from early warnings. Earthscan

Hersh MA (1999) Sustainable decision making: the role of decision support systems. IEEE trans on Systems, Man and Cybernetics, SMC29:395-408

Hobbs BJ (1986) What we can learn from experiments in multiobjective decision analysis. IEEE Trans Systems, Man, Cybernetics, SMC16:384-394

Hokkanen J, Salminen P (1997), Choosing a solid waste management system using multicriteria decision analysis. European J of Operational Research, 98:19-36

Houghton JT, Ding Y, Griggs DJ, Noguer M, van der Linden PJ, Dai X, Maskell K, Johnson CA (2001) Climate change 2001: the scientific basis, contribution of working group I to the third assessment report of the Intergovernmental Panel on Climate Change (IPCC). Cambridge University Press

Janis IL, Mann L (1977) Decision making: a psychological analysis of conflict. In: Choice and Commitment. Free Press, New York

Janssen R, Padilla JE (1999) Preservation or conversion? Valuation and evaluation of a mangrove forest in the Philippines. Environmental and Resource Economics, 14:297-331

Kalika V, Frant S (1992) Environmental aspects of power generation. Energy Sources, 21 (8):687-704

Keeney RL (1980) Siting energy facilities. Academic Press

Keeney RL (1992) Value-focused thinking. Harvard University Press

Keeney RL, Raiffa H (1976) Decisions with multiple objectives: preferences and value tradeoffs. Wiley, New York

Keeney RL, Raiffa H (1993) Decisions with multiple objectives. Cambridge University Press, Cambridge MA

Khrisanov NI, Kerro NI, Kol'nik GA (1990) Comprehensive expert evaluation of the ecological consequences of the construction of hydropower facilities. Hydrotechnical Construction, 24:156-161

Klein GA (1989) Recognition-primed decisions. Advances Man-Mach Syst Res 5:47-92.

Klein GA (1993) A recognition-primed decision model of rapid decision making. In: Klein GA, Orasanu J, Calderwood R, Zsambok CE (eds) Decision making in action: models and methods (cognition and literacy), Norword NJ, Ablex, pp 138-147

Klir GJ, Yuan B (1995) Fuzzy Sets and Fuzzy Logic, Theory and Application., Prentice Hall

Korhonen P, Moskowitz H, Wallenius J (1992) Multiple criteria decision support – a review. Eur J Oper Res, 63:361-375

Linden G, Hanks S, Lesh N (1997) Interactive assessment of user preference models: the automated travel assistant, Proceedings User Modeling '97, http://citeseer.ist.psu.edu/40598.html

Lincoln DR, Rubin ES (1979) Cross-media environmental impacts of coal-fired power plants: an approach using multi-attribute utility theory. IEEE Trans Systems, Man and Cybernetics, SMC9 (5):285-289

Lipschitz R (1993a) Converging themes in the study of decision making in realistic settings. In: Klein GA, Orasanu J, Calderwood R, Zsambok CE (eds) Decision making in action: models and methods (cognition and literacy), Norword NJ, Ablex, pp 103-137

Lipschitz R (1993b) Decision making as argument driven action. In: Klein GA, Orasanu J, Calderwood R, Zsambok CE (eds) Decision making in action: models and methods (cognition and literacy), Norword NJ, Ablex, 172-181

McKenna C (1980) Quantitative methods for public decision making. McGraw Hill, New York

March JG (1994) A primer on decision making. Free Press. New York

Matheson S, Lence B, Fürst J (1997) Distributive fairness considerations in sustainable project selection, Hydrol Sc J, 42(4):531-584

Melton C (1993) Alternatives of lead bearing alloy solder, Proc of IEEE Int Symp on Electronics and the Environment

Mladlineo N, Margeta J (1987) Multicriteria ranking of alternative locations for small scale hydro plants. Eur J Operational Research, 31:215-222

Nagel SS (1988) Evaluation analysis with microcomputers. JAI Press Greenwich CT
Nagel SS, Bievenue L (1989) Simulation to aid in influencing decision-makers. Simulation 53(2):76-77
Nagel SS, Bievenue L (1990a) Education policy and multi-criteria decision-making. In: Walberg HJ Analytic methods for educational productivity, JAI Press, Greenwich CT
Nagel SS, Bievenue L (1990b) Using decision-aiding software for teaching in all fields of knowledge. Collegiate Microcomputer, 8(3):171-182
Nagel SS, Bievenue L (1992) Social science and public policy. University Press of America
O'Riordan T, Cameron J (1994) Interpreting the precautionary principle. Earthscan
Ozernoy VM (1987) A framework for choosing the most appropriate discrete alternative multiple criteria decision-making method in decision support systems and expert systems. In: Savaragai (ed) Toward interactive and intelligent decision support systems, Springer Verlag, Berlin, vol 2, pp 56-64
Ozernoy VM (1992) Choosing the "best" multiple criteria decision-making method. INFO, 30:159-171
Penninton N, Hastie R (1993) A theory of explanation-based decision making. In: Klein GA, Orasanu J, Calderwood R, Zsambok CE (eds) Decision making in action Norword NJ, Ablex, pp 270-287
Pfleiderer RHH, Dieterich M (1995) New roads generate new traffic. World Transport Policy and Practice, 1(1):29-31
Pollak RA (1967) Additive von Neumann-Morgenstern utility functions. Econometrica, 35:485-494
Pomerol J-C, Sergio B-R (1993) Choix Multicritère dans L'Enterprise Principe et Practique. Hermes, Paris, France
Powell J (1996) The evaluation of waste management options. Waste Management and Research, 14:515-526
Renner M (2004) Moving toward a less consumptive economy. Halweil B, Mastny L, Assadourian E, Starke L State of the world 2004, Starke L (ed), WW Norton and Co, pp 96-119
Rogers M, Bruen M (1998) A new system for weighting environmental criteria for use within ELECTRE III. European J of Operational Research, 107:552-563
Rosenthal RE (1985) Concepts, theory and techniques, principles of multiobjective optimization. Decision Sciences, 16:133-152
Roy B (1968) Classement et choix en presence de points de vue multiples, la méthode ELECTRE. RIRO, 2(8):57-75
Roy B (1971) Problems and methods with multiple objectives, Mathematical Programming, 1:239-266
Roy B (1978) ELECTRE III: un algorithme de rangement fondé sur une representation floue des preferences en présence de critères multiples. Cahiers du Centre d'Etudes de Recherche Operationnelle, 20:3-24
Roy B (1985) Mèthodologie multicritère d'aide à la decision. Economica

Roy B (1990) The outranking approach and the foundations of ELECTRE methods. In: Bana e Costa CA (ed) Readings in multiple criteria decision making, Springer, pp 158-183
Roy B, Bouyssou D (1993) Aides multicritère à la decision: methods et cas. Economica, Paris
Saaty TL (1980) The Analytic Hierarchy Process. McGraw Hill, New York
Saaty TL (1990) How to make a decision: the Analytic Hierarchy Process. European J of Operational Research, 48:9-26
Sage AP (1981) Behavioral and organizational considerations in the design of information systems and processes for planning and decision support. IEEE Trans Syst, Man, Cyber SMC11:640-678
Silver MS (1991) Systems that support decision makers, description and analysis. Wiley, New York
Simon HA (1955) A behavioral model of rational choice. Quarterly J of Economics, 69:99-118
Simon HA (1977) A new science of management decision. Revised edition, Prentice-Hall, Englewood Cliffs
Simonovic SP, Burn DH, Lence B (1997) Practical sustainability criteria for decision-making. Int J Sustainable Develop World Ecol, 4(4):231-244
Stanton (1992) Decision aiding algorithms. In: Nagel SN Applications of decision aiding software, Macmillan, New York, pp 170-190
Ting HM (1971) Aggregation of Attributes for multiattributed utility assessment. Technical Report no 66, Operations Research Center, MIT
Tuxill J (1999) Appreciating the benefits of plant biodiversity. In: Brown LR, Flavin C State of the world 1999 progress towards a sustainable society. Starke L (ed) Worldwatch Instiute, Earthscan, pp 96-114
UNESCO (1997) Sustainability criteria for water resources systems. Cambridge University Press, Cambridge, UK
Vetschera R (1986) Sensitivity analysis for the ELECTRE multicriteria method. Zeitschrift Operations Research, 30:B99-B117
Vetschera R (1988) An interactive outranking system for multi-attribute decision making. Comput Operations Research, 15(4):311-322
Vincke P (1989) L'aide multicritère à la decision. SMA, Paris
WCED (1987) Our common future, report of the world commission on environment and development. Oxford University Press
Wenstop FE and Carlsen AJ (1988) Ranking hydroelectric power projects with multicriteria decision analyses. Interfaces, 18:36-48
WWW10 (2005) http:www.endangeredspecie.com/specieprofile.htm
WWW11 (2005) http://www.unido.org/doc/310797.html; http://unfccc.int/resource/conv/
WWW12 (2005) http://www.lamsade.dauphine.fr/English/software.html
Yu P-L (1985) Multi-Criteria decision making concepts, techniques and extensions. Plenum Press
Young M (1992) Sustainable investment and resource use. Parthenon, Park Ridge
Zadeh LA (1994) Fuzzy logic, neural networks and soft computing. Commun ACM 37(3):77-84

Additional Reading for Part II

Ahmad K and Griffin S (1993) Intelligent assistants and engineering decision support: management of water resources. Computing Systems in Engineering, 4:325-335

Alavian V (1993) Integration of environmental management with reservoir and power system operations. Proc Nat Conf on Hydraulic Engineering, 2:205-211

Arrow KJ and Raymond H (1986) Social choice and multicriterion decision-making. MIT Press

Bather J (2000) Decision theory: an introduction to dynamic programming and sequential decisions. John Wiley

Brousseau P (1995) Building decision skills: a unit on ethical decision making. Institute for Global Ethics, Camden ME

Buchanan TJ (1982) Discrete and dynamic decision analysis. John Wiley and Sons

Crain IK (1992) The evolving technology for environmental decision-making – toward the environmental decision support system. Educ Soc, 2:605-614

Dale VH and English MR (1999) Tools to aid environmental decision making. Springer

Fishburn PC (1965) Independence in utility theory with whole product sets. Operations Research, 13:28-45

Fishburn PC (1967) Methods of estimating additive utilities. Management Science, 13 (7):435-453

Fishburn PC (1970) Utility for decision making. Publications in Operations Research, no 18, Wiley, New York

Fletcher R (1987) Practical methods of optimization, 2nd edn. John Wiley and Sons

Gill PE Murray W, Wright MH (1981) Practical optimization. Academic Press

Ginsberg IW, Angelo JA (1990) Earth observations and global change decision making, 1989: a national partnership. Krieger

Hamlett PW (1992) Understanding technological politics: a decision-making approach. Prentice Hall

Hemmati M, Dodds F, Enayati J, McHarry J (2002) Multi-stakeholder processes for governance and sustainability: beyond deadlock and conflict. Earthscan

Jennings D, Wattam S (1998) Decision making: an integrated approach. 2nd edn, Pitman Publishing

Kallman EA and Grillo JP (1993) Ethical decision making and information technology: an introduction with cases. Mitchell McGraw-Hill.

Klein GA (1993) Decision making in action. Norword NJ, Ablex, pp 138-147

Krabs W (1979) Optimization and approximation. John Wiley and Sons

Lance O (1971) Optimal decisions principles of programming. Pergamon Press

LaValle IH (1978) Fundamentals of decision analysis. Holt, Rinehart, and Winston

Leitman G (1976) Multicriteria decision making and differential games. Plenum Press

McDaniel H (1970) Applications of decision tables. Brandon Systems Press Inc

McGosh AM and Scott Morton MS (1978) Management decision support systems. Macmillan Press Ltd

McKeown JJ, Meagan D, Sprevak D (1990) An introduction to unconstrained optimisation. ESM

Nocedal J, Wright SJ (1999) Numerical optimization. Springer

Ossenbruggen PJ (1994) Fundamental principles of systems analysis and decision-making. Wiley

Petitti DB (2000) Meta-Analysis, decision analysis, and cost-effectiveness analysis: methods for quantitative synthesis in medicine, 2nd edn. Oxford University Press

Pollock DSG (1991) Nonlinear optimisation. Queen Mary and Westfield College, University of London

Pratt (1964) Risk aversion in the small and in the large. Econometrica, 32:122-136

Quebral MC (1992) Policy evaluation methods for developing country energy issues. In: Nagel SS (ed) Applications of decision-aiding software. Macmillan, pp 207-233

Robinson EA (1981) Statistical reasoning and decision making. Goose Pond Press

Ryan M (1992) Decision making under uncertainty: regret, disappointment and expected utility theory. HERU, University of Aberdeen

Stanton T (1992) Decision-aiding algorithms. In: Nagel SS (ed) Applications of decision aiding software. Macmillan, pp 170-190

Simonovic SSP, Kundzewicz Z, Rozbjerg D, Takeuchi K (eds) (1995) Modeling and management of sustainable basin-scale water resource systems. IAHS, Wallingford, UK

Stead WE, Stead JG (1996) Management for a small planet: strategic decision making and the environment. 2nd edn. Sage Publications

Targett D (1996) Analytical decision making. Pitman Publishing.

Thrall RM, Coombs CH, Davis RL (1957) Decision processes, 2nd edn. John Wiley and Sons

Truelove P (1992) Decision Making in Transport Planning. Longman Scientific & Technical

Van Delft A, Nijkamp P (1977) Multi-criteria analysis and regional decision-making. Nijhoff Social Sciences Division

Wilson AC, Coelho JD, Macgill SM, Williams HCWL (1981) Optimization in locational and transport analysis. John Wiley and Sons

Wisniewski M (1994) Quantitative methods for decision makers. Pitman.

World Commission on Dams (2000), Dams and development : a new framework for decision-making: the report of the world commission on dams. Earthscan

Zeleny M (1976) Multiple criteria decision making Kyoto 1975. Springer-Verlag

Zitzler E, Deb K, Thiele L, Coello CAC, Corne D (eds) (2001) Evolutionary multi-criterion optimization. First International Conference, Zurich, Switzerland, Springer.

Part III Uncertainty, Imprecision & Fuzzy Systems

III.1 Introduction

Conventional set theory deals with variables with measurable properties which can be unambiguously assigned as members or non-members of a particular set. However real problems, and particularly those associated with sustainable development, are often concerned with messy data, including a mixture of qualitative and quantitative variables, imprecise data and uncertain or inaccurate data. Thus techniques are required to represent and analyse this type of data. A particularly useful approach, presented in this part of the book, is based on fuzzy set techniques and their extension to augmented, intuitionistic and type 2 fuzzy sets.

In a fuzzy set an element is no longer either a member or not a member of a set, but can belong to it to an extent determined by a membership grade. Fuzzy sets can also be represented in linguistic terms (using descriptions rather than numbers). Thus fuzzy set techniques can be used to represent the mixture of qualitative and quantitative and imprecise data occurring in sustainable development problems. Extensions to augmented, intuitionistic and type 2 fuzzy sets can then be used to represent data which is both uncertain and imprecise, which is again important in a sustainable development context.

Part III presents fuzzy set concepts and methods and their application to problems in sustainable development, starting with a basic introduction and moving on to more advanced ideas. It consists of five chapters, (chapters 13-17). Chapter 13 introduces fuzzy set theory and illustrates it by application to the classification of hearing impairment. Chapter 13 also discusses the use of α–cuts and strong α–cuts in obtaining crisp sets from fuzzy sets, numerical and linguistic variables, fuzzy numbers and fuzzification or the process of obtaining fuzzy from crisp sets. Fuzzy relations and conditional statements are also discussed and illustrated by two scenarios for investigating the effects of different strategies for the reduction of carbon dioxide emissions, as well as other global climate change applications.

Chapter 14 introduces the standard definitions of the basic set operations of union, intersection and complement on fuzzy sets and then relates the concepts of fuzzy union, intersection and complement to the classes of t-conorms, t-norms and fuzzy complements. Aggregation and norm operations are also presented. Finally a number of defuzzification methods for obtaining crisp sets from fuzzy sets are considered.

Chapter 15 considers three different approaches to the problem of representing information which is both imprecise (fuzzy) and uncertain or inaccurate, namely augmented, intuitionistic and type 2 fuzzy sets. Definitions of the set operations of union, intersection and complement are introduced for each of these three types of sets. The chapter concludes with two examples of the application of these methodologies to illustrate some of the ways they can be applied to problems in sustainable development and the differences between the approaches. The first example considers the evaluation of the environmental effects of a manufacturing process for producing gears and the second example presents life cycle analysis in the computer industry.

Chapter 16 discusses fuzzy set comparison rules for augmented fuzzy sets and third order augmented fuzzy sets. The comparison rules are illustrated by an example of their application to strategically targeted impact reduction in the use and reuse life cycle stage in the computer industry and evaluation of the effect of a reduction strategy on this life cycle stage. The extension of third order augmented fuzzy sets to allow the first term to have a negative coefficient, which can be used in the representation of costs and benefits, is also considered.

The concluding chapter, chapter 17, illustrates the techniques discussed in Part III of the book by a case study of the application of third order augmented fuzzy sets, intuitionistic fuzzy sets and type 2 fuzzy sets to a decision problem on different strategies for achieving transport aims.

III.2 Learning Objectives

Readers should gain an understanding of fuzzy set theory and techniques and their application to sustainable development. Specific learning objectives of Part III include the following:

- Understanding of basic fuzzy set concepts and the ability to calculate fuzzy set unions, intersections and complements.
- Understanding why fuzzy set techniques are appropriate for the analysis of problems in sustainable development and knowledge of the types of problems they can be applied to.

- Understanding of the relationship between the standard fuzzy set operations and the classes of t-conorms, t-norms and fuzzy complements.
- Knowledge of augmented, intuitionistic and type 2 fuzzy sets and the ability to apply them to problems in sustainable development.
- The ability to apply fuzzy relations and conditional statements to the modelling and analysis of sustainable development problems.
- The ability to apply fuzzification and defuzzification techniques.
- The ability to apply fuzzy techniques, including those based on augmented, type 2 and intuitionistic fuzzy sets to sustainable development problems.

13 Introduction to Fuzzy Sets

13.0 Learning Objectives

Chapter 13 introduces the concept of a fuzzy set and then considers α–cuts, numerical and linguistic variables, fuzzy numbers and fuzzification, as well as fuzzy relations and conditional statements. Illustrative examples include the fuzzy relations between anthropogenic (human generated) emissions of carbon dioxide and sea level rise and two scenarios for investigating the effects of different emissions reduction strategies on global climate change.

Specific learning objectives include the following:

- Understanding of the concept of a fuzzy set.
- Understanding of the differences between numerical and linguistic variables.
- The ability to use α–cuts and strong α–cuts to obtain crisp from fuzzy sets.
- The ability to define fuzzy numbers, particularly in a sustainable development context.
- The ability to apply fuzzification techniques to obtain fuzzy sets from crisp sets.
- An understanding of fuzzy relations and conditional statements and the ability to apply them to modelling global climate change scenarios and other sustainable development problems.

13.1 Introduction

This chapter introduces the concept of fuzzy set, which is defined and illustrated by the example of hearing impairment in section 13.1. Numerical and linguistic variables are presented in section 13.2 and α-cuts, which can be used to obtain crisp sets from fuzzy sets, are introduced in section 13.3. Section 13.4 discusses fuzzy numbers and section 13.5 fuzzification or obtaining fuzzy from crisp sets. Fuzzy and conditional relations are considered in section 13.6. A summary of the chapter is presented in section 13.7.

In conventional set theory a given element is either a member of a set or not a member of the set. Sets with a clearly defined membership, with a given element either a member or non-member, are called *crisp sets*. However in real situations set boundaries are often not clearly defined. For instance the noise from a given computer, printer, car or road may be irritating, but it can still be difficult to decide whether or not it should be classified as 'noisy' either in absolute terms or compared to other computer equipment or other sources of noise such as aeroplanes or pneumatic drills. Most countries do have some legislation about prohibited noise levels in some contexts, though is not necessarily relevant to an individual computer or printer.

Analogously, definitions of 'hazardous' chemicals have changed over time and have generally become more stringent. In addition it has been suggested that the important issue is which chemicals are essential rather than which chemicals are too hazardous to use (O'Brien 2000; Raffensperger and Tickner 1999). However, the term 'essential' is not precisely defined. For instance, a number of years ago lead would have been considered an 'essential' chemical, but the use of lead is being banned by the European Union Reduction of Hazardous Substances (ROHS) Directive from July 2006. Lead-free electronics is widely practiced in Japan and there is some evidence that lead-free solder gives a more reliable joint than the traditional tin-lead one. Therefore both the terms hazardous and essential are not precisely defined and whether or not a particular chemical is considered 'hazardous' and/or 'essential' will depend on a number of factors.

Fuzzy sets (Bandemer and Gottwald 1995; Terano et al 1987; Zadeh 1965) resolve this problem by defining a fuzzy set *membership function* to take all values between 0 and 1, unlike a conventional or *crisp set* membership function which is either 0 (non-member) or 1 (member). This membership function then defines a fuzzy set. Therefore a fuzzy set can be defined as a mapping or membership function from the classical set X of all possible points to the interval $I = [0, 1]$. Each point x in X has a *membership grade* $\mu(x)$ (value taken by the membership function $\mu(.)$ at the point x) which is equal to a real number between 0 and 1 and so a fuzzy set is defined as

$$\mu: \ X \rightarrow I = [0, \ 1] \qquad (13.1a)$$

$$\mu(x) \in I, \ \ \forall x \in X \qquad (13.1b)$$

Therefore, unlike a *crisp set*, where each element is either a member or not a member of the set, in a fuzzy set each element belongs to the set to an extent defined by its membership grade. Thus the fuzzy set membership function gives a measure of the grade of membership of the point x in the fuzzy set defined by μ(X) or the extent to which, for instance, a particular property holds. A fuzzy set can be represented graphically, as shown in fig. 13.1, or mathematically. Figure 13.1 illustrates three fuzzy sets for low, medium and high temperatures. When X, the universal set or space of μ, is finite, the fuzzy set with membership grade given by the value of μ(x) can either be represented as ordered pairs of elements and their membership grades connected by the set theory union operator, represented here by '+' rather than '∪'.

$$\{x_1, \mu(x_1)\} + \{x_2, \mu(x_2)\} + \ldots + \{x_n, \mu(x_n)\} \quad (13.2)$$

or in tabular form as shown in table 13.1.

Table 13.1 Fuzzy set representation

X	x_1	x_2	...	x_n
μ(x)	$\mu(x_1)$	$\mu(x_2)$	...	$\mu(x_n)$

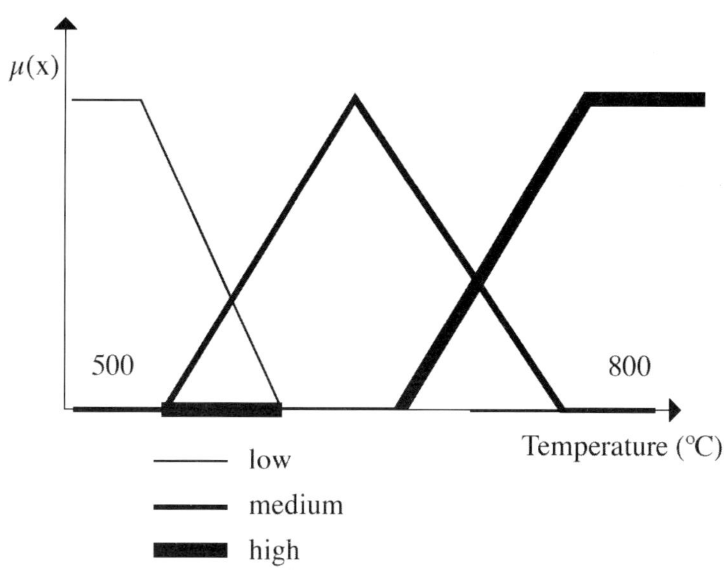

Fig. 13.1 Graphical representation of fuzzy sets representing temperature

A fuzzy set is totally defined by the values taken by its membership function. Therefore the notation $\mu(X)$ will be used to denote the fuzzy set and $\mu(x)$ the values of its membership function. When it is necessary to distinguish between different fuzzy sets, subscripts will be used e.g. μ_1 or μ_A. However, sometimes a fuzzy set is denoted by A and its membership grade by $\mu_A(x)$. When $X = \{x_1, x_2, \ldots, x_n\}$ is a finite set, the following representation is often used:

$$A = \frac{\mu_A(x_1)}{x_1} + \frac{\mu_A(x_2)}{x_2} + \ldots + \frac{\mu_A(x_n)}{x_n} \tag{13.3}$$

where '+' denotes the set theory union operator and the 'denominator' of each term $\frac{\mu_A(x_i)}{x_i}$ denotes a point x in X and the 'numerator' the value of its membership function. μ is a crisp set if it only takes the values 0 and 1. It can then be identified with the classical subset $A = \{x \in X: \mu(x) = 1\}$.

Example 13.1: Classification of Hearing Impairment

Measurement of hearing and hearing impairment is required to inform decisions on rehabilitative and other support measures. Hearing assessment can also contribute to the identification of medical conditions, tumours or other diseases of the auditory system. The measurement of hearing is important in occupational health, particularly for workers in noisy environments. It can be used to determine the effectiveness of occupational hearing conservation programmes, including whether additional measures will be required to protect hearing.

Hearing impairment is measured in decibels hearing level, normally abbreviated as dB HL. These measurements can also be expressed in

Table 13.2 Classification of hearing impairment

Category of Hearing impairment	Threshold in Decibels
No (or minimal) hearing impairment	0-20 dB HL
Slight hearing impairment	20-30 dB HL
Mild hearing impairment	30-45 dB HL
Moderate hearing impairment	45-60 dB HL
Severe hearing impairment	60-75 dB HL
Profound hearing impairment	75-90 dB HL
Extreme hearing impairment	90-110 dB HL
Deafness	> 110 dB HL

words or linguistic terms as shown in table 13.2 (Newby and Popelka 1985)

Thus a fuzzy set *hearing impairment* could be defined. Individuals are assigned membership grades in the fuzzy set *hearing impairment* according to their degree of hearing impairment. This could give, for instance, the membership grades in table 13.3.

Table 13.3 Membership grades for categories of hearing impairment

Category of Hearing impairment	Membership Grade
No (or minimal) hearing impairment	0
Slight hearing impairment	0.2
Mild hearing impairment	0.4
Moderate hearing impairment	0.5
Severe hearing impairment	0.7
Profound hearing impairment	0.8
Extreme hearing impairment	0.9
Deafness	1

Consider the group of people Helga, Leylah, Ahmed, Francois, Miriam, John and Karen, where Karen and Francois are deaf, Helga has no hearing impairment, Ahmed has a mild hearing impairment, Miriam has a moderate hearing impairment, Leylah has a severe hearing impairment and John has a profound hearing impairment. This can be expressed using fuzzy sets as shown in table 13.4.

Table 13.4 Fuzzy set *hearing impairment*

X	x_1	x_2	x_3	x_4	x_5	x_6	x_7
$\mu(x)$	0	0.7	0.4	1	0.5	0.8	1

where $X = \{x_1, \ldots, x_7\}$ and

x_1 – Helga x_2 – Leylah x_3 - Ahmed x_4 – Francois
x_5 – Miriam x_6 – John x_7 - Karen

This fuzzy set could also be expressed as

$$A = \frac{0}{x_1} + \frac{0.7}{x_2} + \frac{0.4}{x_3} + \frac{1}{x_4} + \frac{0.5}{x_5} + \frac{0.8}{x_6} + \frac{1}{x_7}$$

13.2 Numerical and Linguistic Variables

Data obtained from the real world to be used in fuzzy methods is not always fuzzy and real problems may require crisp rather than fuzzy data as the output of fuzzy methods. Therefore methods are required to convert between fuzzy and crisp data. This section and the following three sections will discuss issues associated with the representation of data in fuzzy set form, including fuzzification. Defuzzification or the conversion of fuzzy data into crisp data will be discussed in chapter 14. Raw data can be obtained in two main forms:

- numerical or quantitative variables
- linguistic or qualitative variables

Numerical variables are variables which take quantitative values expressed as numbers in appropriate units. They can be used to give numerical values of, for instance, energy consumption in (kilo)joules and waste generation or resource consumption in kilograms or metric tons. They use crisp i.e. exact numerical data or crisp data together with error bounds. Numerical variables, possibly with error bounds, are generally used in mathematical models and other applications where calculation is required.

Linguistic variables (Zadeh 1975) have values which are expressed in qualitative or linguistic terms i.e. in words rather than numbers. For instance, energy or resource consumption could be measured as high, low, moderately high, negligible or excessive.

Linguistic variables are used in many everyday situations, due to the following:

- Measuring equipment not being available or problems in obtaining or measuring appropriate variables.
- Difficulties in understanding numerical measurements.
- Numerical measurements and variables appearing excessively precise in some contexts.

Linguistic variables are often used in giving instructions in everyday situations. For instance instructions to turn music down are not given in terms of decibels, but to turn the music down a bit, a lot, to a reasonable volume or to turn it off, as well as more idiomatic and rather less polite versions. In this case the use of linguistic variables is more appropriate than the use of numerical variables, which could seem excessively precise and pedantic.

However many variables can be expressed in either numerical or linguistic terms. For instance energy consumption of a particular industrial process can be expressed as high, very high, excessive or moderate or as a number in kilojoules, possibly with error bounds.

The numerical values that could be associated with linguistic variables are frequently context dependent. For instance the term high energy consumption means something different when applied to a house, an industrial process, a city or a country and the concept of safe exposure limits is totally dependent on whether it is exposure to, for instance, oxygen, chlorine or asbestos. Operations on linguistic terms are frequently also context dependent. In many cases the fuzzy union, intersection and complement operators can be used to represent the connectives *OR*, *AND* and *NOT*, but choice of an appropriate fuzzy operator will largely depend on the context. Obtaining the appropriate context dependent meaning for linguistic terms and operations is particularly important when fuzzy set techniques are applied to real problems.

Linguistic variables are obviously 'fuzzy', as they give imprecise measures and can generally be treated as fuzzy numbers. (See section 13.4.) The inclusion of error bounds in expressions for numerical variables gives them a degree of fuzziness. They can also be made fuzzy or *fuzzified*. For instance a numerical measure of a particular environmental impact, such as energy consumption or emissions of carbon dioxide, can be treated as a fuzzy variable if considered as a measure of the severity of that particular impact. Alternatively variables can be fuzzified by assigning linguistic values, such as very low, low, moderate, high and very high, to appropriate ranges of values of the variables. Thus numerical variables can be represented as fuzzy variables by conversion to membership grades. The conversion of crisp variables to fuzzy variables is called *fuzzification* and the reverse process of obtaining a particular crisp output from a fuzzy set is called *defuzzification*.

13.3 α–cuts and Some Definitions

Crisp sets can be obtained from fuzzy sets in a number of ways, including through the use of α–cuts (Bandemer and Gottwald 1995; Kruse et al 1994). For a given fuzzy set μ(x) defined on a space X and any number $\alpha \in [0, 1]$, the *α–cut* ${}^{\alpha}\mu$ and the *strong α–cut* ${}^{\alpha+}\mu$ are the following crisp sets:

$$ {}^{\alpha}\mu = \{x : \mu(x) \geq \alpha\} \tag{13.4a} $$

$$^{\alpha+}\mu = \{x: \mu(x) > \alpha\} \tag{13.4b}$$

Therefore the α–cut of a given set is the crisp set containing all the elements of X with membership grade greater than or equal to the specified value α and the strong α-cut is the crisp set containing all the elements of X with membership grade greater than α. Therefore the strong α-cut can be obtained from the α-cut by removing all the elements with membership grade equal to α. All α-cuts and strong α-cuts form two distinct families of nested crisp sets. The (strong) α-cut of a set with a higher value of α is contained in the (strong) α-cut of the same set for a lower value of α. Therefore the ordering by set inclusion of the corresponding α-cuts or strong α-cuts is monotonic and in the opposite direction from the total ordering of values of α in [0, 1] i.e.

$$^{\alpha 1}\mu \supseteq {}^{\alpha 2}\mu \text{ and } {}^{\alpha 1+}\mu \supseteq {}^{\alpha 2+}\mu \quad \text{for } \alpha_1 \leq \alpha_2 \tag{13.5}$$

The *level set* Λ(μ) of a given fuzzy set μ is the set of all values of α ∈ [0, 1] that represent distinct α–cuts of the fuzzy set μ. Therefore the level set is the set of all the distinct membership grades taken by elements of the fuzzy set, as each distinct α–cut corresponds to a distinct membership grade. Formally

$$\Lambda(\mu) = \{\alpha: \mu(x) = \alpha \text{ for some } x \in X\} \tag{13.6}$$

The *support* S(μ) of a fuzzy set μ in the universal set X is the crisp set that contains all the elements of X with non-zero membership grades in the fuzzy set μ. The support of μ is identical to the strong α–cut of μ for α= 0 and therefore sometimes written $^{0+}\mu$.

The *height* h(μ) of a fuzzy set μ is the largest membership grade obtained by any element in the set. Formally

$$h(\mu) = \sup_{x \in X} \mu(x) \tag{13.7}$$

The *core* of a fuzzy set is the set of elements with the largest degree of membership of the fuzzy set μ.

A fuzzy set is *normal* when its height is equal to one and *subnormal* when its height is less than one. The height of a fuzzy set can also be considered as the highest value of α for which the α–cut is not empty. Therefore a fuzzy set is normal if, but not only if, it contains an element which would be definitely (with membership grade one) in the associated crisp set. Thus having an element with membership grade one is a sufficient, but not a necessary condition for a fuzzy set to be normal and there are sets with height equal to one, but without an element with membership grade equal to one.

For instance the fuzzy set

$$\mu(x) = \frac{x}{x+1} \quad \text{for } x \in X = [0, \infty) \tag{13.8}$$

has h(x) = 1, though there is no element $x \in X$ with h(x) = 1.

A *convex* fuzzy set has convex α–cuts for α in (0, 1]. A crisp set, such as an α–cut for a given value of α, is convex if all points on the straight line segment connecting the points *r* and *s* are in the set whenever the two points *r* and *s* are i.e. if

$$[1-\lambda]r + \lambda s \;\leq\; (1-\lambda)r + \lambda s \tag{13.9}$$

It should be noted that the membership function of a convex fuzzy set is not necessarily convex.

Example 13.2: Fuzzy Set for Noise

Noise pollution is an increasing problem. In addition to disrupting communication and causing stress, exposure to loud noise can lead to hearing impairment. However the term *noisy* is not precisely defined, as well as depending on the context and the subjective perceptions of different individuals about different levels of noise. This example considers the noise levels of different printers and defines a fuzzy set *noisy* which gives a measure of the extent to which a given printer (or other device) is considered to be (unacceptably) noisy. This fuzzy set has the membership function

Table 13.5 Fuzzy set *noisy*

X	x_1	x_2	x_3	x_4	x_5
μ(x):	1	0.5	0.6	0.3	0.2

defined on the space $X = \{x_1, x_2, ..., x_5\}$ of five different types of printers.

The fuzzy set *noisy* shows that, for instance, printer 5 is quiet, whereas printer 1 is very noisy. This set has the following α-cuts and strong α-cuts, as shown in fig. 13.2:

$^{0+}\mu = {}^{0.2}\mu = \{x_1, ..., x_5\} = X \quad {}^{0.2+}\mu = {}^{0.3}\mu = \{x_1, x_2, x_3, x_4\}$
$^{0.3+}\mu = {}^{0.5}\mu = \{x_1, x_2, x_3\} \quad {}^{0.5+}\mu = {}^{0.6}\mu = \{x_1, x_3\} \quad {}^{1}\mu = \{x_1\}$

The level set of this set is $\Lambda(\mu) = \{0.2, 0.3, 0.5, 0.6, 1\}$

The support of this set is $S(\mu) = \{x_1, ..., x_5\} = X$

The height of this set is $h(\mu) = 1$ and it is therefore normal.

13.4 Fuzzy Numbers

Fuzzy numbers or *fuzzy intervals* are fuzzy sets defined on the real interval i.e. with membership functions of the form

$$\mu : \mathcal{R} \rightarrow I \tag{13.10}$$

and which have a quantitative meaning. A fuzzy set which is a fuzzy number has the following properties:

- It is normal i.e. it contains an element with membership grade 1.
- The alpha cuts $^{\alpha}\mu$ are closed intervals for all values of α.
- Its support i.e. the crisp set containing all elements with non-zero membership grades is bounded.

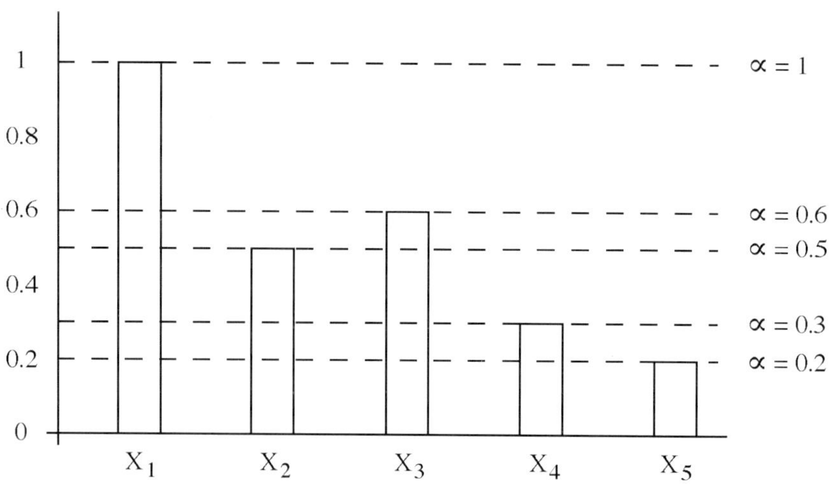

Fig. 13.2 α-cuts and strong α-cuts of the fuzzy set *noisy*

Fuzzy numbers are normal fuzzy sets as, for instance, the fuzzy set of real numbers close to the real number m contains the number m. Fuzzy numbers are convex fuzzy sets, as their α-cuts are closed intervals and therefore equation (3.9) is satisfied, but not all convex fuzzy sets are fuzzy numbers, as they may have α-cuts which are open or half open intervals. Ordinary real numbers and intervals of real numbers can be considered special cases of fuzzy numbers. An interval [a, b] of the real line can be partitioned into a number of different regions representing different linguistic states or fuzzy numbers.

Fuzzy numbers are often represented using triangular functions, as shown in fig. 13.3 or trapezoidal functions, as shown in fig. 13.4. Both triangular and trapezoidal functions are piecewise linear and are therefore easy to program on a computer.

However other shapes can be used and may be preferable in some applications. In particular, in the context of sustainable development it is often necessary to measure the size or significance of the environmental, social and developmental impacts of products and processes. Since variables of this type are monotonically increasing, they can be represented by monotonically increasing linear or non-linear fuzzy numbers, as shown in figures 13.5a and 13.5b. These are both examples of non-symmetric fuzzy numbers. Other non-symmetric membership functions include one-sided or otherwise asymmetric bell-shaped curves and asymmetric

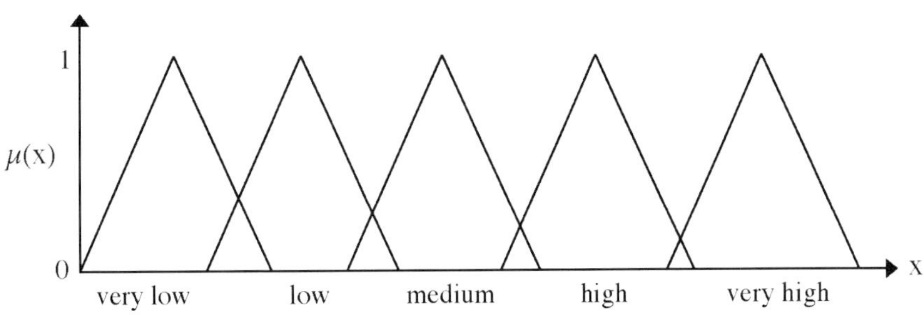

Fig. 13.3 Triangular fuzzy number representation

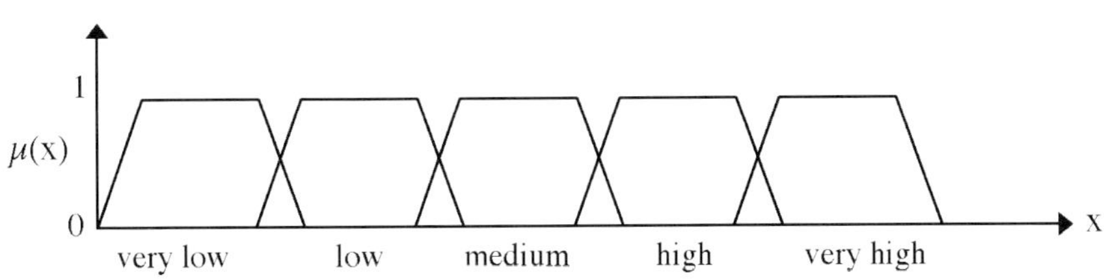

Fig. 13.4 Trapezoidal fuzzy number representation

versions of the triangular and trapezoidal membership functions in figs. 13.3 and 13.4, as well as monotonically decreasing fuzzy numbers, which can be used to represent small numbers. However it should be noted that environmental and social impacts can be represented in other ways than by one monotonically increasing fuzzy number, including by triangular or trapezoidal fuzzy numbers.

It can be shown (Klir and Yuan 1995) that the membership functions of fuzzy numbers are generally piecewise defined. More specifically a fuzzy set μ is a fuzzy number if and only if there exists a closed interval [a, b] ≠ ∅ such that

$$\mu = \begin{cases} 1 & \text{for } x \in [a, b] \\ l(x) & \text{for } x \in (-\infty, a) \\ r(x) & \text{for } x \in (b, \infty) \end{cases} \tag{13.11}$$

where l(x) is a monotonic increasing function from $(-\infty, a)$ to [0, 1] which is continuous from the right and zero for x in $(-\infty, \omega_1)$ and r(x) is a monotonic decreasing function from (b, ∞) to [0, 1] which is continuous from the left and zero for x in (ω_2, ∞). This is illustrated in fig. 13.6.In the case of triangular and trapezoidal fuzzy numbers the functions l(x) and r(x) are linear. For triangular fuzzy numbers a = b = 0.

Example 13.3: Hearing Impairment – Fuzzy Numbers

The categories of hearing impairment discussed in example 13.1 in section 13.1 will now be represented using triangular fuzzy numbers, as illustrated in fig. 13.7, where the abbreviations hi, mod, sev, prof and extr represent hearing impairment, moderate, severe, profound and extreme respectively.

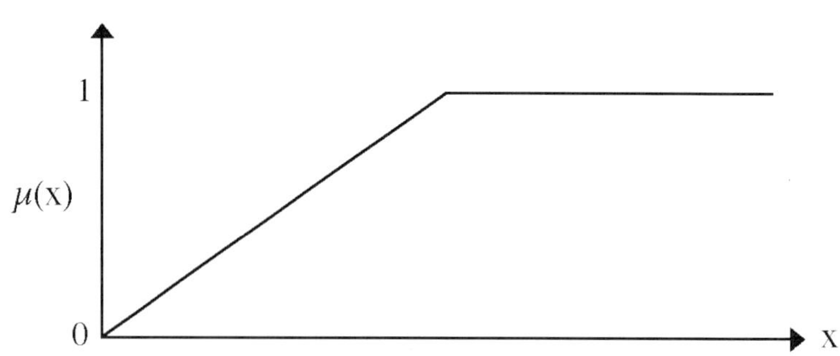

Fig. 13.5a Linear monotonically increasing fuzzy number

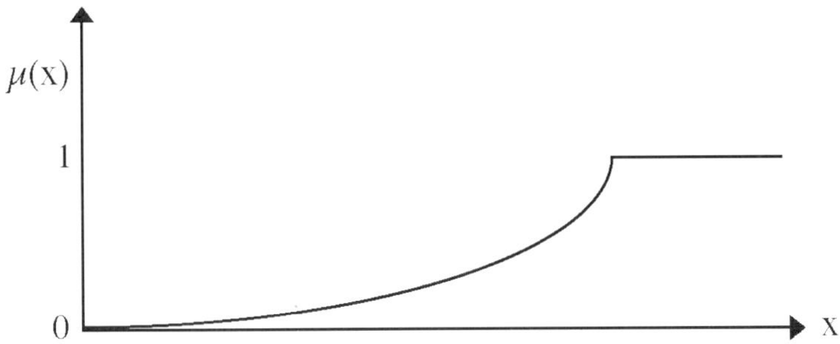

Fig. 13.5b Non-linear monotonically increasing fuzzy number

13.5 Fuzzification

Before fuzzy methods can be applied, the raw data has to be fuzzified or converted into fuzzy set form. There are a number of ways of doing this and different approaches may be required to assign membership grades to numerical and linguistic variables. First a framework consisting of criteria to determine whether or not a particular approach to fuzzification is satisfactory will be discussed. In mathematical terms a fuzzification can be expressed as a mapping $\mathcal{F}$ from a variable X to a fuzzy set μ with membership function μ(X) given by:

$$\mathcal{F}\colon X \rightarrow \mu(X) \tag{13.12a}$$

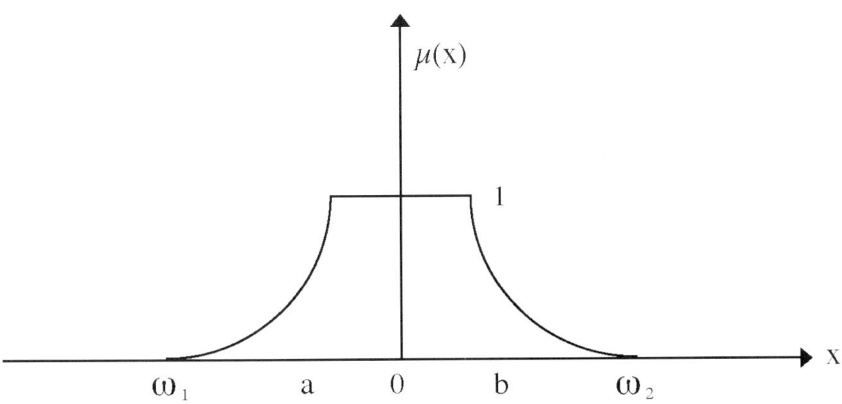

Fig. 13.6 Piecewise continuous fuzzy number

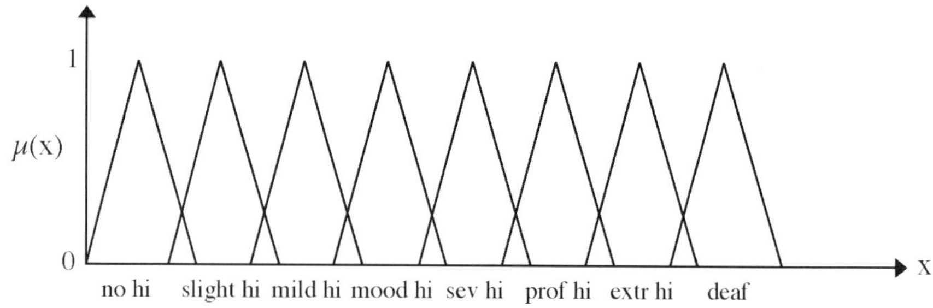

Fig. 13.7 Triangular fuzzy number representation of hearing impairment

The conditions to be satisfied by this fuzzification include the following:

- *(Piecewise) continuity* so that a small change in the value of the crisp data should not give a large change in the resulting membership grade. The condition of piecewise rather than absolute continuity allows fuzzification using fuzzy numbers, where there may be small discontinuities between the membership grades associated with the different fuzzy numbers.
- *Uniqueness* so that each crisp data value has a unique membership grade. However a one-to one mapping between crisp data and fuzzy data is not required and therefore several different crisp data values may have the same membership grade.
- *Computational efficiency* to reduce memory requirements and computational time.
- *Range* so that the values taken by crisp data in a given problem cover the whole [0 1] range of the membership function.
  $\mu(x_1) = 0$ for $x_1 = \min X \quad \mu(x_2) = 1$ for $x_2 = \max X$ (13.12b)
- *Monotonic increasing* so that a higher value of the variable has a greater membership grade.
  $x_1 < x_2 \Rightarrow \mathcal{F}(x_1) < \mathcal{F}(x_2)$ (13.12c)

In the context of sustainable development fuzzy variables which represent the extent to which a particular impact is significant are often particularly useful. Variables of this type can be expressed as linear or non-linear monotonically increasing fuzzy numbers. This generally involves some sort of scaling. There are two possibilities:

- 'local' scaling
- 'global' scaling.

Local scaling only takes account of the values in the problem. It has the advantage of simplicity. *Global scaling* takes account of the wider context and generally considers both:

- The relationship of the particular variable to other variables in the problem.
- The severity or strength of the impact or variable on a global scale compared to occurrences of that variable in other types of problems.

Therefore the use of global scaling gives a measure of both the overall seriousness of the impacts on a global scale and the severity of a given class of impacts relative to other types of impacts. The use of a global scale facilitates the comparison of different types of impacts, but its calculation is often more complicated than that of local scaling and may require additional data, which is not necessarily (easily) available, to give an appropriate global measure.

When global scaling is used, the resulting membership grades may be bunched closely together and the use of a multiplying function may be required to allow the use of a greater part of the range [0, 1] if it is necessary to make comparisons between the different values. When local scaling is used, then relative scaling of the different categories of impacts may be required to allow them to be combined with each other.

One of the simplest approaches to *assigning membership grades* to numerical data is as a monotonically increasing (linear) fuzzy number, with the minimum (or zero) and maximum significant values assigned to 0 and 1 respectively. Linear scaling can then be applied to other values between these limits. However, it may not be easy to determine the minimum and maximum significant values, particularly when global scaling is used. It is also not necessary to assign the data to a linear monotonically increasing fuzzy number. Non-linear fuzzy numbers, with functional forms such as logarithmic can also be used.

Another approach is based on assigning ranges of values of the variables to different fuzzy numbers, such as very low, low, moderate, high and very high. This will generally require appropriate shape(s) to be chosen for the fuzzy numbers, as well as decisions on the number of fuzzy numbers to best represent the variables and facilitate decision making. There will also generally be issues of choices between local and global scaling in the context of using available data to decide what should be considered, for instance, high or moderate values for the particular variable(s) and context. In many cases linguistic or numerical variables cannot be directly converted to fuzzy sets using numerical techniques and the involvement of experts will be required to interpret the linguistic or numerical data. There are a number of different methods of involving experts which can be classified into direct and indirect methods and further into methods involving one expert and methods involving multiple experts, but these will not be discussed further here.

Linguistic variables can generally be arranged in rank order and this facilitates the assignment of membership grades. It is often possible to directly assign linguistic states to fuzzy numbers. However choices still have to be made about the shape of the fuzzy numbers and the assignment of regions of the real axis to different fuzzy numbers, as well as whether the fuzzy numbers should be symmetric and/or all have the same shape. In some cases there may be advantages in an indirect conversion to numerical values first, followed by conversion to membership grades.

Another important decision in deriving fuzzy set models is the appropriate degree of aggregation of the data to give a reasonable number of variables or variable categories on which to base decisions. Aggregation can be carried out using the appropriate fuzzy union operators. It simplifies the data and makes it more accessible, but at the same time may reduce data quality and, if carried out inappropriately, can remove useful information.

Example 13.4: Fuzzification of Carbon Dioxide Emissions

Consider a number of different industrial processes which have the following monthly levels of emissions of the greenhouse gas carbon dioxide: 0.7 kg, 2.5 kg, 6 kg, 12 kg, 15 kg, 16 kg, 29.6 kg and 54 kg. The variable *emissions of carbon dioxide* will first be fuzzified using one fuzzy number representing the seriousness of the emissions.

In this case the spread of the values of the emissions makes the use of linear scaling appropriate. Using local scaling the value of 0 kg is assigned a membership grade of 0 and 60 kg a membership grade of 1. Linear scaling is then carried out between these values and the results are presented in table 13.6.

Table 13.6 Calculation of membership grades using local scaling

Original	Scaled	Membership Grade
0.7 kg	0.012	0
2.5 kg	0.042	0.0
6 kg	0.100	0.1
12 kg	0.200	0.2
15 kg	0.250	0.3
16 kg	0.267	0.3
29.6 kg	0.493	0.5
54 kg	0.900	0.9

The second column gives the values obtained from linear scaling to 3 decimal places and the third column gives rounded values, which can be used as the membership grades.

It will now be assumed that the following levels of emissions are produced either by other industrial processes or in another relevant context:

5000 kg, 21.7 kg, 673 kg, 56.9 kg.

Therefore using global scaling which takes account of these values would again assign the membership grade of 0 to 0kg, but now assign the membership grade of 1 to 5000 kg, as this is the maximum value in this problem. The wide spread of values implies that, in this case, logarithmic scaling i.e. the use of a logarithmic fuzzy number could be more appropriate than linear scaling. Thus scaling so that 5000 kg has the fuzzy set value of 1 can be achieved by taking logs and dividing by log 5000.

However 0.7 has a negative log, but there is no reason for the membership grade of 0.7 kg to be negative. It is just smaller in magnitude than the other values, but not different in sign. This problem can be avoided by first multiplying all values by 10 (second column) before taking logs. The stages of this calculation are shown in table 13.7. The third column gives the log of 10 times the value in the first column. The fourth column gives the membership grade obtained by dividing the value in the third column by log(5000x10) = y. For comparison the last column gives the linear scaling obtained by dividing the value by 5000. In this case linear scaling is not very useful, as all the values are assigned to zero to one decimal place and only two values are non-zero if two decimal places are used.

Table 13.7 Calculation of membership grades using global scaling

x kg	10x	Log(10x)	log10x/y	x/5000
0.7 kg	7	0.8451	0.18	0.00006
2.5 kg	25	1.3979	0.30	0.0005
6 kg	60	1.7782	0.38	0.0012
12 kg	120	2.0792	0.44	0.0024
15 kg	150	2.1761	0.46	0.003
16 kg	160	2.2041	0.47	0.0032
29.6 kg	296	2.4713	0.53	0.0059
54 kg	540	2.7324	0.58	0.011

Converting the data to a logarithmic rather than a linear fuzzy number has both advantages and disadvantages. It discriminates better between the

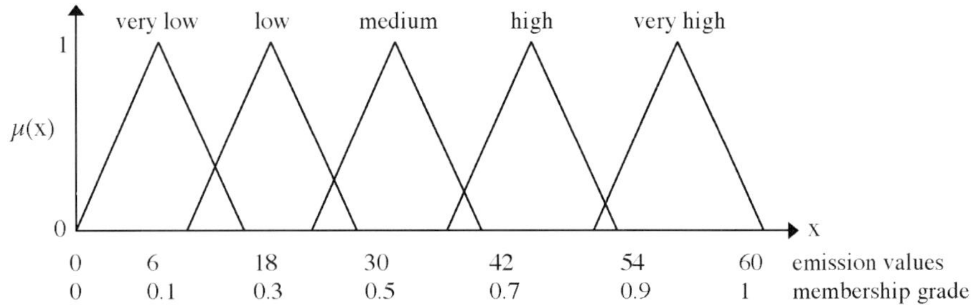

Fig. 13.8 Fuzzification of emissions data using triangular fuzzy numbers

different (low) values, but may give the impression that their impact is greater (in the global sense) than is in fact the case.The values of the emissions can also be fuzzified using a number of different fuzzy numbers. In this case triangular fuzzy numbers representing very low, low, medium, high and very high values of emissions will be used. The values of emissions and membership grades corresponding to the different fuzzy numbers are shown in Fig. 13.8. It can be seen from this figure that the emissions of 0.7, 2.5 and 6 kg correspond to the fuzzy number *very low*, 12, 15 and 16 kg to the fuzzy number *low*, 29.6 kg to the fuzzy number *medium* and 54 kg to the fuzzy number *very high*. Again from the figure, the approximate membership grades would be 0 for 0.7kg, 0.4 for 2.5kg, 0.1 for 6kg, 0.2 for 12 kg, 0.25 for 15 kg, 0.3 for 16kg, 0.5 for 29.6 kg and 0.9 for 54 kg i.e. the same values as previously calculated for local scaling.

13.6 Fuzzy Relations

Crisp relations can be used to represent the existence or absence of a connection or association between the elements of two or more sets. Therefore a relation between the crisp sets X_1 to X_n is a subset of the Cartesian product of the sets $X_1 \times X_2 \times \ldots \times X_n$ and is denoted by $R(X_1, X_2, \ldots, X_n)$. It can be defined by a characteristic function which assigns the values 1 and 0 according to whether or not an n-tuple is in the relation i.e.

$$R(x_1, x_2, \ldots, x_n) = \begin{cases} 1 & \text{if and only if } (x_1, x_2, \ldots, x_n) \in R \\ \text{otherwise} & \end{cases} \tag{13.13}$$

Analogously to crisp sets, crisp relations have the disadvantage of only being able to indicate whether a connection is present and not its strength. For instance consider the problem of reducing the environmental impacts of transport, including emissions, energy consumption and noise. Amongst other factors it is useful to be able to represent the environmental or environmentally damaging features of different vehicles, such as metro, light rail, different types of rail system, buses of varying design, bicycles, electrified bicycles, cars of different makes and size, and freight vehicles. These features could include energy consumption, emissions of carbon monoxide and dioxide, nitrogen oxides and other gases, type of fuel used, fuel consumption per passenger or metric ton of load, and noise levels. One way of doing this is by using relations with one set representing the vehicles and the other sets the levels of the different environmental or environmentally damaging characteristics. However in many cases the data may be qualitative, uncertain and/or based on approximations and estimates, giving a need for measures of the strength of the connections and the use of qualitative or fuzzy set elements rather than crisp values.

In a *fuzzy relation* (Blanchard 1983) a membership grade is used to represent the degree of connection between the elements of the two sets rather than just membership or non-membership of the relation i.e. the existence of a connection between the elements. This is analogous to the use of a membership function to represent the degree of membership in a fuzzy set. A *fuzzy relation* is a fuzzy set defined on the Cartesian product of the crisp or fuzzy sets $X_1, X_2, ..., X_n$. Its membership grade indicates the strength of the relation between the different elements of an n-tuple. A fuzzy relation can be represented by an n-dimensional array, with entries denoting the membership grades of the corresponding n-tuples.

Binary fuzzy relations R(X,Y) are an important and frequently used subclass of n-dimensional fuzzy relations. They give the connection between the elements of two rather than n sets. A binary relation can be represented by a rectangular membership matrix $\mu_R(X,Y)$, with the ijth element representing the degree of connection or the extent to which the relation holds between the ith element x_i of X and the jth element y_j of Y. When X and Y are finite fuzzy sets μ_X and μ_Y, the relation matrix μ_R is given by

$$\mu_R = \begin{bmatrix} \min(\mu_X(u_1), \mu_Y(y_1)) & \dots & \min(\mu_X(u_1), \mu_Y(y_n)) \\ \vdots & \ddots & \vdots \\ \min(\mu_X(u_n), \mu_Y(y_1)) & \cdots & \min(\mu_X(u_n), \mu_Y(y_n)) \end{bmatrix} \qquad (13.14)$$

The *inverse* R^{-1} of a binary fuzzy relation R is a fuzzy relation on X x Y defined by

$$R^{-1}(X,Y) = R(Y,X) \tag{13.15a}$$

i.e. the inverse relation gives the relation between Y and X. When the sets X and Y are fuzzy

$$\mu_R^{-1} = \mu_R^{T} = \begin{bmatrix} \min(\mu_Y(y_1),\mu_X(u_1)) & \cdots & \min(\mu_Y(u_1),\mu_X(y_n)) \\ \vdots & \ddots & \vdots \\ \min(\mu_Y(u_n),\mu_X(y_1)) & \cdots & \min(\mu_Y(u_n),\mu_X(y_n)) \end{bmatrix} \tag{13.15b}$$

Two binary fuzzy relations P(X,Y) and Q(Y,Z) with the set Y common to both can be combined to give a binary fuzzy relation R(X,Z) on X x Z as follows:

$$R(X,Z) = [PoQ] = \max_{y \in Y} \min[P(x,y), Q(y,z)] \text{ for all } x \in X \text{ and } z \in Z \tag{3.16}$$

$$\text{and } \mu_R(X,Z) = \mu_P o \mu_Q = \max_{y \in Y} \min[\mu_P(x,y), \mu_Q(y,z)] \tag{13.17a}$$

$$\text{i.e. } \mu_{R.ij} = \max_k \min(\mu_{P.ik}, \mu_{Q.kj}) \tag{13.17b}$$

This standard or max-min composition for two fuzzy sets is associative

$$[P(X,Y)oQ(Y,Z)]oR(Z,W) = P(X,Y)o[Q(Y,Z)oR(Z,W)] \tag{13.18}$$

but not commutative. Q(Y, Z)oP(X, Y) may not even be defined when $Y \neq Z$. Even when both Q(Y, Z)oP(X, Y) and P(X,Y)oQ(Y,Z) are defined they may not be equal.

Example 13.5: Fuzzy Relations for Climate Change

Greenhouse gases, particularly carbon dioxide, methane, chlorofluorocarbons and nitrous oxide, in the atmosphere absorb heat radiating from the earth's surface and emit some of it downwards, giving an increase in temperature. Without this heat the earth would be about 30° C colder than it is. However human activities have led to increasing atmospheric concentrations of greenhouse gases globally. The fact that anthropogenic (human generated) emissions of greenhouse gases are affecting the global climate is now generally accepted as a scientific fact (Houghton et al 2001), but there are still considerable uncertainties about the timing and magnitude of the resulting temperature changes and effects on the natural environment, including sea levels, and economies. The magnitude of future global warming will depend at least in part on geophysical and biological feedbacks and how they magnify or reduce this warming effect.

The concentration of carbon dioxide, the most important greenhouse gas, has risen by 25% since pre-industrial times. The rate of greenhouse gas build-up will depend heavily on future patterns of economic and technological development. However, stabilising emissions of greenhouse gases at current levels will not stabilise atmospheric concentrations, as once emitted, greenhouse gases remain in the atmosphere for decades to centuries. Therefore drastic reductions, for instance of 50-80% in emissions of carbon dioxide, would be required to stabilise its atmospheric composition.

Under the UN Framework Convention on Climate Change the world's governments have agreed to phased reductions in emissions. However there is considerable reluctance to take the measures necessary to reduce energy consumption and emissions, for instance by a modal shift from private to public transport, more efficient energy use and reduced energy consumption. A doubling of carbon dioxide combined with secondary effects, including increasing atmospheric levels of water vapour and changes in snow and ice cover, would most probably lead to further warming, giving a total global temperature rise of 2-4° C. Although this temperature rise may not seem very large, global warming of even a few degrees represents an enormous change in climate. For instance the total global warming since the peak of the last ice age has only been about 5° C, but it has had very dramatic effects, including shifting the Atlantic ocean inland by about one hundred and sixty kilometres (McCarthy et al 2001).

Climate change involves a number of different relations, for instance between:

- Anthropogenic (human generated) emissions of carbon dioxide and levels of carbon dioxide in the atmosphere.
- Levels of carbon dioxide in the atmosphere and global warming (average temperature rise).
- Global warming and sea level rise.

According to the third report of the International Panel on Climate Change (McCarthy et al 2001)

- Carbon dioxide levels are likely to increase to 540 to 970 parts per million by volume (ppmv) by 2100, though action on emissions could stabilise them at a lower level of 450 ppmv or even lower.
- Global average temperature will increase by 1.4-5.8 degrees Celcius between 1990 and 2100.
- Average sea level is projected to rise by 9-88 cm.

Consider the fuzzy sets μ_X, μ_Y and μ_Z, representing carbon dioxide levels in the atmosphere in parts per million, global average temperature rise in degrees Celsius and average sea level rise in centimetres respectively. To simplify the example it will be assumed that carbon dioxide levels in the atmosphere will not rise beyond 700 ppmv, global average temperature will not rise by more than 3 degrees Celsius and sea level will not rise beyond 40 cm i.e. values somewhere in the middle of the range of projected changes. The three fuzzy sets μ_X, μ_Y and μ_Z are defined in table 13.8.

Table 13.8 Fuzzy sets μ_X, μ_Y and μ_Z

X (ppmv)	450	500	550	600	650	700
$\mu_X(x)$	0.1	0.4	0.7	0.6	0.5	0.5

X (°C)	0.5	1.0	1.5	2.0	2.5	3.0
$\mu_Y(x)$	0.1	0.3	0.6	0.8	0.5	0.3

X (cm)	5	10	15	20	30	40
$\mu_Z(x)$	0.1	0.2	0.4	0.7	0.5	0.4

The matrices $\mu_R(X, Y)$ and $\mu_P(Y, Z)$ for the fuzzy relations between levels of carbon dioxide in the atmosphere and global warming and global warming and sea level rise can be obtained as follows:

$$\mu_R(X, Y) =$$

$$\begin{bmatrix} \min(.1,.1) & \min(.1,.3) & \min(.1,.6) & \min(.1,.8) & \min(.1,.5) & \min(.1,.3) \\ \min(.4,.1) & \min(.4,.3) & \min(.4,.6) & \min(.4,.8) & \min(.4,.5) & \min(.4,.3) \\ \min(.7,.1) & \min(.7,.3) & \min(.7,.6) & \min(.7,.8) & \min(.7,.5) & \min(.7,.3) \\ \min(.6,.1) & \min(.6,.3) & \min(.6,.6) & \min(.6,.8) & \min(.6,.5) & \min(.6,.3) \\ \min(.5,.1) & \min(.5,.3) & \min(.5,.6) & \min(.5.8) & \min(.5,.5) & \min(.5,.3) \\ \min(.5,.1) & \min(.5,.3) & \min(.5,.6) & \min(.5,.8) & \min(.5,.5) & \min(.5,.3) \end{bmatrix}$$

$$\therefore \mu_R(X,Y) = \begin{bmatrix} 0.1 & 0.1 & 0.1 & 0.1 & 0.1 & 0.1 \\ 0.1 & 0.3 & 0.4 & 0.4 & 0.4 & 0.3 \\ 0.1 & 0.3 & 0.6 & 0.7 & 0.5 & 0.3 \\ 0.1 & 0.3 & 0.6 & 0.6 & 0.5 & 0.3 \\ 0.1 & 0.3 & 0.5 & 0.5 & 0.5 & 0.3 \\ 0.1 & 0.3 & 0.5 & 0.5 & 0.5 & 0.3 \end{bmatrix}$$

$$\mu_P(Y,Z) = \begin{bmatrix} 0.1 & 0.1 & 0.1 & 0.1 & 0.1 & 0.1 \\ 0.1 & 0.2 & 0.3 & 0.3 & 0.3 & 0.3 \\ 0.1 & 0.2 & 0.4 & 0.6 & 0.5 & 0.4 \\ 0.1 & 0.2 & 0.4 & 0.7 & 0.5 & 0.4 \\ 0.1 & 0.2 & 0.4 & 0.5 & 0.5 & 0.4 \\ 0.1 & 0.2 & 0.3 & 0.3 & 0.3 & 0.3 \end{bmatrix}$$

The matrix $\mu_R(X, Y)$ gives the relationship between particular levels of global warming (in degrees Celsius) and atmospheric levels of carbon dioxide (in parts per million). The first row of this matrix is the fuzzy set of the (predicted) global temperature rise when the atmospheric concentration of carbon dioxide is 450 parts per million. The fourth row is the fuzzy set of the (predicted) global temperature rise when the atmospheric concentration of carbon dioxide is 600 parts per million.

The matrix $\mu_P(Y, Z)$ gives the relationship between average sea level rise (in centimeters) and global temperature rise (in degrees Celsius). The first row is the fuzzy set of the (predicted) average sea level rise when the average global temperature rise is 0.5 °C. The third row is the fuzzy set of the (predicted) average sea level rise when the average global temperature rise is 1.5 °C. The inverse relation between global warming and atmospheric levels of carbon dioxide can be obtained as

$$\mu_R^{-1}(X,Y) = \begin{bmatrix} 0.1 & 0.1 & 0.1 & 0.1 & 0.1 & 0.1 \\ 0.1 & 0.3 & 0.3 & 0.3 & 0.3 & 0.3 \\ 0.1 & 0.4 & 0.6 & 0.6 & 0.5 & 0.5 \\ 0.1 & 0.4 & 0.7 & 0.6 & 0.5 & 0.5 \\ 0.1 & 0.4 & 0.5 & 0.5 & 0.5 & 0.5 \\ 0.1 & 0.3 & 0.3 & 0.3 & 0.3 & 0.3 \end{bmatrix}$$

This matrix gives the relationship between average global temperature rise and atmospheric levels of carbon dioxide. The first row gives the fuzzy set for the atmospheric level of carbon dioxide which is expected to give an average global temperature rise of 0.5 °C.

The relation between emissions and sea level rise can now be calculated. To illustrate how this relation matrix is obtained detailed calculations will be given for the elements $\mu_{Q.23}(X, Z)$ and $\mu_{Q.34}(X, Z)$.

$$\mu_{Q.23}(X, Z) = \max[\min\{\mu_{R.21}(X), \mu_{P.13}(Y)\}, \min\{\mu_{R.22}(X), \mu_{P.23}(Y)\}, \ldots, \min\{\mu_{R.26}(X), \mu_{P.63}(Y)\}]$$

$$= \max[\min(.1, .1), \min(.3,.3), \min(.4,.4), \min(.4,.4), \min(.4,.4), \min(.3,.3)]$$

$$= \max[0.1, 0.3, 0.4, 0.4, 0.4, 0.3]$$

$$= 0.4$$

$$\mu_{Q.34}(X, Z) = \max[\min\{\mu_{R.31}(X),\mu_{P.14}(Y)\}, \min\{\mu_{R.32}(X), \mu_{P.24}(Y)\}, \ldots, \min\{\mu_{R.36}(X),\mu_{P.64}(Y)\}]$$

$$= \max[\min(.1,.1), \min(.3,.3), \min(.6,.6), \min(.7,.7), \min(.5,.5), \min(.3,.3)]$$

$$= \max[0.1, 0.3, 0.6, 0.7, 0.5, 0.3]$$

$$= 0.7$$

The other terms can be calculated similarly to give

$$\mu_Q(X,Z) = \begin{bmatrix} 0.1 & 0.1 & 0.1 & 0.1 & 0.1 & 0.1 \\ 0.1 & 0.2 & 0.4 & 0.4 & 0.4 & 0.4 \\ 0.1 & 0.2 & 0.4 & 0.7 & 0.5 & 0.4 \\ 0.1 & 0.2 & 0.4 & 0.6 & 0.5 & 0.4 \\ 0.1 & 0.2 & 0.4 & 0.5 & 0.5 & 0.4 \\ 0.1 & 0.2 & 0.4 & 0.5 & 0.5 & 0.4 \end{bmatrix}$$

The matrix $\mu_Q(X, Z)$ gives the relationship between the average sea level rise (in centimeters) and the atmospheric concentration of carbon dioxide (in parts per million). In particular the second row is the fuzzy set for the (predicted) sea level rise when the atmospheric concentration of carbon dioxide is 500 parts per million.

13.6.1 Fuzzy Conditional Statements

Fuzzy relations are used in a number of different contexts, including fuzzy control systems and fuzzy logic. The relationship between system outputs and inputs is described by a *fuzzy conditional statement* or relation R of the form {input condition with membership function μ_1} implies {output condition with membership function μ_2}.

Fuzzy conditional relations can connect more than two fuzzy sets e.g.
IF {emissions are high with membership functions μ_1} THEN {atmospheric levels of carbon dioxide will rise with membership function μ_2} THEN {sea levels will rise with membership functions μ_3}.

Example: 13.6 Climate Change - Fuzzy Conditional Statements

A commonly used technique for investigating and presenting the effects of different emissions reduction strategies on global climate change is the use of scenarios. Conditional statements can be used to represent these scenarios and fuzzy conditional statements are particularly appropriate due to the existence of several different types of imprecision and uncertainty, including in the likely growth in anthropogenic emissions and the resulting effects on the global climate and the environment.

The different scenarios are based on different assumptions about the rate of population and GDP (gross domestic product) growth, the rate of technological change and the existence and the nature of policy measures to stabilise or reduce greenhouse gas emissions.

In this example the following two scenarios (Nakicenovic et al 2000) are considered: a rapidly changing world with rapid emissions reductions and a rapidly changing world with fairly rapidly growing emissions. These scenarios will be denoted scenario 1 and scenario 2 respectively. In scenario 1 it is assumed that there is a carbon tax, high efficiency cars, buildings and power plants, high biomass penetration and rapid reforestation. In scenario 2 it is assumed that there are slow efficiency improvements, high deforestation, high cost solar energy, high oil and gas prices and cheap coal and synfuels. These are just two of the many possible scenarios. There are also questions as to the best way to achieve high emissions reductions and how realistic the assumptions are of, for instance, the gain in energy from high efficiency cars not being lost through increases in the distance travelled. A more sustainable approach might involve a modal shift to public transport, walking and cycling or town planning to reduce the need to travel to use services and facilities, though there would again be questions as to how and whether either of these proposals could be achieved.

Fuzzy set techniques will be used to express the scenarios in terms of fuzzy conditional statements. These conditional statements will then be combined to obtain fuzzy relations. Graphical illustrations of the fuzzy conditional statements will also be given.

Scenarios 1 and 2 can be expressed in terms of the following fuzzy conditional statements:

1. IF {carbon dioxide emissions are reduced significantly} THEN {global temperatures will rise relatively slowly}
2. IF {carbon dioxide emissions rise significantly} THEN {global temperatures will rise significantly}

Fuzzy sets for these conditional statements can be defined on the spaces:

X - global emissions of carbon dioxide in billion metric tons of carbon
Y - global warming in °C, where

$$X = \{0, 0.5, 1, 2, 5, 10, 15, 20, 25, 30, 35, 40, 45\} \quad (13.19a)$$

$$Y = \{0, 0.5, 1, 1.5, 2, 2.5, 3, 4, 5, 6\} \quad (13.19b)$$

The fuzzy conditional statements for relations 1 and 2 for the year 2050 are stated in tables 13.9 and 13.10 respectively.

Table 13.9a Fuzzy set for *carbon dioxide emissions are reduced significantly*

X	0	0.5	1	2	5	10	15	20	25	30	35	40	45
$\mu_1(x)$	0.1	0.4	0.7	0.8	0.5	0.1	0	0	0	0	0	0	0

Table 13.9b Fuzzy set for *global temperature will rise relatively slowly*

Y	0	0.5	1	1.5	2	2.5	3	4	5	6
$\mu_2(y)$	0.2	0.5	0.8	0.5	0.3	0.1	0	0	0	0

Table 13.10a Fuzzy set for *carbon dioxide emissions rise significantly*

X	0	0.5	1	2	5	10	15	20	25	30	35	40	45
$\mu_1(x)$	0	0	0	0	0	0	0.2	0.4	0.6	0.8	0.6	0.4	0.2

Table 13.10b Fuzzy set for *global temperature will rise significantly*

Y	0	0.5	1	1.5	2	2.5	3	4	5	6
$\mu_1(y)$	0	0.1	0.2	0.3	0.5	0.7	0.8	0.6	0.4	0.2

The relation matrices μ_{R1} and μ_{R2} for scenarios 1 and 2 can be obtained by combining the fuzzy sets in tables 13.9a and b and 13.10a and b respectively. For illustration the working is shown in the calculation of μ_{R1}, but not μ_{R2}.

$$\mu_{R1} =$$

$$\begin{bmatrix}
\min(.1,.2) & \min(.1,.5) & \min(.1,.8) & \min(.1,.5) & \min(.1,.3) & \min(.1,.1) & 0 & 0 & 0 & 0 \\
\min(.4,.2) & \min(.4,.5) & \min(.4,.8) & \min(.4,.5) & \min(.4,.3) & \min(.4,.1) & 0 & 0 & 0 & 0 \\
\min(.7,.2) & \min(.7,.5) & \min(.7,.8) & \min(.7,.5) & \min(.7,.3) & \min(.7,.1) & 0 & 0 & 0 & 0 \\
\min(.8,.2) & \min(.8,.5) & \min(.8,.8) & \min(.8,.5) & \min(.8,.3) & \min(.8,.1) & 0 & 0 & 0 & 0 \\
\min(.5,.2) & \min(.5,.5) & \min(.5,.8) & \min(.5,.5) & \min(.5,.3) & \min(.5,.1) & 0 & 0 & 0 & 0 \\
\min(.1,.2) & \min(.1,.5) & \min(.1,.8) & \min(.1,.5) & \min(.1,.3) & \min(.1,.1) & 0 & 0 & 0 & 0 \\
0 & 0 & 0 & 0 & 0 & 0 & 0 & 0 & 0 & 0 \\
0 & 0 & 0 & 0 & 0 & 0 & 0 & 0 & 0 & 0 \\
0 & 0 & 0 & 0 & 0 & 0 & 0 & 0 & 0 & 0 \\
0 & 0 & 0 & 0 & 0 & 0 & 0 & 0 & 0 & 0 \\
0 & 0 & 0 & 0 & 0 & 0 & 0 & 0 & 0 & 0 \\
0 & 0 & 0 & 0 & 0 & 0 & 0 & 0 & 0 & 0 \\
0 & 0 & 0 & 0 & 0 & 0 & 0 & 0 & 0 & 0
\end{bmatrix}$$

$$= \begin{bmatrix}
0.1 & 0.1 & 0.1 & 0.1 & 0.1 & 0.1 & 0 & 0 & 0 & 0 \\
0.2 & 0.4 & 0.4 & 0.4 & 0.3 & 0.1 & 0 & 0 & 0 & 0 \\
0.2 & 0.5 & 0.7 & 0.5 & 0.3 & 0.1 & 0 & 0 & 0 & 0 \\
0.2 & 0.5 & 0.8 & 0.5 & 0.3 & 0.1 & 0 & 0 & 0 & 0 \\
0.2 & 0.5 & 0.5 & 0.5 & 0.3 & 0.1 & 0 & 0 & 0 & 0 \\
0.1 & 0.1 & 0.1 & 0.1 & 0.1 & 0.1 & 0 & 0 & 0 & 0 \\
0 & 0 & 0 & 0 & 0 & 0 & 0 & 0 & 0 & 0 \\
0 & 0 & 0 & 0 & 0 & 0 & 0 & 0 & 0 & 0 \\
0 & 0 & 0 & 0 & 0 & 0 & 0 & 0 & 0 & 0 \\
0 & 0 & 0 & 0 & 0 & 0 & 0 & 0 & 0 & 0 \\
0 & 0 & 0 & 0 & 0 & 0 & 0 & 0 & 0 & 0 \\
0 & 0 & 0 & 0 & 0 & 0 & 0 & 0 & 0 & 0 \\
0 & 0 & 0 & 0 & 0 & 0 & 0 & 0 & 0 & 0
\end{bmatrix}$$

$$\mu_{R2} = \begin{bmatrix} 0 & 0 & 0 & 0 & 0 & 0 & 0 & 0 & 0 & 0 \\ 0 & 0 & 0 & 0 & 0 & 0 & 0 & 0 & 0 & 0 \\ 0 & 0 & 0 & 0 & 0 & 0 & 0 & 0 & 0 & 0 \\ 0 & 0 & 0 & 0 & 0 & 0 & 0 & 0 & 0 & 0 \\ 0 & 0 & 0 & 0 & 0 & 0 & 0 & 0 & 0 & 0 \\ 0 & 0 & 0 & 0 & 0 & 0 & 0 & 0 & 0 & 0 \\ 0 & 0.1 & 0.2 & 0.2 & 0.2 & 0.2 & 0.2 & 0.2 & 0.2 & 0.2 \\ 0 & 0.1 & 0.2 & 0.3 & 0.4 & 0.4 & 0.4 & 0.4 & 0.4 & 0.2 \\ 0 & 0.1 & 0.2 & 0.3 & 0.5 & 0.6 & 0.6 & 0.6 & 0.4 & 0.2 \\ 0 & 0.1 & 0.2 & 0.3 & 0.5 & 0.7 & 0.8 & 0.6 & 0.4 & 0.2 \\ 0 & 0.1 & 0.2 & 0.3 & 0.5 & 0.6 & 0.6 & 0.6 & 0.4 & 0.2 \\ 0 & 0.1 & 0.2 & 0.3 & 0.4 & 0.4 & 0.4 & 0.4 & 0.4 & 0.2 \\ 0 & 0.1 & 0.2 & 0.2 & 0.2 & 0.2 & 0.2 & 0.2 & 0.2 & 0.2 \end{bmatrix}$$

The fuzzy sets in tables 13.9a and b and 13.10a and b are illustrated in bar chart form in figures 13.9a and b and 13.10a and b respectively.

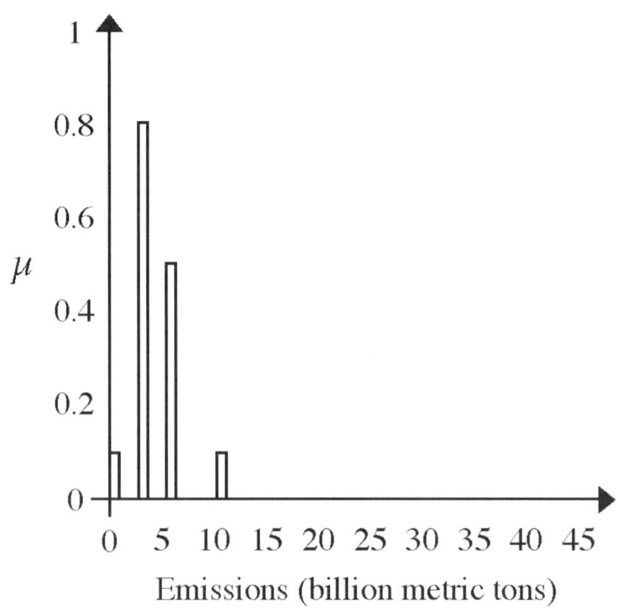

Fig. 13.9a Fuzzy set *for carbon dioxide emissions are reduced significantly*

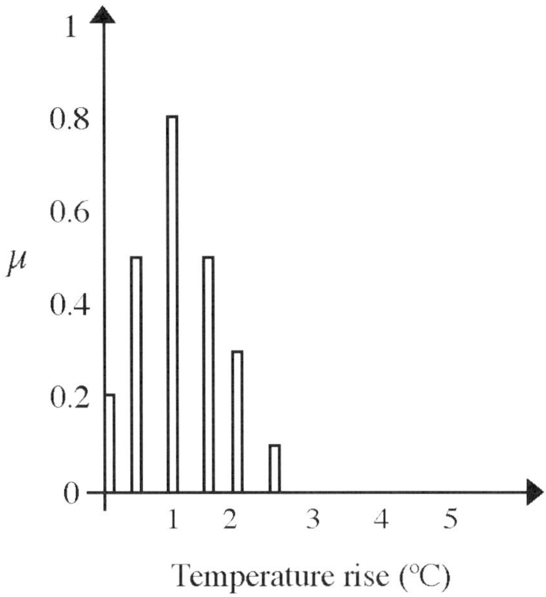

Fig. 13.9b Fuzzy set for global temperature will rise relatively slowly

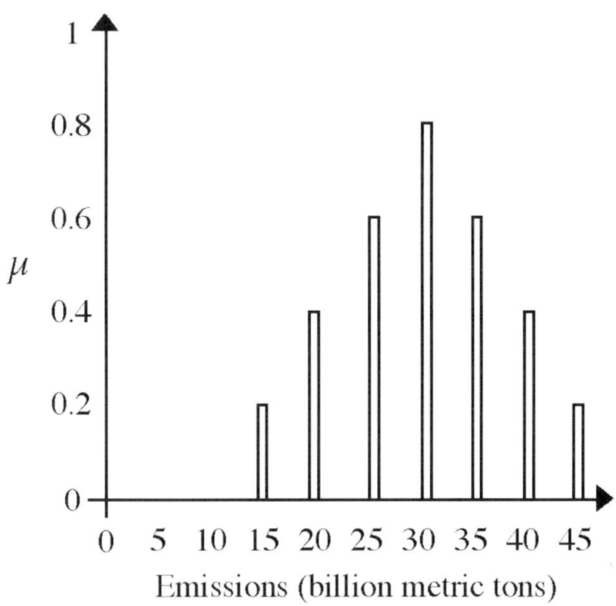

Fig. 13.10a Fuzzy set for *carbon dioxide emissions rise significantly*

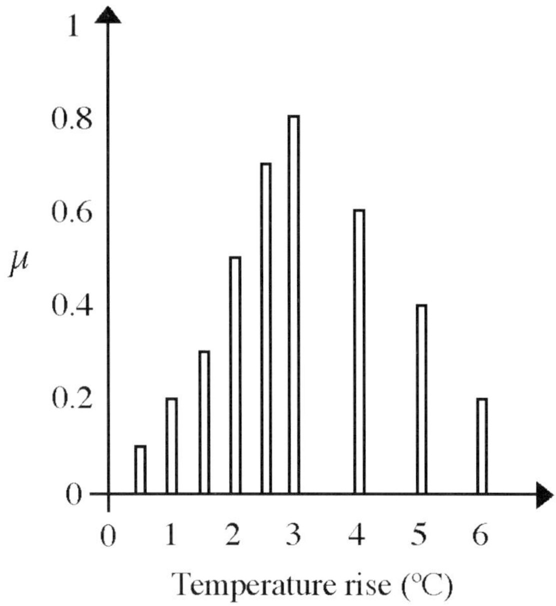

Fig. 13.10b Fuzzy set for *global temperature will rise rapidly*

It should be noted that both the relation matrices μ_{R1} and μ_{R2} are for the year 2050. Similar techniques could be used to obtain relation matrices for other years or for the changes over a period of years. Thus this approach can be used to show the effects of uncertainty on the conditional relation, either at different points in time or over a period. However, unlike some graphical representations, a conditional representation cannot be used to give a continuous picture of the changes over time.

13.7 Roundup of Chapter 13

Chapter 13 has introduced fuzzy sets and discussed numerical and linguistic variables, fuzzification and fuzzy relations and conditional statements. Unlike crisp sets, where an element is either a member or non-member of the set, fuzzy sets are defined by membership functions which take values between 0 and 1 and define the extent to which an element is in a set. They are particularly useful in representing imprecise quantities, such as noisy or hazardous, which frequently occur in sustainable development problems.

Numerical or quantitative variables are expressed as numbers in appropriate units. Linguistic or qualitative variables are expressed in words. Sustainable development problems often include both numerical and linguistic variables. Fuzzy numbers are fuzzy sets which are defined on the real interval and have particular properties. Commonly used fuzzy numbers include triangular, trapezoidal and monotonically increasing linear and non-linear fuzzy numbers.

Fuzzification is the conversion of crisp data to a fuzzy set. It can involve the use of fuzzy numbers and global or local scaling. α–cuts and strong α–cuts can be used to obtain crisp from fuzzy sets and contain all the elements with values greater than (or equal) to α.

Crisp relations can be used to represent the existence or absence of a connection between the members of two or more sets. Fuzzy relations allow the strength or tenuousness of this connection to be indicated using membership grades with values between 0 and 1.

Chapter 14 continues the introduction to fuzzy sets presented in this chapter through the presentation of set operations for fuzzy sets and a discussion of their relationship to t-norms and t-conorms.

14 Fuzzy Set Operations

14.0 Learning Objectives

This chapter draws on and further enlarges on the basic concepts in fuzzy set theory and illustrates them by examples of the environmental impacts of information technology. Specific learning objectives include the following:

- The ability to apply the standard definitions of the basic set operations of union, intersection and complement for fuzzy sets.
- Understanding of the terms t-norm and t-conorm and knowledge of the conditions satisfied by t-norms and t-conorms.
- Understanding of the relationship between the standard fuzzy set operations and the classes of t-conorms, t-norms and fuzzy complements.
- Knowledge of aggregation and norm operations.
- Knowledge of the criteria a defuzzification method should meet.
- The ability to apply defuzzification techniques to obtain crisp sets from fuzzy sets.

14.1 Introduction

The previous chapter introduced the concept of a fuzzy set and showed how fuzzification techniques can be used to obtain fuzzy from crisp sets. This chapter discusses the fuzzy set operations of intersection, union and complement in section 14.2 and relates them to the classes of t-norms, t-conorms and fuzzy complements in section 14.3. The fuzzy set operations are illustrated by the fuzzy sets *high power consumption* and *noisy* defined on a space of five printers. Section 14.4 presents defuzzification techniques for obtaining crisp from fuzzy sets and section 14.5 gives a summary of the chapter.

14.2 Fuzzy Set Operations

14.2.1 Aggregation Operations

It is useful to be able to combine fuzzy sets, for instance to determine the total environmental impacts at a particular life cycle stage or the total impact of a specific factor over all life cycle stages. Several fuzzy sets can be combined to produce a single fuzzy set using aggregation operations, of which the fuzzy union and intersection are the most commonly used. A general *aggregation operation* can combine n fuzzy sets $\mu_1(x), ..., \mu_n(x)$ and is therefore a function from the n times cross product of the unit interval into the unit interval. This n times cross product is defined as

$$[0, 1]^n = (\alpha_1, \alpha_2, ..., \alpha_n) \quad \text{for } \alpha_1, ..., \alpha_n \in [0,1] \tag{14.1}$$

A general aggregation operator can be defined as follows:

$$h: [0, 1]^n \rightarrow [0, 1] \tag{14.2a}$$
$$h = h(\alpha_1, \alpha_2, ..., \alpha_n) \tag{14.2b}$$

However not all functions of this type give useful aggregation operations and functions are required to satisfy at least the following three properties to be accepted as aggregation operations (Klir and Yuan 1995):

A1. $h(0, 0, ..., 0) = 0$ and $h(1, 1, ..., 1) = 1$ (*boundary* condition).
A2. $h(\alpha_1, \alpha_2, ..., \alpha_n) \leq h(\beta_1, \beta_2, ..., \beta_n)$ if $\alpha_i \leq \beta_i$ for all $\alpha_i, \beta_i \in R$ $i=1...n$ (*monotonic increasing* in all arguments).
A3. $h(., ..., .)$ is a continuous function (*continuity*).

This gives functions which are continuous, monotonic in all arguments and bounded by 0 and 1. Many fuzzy set aggregation operations also satisfy the following two conditions:

A4. $h(\alpha_1, \alpha_2, ..., \alpha_n) = h(\alpha_{p(1)}, \alpha_{p(2)}, ..., \alpha_{p(n)})$ for any permutation p on $\mathcal{N}_n$ (where $\mathcal{N}_n = \{1, 2, ..., n\}$) (*symmetry* in all arguments).
A5. $h(\alpha, \alpha, ..., \alpha) = \alpha$ for all $\alpha \in [0, 1]$ (*idempotency*).

Property A4 assumes that each of the aggregated fuzzy sets is equally important and should be omitted if this is not the case. When property A5 is satisfied combining a fuzzy set multiple times with itself results in the original fuzzy set. This is only possible if the range of the function $h(., ..., .)$ is [0, 1] and therefore property A5 subsumes property A1. Consequently property A1 is not required if property A5 holds.

14.2.2 Fuzzy Union, Intersection and Complement

The union, intersection and complement of fuzzy sets will be considered in this subsection, though it should be noted that the complement is not really an aggregation operation. Unlike in the crisp set case, there are a number of different definitions of all three of these operations. What are often called the standard definitions will first be presented and more general definitions will be given in section 14.3. In the standard definition the *union* of two fuzzy sets μ_1 and μ_2 corresponds to the connective OR and the membership grade of each element in the union is given by the maximum of the two membership grades i.e.

$$\mu_{sup} = \mu_1 \vee \mu_2 \tag{14.3a}$$

$$\text{with } (\mu_1 \vee \mu)x) = \max[\mu_1(x), \ \mu_2(x)] \tag{14.3b}$$

The standard definition of the *intersection* of two fuzzy sets with membership functions μ_1 and μ_2 corresponds to the connective AND, is given by the minimum of the two membership functions and has membership function

$$\mu_{inf} = \mu_1 \wedge \mu_2 \tag{14.4a}$$

$$\text{with } (\mu_1 \wedge \mu_2)(x) = \min[\mu_1(x), \ \mu_2(x)] \tag{14.4b}$$

The standard definition of the *complement* of a fuzzy set with membership function μ corresponds to the connective NOT and has membership function

$$\bar{\mu}(x) = 1 - \mu(x) \tag{14.5}$$

The union of two fuzzy sets can be used to give a measure of the extent to which one or other of two properties holds. For instance the union of the two fuzzy sets *quiet* and *low energy consumption* on a space of different types of printers gives a measure of the extent to which the printers are quiet *or* have low energy consumption. Similarly the intersection of two fuzzy sets can be used to give a measure of the extent to which both of two properties hold. For instance the intersection of two fuzzy sets representing *quiet* and *low energy consumption* on a space of different types of printers gives a measure of the extent to which the printers are *both* quiet *and* have low energy consumption. The complement gives a measure of the extent to which a property does not hold, so that the complement of the fuzzy set *low energy consumption* is '*not low*' i.e. medium or high energy consumption.

Example 14.1: Printer Power Consumption and Noise I

Consider the space X = {$x_1, x_2, ..., x_5$} of five printers, printer 1 to printer 5. The fuzzy sets *high power consumption* (HPC) and *noisy* are defined on the space X.

noisy has membership function

X	x_1	x_2	x_3	x_4	x_5
μ(x)	1	0.5	0.6	0.3	0.2

HPC has membership function

X	x_1	x_2	x_3	x_4	x_5
μ(x)	0.6	0.35	0.4	0.7	0.6

noisy AND *HPC*:

X	x_1	x_2	x_3	x_4	x_5
μ(x)	min(1, .6)	min(.5, .35)	min(.6, .4)	min(.3, .7)	min(.2, .6)

=

X	x_1	x_2	x_3	x_4	x_5
μ(x)	0.6	0.35	0.4	0.3	0.2

noisy OR *HPC*:

X	x_1	x_2	x_3	x_4	x_5
μ(x)	max(1,.6)	max(.5, .35)	max(.6, .4)	max(.3, .7)	max(.2, .6)

=

X	x_1	x_2	x_3	x_4	x_5
μ(x)	1	0.5	0.6	0.7	0.6

NOT *HPC*:

X	x_1	x_2	x_3	x_4	x_5
μ(x)	1-0.6	1-0.35	1-0.4	1-0.7	1-0.6

∴ NOT *HPC* =

X	x_1	x_2	x_3	x_4	x_5
$\mu(x)$	0.4	0.65	0.6	0.3	0.4

14.3 t-Norms, t-Conorms and Fuzzy Complements

The standard definitions of the fuzzy operations of union, intersection and complement were presented in the previous section. The standard operations have a number of particularly useful properties. The standard fuzzy intersection or min operator gives the largest fuzzy set of all possible fuzzy intersections (t-norms), whereas the standard fuzzy union or max operator produces the smallest fuzzy set of all possible fuzzy unions (t-conorms). Therefore the fuzzy set obtained from the min operator contains the fuzzy sets obtained from all other possible fuzzy intersections, whereas the fuzzy set obtained from the max operator is contained in the fuzzy sets obtained from all other possible fuzzy unions. In this section general classes of fuzzy intersections based on t-norms, fuzzy unions based on t-conorms and fuzzy complements will be introduced.

14.3.1 t-Norms

The *intersection* of two fuzzy sets μ_1 and μ_2 is a binary operation on the unit interval i.e. a function of the form

$$\mu_1 \cap \mu_2 = i(\mu_1, \mu_2) \tag{14.6a}$$

$$i: [0, 1] \times [0, 1] \rightarrow [0, 1] \tag{14.6b}$$

This function depends only on the values of μ_1 and μ_2 and not on x. It has been found that a meaningful fuzzy intersection for any pair of fuzzy sets can be obtained from the class of functions called t-norms. Therefore the terms t-norm and fuzzy intersection are often used interchangeably. A function is a *t-norm* if it has the following four properties (Klir and Yuan 1995; Kruse et al 1994) for all α, β and γ ∈ [0, 1]:

I1. $i(\alpha, 1) = \alpha$ (*boundary* condition).
I2. $\beta \leq \gamma \Rightarrow i(a, \beta) \leq i(\alpha, \gamma)$ (*monotonic increasing*).
I3. $i(\alpha, \beta) = i(\beta, \alpha)$ (*commutativity*).
I4. $i[\alpha, i(\beta, \gamma)] = i[i(\alpha, \beta), \gamma]$ (*associativity*).

The first three properties can be used to show that the fuzzy intersection defined by equations (14.6) is the standard set intersection when the two sets are crisp. Monotonicity ensures that the membership function of the intersection does not increase when the membership functions of the two sets decrease. Commutativity means that the intersection is independent of the order of the two sets and associativity that the intersection of a number of sets can be obtained from any pairwise grouping of the sets. This allows the definition of the fuzzy intersection to be extended to more than two sets.

It is sometimes useful to restrict the class of fuzzy intersections by considering additional properties, such as the following (Klir and Yuan 1995):

I5. i is a continuous function (*continuity*).

I6. $i(\alpha, \alpha) < \alpha$ (*subidempotency*).

I7. $\alpha_1 < \alpha_2$ and $\beta_1 < \beta_2 \Rightarrow i(\alpha_1, \beta_1) < i(\alpha_2, \beta_2)$ (*strict monotonicity*).

Continuity ensures that small changes in the membership grade of either set do not produce large (discontinuous) changes in the intersection. t-norms or fuzzy intersections that satisfy properties I5-I7 are called strict *Archimedean t-norms*, whereas those that satisfy properties I5 and I6, but not I7, are *Archimedean t-norms*. Subidempotency prevents the membership function of the intersection of two identical sets exceeding the membership function of the set on its own.

t-norms which are used as fuzzy intersections include the following (Klir and Yuan 1995; Kruse et al 1994):

Standard intersection: $i(\mu_a, \mu_b) = \min(\mu_a, \mu_b)$

Algebraic product: $i(\mu_a, \mu_b) = \mu_a\mu_b$

Bounded difference: $i(\mu_a, \mu_b) = \max(0, \mu_a + \mu_b - 1)$

Drastic intersection: $$i_{min} = \begin{cases} \mu_a & \text{when } \mu_b = 1 \\ \mu_b & \text{when } \mu_b = 1 \\ 0 & \text{otherwise} \end{cases}$$

It can be shown that all fuzzy intersections $i(\mu_a, \mu_b)$ take values between the drastic intersection $i_{min}(\mu_a, \mu_b)$ and the standard intersection $\min(\mu_a, \mu_b)$

i.e. $$i_{min}(\mu_a, \mu_b) \leq i(\mu_a, \mu_b) \leq \min(\mu_a, \mu_b) \tag{14.7}$$

The standard fuzzy intersection or minimum function is the only *idempotent* t-norm i.e. the only t-norm for which the intersection of two identical sets is equal to the membership function of the set on its own.

Example 14.2: Printer Power Consumption and Noise II

Consider the fuzzy sets *high power consumption* (HPC) and *noisy* defined in example 14.1 in section 14.2.2 and which are repeated below for ease of reference:

HPC

x:	x_1	x_2	x_3	x_4	x_5
$\mu(x)$:	0.6	0.35	0.4	0.7	0.6

noisy

x:	x_1	x_2	x_3	x_4	x_5
$\mu(x)$:	1	0.5	0.6	0.3	0.2

The *standard fuzzy intersection* has already been shown to be given by:

x:	x_1	x_2	x_3	x_4	x_5
$\mu(x)$:	0.6	0.35	0.4	0.3	0.2

The *algebraic product intersection* is given by:

x:	x_1	x_2	x_3	x_4	x_5
μ (x):	1 x 0.6	0.5 x 0.35	0.6 x 0.4	0.3 x 0.7	0.2 x 0.6

= x:	x_1	x_2	x_3	x_4	x_5
μ (x):	0.6	0.175	0.24	0.21	0.12

The *bounded difference intersection* is given by:

x:	x_1	x_2	x_3	x_4	x_5
$\mu(x)$:	max(0,1+.6-1)	max(0, .5+.35-1)	max(0, .6+.4-1)	max(0, .3+.7-1)	max(0, .2+.6-1)
= x:	x_1	x_2	x_3	x_4	x_5
$\mu(x)$	0.6	0	0	0	0

The *drastic intersection* is given by:

x:	x_1	x_2	x_3	x_4	x_5
μ (x):	0.6	0	0	0	0

Then it can easily be seen that the values of both the bounded difference and algebraic product intersections are between those of the drastic and standard intersections. The relationship between the different fuzzy intersections is illustrated in fig. 14.1, where the subscripts 1-5 indicate the variables and the subscripts a, b, c and d indicate the standard, algebraic product, bounded difference and drastic intersections respectively. Thus, for instance i_{3a} is the standard fuzzy intersection for x_3 and y_3 and i_{45cd} is the bounded difference and drastic intersection for both x_4 and y_4 and x_5 and y_5.

14.3.2 t-Conorms

The *union* of two fuzzy sets μ_1 and μ_2 is a binary operation on the unit interval i.e. a function of the form

$$\mu_1 \cup \mu_2 = u(\mu_1, \mu_2) \tag{14.8a}$$
$$u: [0, 1] \times [0, 1] \rightarrow [0, 1] \tag{14.8b}$$

It has been found that the class of functions called t-conorms has appropriate properties to give a meaningful fuzzy union for any pair of fuzzy sets. Therefore the terms t-conorm and fuzzy union are often used

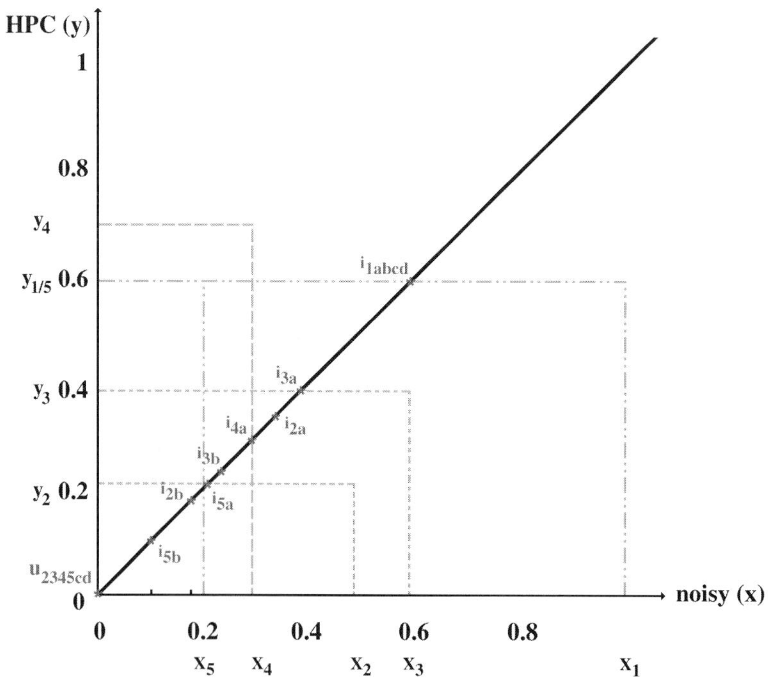

Fig. 14.1 Relationship between different fuzzy intersections

interchangeably. A function is a *t-conorm* if it has the following four properties (Klir and Yuan 1995) for all α, β and $\gamma \in [0, 1]$:

U1. $u(\alpha, 0) = \alpha$ (*boundary* condition).
U2. $\beta \leq \gamma \Rightarrow u(\alpha, \beta) \leq u(\alpha, \gamma)$ (*monotonic increasing*).
U3. $u(\alpha, \beta) = u(\beta, \alpha)$ (*commutativity*).
U4. $u[\alpha, u(\beta, \gamma)] = u[u(\alpha, \beta), \gamma]$ (*associativity*).

The only difference between these properties and the analogous properties for fuzzy intersections is in the boundary condition. In the case of the t-norm the boundary condition states that any fuzzy set is unaffected by intersection with the element 1, whereas in the case of the t-conorm the boundary condition states that any element is unaffected by union with the element 0. The first three properties can be used to show that the fuzzy union defined by equations (14.8) is the standard set union when the two sets are crisp. It is sometimes useful to restrict the class of fuzzy unions by considering additional properties, such as the following (Klir and Yuan 1995; Kruse et al 1994):

U5. u is a continuous function (*continuity*).
U6. $u(\alpha, \alpha) > \alpha$ (*superidempotency*).
U7. $\alpha_1 < \alpha_2 \,\&\, \beta_1 < \beta_2 \Rightarrow u(\alpha_1, \beta_1) < u(\alpha_2, \beta_2)$ (*strict monotonicity*).

t-conorms that satisfy properties U5-U7 are called *strict Archimedean t-conorms*, whereas those that satisfy properties U5 and U6, but not U7 are *Archimedean t-conorms*. The properties of Archimedian t-conorms are analogous to those of t-norms, with superidempotency replacing subidempotency, and different boundary conditions.

Thus the main difference between t-conorms and t-norms is in the *different boundary conditions*. For Archimedean t-conorms and t-norms which satisfy the additional properties I5-I7 or U5-U7 there is an additional difference in t-conorms being subidempotent and t-norms superidempotent.

t-conorms which are used as fuzzy unions include the following (Klir and Yuan 1995; Kruse et al 1994):

Standard union: $u(\mu_a, \mu_b) = \max(\mu_a, \mu_b)$

Algebraic sum: $u(\mu_a, \mu_b) = \mu_a + \mu_b - \mu_a\mu_b$

Bounded sum: $u(\mu_a, \mu_b) = \min(1, \mu_a + \mu_b)$

Drastic union:
$$u_{max} = \begin{cases} \mu_a & \text{when } \mu_b = 0 \\ \mu_b & \text{when } \mu_b = 0 \\ 1 & \text{otherwise} \end{cases}$$

It can be shown that all fuzzy unions $u(\mu_a, \mu_b)$ take values between the standard union $\max(\mu_a, \mu_b)$ and the drastic union $u_{max}(\mu_a, \mu_b)$ i.e.

$$\max(\mu_a, \mu_b) \leq u(\mu_a, \mu_b) \leq u_{max}(\mu_a, \mu_b) \tag{14.9}$$

The standard fuzzy union or maximum function is the only idempotent t-conorm for which the union of two identical sets is equal to the membership function of each set on its own.

Example 14.3: Printer Power Consumption and Noise III

Consider the fuzzy sets *high power consumption* (HPC) and *noisy* defined in example 14.1 in section 14.2.2 and which are repeated below for ease of reference:

HPC

x:	x_1	x_2	x_3	x_4	x_5
$\mu(x)$:	0.6	0.35	0.4	0.7	0.6

noisy

x:	x_1	x_2	x_3	x_4	x_5
$\mu(x)$:	1	0.5	0.6	0.3	0.2

The *standard fuzzy union* has already been shown to be given by:

x:	x_1	x_2	x_3	x_4	x_5
$\mu(x)$	1	0.5	0.6	0.7	0.6

The *algebraic sum union* is given by:

x:	x_1	x_2	x_3	x_4	x_5
$\mu(x)$:	1+0.6-0.6	0.5+0.35-0.175	0.6+0.4-0.24	0.3+0.7-0.21	0.2+0.6-0.12

= x:	x_1	x_2	x_3	x_4	x_5
μ (x):	1	0.675	0.76	0.79	0.68

The *bounded sum union* is given by:

x:	x_1	x_2	x_3	x_4	x_5
$\mu(x)$:	min(1,1+.6)	min(1, .5+.35)	min(1, .6+.4)	min(1, .3+.7)	min(1, .2+.6)

= x:	x_1	x_2	x_3	x_4	x_5
$\mu(x)$	1	0.85	1	1	0.8

The *drastic union* is given by:

x:	x_1	x_2	x_3	x_4	x_5
μ (x)	1	1	1	1	1

Then it can easily be seen that the values of both the bounded sum and algebraic sum unions are between those of the standard and drastic unions. The relationship between the different fuzzy unions is shown in fig. 14.2 where the subscripts 1-5 indicate the variables and the subscripts a,b,c and d indicate the standard, algebraic product, algebraic sum and drastic fuzzy unions respectively. Thus, for instance u_{2a} is the standard union for x_2 and y_2 and u_{34cd} is the algebraic sum and drastic union for both x_3 and y_3 and x_4 and y_4.

14.3.3 Fuzzy Complements

The *fuzzy complement* $c(\mu_a)$ of a fuzzy set μ_a can be interpreted both as the extent to which an element x belongs to the fuzzy complement $c(\mu_a)$

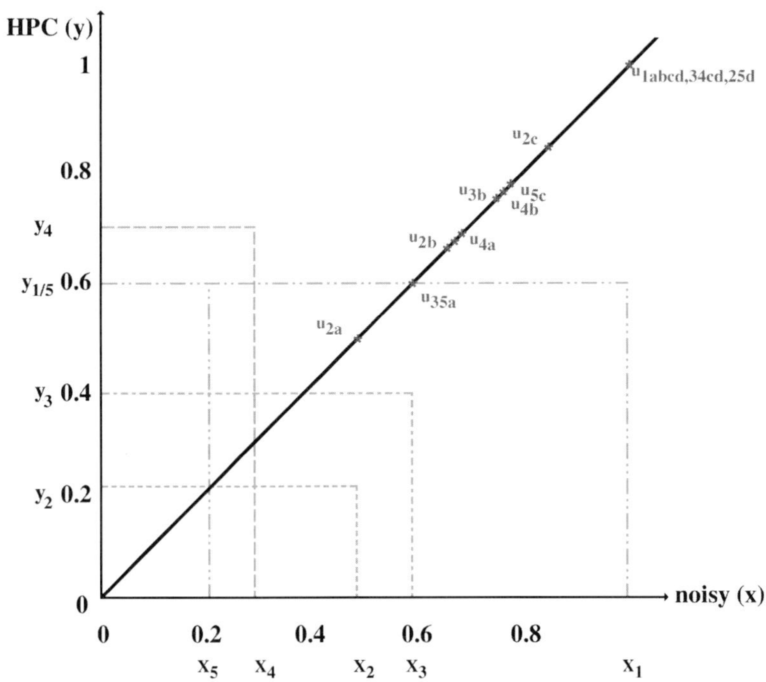

Fig. 14.2 The relationship between the different fuzzy unions

and the extent to which it does not belong to the fuzzy set μ_a. A fuzzy complement can be defined as a function c from the unit interval to the unit interval i.e.

$$c: [0, 1] \rightarrow [0, 1] \tag{14.10}$$

It assigns a membership grade $c(\mu_a(x))$ to each element x of μ_a. Functions c which are meaningful complements of fuzzy sets have the following properties (Klir and Yuan 1995):

C1. $c(0) = 1$ and $c(1) = 0$ (*boundary* condition).
C2. If $\alpha \leq \beta$, then $c(\alpha) \geq c(\beta)$ for all $\alpha, \beta \in [0, 1]$ (*monotonicity*).

Thus fuzzy complements take values between 0 and 1 and decrease in value when the associated fuzzy membership grades increases. The functions with these properties form the most general class of fuzzy complements. Most useful fuzzy complements also have the following two properties:

C3. c is a continuous function (*continuity*).
C4. $c(c(\alpha)) = \alpha$ (*involutive* condition) i.e. the complement is its own inverse.

It can be shown that functions with properties C2 and C4 also have properties C1 and C3 and are bijective or one-to-one and onto. *Involutive complements* form a subclass of continuous complements, and *continuous complements* are a subclass of all fuzzy complements. Classes of involutive fuzzy complements include the Sugeno and Yager classes. The *Sugeno* fuzzy complement is given by (Klir and Yuan 1995)

$$c_\lambda(\mu_a) = \frac{1 - \mu_a}{1 + \lambda\mu_a} \tag{14.11}$$

where $\lambda \in (-1, \infty)$ and a particular involutive fuzzy complement is obtained for each value of λ. Fig. 14.3 shows the Sugeno class of fuzzy complements for values of λ between –1 and 20 and three different values of μ_a, equal to 0.1, 0.5 and 0.9 on the upper, middle and lower curves respectively.

The *Yager* class of fuzzy complements is given by (Klir and Yuan 1995)

$$c_w(\mu_a) = (1 - \mu_a^{\,w})^{1/w} \tag{14.12}$$

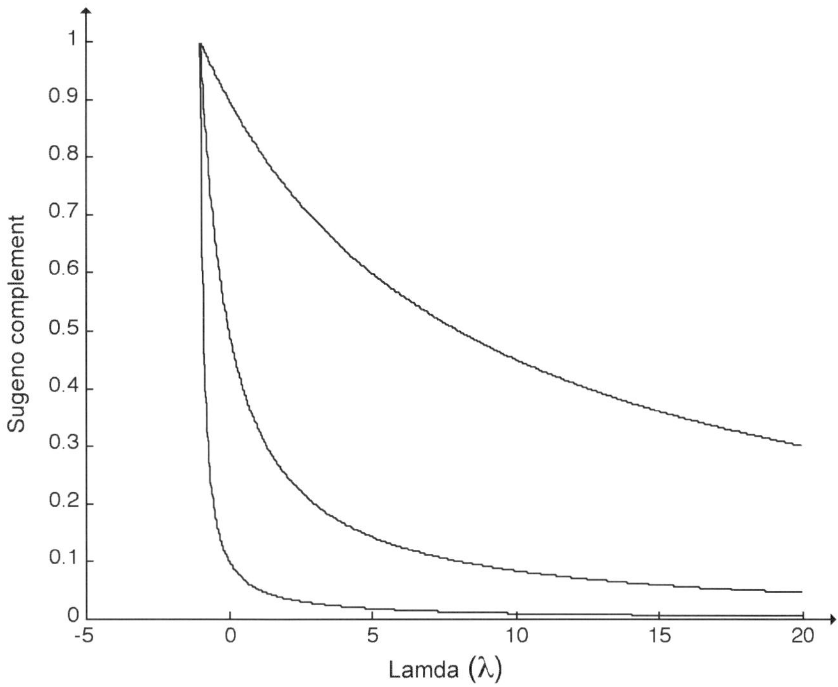

Fig. 14.3 Sugeno fuzzy complement for values of λ between –1 and 20

where w $\in (0, \infty)$ and a particular involutive fuzzy complement is obtained for each value of w. Fig. 14.4 shows the Yager class of fuzzy complements for values of w between 0 and 20 and three different values of μ_a, equal to 0.1, 0.5 and 0.9 on the upper, middle and lower curves respectively.

Example 14.4: Printer Power Consumption

Consider the fuzzy set *high power consumption* (HPC) defined in section 14.2.2, which is repeated below for ease of reference:

x:	x_1	x_2	x_3	x_4	x_5
$\mu(x)$:	0.6	0.35	0.4	0.7	0.6

The *standard fuzzy complement* has already been shown to be given by:

x:	x_1	x_2	x_3	x_4	x_5
$\overline{\mu}(x)$:	0.4	0.65	0.6	0.3	0.4

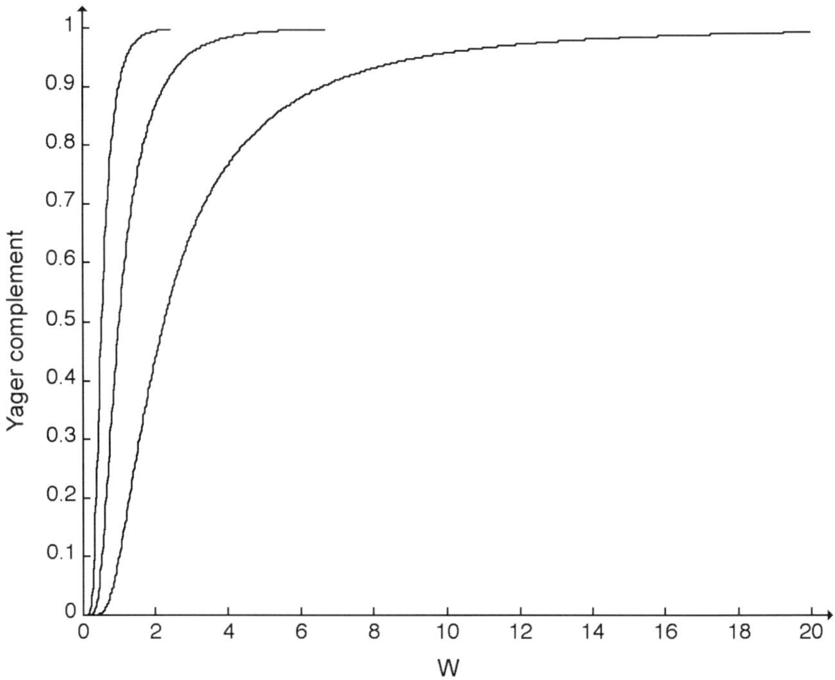

Fig. 14.4 Yager fuzzy complement for values of w between 0 and 20

The *standard fuzzy complement* has already been shown to be given by:

x:	x_1	x_2	x_3	x_4	x_5
$\overline{\mu}(x)$:	0.4	0.65	0.6	0.3	0.4

The *Sugeno* class of fuzzy complements is given by

x:	x_1	x_2	x_3	x_4	x_5
$\overline{\mu}(x)$:	$\frac{1-0.6}{1+\lambda 0.6}$	$\frac{1-0.35}{1+\lambda 0.35}$	$\frac{1-0.4}{1+\lambda 0.4}$	$\frac{1-0.7}{1+\lambda 0.7}$	$\frac{1-0.6}{1+\lambda 0.6}$

= x:	x_1	x_2	x_3	x_4	x_5
$\overline{\mu}(x)$:	$\frac{0.4}{1+\lambda 0.6}$	$\frac{0.65}{1+\lambda 0.35}$	$\frac{0.6}{1+\lambda 0.4}$	$\frac{0.3}{1+\lambda 0.7}$	$\frac{0.4}{1+\lambda 0.6}$

For $\lambda = 2$ the Sugeno fuzzy complement is

x:	x_1	x_2	x_3	x_4	x_5
$\overline{\mu}(x)$:	$\frac{0.4}{1+2x0.6}$	$\frac{0.65}{1+2x0.35}$	$\frac{0.6}{1+2x0.4}$	$\frac{0.3}{1+2x0.7}$	$\frac{0.4}{1+2x0.6}$

$\therefore$ for $\lambda = 2$ the Sugeno fuzzy complement is

x:	x_1	x_2	x_3	x_4	x_5
$\bar{\mu}(x)$:	0.18	0.38	0.33	0.13	0.18

The *Yager* class of fuzzy complements is given by:

x:	x_1	x_2	x_3	x_4	x_5
$\bar{\mu}(x)$:	$(1\text{-}0.6^w)^{1/w}$	$(1\text{-}0.35^w)^{1/w}$	$(1–0.4^w)^{1/w}$	$(1–0.7^w)^{1/w}$	$(1\text{-}0.6^w)^{1/w}$

For w = 2 the Yager fuzzy complement is

x:	x_1	x_2	x_3	x_4	x_5
$\bar{\mu}(x)$:	$(1\text{-}0.6^2)^{1/2}$	$(1\text{-}0.35^2)^{1/2}$	$(1–0.4^2)^{1/2}$	$(1–0.7^2)^{1/2}$	$(1\text{-}0.6^2)^{1/2}$

=

x:	x_1	x_2	x_3	x_4	x_5
$\bar{\mu}(x)$:	0.8	0.94	0.92	0.71	0.8

Thus, comparing the values taken by the standard fuzzy complement, the Sugeno fuzzy complement for $\lambda = 2$ and the Yager fuzzy complement for w = 2 indicates that the membership grades for the three different fuzzy complements take very different values, but that the rank order of the membership grades for the different elements is the same for each of these fuzzy complements.

14.3.4 Averaging Operations

Fuzzy intersections and unions are examples of *aggregation operations.* These standard operations bound all the aggregation operations with the monotonicity and idempotency properties A2 and A5 defined in section 14.2.1 i.e. all aggregation operations with properties A2 and A5 satisfy the following inequalities for all n tuples $(\alpha_1, \alpha_2, ..., \alpha_n) \in [0, 1]^n$

$$\min(\alpha_1, \alpha_2, ..., \alpha_n) \le h(\alpha_1, \alpha_2, ..., \alpha_n) \le \max(\alpha_1, \alpha_2, ..., \alpha_n) \quad (14.13)$$

It can also easily be shown that all aggregation operations which satisfy inequalities (14.13) are also *idempotent* (Klir and Yuan 1995) i.e.

$$h(\alpha, \alpha, ..., \alpha) = \alpha \quad (14.14)$$

and that in addition they are the only idempotent aggregation operations i.e. all idempotent aggregation operation satisfy inequalities (14.13).

These idempotent aggregation operations which take values between the standard fuzzy intersection (min) and fuzzy union (max) are generally called *averaging operations*. Thus averaging operations are idempotent aggregation operations. The *generalised means* are a class of averaging operations that cover the entire interval between the min and max operations. They are defined as follows (Klir and Yuan 1995)

$$h_\beta(\alpha_1, \alpha_2, \ldots, \alpha_n) = \left(\frac{\alpha_1^{\beta} + \alpha_2^{\beta} + \ldots + \alpha_n^{\beta}}{n} \right)^{1/\beta} \qquad (14.15)$$

where $\beta \neq 0$ and $\beta \in \mathcal{R}$ and $\alpha_i \neq 0$ for all i. The parameter β distinguishes the different means. For $\beta \to 0$, the function h converges to the geometric mean $(\alpha_1\alpha_2\ldots\alpha_n)^{1/n}$. The class of generalised means satisfies properties A1 to A5 in section 14.2.1 and is therefore a parametric class of continuous, symmetric and idempotent aggregation operations.

Binary aggregation operations (i.e. two arguments) that satisfy the properties of monotonicity, commutativity and associativity and the boundary conditions

$$h(0, 0) = 0 \text{ and } h(1, 1) = 1 \qquad (14.16)$$

are called *norm operations*. The associativity property can be used to extend the definition of these operations to any finite number of arguments. t-norms are norm operations which have the property $h(\alpha, 1) = \alpha$ and t-conorms those with the property $h(\alpha, 0) = \alpha$ (Klir and Yuan 1995).

Example 14.5: Printer Power Consumption and Noise IV

Consider the fuzzy sets *high power consumption* (HPC) and *noisy* defined in example 14.1 in section 14.2.2 and which are repeated below for ease of reference:

HPC

x:	x_1	x_2	x_3	x_4	x_5
$\mu(x)$:	0.6	0.35	0.4	0.7	0.6

noisy

x:	x_1	x_2	x_3	x_4	x_5
$\mu(x)$:	1	0.5	0.6	0.3	0.2

For these two fuzzy sets, the generalised means are given by

x:	x_1	x_2	x_3	x_4	x_5
μ(x):	$\left(\frac{1^\alpha+0.6^\alpha}{2}\right)^{1/\alpha}$	$\left(\frac{0.5^\alpha+0.35^\alpha}{2}\right)^{1/\alpha}$	$\left(\frac{0.6^\alpha+0.4^\alpha}{2}\right)^{1/\alpha}$	$\left(\frac{0.3^\alpha+0.7^\alpha}{2}\right)^{1/\alpha}$	$\left(\frac{0.2^\alpha+0.6^\alpha}{2}\right)^{1/\alpha}$

For α=2 this gives

x:	x_1	x_2	x_3	x_4	x_5
μ(x):	$\left(\frac{1^2+0.6^2}{2}\right)^{1/2}$	$\left(\frac{0.5^2+0.35^2}{2}\right)^{1/2}$	$\left(\frac{0.6^2+0.4^2}{2}\right)^{1/2}$	$\left(\frac{0.3^2+0.7^2}{2}\right)^{1/2}$	$\left(\frac{0.2^2+0.6^2}{2}\right)^{1/2}$

= x:	x_1	x_2	x_3	x_4	x_5
μ(x)	0.82	0.43	0.51	0.54	0.45

14.4 Defuzzification

Defuzzification is the process of obtaining a single crisp value or a vector of crisp values from a fuzzy set or sets, so that this crisp value or vector best represents the fuzzy set or sets in an appropriate sense. Fuzzy systems have been divided into two broad categories (van Leekwijck and Kerre 1999): *fuzzy knowledge* and *fuzzy control* systems. Of the two categories, it is fuzzy knowledge rather than fuzzy control systems which are more generally used in sustainable development problems. Defuzzification is generally required in the fuzzy control case, but not always in the fuzzy knowledge case. Criteria are required to determine whether or not a defuzzification method performs satisfactorily. These criteria (van Leekwijck and Kerre 1999) include the following:

- *Continuity* i.e. a small change in membership grade should not give a large change in the value of the crisp data.
- *Translation and scaling* criteria, which are relevant when the definition of fuzzy sets is extended to allow the membership function to take a wider range of values than the unit interval.
- *Core selection* i.e. selecting an element which is in the core or set of elements with the maximum possible degree of membership.
- *Computational efficiency* i.e. fast computational time and low memory requirements.

It is generally necessary to obtain numerical representations of linguistic variables before a fuzzy set containing linguistic variables can be defuzzified. Defuzzification methods can be classified as general and specific. *General* methods defuzzify a single fuzzy output set produced by the fuzzy system. *Specific* methods, which are used particularly in fuzzy control, combine defuzzification with other operations to make the control algorithm more computationally efficient. Defuzzification methods can further be classified as (Leekwijck and Kerre 1999):

- *Maxima methods*, which select an element from the set of values, called the core, for which the membership grade $\mu(x)$ is equal to its height i.e. the maximum value. They satisfy the core, translation and scaling criteria and are very computationally efficient, but do not satisfy continuity. They are appropriate for fuzzy knowledge systems and therefore for many sustainable development problems, but less so for fuzzy controllers.
- *Distribution methods*, which take into account the way the values are distributed by converting the membership function into a probability distribution and then computing the expected value. This group of methods is particular appropriate for fuzzy controllers, as it satisfies continuity, but not the core criterion. Many of the translation and scaling criteria are not satisfied. Some of the methods are very computationally efficient and others less so.
- *Area methods*, which use the area under the membership function to determine the crisp value for the fuzzy set. This group of methods is mainly appropriate for fuzzy controllers. The methods satisfy continuity, but not core selection, and some of the translation and scaling properties.

The methods stated below are all for a fuzzy set $\mu(X)$ defined on a space $X = \{x_1, x_2, ..., x_n\}$. Maxima methods include:

- *First of maxima*: This selects the smallest element of the core of $\mu(X)$: $\text{FOM}[\mu(X)] = \min \text{core}[\mu(X)]$

- *Last of maxima*: This selects the largest element of the core of $\mu(X)$: $\text{LOM}[\mu(X)] = \max \text{core}[\mu(X)]$

- *Random choice of maxima* (RCOM): this method picks an arbitrary element from the core of $\mu(X)$.

Distribution methods include the centre of area method in which the defuzzified value $\text{COA}(\mu(X))$ is defined as a weighted average over all points in the space, given by

$$\mathrm{COA}[\mu(X)] = \frac{\sum_{i=1}^{n} x_i\,\mu(x_i)}{\sum_{i=1}^{n} \mu(x_i)} \tag{14.17}$$

If the value obtained is not equal to any value in the space X, the value closest to it is used.

Example 14.6: Emissions of Industrial Processes

There is considerable uncertainty in the measurements of emissions from a particular industrial process and therefore these emissions are best represented using fuzzy methods, as follows (with values of x in kg per month).

X	0.5	1	2	5	10	15	20	25	30
μ(x)	0.2	0.4	0.5	1.0	1.0	0.8	1.0	0.7	0.5

The height is equal to 1 and the elements of the core are 5, 10 and 20. Therefore:

FOM(μ(x)) = 5
LOM(μ(x)) = 20
RCOM(μ(x)) = 5, 10 or 20

The centre of area method gives the defuzzified value

$$\mathrm{COA}[\mu(X)] = \frac{\sum_{i=1}^{n} x_i\,\mu(x_i)}{\sum_{i=1}^{n} \mu(x_i)}$$

$$= \frac{0.5\mathrm{x}0.2 + 1\mathrm{x}0.4 + 2\mathrm{x}0.5 + 5\mathrm{x}1 + 10\mathrm{x}1 + 15\mathrm{x}0.8 + 20\mathrm{x}1 + 25\mathrm{x}0.7 + 30\mathrm{x}0.5}{0.2 + 0.4 + 0.5 + 1 + 1 + 0.8 + 1 + 0.7 + 0.5}$$

= 13.28

The closest value in the space X to D_{CA} is 15. Therefore the different methods give quite different results.

14.5 Roundup of Chapter 14

Chapter 14 has continued the introduction to fuzzy set theory commenced in chapter 13 through the presentation of fuzzy union, intersection and complement operators and discussion of their relationship to the classes of t-conorms, t-norms and fuzzy complements. The standard fuzzy union and intersection have membership functions obtained by taking the maximum and minimum respectively of the membership functions of the fuzzy sets being combined. The complement of a fuzzy set has membership function equal to one minus the membership function of the original fuzzy set.

Classes of fuzzy unions and intersections can be defined using t-conorms and t-norms respectively. These are classes of binary operators which satisfy a number of properties, in particular monotonicity, commutativity and associativity. The classes of t-norms and t-conorms differ from each other in having different boundary conditions. Classes of fuzzy complements which satisfy particular conditions can also be defined.

Techniques for defuzzification or obtaining crisp from fuzzy sets have been discussed and properties to be satisfied by these methods have been stated, including continuity, computational efficiency and core selection i.e. selecting an element with the maximum degree of membership. Deffuzification methods can be classified as maxima methods, distribution methods and area methods.

Many sustainable development problems involve data which is inaccurate or uncertain as well as imprecise. They can be analysed using the augmented, intuitionistic and type 2 fuzzy sets introduced in chapter 15.

15 Augmented, Intuitionistic & Type 2 Fuzzy Sets

15.0 Learning Objectives

This chapter presents three different approaches to the problem of representing information which is both imprecise (fuzzy) and uncertain or inaccurate, namely augmented, intuitionistic and type 2 fuzzy sets. Specific learning objectives include the following:

- The ability to define augmented, intuitionistic and type 2 fuzzy sets and understanding of the differences between them.
- Understanding of some of the applications of augmented, intuitionistic and type 2 fuzzy sets to problems in sustainable development.
- The ability to calculate the union, intersection and complement for augmented and intuitionistic fuzzy sets and the union and intersection for type 2 fuzzy sets.

15.1 Two Different Types of Uncertainty: Imprecision and Lack of Accuracy or Reliability

The previous two chapters considered the type of uncertainty associated with imprecision and quantities that are not precisely defined, such as warmth or degree of hearing impairment. This type of uncertainty or imprecision can be represented using fuzzy variables and fuzzy analysis techniques. However there are also other types of uncertainty, in particular the type of uncertainty associated with a lack of information or knowledge, leading to a lack of reliability and/or accuracy of the data. There are a number of different ways of treating this type of uncertainty, either on its own or in combination with the type of uncertainty resulting from imprecision.

There are two main types of uncertainty in membership grades resulting from a lack of information or reliable data:

- Uncertainty due to measurement problems.

- Uncertainty due to problems in assigning membership grades to measurements or estimates.

The first type of uncertainty arises because it may be difficult or impossible to measure variables and/or there may be large errors in the measurements. In some cases only estimates may be available, possibly derived from measurements of another variable, and there may be limited information about how accurate these estimates are. The second type of uncertainty arises due to a lack of information about the fuzzy set and the correspondence between measured values, whether qualitative (e.g. high or low) or numerical (e.g. 3.5 cm) and membership grades. A number of different techniques have been developed for representing information which is uncertain, inaccurate and/or inconsistent as well as imprecise, including augmented, intuitionistic and type 2 fuzzy sets. These approaches give additional information, but at the price of slightly greater complexity. There are also techniques, such as evidence theory, for analysing lack of knowledge on its own, without also taking account of imprecision (Klir and Yuan 1995), as well as probability theory and stochastic processes, which consider the type of uncertainty due to randomness.

This chapter presents augmented, intuitionistic and type 2 fuzzy sets in sections 15.2, 15.3 and 15.4 respectively. This is followed by two examples of the application of these methods in section 15.5 and 15.6 and a summary of the chapter in section 15.7.

15.2 Augmented Fuzzy Sets

Augmented fuzzy sets (Hersh 1999a Hersh et al 1998) resolve the problem of representing information which is both imprecise and uncertain by using two parameters rather than one i.e. an ordered pair of numbers, both of which take values between 0 and 1. The second parameter represents the degree of certainty or reliability of the information represented by the first value.

An augmented fuzzy set can be defined as a mapping from the space X of all possible points to an ordered pair of numbers on the unit interval I = [0, 1] i.e.

$$f: X \rightarrow I \times I \tag{15.1a}$$

$$f(x) = (\mu(x), \nu(x)), \ \forall x \in X, \mu, \nu: X \rightarrow I \tag{15.1b}$$

There are two different cases. The first element μ(x) of the ordered pair is always a fuzzy set membership function. The second value ν(x) represents the degree of certainty in the assessment of the first value μ(x) and can be treated as either

- a (fuzzy set) membership function
- a probability.

In the membership function case ν(x) is the fuzzy set: 'the value of the membership function is known with certainty' (and therefore can be represented by a monotonically increasing fuzzy number). In the probabilistic case ν(x) represents the probability of the membership function taking the value μ(x).

This gives rise to the question of how values of ν(x) should be measured or estimated. Any available information should be used, but in many cases (subjective) judgements will also be required. Therefore the resulting value is likely to itself have some associated uncertainty, but in general the additional complexity required to measure this uncertainty makes it not worthwhile. In many cases the value of ν(x) is most easily obtained by first expressing it as a linguistic variable and then converting the linguistic value to the membership grade of a fuzzy set, using the scale in table 15.1.

Table 15.1 Scale for converting linguistic values of certainty to a fuzzy set

Linguistic value	Membership grade
No knowledge of very low certainty	0 - 0.2
Low certainty	0.2 - 0.4
Moderate certainty	0.4 - 0.6
High certainty	0.6 - 0.8
Very high or full certainty	0.8 – 1

Alternatively if error bounds are available they can be used to calculate values of the degree of certainty ν(x). In general the value of ν(x) should decrease with or even be directly inversely proportional to the ratio of the error bound to the value of the numerical variable.

For instance, consider the fuzzy set *'acceptable' levels of emissions of greenhouse gases*. The term 'acceptable' relates to the resulting climate and other changes, including rises in sea level, in response to particular levels of emissions. However, as already discussed (for instance in section 13.6), there is still considerable uncertainty as to the exact response to any given levels of emissions, as well as the time span over which these effects will occur. Thus the certainty associated with the response to a particular level of emissions can be classified as 'low'. There is also further uncertainty as to how the term 'acceptable' levels of emissions should be interpreted in the context of a particular industrial process, for instance with regards to the proportion of the total that should be 'allocated' to it or that it is in some sense reasonable for it to produce. This uncertainty can be classified as 'moderate'. Combining the two uncertainties and noting that uncertainties are unlikely to cancel out, gives 'low' certainty or a value of 0.2 - 0.4 for the degree of certainty $\nu(x)$.

When X is finite, an augmented fuzzy set can be described in tabular form, as shown in table 15.2.

Table 15.2 A finite augmented fuzzy set

X	x_1	x_2	...	x_n
$\mu(x)$	$\mu(x_1)$	$\mu(x_2)$	...	$\mu(x_n)$
$\nu(x)$	$\nu(x_1)$	$\nu(x_2)$	...	$\nu(x_n)$

15.2.1 Set Operations for Augmented Fuzzy Sets: Union

There are significant differences in the forms of the expressions for the set operations of union and intersection according to whether $\nu(x)$ is a membership function or a probability. Only the membership function case will be considered here and the definitions presented will be based on the standard (max and min operations) for the union and intersection. However it is again possible to obtain wider classes of unions and intersections of augmented fuzzy sets based on t-conorms and t-norms.

The *definition* of the augmented fuzzy set union operator can be *motivated* by considering that the union of a number of crisp sets is defined to include all the elements of any of the component sets. In the case of fuzzy sets, membership is replaced by membership functions and the standard generalisation of the union operator to fuzzy sets is through the maximisation operator, with each element in the union of fuzzy sets having the maximum membership function taken over all the fuzzy sets in the union.

A similar principle can be applied to augmented fuzzy sets, except that there are now two values, the data itself and the degree of certainty with which it is known, which have to be taken into account, so that a simple maximum can no longer be used. Thus an appropriate definition of the union operator for the augmented fuzzy set case will be an appropriate generalisation of maximisation of an ordered pair of functions. This can be treated analogously to maximisation of a function of two variables or bicriteria optimisation.

There are many different ways of doing this, some of which give fairly complicated algorithms. The particular choice taken here is both computationally simple and relatively general. It is based on comparing the performance of the augmented fuzzy set options (a) and (b) to the default option (c), where the three options are as follows:

(a) Maximisation of a simple function of μ and ν, followed by maximising over μ

(b) Maximisation over ν and then over μ

(c) Maximisation over μ and then over ν

Option (a) is the preferred option, but is only accepted if performance is satisfactory relative to the default option (c) in terms of constants δ and ε. These constants can be chosen by the user to give the relative importance of maximising over μ or ν or over both together. When maximising over μ is more important, then δ should be smaller than ε and conversely, when maximising over ν is more important, then ε should be smaller than δ.

If option (a) is not found to satisfy appropriate conditions, option (b) is then tested in an analogous way. If neither option is found to be satisfactory, then option (c) is accepted. Sometimes it is useful to put additional conditions on the maximising values to ensure that the best overall values are obtained. The product $\mu\nu$ has been chosen as the simple function of μ and ν rather than the sum $\mu + \nu$ for the following reasons:

- To avoid the case where the highest overall sum is obtained with either of μ or ν zero.
- To give greater discrimination between different augmented fuzzy sets, since there are fewer pairs of numbers on the unit interval with a given product than a given sum. For instance (considering values to one decimal place) only (1,0.6) and (0.6, 1) have a product of 0.6, whereas (0, 0.6), (0.1, 0.5), (0.2, 0.4), (0.3, 0.3), (0.4, 0.2), (0.5, 0.1) and (0.6, 0) all have a sum of 0.6.

The *union operator* for *augmented fuzzy sets* can now be defined. The union of n augmented fuzzy sets $f_1 = (\mu_1, \nu_1)$, $f_2 = (\mu_2, \nu_2)$, ..., $f_n = (\mu_n, \nu_n)$ has membership function

$$f_{sup}(x) = \bigvee_{i=1}^{n} f_i(x) : X \to I \times I \qquad (15.2a)$$

$$f_{sup}(x) = (\mu_{sup}(x), \nu_{sup}(x)), \ \forall x \in X, \mu_{sup}(x), \nu_{sup}(x) \in I \qquad (15.2b)$$

with (omitting the argument x to simplify the notation),

$\mu_{sup} = \mu_u, \ \nu_{sup} = \nu_u$
if (ai) $\mu_u + \delta \geq \mu_h$, (aii) $\nu_u - \varepsilon \geq \nu_h$ or (aiii) $\mu_u = \mu_h, \quad \nu_u = \nu_h$

$\mu_{sup} = \mu_m, \ \nu_{sup} = \nu_m$
if (ai) but not (aii) is satisfied, (bi) $\mu_m + \delta \geq \mu_h$, (bii) $\nu_m - \varepsilon \geq \nu_h$

$\mu_{sup} = \mu_h, \ \nu_{sup} = \nu_h$ otherwise

where $\delta, \varepsilon \geq 0$ are chosen by the user

$$h = \min j \in J, \quad J = \{j: \nu_j = \max_{k \in K} \nu_k\}, \quad K = \{k: \mu_k = \max_{i=1...n} \mu_i\} \qquad (15.3a)$$

$$m = \min p \in P, \quad P = \{p: \mu_p = \max_{r \in R} \mu_r\}, \quad R = \{r: \nu_r = \max_{i=1...n} \nu_i\} \qquad (15.3b)$$

$$u = \min w \in W, \quad W = \{w: \mu_w = \max_{z \in Z} \mu_z\},$$

$$Z = \{z: \mu_z \nu_z = \max_{i=1...n} (\mu_i \nu_i)\} \qquad (15.3c)$$

As already indicated, the constants δ and ε are chosen by the user. Allowing the user to choose these values gives increased flexibility, but some users, particularly students, may prefer not to have this flexibility! Therefore values of 0.1 or 0.2 are suggested, or 0.03-0.05 if all the values of μ are expected to be small e.g. less than 0.2. It should also be noted that different values of δ and ε can be chosen for the different elements of the augmented fuzzy set. There are also a number of special cases. For instance taking $\delta = 0, \quad 0 < \varepsilon < 1$ gives the normal fuzzy set operator with value $\mu_{sup} = \max_{i=1...n} \mu_i$.

The associated certainty ν can then be obtained as $\nu_{sup} = \max_{j \in J} \nu_j$ for $J = \{j: \mu_j = \max_{i=1...n} \mu_i\}$.

It will now be shown that the augmented fuzzy union operator satisfies appropriate *extensions of the t-conorm properties* U1-U7 in section 14.3.2. The zero augmented fuzzy set is equal to (0, 1). In the case of the union of this fuzzy set and any other fuzzy set (μ_a, ν_a) condition (aiii) clearly holds and satisfaction of the boundary condition U1 follows. There are a number of possible definitions of (strict) inequality for augmented fuzzy sets. However consideration will be restricted here to the case when each element of the ordered pair (μ_b, ν_b) is (strictly) greater than the associated element of the ordered pair (μ_d, ν_d). In this case both the monotonicity and strict monotonicity properties U2 and U7 are satisfied in the sense that each element of $u[(\mu_a, \nu_a), (\mu_d, \nu_d)]$ is greater than or equal to the corresponding element of $u[(\mu_a, \nu_a), (\mu_b, \nu_b)]$ i.e.

$$\mu_b \leq \mu_d \text{ and } \nu_b \leq \nu_d \Rightarrow u[(\mu_a, \nu_a), (\mu_b, \nu_b)] \leq u[(\mu_a, \nu_a), (\mu_d, \nu_d)] \quad (15.4)$$

with an analogous expression for the strict monotonicity property U7.

These properties follow directly from the definitions and the (strict) monotonicity of the max operator. The commutativity property U3 follows directly from the commutativity of the max operator. The associativity property U4 can be demonstrated by considering the different possible cases for satisfaction of the properties (a) and (b) when three fuzzy sets are combined in different orders, but the details will not be given here.

Piecewise continuity of the augmented fuzzy union operator and hence partial satisfaction of property U5 follows from the continuity of the max operator and the definition (15.2) - (15.3). The idempotency property U6 follows directly from the idempotency property of the max operator.

15.2.2 Intersection & Complement for Augmented Fuzzy Sets

The *intersection* of n *augmented fuzzy sets* can be defined from the min operator in an analogous way to the definition of the union from the max operator. The membership function is

$$f_{inf}(x) = \bigwedge_{i=1}^{n} f_i(x) : X \rightarrow I \times I \quad (15.5a)$$

$$f_{inf}(x) = (\mu_{inf}(x), \nu_{inf}(x)), \ \forall x \in X, \mu_{inf}(x), \nu_{inf}(x) \in I \quad (15.5b)$$

where, omitting the argument x to simplify notation

$$\mu_{inf} = \mu_y, \ \nu_{inf} = \nu_y$$

if (ai) $\mu_y - \delta \leq \mu_g$, (aii) $\nu_y + \varepsilon \leq \nu_g$ or (aiii) $\mu_y = \mu_g$, $\nu_y = \nu_g$

$\mu_{inf} = \mu_n, \ \nu_{inf} = \nu_n$

if (ai) but not (aii) is satisfied, (bi) $\mu_n - \delta \leq \mu_g$, (bii) $\nu_m + \varepsilon \leq \nu_g$

$\mu_{inf} = \mu_g, \ \nu_{inf} = \nu_g$, otherwise

where δ and ε are given and

$$g = \min j \in J, \quad J = \{j: \nu_j = \min_{k \in K} \nu_k\}, \quad K = \{k: \mu_k = \min_{i=1...n} \mu_i\}$$

$$n = \min p \in P, \quad P = \{p: \mu_p = \min_{q \in Q} \mu_q\}, \quad Q = \{q: \nu_q = \min_{i=1...n} \nu_i\}$$

$$y = \min w \in W, \quad W = \{w: \mu_w = \min_{z \in Z} \mu_z\},$$

$$Z = \{z : (\mu_z, \nu_z) = \min_{i \in V}(\mu_i \nu_i)\} \text{ if } \min_{i \in V}(\mu_i \nu_i) \neq 0$$
$$\text{or } \min_{i \in V}(\mu_i \nu_i) = 0 \text{ and } \mu_z = \mu_g, \nu_z = 0, \text{ or } \mu_z = 0, \nu_z = \nu_n\}$$

$$Z = \{z: (\mu_z, \nu_z) = \min_{i \in V}(\mu_i + \nu_i)\} \text{ otherwise}$$

This definition preferentially uses the product rather than the sum of μ and ν, as this gives greater discrimination between different augmented fuzzy sets. However it uses the sum when the minimum product is zero and the non-zero element of the minimising pair is not at the minimum value μ_g or ν_n. This avoids a minimum being obtained with either μ or ν zero and a high value on the other element of the ordered pair. Appropriate generalisations of the comments for the union operator in section 15.2.1 hold for the intersection operator.

Satisfaction of properties I1-I7 in section 14.3.1 can be demonstrated analogously to the demonstration of properties U1-U7 for the union operator for augmented fuzzy sets, again noting that appropriate definitions of (strict) inequality are required and property I5 is satisfied in terms of piecewise continuity.

The *complement* $\bar{f}(x)$ of an *augmented fuzzy set* with membership function $f(x) = (\mu(x), \nu(x))$ has membership function

$$\bar{f}(x): X \rightarrow I \times I \tag{15.6a}$$

$$\bar{f}(x) = (1-\mu(x), \nu(x)) \tag{15.6b}$$

Noting that the unit and zero of the augmented fuzzy set are given by (1, 1) and (0, 1) respectively, then the boundary condition C1 in section 14.3.3 follows immediately from the definitions (15.6). Properties C2-C4 in section 14.3.3 follow from the definition and the properties of the minus function.

15.3 Intuitionistic Fuzzy Sets

Intuitionistic fuzzy sets can be used to represent the *consistency* rather than degree of certainty or accuracy of the information. They were first introduced by Atanassov (1986 1989 1999) and later generalised to intuitionistic L-fuzzy sets with range [0, L] rather than [0, 1] and interval-valued fuzzy sets (Atanassov and Gargov 1989; Bustince and Burillo 1995), where the membership and non-membership functions take interval values contained in the unit interval. Intuitionistic fuzzy sets give a measure of *non-membership*, as well as *membership* of the set, with the sum of the two measures less than or equal to one. They are related to the bilattice approach which is intended to give the relationship between incomplete and contradictory information.

Intuitionistic fuzzy sets have the form

$$A = \{x, \mu(x), \gamma(x): x \in X\}, \quad \mu(x), \gamma(x): X \to I \qquad (15.7a)$$

where $\mu(x)$ and $\gamma(x)$ denote respectively the degree of membership and non-membership of an element x in the fuzzy set and

$$0 \leq \mu(x) + \gamma(x) \leq 1 \qquad (15.7b)$$

Ordinary fuzzy sets can easily be expressed as intuitionistic fuzzy sets by taking

$$\gamma(x) = 1 - \mu(x) \qquad (15.8a)$$

giving the set

$$A = \{x, \mu(x), 1\text{-}\mu(x): x \in X\}. \qquad (15.8b)$$

However in general $\gamma(x) \neq 1 - \mu(x)$, so that $\mu(x) + \gamma(x) < 1$

The *intuitionistic index* of the element x in the set A is given by

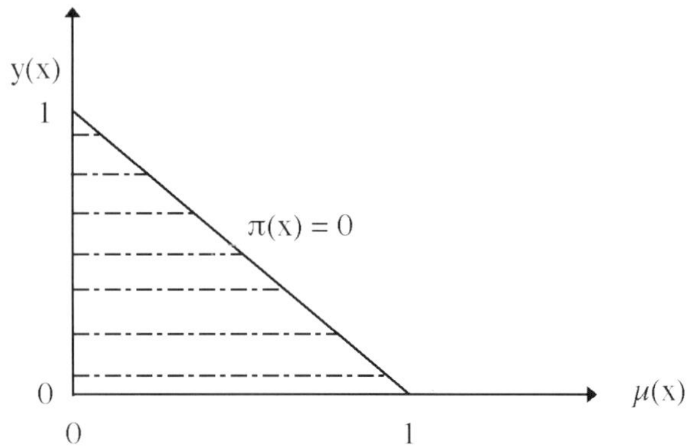

Fig. 15.1 Intuitionistic fuzzy sets

$$\pi(x) = 1 - \mu(x) - \gamma(x) \tag{15.9a}$$

$$\text{so that} \quad 0 \leq \pi(x) \leq 1 \quad \text{for all x} \tag{15.9b}$$

Therefore the intuitionistic index gives a measure of the inconsistency in the information, with total consistency when $\pi(x)$ is equal to zero and total inconsistency when it is equal to 1. In the latter case the set is *completely intuitionistic* and $\mu(x) = \gamma(x) = 0$. This is illustrated in fig. 15.1, where the coordinate axes are the membership and non-membership grades and the diagonal line represents totally consistent intuitionistic fuzzy sets with $\pi(x)$ equal to 0. Intuitionistic fuzzy sets which satisfy condition (15.7b) and are not totally consistent are in the shaded region below this line.

Obtaining an intuitionistic fuzzy sets requires values of $\gamma(x)$ as well as $\mu(x)$. Similar techniques to those in section 13.4 can in principle be used to obtain $\mu(x)$. However, the calculation is complicated by the fact that $\mu(x)$ and $\gamma(x)$ need to satisfy inequalities (15.7b) and equality (15.9a). It is therefore generally appropriate to first calculate the consistency and then the values of membership, $\mu(x)$, and non-membership, $\gamma(x)$. If information on the consistency is available it may be easiest to first express it as a linguistic variable and then convert the linguistic value to a numerical value for the consistency $\pi(x)$, using the scale in table 15.3. It should be noted that a high degree of consistency corresponds to a low value of $\pi(x)$.

Fuzzification techniques similar to those in section 13.5 can then be used to apply local or global scaling to obtain a membership function $\mu''(.)$ which then has to be scaled by multiplying by $[1 - \pi(x)]$ to ensure that both $\mu(x)$ and $\gamma(x)$ are non-negative. If this scaling is not carried out, negative values of $\gamma(x)$ could be obtained. For instance, $\mu(x) = 0.7$ and $\pi(x) = 0.4$

Table 15.3 Conversion of consistency expressions to membership grades

Consistency	$\pi(x)$
No or very low consistency	0.8 – 1
Low consistency	0.6 – 0.8
Moderate consistency	0.4 – 0.6
High consistency	0.2 – 0.4
Very high or total consistency	0 – 0.2

would give $\gamma(x) = -0.1$, which is negative, whereas scaling $\mu''(x) = 0.7$ by $[1 - \pi(x)]$ would give $\mu(x) = 0.42$ and $\nu(x) = 0.18$, both of which are positive. Therefore, using equation (15.9a) for $\nu(x)$, the values of the membership function $\mu(x)$ and non-membership function $\nu(x)$ can be obtained as follows:

$$\mu(x) = \mu''(x)[1-\pi(x] \qquad (15.10a)$$

so that, using equation (15.9a),

$$\begin{aligned}\gamma(x) &= 1 - \mu(x) - \pi(x) = 1 - \mu''(x)[1-\pi(x)] - \pi(x) \\ &= [1 - \mu''(x)][1-\pi(x)] \qquad (15.10b)\end{aligned}$$

The values of $\mu(x)$ should be rounded to the desired number of decimal places (generally one) before calculating $\gamma(x)$ to ensure that inequality (15.7b) and equation (15.9a) are satisfied.

Example 15.1: Fuzzification of Carbon Dioxide Emissions

Consider the monthly emissions of carbon dioxide of the industrial processes in example 13.4 in section 13.5. As in example 13.4, local scaling can be applied to obtain the modified membership grades $\mu''(x)$ given in table 13.6 and restated in table 15.4. It will be assumed that the consistency of this data is moderate, so that, using table 15.3, a value between 0.4 and 0.6 should be used for $\pi(x)$. In this case $\pi(x)$ will be put equal to 0.4.

$\mu(x)$ and $\gamma(x)$ can be calculated from $\mu''(x)$ and $\pi(x)$ using equation (15.10a) and (15.10b) respectively.

Table 15.4 Calculation of membership grades using local scaling

Original Data	$\mu''(x)$
0.7 kg	0
2.5 kg	0.0
6 kg	0.1
12 kg	0.2
15 kg	0.3
16 kg	0.3
29.6 kg	0.5
54 kg	0.9

For instance, for $\mu''(x) = 0.2$,

$$\mu(x) = \mu''(x)[1-\pi(x)] = 0.2[1-0.4] = 0.1 \text{ to 1 d.p.}$$

$$\gamma(x) = 1 - \mu(x) - \pi(x) = 1 - 0.1 - 0.4 = 0.5$$

For $\mu''(x) = 0.5$,

$$\mu(x) = \mu''(x)[1-\pi(x)] = 0.5[1-0.4] = 0.3$$

$$\gamma(x) = 1 - \mu(x) - \pi(x) = 1 - 0.3 - 0.4 = 0.3$$

The other results can be calculated similarly and are stated in table 15.5.

Table 15.5 Fuzzification calculation for intuitionistic fuzzy set

Original Data	$\mu''(x)$	$\mu(x)$	$\gamma(x)$
0.7 kg	0	0	0.6
2.5 kg	0.0	0.0	0.6
6 kg	0.1	0.1	0.5
12 kg	0.2	0.1	0.5
15 kg	0.3	0.2	0.4
16 kg	0.3	0.2	0.4
29.6 kg	0.5	0.3	0.3
54 kg	0.9	0.5	0.1

15.3.1 Aggregation Operations for Intuitionistic Fuzzy Sets

Set combination operations can be defined for intuitionistic fuzzy sets as follows. The *union* of two *intuitionistic* fuzzy sets with membership functions $f_1 = (\mu_1, \gamma_1)$ and $f_2 = (\mu_2, \gamma_2)$ has membership function

$$f_{sup} = f_1 \vee f_2 = [\max(\mu_1, \mu_2), \min(\gamma_1, \gamma_2)] \tag{15.11}$$

The *intersection* of two *intuitionistic* fuzzy sets with membership functions $f_1 = (\mu_1, \gamma_1)$ and $f_2 = (\mu_2, \gamma_2)$ has membership function

$$f_{inf} = f_1 \wedge f_2 = [\min(\mu_1, \mu_2), \max(\gamma_1, \gamma_2)] \tag{15.12}$$

These definitions can be generalised to n intuitionistic fuzzy sets with membership functions $f_1 = (\mu_1, \gamma_1)$, $f_1 = (\mu_2, \gamma_2)$, ..., $f_n = (\mu_n, \gamma_n)$, as follows. The union of n intuitionistic fuzzy sets is given by:

$$f_{sup}(x) = \bigvee_{i=1}^{n} f_i(x) : X \to I \times I \tag{15.13a}$$

$$f_{sup}(x) = (\mu_{sup}(x), \gamma_{sup}(x)), \ \forall\ x \in X, \ \mu_{sup}(x) + \gamma_{sup}(x) \leq 1, \quad \mu_{sup}(x), \gamma_{sup}(x) \geq 0 \tag{15.13b}$$

$$\mu_{sup}(x) = \max_i \mu_i(x), \quad \gamma_{sup}(x) = \min_i \gamma_i(x) \tag{15.13c}$$

The intersection of n intuitionistic fuzzy sets is given by:

$$f_{inf}(x) = \bigwedge_{i=1}^{n} f_i(x) : X \to I \times I \tag{15.14a}$$

$$f_{inf}(x) = (\mu_{inf}(x), \gamma_{inf}(x)), \ \forall\ x \in X, \ \mu_{inf}(x) + \gamma_{inf}(x) \leq 1, \quad \mu_{inf}(x), \gamma_{inf}(x) \geq 0 \tag{15.14b}$$

$$\mu_{inf}(x) = \min_i \mu_i(x), \quad \gamma_{inf}(x) = \max_i \gamma_i(x) \tag{15.14c}$$

It can easily be shown that the expressions for the union and intersection are equivalent to the standard fuzzy set expressions when $\gamma_i(x) = 1 - \mu_i(x)$ (i=1...n) and that they satisfy

$$\mu_{sup}(x) + \gamma_{sup}(x) \leq 1 \quad (15.15a)$$

$$\mu_{inf}(x) + \gamma_{inf}(x) \leq 1 \quad (15.15b)$$

When $\gamma_i(x) = 1 - \mu_i(x)$ (i=1...n) (15.16)

Then

$$(\mu_{sup}(x), \gamma_{sup}(x)) = \left(\max_i \mu_i(x),\ \min_i \{1 - \mu_i(x)\} \right)$$
$$= \left(\max_i \mu_i(x), 1 - \max_i \mu_i(x) \right) \quad (15.17)$$

and

$$(\mu_{inf}(x), \gamma_{inf}(x)) = \left(\min_i \mu_i(x),\ \max_i \{1 - \mu_i(x)\} \right)$$
$$= \left(\min_i \mu_i(x), 1 - \min_i \mu_i(x) \right) \quad (15.18)$$

$$\therefore\ \mu_{sup}(x) + \gamma_{sup}(x) = \max_i \mu_i(x) + \min_i \gamma_i(x)$$
$$\leq \max_i \mu_i(x) + 1 - \max_i \mu_i(x) \leq 1 \text{ (using (5.17))}$$
(15.19a)

and

$$\mu_{inf}(x) + \gamma_{inf}(x) = \min_i \mu_i(x) + \max_i \gamma_i(x)$$
$$\leq \min_i \mu_i(x) + 1 - \min_i \mu_i(x) \leq 1 \text{ (from (15.18))}$$
(15.19b)

Combining equality (15.17) with inequality (15.19a) and equality (15.18) with inequality (5.19b) shows that the expressions for the union and intersection of intuitionistic fuzzy sets are equivalent to the union and intersection of ordinary fuzzy sets when conditions (5.16) holds.

Satisfaction of appropriate extensions of the t-conorm properties U1-U7 in section 14.3.2 for the union of two intuitionistic fuzzy sets will now be considered. Defining the zero of an intuitionistic fuzzy set as (0, 1), the boundary condition U1 clearly holds. The monotonicity and strict monotonicity properties U2 and U7 follow from the properties of the max and min operators for inequality of two intuitionistic fuzzy sets (μ_b, γ_b) and (μ_d, γ_d) defined as follows

$$\mu_b \leq \mu_d \text{ and } \gamma_b \geq \gamma_d \tag{15.20}$$

The commutativity, associativity and continuity properties U3, U4 and U5 follow directly from the properties of the min and max operators, since the two components of the intuitionistic fuzzy set are defined independently of each other. Similarly idempotency follows from idempotency of both the max and min operators. The properties I1-I7 in section 14.3.1 can analogously be shown to hold for the intersection of two intuitionistic fuzzy sets.

The *complement* of an *intuitionistic* fuzzy set with membership function $f = (\mu, \gamma)$ has membership function

$$\bar{f} = (\gamma, \mu) \tag{15.21}$$

When $\gamma = 1-\mu$, the intuitionistic fuzzy set is $f = (\mu, 1-\mu)$ and the complement is $\bar{f} = (1-\mu, \mu)$, which is the same as in the ordinary fuzzy set case. Defining the unit of an intuitionistic fuzzy set as (1, 0) and using the zero defined as (0, 1), properties C1-C4 in section 14.3.3 follow from the properties of the max and min operators.

15.4 Type 2 Fuzzy Sets

A *type 2 fuzzy set* (John 1998; Karnik and Karnik 1998; Mendel 2001; Mizumoto and Tanaka 1976; Yager 1980) has a fuzzy membership function f(x) which is a fuzzy set (sometimes called a *type 1 fuzzy set*) on the unit interval [0 1], rather than a point in the unit interval. Therefore the membership function can be expressed as

$$f(x) = \{x_1, \mu(x_1)\} + \{x_2, \mu(x_2)\} + ... + \{x_n, \mu(x_n)\} = \sum_{i=1}^{n} \{x_i, \mu(x_i)\} \tag{15.22}$$

where the $\mu(x_i)$, i = 1...n are fuzzy set membership functions which may be expressed as linguistic expressions or functions from the unit interval into the unit interval.

Example 15.2: Type 2 Fuzzy Set for Hearing Impairment

Consider the fuzzy set *hearing impairment* defined on the space of people $X_1 = \{x_1, ..., x_7\}$ introduced in section 13.1, with

x_1 – Helga x_2 – Leylah x_3 - Ahmed x_4 – Francois
x_5 – Miriam x_6 – John x_7 - Karen

This fuzzy set can be made into a type 2 fuzzy set by replacing the membership grade of each element by a type 1 fuzzy set. This results in the membership function given in table 15.6.

Table 15.6 Type 2 Fuzzy set *hearing impairment*

X	x_1	x_2	x_3	x_4	x_5	x_6	x_7
f(x)	no hi	sev hi	mild hi	deaf	mod hi	prof hi	deaf

where the abbreviations hi, sev, mod and prof stand for hearing impairment, severe, moderate and profound respectively. For illustrative purposes, the fuzzy sets *mild hearing impairment, moderate hearing impairment* and *profound hearing impairment* are defined in table 15.7. The definition of the fuzzy sets *no hearing impairment* and *deaf* is left as an exercise.

Table 15.7a Fuzzy set *mild hearing impairment*

X	0.1	0.2	0.3	0.4	0.5
μ (x)	0.3	0.7	1	0.6	0.2

Table 15.7b Fuzzy set *moderate hearing impairment*

X	0.3	0.4	0.5	0.6	0.7
μ(x)	0.3	0.6	1	0.6	0.3

Table 15.7c Fuzzy set *profound hearing impairment*

X	0.6	0.7	0.8	0.9	1
μ(x)	0.2	0.6	1	0.7	0.3

In the case of standard (type 1) fuzzy sets there is considerable flexibility in the description of the sets, in terms of the number of different linguistic classifiers used or whether, for instance, the membership grades are expressed to one or two decimal places or to 0.2. The same is true for type 2 fuzzy sets. The number of type 1 fuzzy sets required depends on the application. There are a number of different definitions of union and intersection for type 2 fuzzy sets.

The most widely used definitions are due to Zadeh (1975) and are based on the extension principle. The union of two type 2 fuzzy sets:

$f_A = \{x_1, \mu_A(x_1)\} + \{x_2, \mu_A(x_2)\} + \ldots + \{x_n, \mu_A(x_n)\}$ and

$f_B = \{x_1, \mu_B(x_1)\} + \{x_2, \mu_B(x_2)\} + \ldots + \{x_m, \mu_B(x_m)\}$ has membership function

$$f_{sup}(x) = f_A(x) \vee f_B(x): \; X \rightarrow I \times I \tag{15.23a}$$

$$f_{sup} = \bigvee_i f_i(x) = \sum_{i,j} \left(x_i \vee x_j, \; \mu_A(x_i) \wedge \mu_B(x_j) \right) \tag{15.23b}$$

The intersection of these sets is given by

$$f_{inf}(x) = f_A(x) \wedge f_B(x): \; X \rightarrow I \times I \tag{15.24a}$$

$$f_{inf} = \bigwedge_i f_i(x) = \sum_{i,j} \left(x_i \wedge x_j, \; \mu_A(x_i) \wedge \mu_B(x_j) \right) \tag{15.24b}$$

Similarly the union and intersection of n type 2 fuzzy sets

$f_1 = \{x_1, \mu_1(x_1)\} + \{x_2, \mu_1(x_2)\} + \ldots + \{x_{m_1}, \mu_1(x_{m_1})\}$,

$f_2 = \{x_1, \mu_2(x_1)\} + \{x_2, \mu_2(x_2)\} + \ldots + \{x_{m_2}, \mu_2(x_{m_2})\}, \ldots$,

$f_n = \{x_1, \mu_n(x_1)\} + \{x_2, \mu_n(x_2)\} + \ldots + \{x_{m_n}, \mu_n(x_{m_n})\}$ are given by

$$f_{sup} = \bigvee_i f_i(x) = \sum_{\alpha_1 \ldots \alpha_m} \left(\bigvee_{\alpha_j} x_{\alpha_j}, \; \bigwedge_i \mu_i(x_{\alpha_j}) \right) \tag{15.25}$$

$$f_{inf} = \bigwedge_i f_i(x) = \sum_{\alpha_1 \ldots \alpha_m} \left(\bigwedge_{\alpha_j} x_{\alpha_j}, \; \bigwedge_i \mu_i(x_{\alpha_j}) \right) \tag{15.26}$$

for the union and intersection respectively, with

$$\alpha_1, \ldots, \alpha_m \in \mathcal{N}_n = \{1,2, \ldots, n\}, \quad m = \max_i m \tag{15.27}$$

It should be noted that, with the definitions given, the intersection and union of type 2 fuzzy sets possess the commutative, associative and distributive properties (Mendel 2001). Although programmable expressions have been given for the union and intersection of n type 2 fuzzy sets, it can be easier to use the commutative and associative properties to obtain the union and intersection from the expressions for the union or intersection of two type 2 fuzzy sets. This is illustrated in the life cycle analysis example in section 15.6, with calculation of the union and intersection in sections 15.6.3 and 15.6.4 respectively.

15.5 Example: Evaluation of the Environmental Effects of Manufacturing Processes for Producing Gears

In this section an example from manufacturing industry is considered, in this case for the production of gears. Even for this relatively simple product, there are a number of different possible manufacturing sequences, including the following:

- centering, rotating, milling of gearing, boring, hardening, grinding
- converting, welding, hardening, hard-rotating, grinding.

Fuzzy set techniques can be used to give a measure of the environmental impacts of the manufacturing sequences, both for each of the different stages and overall. In this case, five different impacts will be considered, giving the following space of environmental impacts:

$W = \{x_1, x_2, x_3, x_4, x_5\}$, where

x_1 - environmental overheads x_2 - waste production x_3 - toxicity
x_4 - resource consumption x_5 - energy consumption

Environmental overheads represent the negative impacts on the environment that occur even when no production processes are running. The 6-stage process will be considered here and its environmental impacts analysed using augmented and intuitionistic fuzzy set techniques.

15.5.1 Total Process Impacts Using Augmented Fuzzy Sets

The environmental impacts at each of the six stages (centering, rotating, milling of gearing, boring, hardening and grinding) of the manufacturing process can be represented in tabular form as shown in table 15.8. Here $\mu_{i.j}$ represents the degree of concern or significance of the environmental impact x_j at the ith stage of the manufacturing process and $\nu_{i.j}$ represents the degree of certainty in the assessment. For instance $\mu_{2.3}$ represents the degree of significance of the impact x_3 (toxicity) at the second (rotating) stage of the manufacturing process. Values obtained from a number of experts can then be used to obtain the augmented fuzzy sets for the environmental impacts at the six stages of the manufacturing process. These fuzzy sets are given in table 15.9.

Table 15.8 Augmented fuzzy set for environmental impacts of gear manufacture

X	Environmental overheads x_1	Waste production x_2	Toxicity x_3	Resources x_4	Energy x_5
$\mu_{i.}$	$\mu_{i.1}$	$\mu_{i.2}$	$\mu_{i.3}$	$\mu_{i.4}$	$\mu_{i.5}$
$\nu_{i.}$	$\nu_{i.1}$	$\nu_{i.2}$	$\nu_{i.3}$	$\nu_{i.4}$	$\nu_{i.5}$

Table 15.9a Augmented fuzzy set for centering stage of gear manufacture

X	Environmental Overheads x_1	Waste production x_2	Toxicity x_3	Resources x_4	Energy x_5
μ_1	1	0.9	0	1	0.7
ν_1	0.5	0.3	1	0.5	0.5

Table 15.9b Augmented fuzzy set for rotating stage of gear manufacture

X	Environmental Overheads x_1	Waste production x_2	Toxicity x_3	Resources x_4	Energy x_5
μ_1	1	0.8	0	0.3	0
ν_1	0.5	0.6	0.8	0.5	0.5

Table 15.9c Augmented fuzzy set for milling of gearing stage

X	Environmental Overheads x_1	Waste production x_2	Toxicity x_3	Resources x_4	Energy x_5
μ_1	1	0.5	0	1	0.7
ν_1	0.5	0.8	0.8	0.8	0.8

Table 15.9d Augmented fuzzy set for boring stage of gear manufacture

X	Environmental Overheads x_1	Waste production x_2	Toxicity x_3	Resources x_4	Energy X_5
μ_1	1	0.8	0.7	1	0
ν_1	1	0.5	0.8	0.5	1

Table 15.9e Augmented fuzzy set for hardening stage of gear manufacture

X	Environmental Overheads x_1	Waste production x_2	Toxicity x_3	Resources x_4	Energy x_5
μ_1	1	0.6	0	1	0
ν_1	0.5	0.8	1	0.8	0.8

Table 15.9f Augmented fuzzy set for grinding stage of gear manufacture

X	Environmental Overheads x_1	Waste production x_2	Toxicity x_3	Resources x_4	Energy x_5
μ_1	0.7	0.9	1	0.7	0.7
ν_1	0.5	0.3	0.75	0.5	0.5

The total environmental impact for the manufacturing sequence of each of the five variables x_1 - x_5 can be obtained by aggregating the six augmented fuzzy sets for the impacts at the different manufacturing stages using the augmented fuzzy set union operation defined in section 15.2.1. The conditions (ai), (aii), (aiii), (bi) and (bii) referred to below are stated in this section. From the definitions the values of μ_h, μ_m, μ_u, ν_h, ν_u, ν_u can be obtained and are presented in table 15.10.

Table 15.10 Calculation of total impacts

X	Environmental Overheads x_1	Waste production x_2	Toxicity x_3	Resources x_4	Energy x_5
μ_h	1	0.9	1	1	0.7
ν_h	1	0.3	0.8	0.8	0.8
μ_m	1	0.6	0	1	0
ν_m	1	0.8	1	0.8	1
μ_u	1	0.8	1	1	0.7
ν_u	1	0.6	0.8	0.8	0.8

$f_{sup} = (\mu_{sup}, \nu_{sup})$ does not depend on δ and ε for the impacts x_1, x_3, x_4 and x_5, since condition (aiii) is satisfied for these variables (for all values of δ and ε) and so

$$\mu_{1.sup} = 1, \qquad \nu_{1.sup} = 1$$
$$\mu_{3.sup} = 1, \qquad \nu_{3.sup} = 0.8$$
$$\mu_{4.sup} = 1, \qquad \nu_{4.sup} = 0.8$$
$$\mu_{5.sup} = 0.7, \qquad \nu_{5.sup} = 0.8$$

The effect of varying δ and ε can be investigated for x_2. Conditions (ai) and (aii) are satisfied for $\delta = \varepsilon = 0.1$, giving $\mu_{2.sup} = \mu_{2.u} = 0.8$, $\nu_{2.sup} = \nu_{2.u} = 0.6$.

For $\delta = \varepsilon = 0$, conditions (ai) and (bi) are not satisfied, so $\mu_{2.sup} = \mu_{2.h} = 0.9$, $\nu_{2.sup} = \nu_{2.h} = 0.3$.

For $\delta = 0.3, \varepsilon = 0.4$, conditions (ai), (bi) and (bii), but not (aii) are satisfied, so $\mu_{2.sup} = \mu_{2.m} = 0.6$, $\nu_{2.sup} = \nu_{2.m} = 0.8$.

The results can be combined in a table of total environmental impacts for the whole manufacturing sequence. Table 15.11 gives the total impacts for $\delta = \varepsilon = 0.1$.

Table 15.11 Augmented fuzzy set of total environmental impacts

X	Environmental Overheads x_1	Waste production x_2	Toxicity x_3	Resources x_4	Energy x_5
μ_{sup}	1	0.8	1	1	0.7
ν_{sup}	1	0.6	0.8	0.8	0.8

This table gives a good representation of the different categories of environmental impacts over the whole manufacturing sequence. The approach could be used to help a design team improve the manufacturing processes or choose between different processes to obtain the one with the least environmental impact. Table 5.11 indicates that this manufacturing sequence places a heavy burden on the environment in each impact category with a high degree of certainty, indicating that measures will be required to reduce all categories of impacts. The environmental overheads are particularly high, indicating that the basic structure of the operations needs to be examined, as a high load is being placed on the environment even when nothing is being produced. Augmented fuzzy sets could be used to compare the environmental impacts for the five and six stage manufacturing processes.

15.5.2 Total Process Impacts: Intuitionistic Fuzzy Sets

The five environmental impacts at each of the six stages of the manufacturing process can be represented in tabular form in table 15.12.

Table 15.12a Intuitionistic fuzzy set for centering stage of gear manufacturing

X	Environmental overheads x_1	Waste production x_2	Toxicity x_3	Resources x_4	Energy x_5
μ_1	1	0.9	0	1	0.7
γ_1	0	0	0.8	0	0.2

Table 15.12b Intuitionistic fuzzy set for rotating stage of gear manufacturing

X	Environmental overheads x_1	Waste production x_2	Toxicity x_3	Resources x_4	Energy x_5
μ_2	1	0.8	0	0.3	0
γ_2	0	0	0.7	0.4	0.6

Table 15.12c Intuitionistic fuzzy set for milling of gearing stage

X	Environmental overheads x_1	Waste production x_2	Toxicity x_3	Resources x_4	Energy x_5
μ_3	1	0.5	0	1	0.7
γ_3	0	0.5	0.7	0	0.3

Table 15.12d Intuitionistic fuzzy set for boring stage of gear manufacturing

X	Environmental overheads x_1	Waste production x_2	Toxicity x_3	Resources x_4	Energy x_5
μ_4	1	0.8	0.7	1	0
γ_4	0	0	0.3	0	0.7

Table 15.12e Intuitionistic fuzzy set for hardening stage of gear manufacturing

X	Environmental overheads x_1	Waste production x_2	Toxicity x_3	Resources x_4	Energy x_5
μ_5	1	0.6	0	1	0
γ_5	0	0.3	1	0	0.8

Table 15.12f Intuitionistic fuzzy set for grinding stage of gear manufacturing

X	Environmental overheads x_1	Waste production x_2	Toxicity x_3	Resources x_4	Energy x_5
μ_6	0.7	0.9	1	0.7	0.7
γ_6	0.2	0	0	0.2	0.2

The total environmental impact for the manufacturing sequence in each of the five categories can be obtained by aggregating these results using the union operator for intuitionistic fuzzy sets defined in section 15.3.1. The resulting impacts for the total process are given in table 15.13.

Table 15.13 Intuitionistic fuzzy set of total impacts of gear manufacturing

X	Environmental overheads x_1	Waste production x_2	Toxicity x_3	Resources x_4	Energy x_5
μ_{sup}	1	0.9	1	1	0.7
γ_{sup}	0	0	0	0	0.2
π_{sup}	0	0.1	0	0	0.1

This result is fairly similar to that obtained in the augmented fuzzy set case in table 15.14, except that it gives a measure of the consistency rather than the degree of certainty of the information. It again shows that all the impacts are high, this time with a high degree of consistency rather than certainty, as all the values of π are small.

15.5.3 The Extent to which Impacts hold over all the Manufacturing Stages

In some cases knowledge of whether or to what extent several environmental impacts hold at a particular life cycle stage or one impact holds over all the life cycle or process stages is required. This can be obtained using the appropriate fuzzy set intersection operator.

In this example the extent to which all of the impacts are present at all life cycle stages will be considered. Values will be obtained first for the second order augmented fuzzy set. The values of μ_g, μ_y, μ_n, ν_g, ν_y and ν_n defined in section 15.2.2 are given in table 15.13. The minimising values

Table 5.14 Calculation: impacts over all manufacturing stages

	Environmental overheads x_1	Waste production x_2	Toxicity x_3	Resources x_4	Energy x_5
μ_g	0.7	0.5	0	0.3	0
ν_g	0.5	0.8	0.8	0.5	0.5
μ_n	0.7	0.9	0	0.3	0
ν_n	0.5	0.3	0.8	0.5	0.5
μ_y	0.7	0.9	0	0.3	0
ν_y	0.5	0.3	0.8	0.5	0.5

of μ_y and ν_y for x_3 have been obtained from minimising the product, as the zero value for the pair (0,0.8) is obtained with ν taking its minimum value.

From the table it can easily by seen that for all values of δ and ε

$\mu_{1.inf} = 0.7$ $\nu_{1.inf} = 0.5$ $\mu_{3.inf} = 0$ $\nu_{3.inf} = 0.8$

$\mu_{4.inf} = 0.3$ $\nu_{4.inf} = 0.5$ $\mu_{5.inf} = 0$ $\nu_{5.inf} = 0.5$

For x_2 $\mu_n = \mu_y = 0.9$, $\nu_n = \nu_y = 0.3$ and $\mu_g = 0.5$, $\nu_g = 0.8$, so conditions (a) and (b) in section 15.2.2 are not satisfied for $\delta < 0.4$ or $\varepsilon < 0.5$. Therefore $\mu_{2.inf} = 0.9$, $\mu_{2.inf} = 0.3$. The impacts over all manufacturing stages are presented in table 15.15. The first parameter μ represents the degree to which the impacts are present at all life cycle stages and the second parameter ν represents a lower bound on the degree of certainty with which the impacts are known to be present at all life cycle stages.

Table 15.15 Augmented fuzzy set for impacts over all manufacturing stages

X	Environmental overheads x_1	Waste production x_2	Toxicity x_3	Resources x_4	Energy x_5
μ_{inf}	0.7	0.9	0	0.3	0
ν_{inf}	0.5	0.3	0.8	0.5	0.5

The intuitionistic fuzzy set case will be considered now. In this case the results in table 15.16 are obtained from the definitions (15.14) in section 15.3.1 and give a lower bound on the extent to which all the impacts are present, an upper bound on the extent to which they are not present and an approximate value for the degree of consistency.

Table 15.16 Intuitionistic fuzzy set for impacts over all manufacturing stages

X	Environmental overheads x_1	Waste production x_2	Toxicity x_3	Resources x_4	Energy x_5
μ_{inf}	0.7	0.5	0	0.3	0
γ_{inf}	0.2	0.5	1	0.4	0.8
π_{inf}	0.1	0.0	0	0.3	0.2

There is again a high degree of consistency, though slightly lower than for the aggregation of the impacts over all stages of the manufacturing process. Both types of fuzzy sets give similar, though not identical values for the impacts over all stages of the manufacturing process. Intuitionistic fuzzy sets can also be used to help a design team reduce the environmental impacts of a particular process or compare the impacts of different processes for the same product.

15.6 Example: Life Cycle Analysis in Computer Industry

In this section the environmental impacts of computer production and use will be considered, using augmented, intuitionistic and type 2 fuzzy set techniques. The space of environmental impacts of importance in the computer industry is given by (Hersh 1998):

$$EI = \{x_1, x_{2,\dots,} x_8\} \quad \text{where}$$

x_1 - energy consumption
x_2 - resource consumption
x_3 - landfill exhaustion
x_4 - local air impacts
x_5 - water impacts
x_6 - atmospheric impacts
x_7 - waste
x_8 - forest destruction

The computer life cycle can be divided into the following six life cycle stages:

- material acquisition
- material processing
- manufacture
- transport and distribution
- use and reuse
- disposal and recycling

15.6.1 Total Life Cycle Impact with Augmented Fuzzy Sets

For each of the six life cycle stages a fuzzy set can be defined to give the extent of the environmental impact at that life cycle stage, as shown in table 15.17.

Table 15.17 Augmented fuzzy sets for the impacts at the six life cycle stages

X	Energy x_1	Resources x_2	Land-fill x_3	Local air x_4	Water x_5	Atmos-phere x_6	Waste x_7	Forest loss x_8
μ_i	$\mu_{i.1}$	$\mu_{i.2}$	$\mu_{i.3}$	$\mu_{i.4}$	$\mu_{i.5}$	$\mu_{i.6}$	$\mu_{i.7}$	$\mu_{i.8}$
ν_i	$\nu_{i.1}$	$\nu_{i.2}$	$\nu_{i.3}$	$\nu_{i.4}$	$\nu_{i.5}$	$\nu_{i.6}$	$\nu_{i.7}$	$\nu_{i.8}$

Here $\mu_{i,j}$ and $v_{i,j}$ represent the degree of concern or severity of the environmental impact x_j at the ith life cycle stage and the degree of certainty in the assessment respectively. The fuzzy sets LC_i (i = 1...6) can be obtained for the impacts at the six life cycle stages, as shown in table 15.18.

Table 15.18a Augmented fuzzy set for material acquisition

X	x_1	x_2	x_3	x_4	x_5	x_6	x_7	x_8
μ_1	1	1	0	1	0.66	0.66	1	0.66
v_1	0.50	0.50	1	0.50	0.50	0.50	0.50	0.50

Table 15.18b Augmented fuzzy set for material processing

X	x_1	x_2	x_3	x_4	x_5	x_6	x_7	x_8
μ_2	1	1	0	0.33	0	0.33	0.33	0
v_2	0.50	0.50	0.75	0.50	0.50	0.50	0.50	0.75

Table 15.18c Augmented fuzzy set for manufacture

X	x_1	x_2	x_3	x_4	x_5	x_6	x_7	x_8
μ_3	1	1	0	1	0.66	0.66	0.66	0
v_3	0.50	0.50	0.75	0.75	0.75	0.75	0.75	0.75

Table 15.18d Augmented fuzzy set for transport and distribution

X	x_1	x_2	x_3	x_4	x_5	x_6	x_7	x_8
μ_4	1	0.66	0	1	0	1	1	1
v_4	0.50	0.50	1	0.75	0.75	0.75	0.50	0.25

Table 15.18e Augmented fuzzy set for use and reuse

X	x_1	x_2	x_3	x_4	x_5	x_6	x_7	x_8
μ_5	1	1	0.66	1	0	0.66	1	1
v_5	1	0.75	0.75	0.50	1	0.50	1	0.75

Table 15.18f Augmented fuzzy set for disposal and recycling

X	x_1	x_2	x_3	x_4	x_5	x_6	x_7	x_8
μ_6	0.66	1	1	0.66	0.66	0.33	0.66	0
v_6	0.50	0.75	0.75	0.50	0.5	0.50	0.50	1

The overall environmental impact can be obtained by combining these fuzzy sets, using the fuzzy union operator defined in section 15.2.1, and stated in table 15.19. Condition (aiii) is satisfied for all values of δ and ε, as the life cycle stage with maximum impact for each variable also has the one the greatest product of impact and degree of certainty. A coarse scale was used here for the values of the $\mu_{i.j}$ and $v_{i.j}$ due to the scale of the available data. However a finer scale should be used when this will not give a false impression of the accuracy of the data, since it allows greater discrimination between the values of different impacts.

Table 15.19 Overall environmental impacts

X	Energy x_1	Resources x_2	Landfill x_3	Local air x_4	Water x_5	Atmosphere x_6	Waste x_7	Forest loss x_8
μ_{sup}	1	1	1	1	0.66	1	1	1
v_{sup}	1	0.75	0.75	0.75	0.75	0.75	1	0.75

15.6.2 Total Life Cycle Impact using Intuitionistic Fuzzy Sets

The eight environmental impacts at each of the six stages of the manufacturing process can be represented in tabular form as shown in table 15.20.

Table 15.20a Intuitionistic fuzzy set for material acquisition

X	Energy x_1	Resources x_2	Landfill x_3	Local air x_4	Water x_5	Atmosphere x_6	Waste x_7	Forest loss x_8
μ_{sup}	1	1	0	1	0.66	0.66	1	0.66
v_{sup}	0	0	1	0	0.2	0.2	0	0.2

Table 15.20b Intuitionistic fuzzy set for material processing

X	x_1	x_2	x_3	x_4	x_5	x_6	x_7	x_8
μ_2	1	1	0	0.33	0	0.33	0.33	0
γ_2	0	0	0.8	0.3	0.5	0.5	0.4	0.8

Table 15.20c Intuitionistic fuzzy set for manufacture

X	x_1	x_2	x_3	x_4	x_5	x_6	x_7	x_8
μ_3	1	1	0	1	0.66	0.66	0.66	0
γ_3	0	0	0.6	0	0.3	0.2	0.2	0.6

Table 15.20d Intuitionistic fuzzy set for transport and distribution

X	x_1	x_2	x_3	x_4	x_5	x_6	x_7	x_8
μ_4	1	0.66	0	1	0	1	1	1
γ_4	0	0.2	0.9	0	0.6	0	0	0

Table 15.20e Intuitionistic fuzzy set for use and reuse

X	x_1	x_2	x_3	x_4	x_5	x_6	x_7	x_8
μ_5	1	1	0.66	1	0	0.66	1	1
γ_5	0	0	0.2	0	0.7	0.2	0	0

Table 15.20f Intuitionistic fuzzy set for disposal and recycling

X	x_1	x_2	x_3	x_4	x_5	x_6	x_7	x_8
μ_6	0.66	1	1	0.66	0.66	0.33	0.66	0
γ_6	0.2	0	0	0.2	0.2	0.5	0.2	0.9

The total environmental impact over the manufacturing sequence for each of the eight variables can be obtained by aggregating these results using the union operator for intuitionistic fuzzy sets presented in section 15.3.1. This gives the impacts for the total process shown in table 15.21.

Table 15.21 Environmental impacts for the whole process

X	Ener-gy x_1	Resou-rces x_2	Land-fill x_3	Local air x_4	Water x_5	Atmos-phere x_6	Waste x_7	Forest loss x_8
μ_{sup}	1	1	1	1	0.66	1	1	1
ν_{sup}	0	0	0	0	0.2	0	0	0
π_{sup}	0	0	0	0	0.14	0	0	0

The results are again similar to those obtained in the augmented fuzzy set case, with high impacts in all categories and a high degree of consistency.

15.6.3 Total Life Cycle Impact using Type 2 Fuzzy Sets

The eight environmental impacts at each of the six stages of the manufacturing process can be represented in tabular form as shown in table 15.22, with the associated type 1 fuzzy sets defined in table 15.23.

Table 15.22 Type 2 fuzzy sets for the environmental impacts

X	x_1	x_2	x_3	x_4	x_5	x_6	x_7	x_8
μ_1	pos sig	pos sig	none	pos sig	pos mod	pos mod	pos sig	pos mod

X	x_1	x_2	x_3	x_4	x_5	x_6	x_7	x_8
μ_2	pos sig	pos sig	prob none	pos min	pos none	pos min	pos min	prob none

X	x_1	x_2	x_3	x_4	x_5	x_6	x_7	x_8
μ_3	pos sig	pos sig	prob none	prob sig	prob mod	prob mod	prob mod	prob none

X	x_1	x_2	x_3	x_4	x_5	x_6	x_7	x_8
μ_4	pos sig	pos mod	none	prob sig	prob none	prob sig	pos sig	vps sig

X	x_1	x_2	x_3	x_4	x_5	x_6	x_7	x_8
μ_5	sig	prob sig	prob mod	pos sig	none	pos mod	sig	prob sig

X	x_1	x_2	x_3	x_4	x_5	x_6	x_7	x_8
μ_6	pos mod	prob sig	prob sig	pos mod	pos mod	pos min	pos mod	none

Table 15.23a Type 1 fuzzy set: *possibly none* (pos none)

X	0	0.1	0.2
μ	1.0	0.7	0.3

Table 15.23b Type 1 fuzzy set: *probably none* (prob none)

X	0	0.1
μ	1.0	0.5

Table 15.23c Type 1 fuzzy set: *none*

X	0
μ	1.0

Table 15.23d Type 1 fuzzy set: *possibly minor* (pos min)

X	0.1	0.2	0.3	0.4	0.5	0.6
μ	0.2	0.5	1	0.7	0.4	0.1

Table 15.23e Type 1 fuzzy set: *minor* (min)

X	0.3	0.4
μ	1	0.5

Table 15.23f Type 1 fuzzy set: *possibly moderate* (pos mod)

X	0.4	0.5	0.6	0.7	0.8	0.9
μ	0.3	0.6	0.7	1	0.5	0.2

Table 15.23g Type 1 fuzzy set: *probably moderate* (prob mod)

X	0.5	0.6	0.7	0.8
μ	0.3	0.6	1	0.4

Table 15.23h Type 1 fuzzy set: *moderate* (mod)

X	0.6	0.7
μ	0.6	1

Table 15.23i Type 1 fuzzy set: *very possibly significant* (vps sig)

X	0.7	0.8	0.9	1
μ	0.2	0.6	0.7	1

Table 15.23j Type 1 fuzzy set: *possibly significant* (pos sig)

X	0.8	0.9	1
μ	0.4	0.7	1

Table 15.23k Type 1 fuzzy set: *probably significant* (prob sig)

X	0.9	1
μ	0.5	1

Table 15.23l Type 1 fuzzy set: *significant* (sig)

X	1
μ	1

The total environmental impact over the manufacturing sequence for each of the eight variables can be obtained by aggregating these results using the union operator for type 2 fuzzy sets presented in equation (15.25) in section 15.4. To illustrate the methodology full working is given for the first three environmental impacts. In the case of x_1 there are three distinct linguistic values, namely possibly significant, significant and moderate, over the six life cycle stages and so

$$f_{sup}(x_1) = \text{possibly significant V significant V possibly moderate}$$

possibly significant V significant

$$= \{(0.8,0.4) + (0.9,0.7) + (1,1)\} \text{ V } (1,1)$$
$$= (\max(0.8,1), \min(0.4,1)) + (\max(0.9,1), \min(0.7,1))$$
$$+ (\max(1,1), \min(1,1))$$
$$= (1,0.4) + (1,0.7) + (1,1) = (1,1)$$

$$\therefore f_{sup}(x_1) = (1,1)$$
$$\text{V}\{(0.4,0.3)+(0.5,0.6)+(0.6,0.7)+(0.7,1)+(0.8,0.5)+(0.9, 0.2)\}$$
$$= (1,1)$$

$$f_{sup}(x_2) = \text{possibly significantVpossibly moderate V probably significant}$$

possibly significant V possibly moderate

$$= \{(0.8,0.4) + (0.9,0.7) + (1,1)\}$$
$$\text{V } \{0.4,0.3) + (0.5,0.6) + (0.6,0.7) + (0.7,1) + (0.8,0.5) + (0.9, 0.2)\}$$
$$= (\max(0.8,0.4), \min(0.4,0.3)) + (\max(0.8,0.5), \min(0.4,0.6))$$
$$+ (\max(0.8,0.6), \min(0.4,0.7)) + (\max(0.8,0.7), \min(0.4,1))$$
$$+ (\max(0.8,0.8), \min(0.4,0.5)) + (\max(0.8,0.9), \min(0.4,0.2))$$
$$+ (\max(0.9,0.4), \min(0.7,0.3)) + (\max(0.9,0.5), \min(0.7,0.6))$$
$$+ (\max(0.9,0.6), \min(0.7,0.7)) + (\max(0.9,0.7), \min(0.7,1))$$
$$+ (\max(0.9,0.8), \min(0.7,0.5)) + (\max(0.9,0.9), \min(0.7,0.2))$$
$$+ (\max(1,0.4), \min(1,0.3)) + (\max(1,0.5), \min(1,0.6))$$
$$+ (\max(1,0.6), \min(1,0.7)) + (\max(1,0.7), \min(1,1))$$
$$+ (\max(1,0.8), \min(1,0.5)) + (\max(1,0.9), \min(1,0.2))$$

$$\begin{aligned}\Rightarrow f_{sup}(x_2) &= (0.8,0.3)+(0.8,0.4)+(0.8,0.4)+(0.8,0.4)+(0.8,0.4)\\ &\quad+(0.9,0.2)+(0.9,0.3)+(0.9,0.6)+(0.9,0.7)+(0.9,0.7)\\ &\quad+(0.9,0.5)+(0.9,0.2)+(1,0.3)+(1,0.6)+(1,0.7)\\ &\quad+(1,1)+(1,0.5)+(1,0.2)\\ &= (0.8,0.4)+(0.9,0.7)+(1,1)\end{aligned}$$

$$\begin{aligned}\therefore f_{sup}(x_2) &= \{(0.8,0.4)+(0.9,0.7)+(1,1)\} \vee \{(0.9,0.5)+(1,1)\}\\ &= (\max(0.8,0.9), \min(0.4,0.5)) + (\max(0.9,0.9), \min(0.7,0.5))\\ &\quad+(\max(1,0.9), \min(1,0.5)) + (\max(0.8,1), \min(0.4,1))\\ &\quad+(\max(0.9,1), \min(0.7,1) + (\max(1,1), \min(1,1))\\ &= (0.9,0.4)+(0.9,0.5)+(1,0.5)+(1,0.4)+(1,0.7)+(1,1)\\ &= (0.9,0.5)+(1,1)\end{aligned}$$

$f_{sup}(x_3)$ = none V probably none V probably mod. V probably signific.

$$\begin{aligned}\text{none V probably none} &= (0,1) \vee \{(0,1)+(0.1,0.5)\}\\ &= (0,1)+(0.1,0.5)\end{aligned}$$

probably moderate V probably significant

$$\begin{aligned}&= \{(0.5,0.3)+(0.6,0.6)+(0.7,1)+(0.8,0.4)\} \vee (\{(0.9,0.5)+(1,1)\}\\ &= (\max(0.5,0.9), \min(0.3,0.5)) + (\max(0.5,1), \min(0.3,1))\\ &\quad+(\max(0.6,0.9), \min(0.6,0.5)) + (\max(0.6,1), \min(0.6,1))\\ &\quad+(\max(0.7,0.9), \min(1,0.5)) + (\max(0.7,1), \min(1,1))\\ &\quad+(\max(0.8,0.9), \min(0.4,0.5)) + (\max(0.8,1), \min(0.4,1))\\ &= (0.9,0.3)+(1,0.3)+(0.9,0.5)+(1,0.6)+(0.9,0.5)+(1,1)+(0.9,0.4)\\ &\quad+(1,0.4)\\ &= (0.9,0.5)+(1,1)\end{aligned}$$

$$\begin{aligned}\therefore f_{sup}(x_3) &= \{(0,1)+(0.1,0.5)\} \vee \{(0.9,0.5)+(1,1)\}\\ &= (\max(0,0.9), \min(1,0.5)) + (\max(0,1), \min(1,1))\\ &\quad+(\max(0.1,0.9), \min(0.5,0.5)) + (\max(0.1,1), \min(0.5,1))\\ &= (0.9,0.5)+(1,1)\end{aligned}$$

The total impacts over the six life cycle stages can be calculated similarly for the other environmental impacts to give the following expressions:

$f_{sup}(x_4)$ = (0.9,0.5) + (1,1)
$f_{sup}(x_5)$ = (0.5,0.3) + (0.6,0.6) + (0.7,1) + (0.8,0.5) + (0.9,0.2)
$f_{sup}(x_6)$ = (0.9,0.5) + (1,1)
$f_{sup}(x_7)$ = (0.8,0.4) +(0.9, 0.7) +(1,0.1)
$f_{sup}(x_8)$ = (0.9,0.5) + (1,1)

Analogously to the intuitionistic and augmented fuzzy set analysis, type 2 fuzzy set analysis has shown that the impacts over the six life cycle stages are high with a high degree of certainty for all impact categories except water impacts (x_5), which take a moderately high value. The relatively low spread of values (0.8 to 1 for most impact categories) indicates a high degree of certainty in the values.

Therefore all three approaches give similar results of high impacts in each category over the total life cycle with a high degree of certainty in the augmented and type 2 fuzzy set case and a high degree of consistency in the intuitionistic fuzzy set case. In the augmented fuzzy set case the degree of certainty is expressed as a second parameter on [0, 1], whereas in the type 2 fuzzy set case it is implicitly expressed by the spread of the non-zero membership grades, with a small spread equivalent to a high degree of certainty.

15.6.4 Total Impacts over Each Life Cycle Stage

Next the extent to which all the impacts are present in each life cycle stage will be considered using the three techniques. The notation LC_i (i=1...6) will be used to denote the six life cycle stages. The second order augmented fuzzy set case well be considered first. The augmented fuzzy sets are given in table 15.18 in section 15.6.1. From the definitions in section 15.2.2 the values of μ_g, μ_n, μ_y, ν_g, ν_n, ν_y can be obtained and are presented in table 15.24.

Table 15.24 Values of μ_g, ν_g, μ_n, ν_n, μ_y and ν_y

	LC_1	LC_2	LC_3	LC_4	LC_5	LC_6
μ_g	0	0	0	0	0	0
ν_g	1	0.50	0.75	0.75	1	1
μ_n	0.66	0	1	1	0.66	0.33
ν_n	0.50	0.50	0.50	0.25	0.50	0.50
μ_y	0	0	0	0	0	0.33
ν_y	1	0.50	0.75	0.75	1	0.50

For the life cycle stage LC_2 the values of μ_y and ν_y are obtained by minimising the product of μ and ν, as the minimum product of zero is obtained for μ equal to its minimising value μ_g. In the case of the other five life cycle stages the minimum value of the product $\mu\nu$ is again zero, but this value is not obtained for the minimising values of either μ or ν. Therefore the values of μ_y and ν_y are obtained by minimising the sum of μ and ν. However, minimising the product and minimising the sum give the same result for the life cycle stages LC_1, LC_3, LC_4 and LC_5.

The results for $LC_1 - LC_5$ are independent of δ and ε, as $\mu_g = \mu_y$ and $\nu_g = \nu_y$ for these life cycle stages. For LG_6, conditions (a1) and (b1) require $\delta \geq 0.33$. In view of the spread of values of μ_5, this value of δ is not particularly large, giving $\mu_{6.inf} = 0.33$, $\nu_{6.inf} = 0.5$. The extent to which all the impacts hold over each life cycle stage can now be stated in table 15.25.

Table 15.25 Total augmented fuzzy set impact for each life cycle stage

X	LC_1	LC_2	LC_3	LC_4	LC_5	LC_6
μ_{inf}	0	0	0	0	0	0.33
ν_{inf}	1	0.50	0.75	0.75	1	0.5

The intuitionistic fuzzy set case will be considered now, using the definitions in section 15.3.1. The intuitionistic fuzzy sets for the problem are presented in table 15.20. In this case the results in table 15.26 give a lower bound on the extent to which all the impacts are present, an upper bound on the extent to which they are not present and an approximate value for the degree of consistency.

Table 15.26 Total intuitionistic fuzzy set impact for each life cycle stage

X	LC_1	LC_2	LC_3	LC_4	LC_5	LC_6
μ_{inf}	0	0	0	0	0	0
γ_{inf}	1	0.8	0.6	0.9	0.7	0.9
π_{inf}	0	0.2	0.4	0.1	0.3	0.1

The type 2 fuzzy set case will be considered now. The type 2 fuzzy sets for the problem are presented in tables 15.22 and 15.23.

$f_{inf}(LC_1)$ = possibly significant Λ none Λ possibly moderate possibly significant Λ none $= \{(0.8,0.4) + (0.9,0.7) + (1,1)\}\ \Lambda\ (0,1)$

$= (\min(0.8,0), \min(0.4,1)) + (\min(0.9,0), \min(0.7,1))$
$\quad + (\min(1,0), \min(1,1))$

$= (0,0.4) + (0,0.7) + (0,1) = (0,1)$

$\therefore\ f_{inf}(LC_1) = (0,1)$

$\Lambda\{(0.4,0.3) + (0.5,0.6) + (0.6,0.7) + (0.7,1) + (0.8,0.5) + (0.9, 0.2)\}$

$= (\min(0,0.4), \min(1,0.3)) + (\min(0,0.5), \min(1,0.6))$
$\quad + (\min(0,0.6), \min(1,0.7)) + (\min(0,0.7)\ \min(1,1))$
$\quad + (\min(0,0.8), \min(1,0.5)) + (\min(0,0.9)\ \min(1,0.2))$

$= (0,0.3) + (0,0.6) + (0,0.7) + (0,1) + (0,0.5) + (0,0.2)$

$= (0,1)$

$f_{inf}(LC_2)$ = possibly significant Λ probably none Λ possibly minor Λ possibly none

possibly significant Λ probably none

$= \{(0.8,0.4) + (0.9,0.7) + (1,1)\}\ \Lambda\ \{(0,1) + (0.1,0.5)\}$

$= (\min(0.8,0), \min(0.4,1)) + (\min(0.8,0.1), \min(0.4,0.5))$
$\quad + (\min(0.9,0), \min(0.7,1)) + (\min(0.9,0.1), \min(0.7,0.5))$
$\quad + (\min(1,0), \min(1,1)) + (\min(1,0.1), \min(1,0.5))$

$= (0,0.4) + (0.1,0.4) + (0,0.7) + (0.1,0.5) + (0,1) + (0.1,0.5)$

$= (0,1) + (0.1,0.5)$

possibly minor Λ possibly none

$= \{(0.1,0.2) + (0.2,0.5) + (0.3,1) + (0.4,0.7) + (0.5,0.4) + (0.6,0.1)\}$
$\quad \Lambda\ \{(0,1) + (0.1,0.7) + (0.2,0.3)\}$

$= (\min(0.1,0), \min(0.2,1)) + (\min(0.1,0.1), \min(0.2,0.7))$
$\quad + (\min(0.1,0.2), \min(0.2,0.3)) + (\min(0.2,0), \min(0.5,1))$
$\quad + (\min(0.2,0.1), \min(0.5,0.7)) + (\min(0.2,0.2), \min(0.5,0.3))$
$\quad + (\min(0.3,0), \min(1,1)) + (\min(0.3,0.1), \min(1,0.7))$
$\quad + (\min(0.3,0.2), \min(1,0.3)) + (\min(0.4,0), \min(0.7,1))$
$\quad + (\min(0.4,0.1), \min(0.7,0.7)) + (\min(0.4,0.2), \min(0.7,0.3))$
$\quad + (\min(0.5,0), \min(0.4,1)) + (\min(0.5,0.1), \min(0.4,0.7))$
$\quad + (\min(0.5,0.2), \min(0.4,0.3)) + (\min(0.6,0), \min(0.1,1))$
$\quad + (\min(0.6,0.1), \min(0.1,0.7)) + (\min(0.6,0.2), \min(0.1,0.3))$

$\therefore$ possibly minor Λ possibly none

$$= (0,0.2) + (0.1,0.2) + (0.1,0.2) + (0,0.5) + (0.1,0.5) + (0.2,0.3) + (0,1) + (0.1,0.7) + (0.2,0.3) + (0,0.7) + (0.1,0.7) + (0.2,0.3) + (0,0.4) + (0.1,0.4) + (0.2,0.3) + (0,0.1) + (0.1,0.1) + (0.2,0.1)$$

$$= (0,1) + (0.1,0.7) + (0.2,0.3)$$

$$\therefore\ f_{inf}(LC_2) = \{(0,1) + (0.1,0.5)\}\ \Lambda\ \{(0,1) + (0.1,0.7) + (0.2,0.3)\}$$

$$= (\min(0,0), \min(1,1)) + (\min(0,0.1), \min(1,0.7)) + (\min(0,0.2), \min(1,0.3)) + (\min(0.1,0), \min(0.5,1)) + (\min(0.1,0.1), \min(0.5,0.7)) + (\min(0.1,0.2), \min(0.5,0.3))$$

$$= (0,1) + (0,0.7) + (0,0.3) + (0,0.5) + (0.1,0.5) + (0.1,0.3)$$

$$= (0,1) + (0.1,0.5)$$

$f_{inf}(LC_3)$ = possibly significant Λ probably none Λ probably significant Λ probably moderate

probably significant Λ probably moderate

$$= (\{(0.9,0.5) + (1,1)\}\ \Lambda\ \{(0.5,0.3) + (0.6,0.6) + (0.7,1) + (0.8,0.4)\}$$

$$= (\min(0.9,0.5), \min(0.5,0.3)) + (\min(0.9,0.6), \min(0.5,0.6)) + (\min(0.9,0.7), \min(0.5,1)) + (\min(0.9,0.8), \min(0.5,0.4)) + (\min(1,0.5), \min(1,0.3)) + (\min(1,0.6), \min(1,0.6)) + (\min(1,0.7), \min(1,1)) + (\min(1,0.8), \min(1,0.4))$$

$$= (0.5,0.3) + (0.6,0.5) + (0.7,0.5) + (0.8,0.4) + (0.5,0.3) + (0.6,0.6) + (0.7,1) + (0.8,0.4)$$

$$= (0.5,0.3) + (0.6,0.6) + (0.7,1) + (0.8,0.4)$$

$$\therefore\ f_{inf}(LC_3) = \{(0,1) + (0.1,0.5)\}\ \Lambda\ \{((0.5,0.3) + (0.6,0.6) + (0.7,1) + (0.8,0.4)\}$$

$$= (\min(0,0.5), \min(1,0.3)) + (\min(0,0.6), \min(1,0.6)) + (\min(0,0.7), \min(1,1)) + (\min(0,0.8), \min(1,0.4)) + (\min(0.1,0.5), \min(0.5,0.3)) + (\min(0.1,0.6), \min(0.5,0.6) + (\min(0.1,0.7), \min(0.5,1)) + (\min(0.1,0.8), \min(0.5,0.4))$$

$$= (0,0.3) + (0,0.6) + (0,1) + (0,0.4) + (0.1,0.3) + (0.1,0.5) + (0.1,0.5) + (0.1,0.4)$$

$$= (0,1) + (0.1,0.5)$$

The extent to which each impact holds over the six life cycle stages can be calculated similarly for the other categories of impact to give the following expressions:

$f_{inf}(LC_4) = (0,1)$

$f_{inf}(LC_5) = (0,1)$

$f_{inf}(LC_6) = (0,1)$

Thus all three of the fuzzy set approaches indicate that the extent to which all the impacts hold over any of the life cycle stages is zero or minimal. However this does not contradict the previously obtained result that most of the total impacts over all life cycle stages are very high. The example also illustrates the fact that the calculation of the total impact is quickest for intuitionistic fuzzy sets and slowest for type 2 fuzzy sets.

15.7 Roundup of Chapter 15

Chapter 15 has presented three different approaches to the problem of representing data which is both imprecise (fuzzy) and uncertain or inaccurate, namely augmented, intuitionistic and type 2 fuzzy sets.

Both augmented and intuitionistic fuzzy sets have two parameters, the first of which is the membership function representing the fuzzy data. In the case of augmented fuzzy sets the second parameter represents the degree of certainty with which the data is known, whereas the second parameter in the intuitionistic fuzzy set case is the non-membership of the variable in the fuzzy set. Both parameters in the augmented and intuitionistic fuzzy set cases take values between 0 and 1, and their sum is also between 0 and 1 in the intuitionistic fuzzy set case. The intuitionistic index is the difference between the sum of the two parameters and 1 and gives a measure of the inconsistency of the information. A type 2 fuzzy set has a membership function which is itself a fuzzy set (sometimes called a type 1 fuzzy set) rather than taking values between 0 and 1. Union, intersection and complement operators have been given for all three types of fuzzy set.

Chapter 16 builds on the material presented in this chapter through the presentation of comparison rules for augmented fuzzy sets and the definition of third order augmented fuzzy sets.

16 Augmented Fuzzy Set Ordering Algorithm and Third Order Augmented Fuzzy Sets

16.0 Learning Objectives

Due to the additional parameter it is not possible to compare the values of two different augmented and intuitionistic fuzzy sets simply using standard inequalities and therefore comparison algorithms are required. Chapter 16 discusses augmented fuzzy set comparison rules and illustrates them by application to strategically targeted impact reduction in the computer industry. It also introduces third order augmented fuzzy sets, including those with negative coefficients, which have a particular role in sustainable decision making. Specific learning objectives include the following:

- Knowledge of fuzzy set comparison rules for augmented and third order augmented fuzzy sets
- The ability to apply fuzzy set comparison rules to the strategically targeted impact reduction of products and processes.
- The ability to use augmented fuzzy sets to evaluate the effects of impact reduction strategies.
- Knowledge of third order augmented fuzzy sets, including with a negative first coefficient.
- The ability to apply third order augmented fuzzy sets with the first term having a negative coefficient to the representation of costs and benefits in sustainable development problems.

16.1 Introduction

This chapter further extends the techniques for modelling imprecision and uncertainty presented in chapter 15. In particular section 16.2 presents an ordering algorithm for augmented fuzzy sets which can be used to obtain a strategy for targeting the reduction of the environmental and social impacts of industrial processes. This algorithm is illustrated by application to computer life cycle analysis. Third order augmented fuzzy sets which can be used to represent the priority or significance of information in addition to its imprecise value and degree of certainty are presented in section 16.3. Third order augmented fuzzy sets have particular applications in sustainable decision making problems. They are extended in section 16.4 to third order augmented fuzzy sets which can take negative as well as positive coefficients. These fuzzy sets can be applied to representing costs and benefits or advantages and disadvantages in sustainable design and decision making. A summary of the chapter is presented in section 16.5.

16.2 Augmented Fuzzy Set Ordering Algorithm

16.2.1 Introduction

It is frequently useful to be able to order augmented or intuitionistic fuzzy sets. For instance, in order to improve the environmental and social aspects of design it is necessary to identify the impacts and the life cycle stages which are the most damaging. This allows effort to be concentrated on minimising the most damaging impacts at the appropriate life cycle stages. A strategy for targeting impact reduction at different life cycle stages can be obtained from an ordering of all the impacts over all the life cycle stages (Hersh 1999ab). This section will present an algorithm for ordering augmented fuzzy sets.

Ordinary inequality relations can often be used to order ordinary fuzzy sets, but cannot be used to order augmented (or intuitionistic) fuzzy sets as the degree of certainty (or non-membership) has to be taken into account. For instance variables or impacts with low degrees of certainty may in actual fact have membership grades very different from the given values. This can lead to changes in the order of severity of the impacts, so that, for instance, what seems a moderate impact with low degree of certainty can in actual fact be more damaging than an apparently high impact. Consider the two augmented fuzzy sets (0.7, 0.3) and (0.5, 0.8) representing two environmental impacts. Since the first augmented fuzzy set has a very low value of certainty, it is quite likely that the 'true' membership grade for the significance of the impact is not actually 0.7, whereas the high value of certainty for the second augmented fuzzy set means that its 'true' membership grade is likely to be close to 0.5. Therefore it is possible that the 'true' membership grade for the significance of the first impact is, for instance, 0.4, which is less than the membership grade for the significance of the second impact.

Augmented fuzzy sets can be *ordered* by using an iterative procedure based on analogous δ–ε conditions to those used in the case of the augmented fuzzy set union operator. Suppose there are n impacts or other variables which can be represented by the following augmented fuzzy sets $f_1 = (\mu_1, \nu_1)$, $f_2 = (\mu_2, \nu_2)$, ... $f_n = (\mu_n, \nu_n)$. The fuzzy set union operator in section 15.2.1, which is equivalent to the maximisation operator, can be used to find the 'largest' augmented fuzzy set which becomes the first term in the ordering. This operator is then used to find the 'largest' of the remaining n-1 augmented fuzzy sets, which becomes the second term in the ordering. It is then used another n-3 times to find the order of the remaining n-2 augmented fuzzy sets. Thus ordering all n elements is an iterative process which requires the fuzzy set union operator to be used n-1 times.

If only the values of μ were considered, then augmented fuzzy sets with higher values of μ would always be ordered above those with lower values. However consideration of the value of ν as well as μ may change this ordering. The second element ν gives a measure of the likelihood that the impact (or other variable) will have the given membership grade μ rather than another value. Therefore for moderate to high values of the membership grade, an augmented fuzzy set with a higher degree of certainty is ordered above one with a lower degree of certainty, as there is greater likelihood of the impact (or other variable) actually having this high value in the first than the second case i.e. (0.7, 0.9) is ordered above (0.7, 0.6).

However the reverse holds for the case of low values of membership grades (impacts), as it is more likely that this low value of membership grade (impact) is close to the 'true' value if the associated degree of certainty is high than if it is low. It is therefore possible that a low membership grade (impact or other variable) with low degree of certainty in actual fact corresponds to a moderate or even high membership grade. Consequently an augmented fuzzy set with a low membership grade and a low degree of certainty should be ordered above one with a low membership grade and a high degree of certainty. For instance (0.2, 0.4) is ordered above (0.2, 0.8), as it is more likely that the 'true' membership grade of the second augmented fuzzy set is 0.2, whereas it is quite possible that the 'true' membership grade of the second set is higher. It could also be less than 0.2, but the fact that the range of values above 0.2 is greater than those below 0.2 makes it more likely that the 'true' value will be greater than 0.2.

To take account of this, for 'low' membership grades the associated degree of certainty ν is replaced by $\nu' = 1 - \nu$ and the augmented fuzzy set

(μ, ν') rather than (μ, ν) is used in the algorithm. What is considered a 'low' membership grade is determined by a third parameter η chosen by the user, with $0 \leq \eta \leq 0.33$. At each stage in the algorithm the union operator is applied to obtain the augmented fuzzy set with the 'maximum' value at this stage. This fuzzy set is placed next in the ordering and the remaining elements are then reordered to group together all the terms which have not yet been 'chosen'. The algorithm can now be stated and is illustrated by the flow diagram in fig. 16.1.

16.2.2 Augmented Fuzzy Set Ordering Algorithm

Preliminaries

Put $f'_{1.i} = (\mu_{1.i}, \nu'_{1.i})$, $i = 1 \ldots n$,
with $\mu_{1.i} = \mu_i$, $\nu'_{1.i} = \nu_i$ if $\mu_i \geq \eta$, $\nu'_{1.i} = 1 - \nu_i$ if $\mu_i < \eta$

The algorithm then consists of repeating the following three loop steps for i=1 to n-1:

Loop step 1

Obtain $(\mu_{n.i}, \nu'_{n.i}) = (\mu_{i.sup}, \nu'_{i.sup})$ from

$$(\mu_{i.sup}, \nu'_{i.sup}) = (\mu_{k_i}, \nu'_{k_i}) = f'_{1.sup} = \bigvee_{j=i}^{n} f'_{1.j} : X \to I \times I$$

using the augmented fuzzy set union operator defined in section 15.2.1.

Loop step 2

Put $\nu'_{n.i} = \nu_{n.i}$ if $\mu_{n.i} \geq \eta$, $\nu_{n.1} = 1 - \nu'_{n.i}$ if $\mu_{n.i} < \eta$.

Loop step 3

Reorder the remaining elements as $f'_{i+1.j} = f'_{i.j-1}$, for $j = i+1 \ldots k_i - 1$
$f'_{i+1.j} = f'_{i.j}$ for $j = k_i + 1 \ldots n$

This leaves one element which becomes $(\mu_{n.n}, \nu'_{n.n})$. Loop step 2 is then carried out to obtain $(\mu_{n.n}, \nu_{n.n})$ from it.

This gives the ordering $(\mu_{n.1}, \nu_{n.1})$, $(\mu_{n.2}, \nu_{n.2})$, ..., $(\mu_{n.n}, \nu_{n.n})$

16.2.3 Relationships Between the Terms after Ordering

For (crisp) numbers or ordinary fuzzy sets, what is meant by ordering is very clear i.e. satisfying an inequality relationship. However, the fact that augmented fuzzy sets have two components means that, analogously to the case of multivariable optimisation, there are a number of different ways of ordering them. The inequality satisfied by consecutive pairs of terms

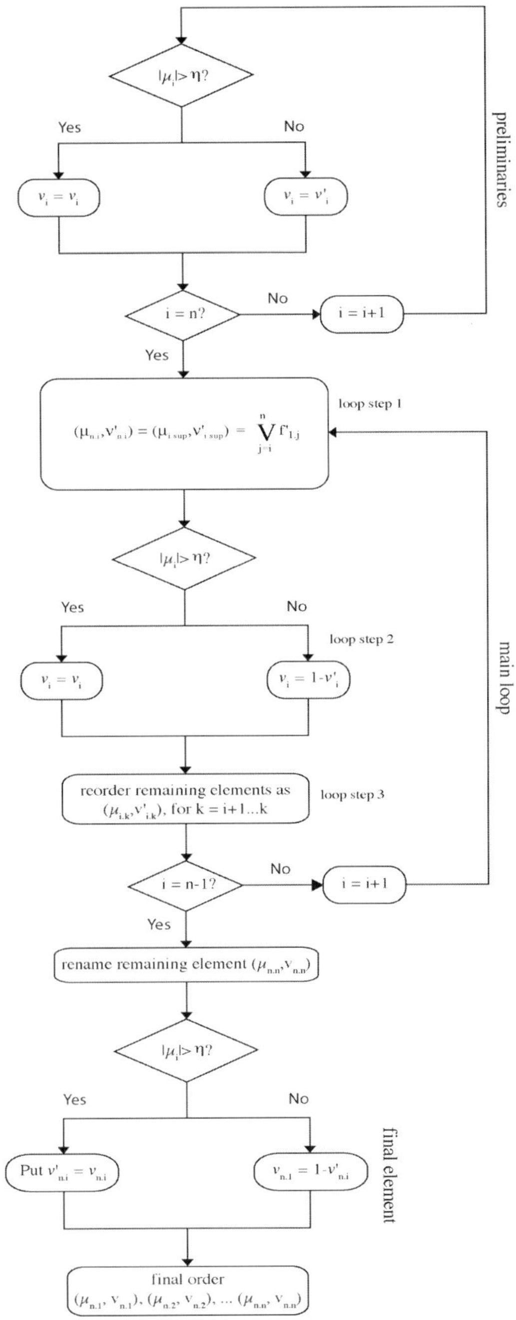

Fig. 16.1, Flow diagram for ordering algorithm

ordered by the algorithm is stronger than that satisfied by non-consecutive terms. This case will be considered first.

Consecutive pairs of terms ordered by the algorithm i.e. for which

$$(\mu_{n.s+1}, \nu_{n.s+1}) \leq (\mu_{n.s}, \nu_{n.s}) \tag{16.1}$$

satisfy one of the following relationships for given values of δ and ε:

a. Both the μ and ν components are greater than the corresponding terms of the succeeding augmented fuzzy set, as illustrated in fig. 16.2a, i.e.

$$\mu_{n.s+1} \leq \mu_{n.s}, \quad \nu_{n.s+1} \leq \nu_{n.s} \tag{16.2a}$$

b. The μ component is greater than the μ component of the succeeding augmented fuzzy set by at least δ and the ν component is not less than the ν component of the next augmented fuzzy set by more than ε, as given by inequality (16.2b). This is illustrated in fig. 16.2b, where the shaded region satisfies these conditions.

$$\mu_{n.s+1} \leq \mu_{n.s} - \delta, \quad \nu_{n.s+1} \leq \nu_{n.s} + \varepsilon \tag{16.2b}$$

c. The ν component is greater than the ν component of the next augmented fuzzy set by at least ε and the μ component is not less than the μ component of the next augmented fuzzy set by more than δ, as given by inequality (16.2c). This is illustrated in fig. 16.2c, where the shaded region satisfies these conditions.

$$\mu_{n.s+1} \leq \mu_{n.s} + \delta, \quad \nu_{n.s+1} \leq \nu_{n.s} - \varepsilon \tag{16.2c}$$

Thus the ordering given by the algorithm is such that (other than the fairly rare cases in which both components of the augmented fuzzy set are greater than their successors) each augmented fuzzy set is 'greater' than the next set only in a δ–ε sense. As indicated by the notation this relation only holds for consecutive pairs of terms in the ordering. It is not transitive. Therefore this relation will not necessarily hold for the terms s and s+2 when it holds for two consecutive pairs of terms in the ordering, s and s+1, and s+1 and s+2.

Under two not very restrictive conditions a slightly weaker inequality relationship holds between a term and *all succeeding terms* in the series:

- The μ component is not reduced by more than δ relative to the μ components of all succeeding terms in the ordering.
- The ν component is not reduced by more than ε relative to the ν components of all succeeding terms in the ordering.

i.e. $\mu_{n.j} \leq \mu_{n.i} + \delta, \quad i=1...n, j \leq i$ (16.3a)

$\nu_{n.j} \leq \nu_{n.i} + \varepsilon \quad i=1...n, j \leq i$ (16.3b)

These relationships are illustrated in fig. 16.2d, where the shaded region satisfies inequalities (16.3). It should be noted that the shaded regions in fig. 16.2d and 16.2a respectively contain the union and intersection of the shaded regions in figs. 16.2b and 16.2c. The conditions for both the inequalities (16.3) to hold are that there do not exist sets of at least three terms for which, after reordering as

$$(\mu_{r_i}, \nu_{r_i}),\ (\mu_{r_{i+1}}, \nu_{r_{i+1}}),\ ...,\ (\mu_{r_{i+k}}, \nu_{r_{i+k}}),\quad k \geq 2$$

$$\mu_{r_i} \geq \mu_{r_{i+1}} \geq ... \geq \mu_{r_{i+k}},\ \nu_{r_i} \leq \nu_{r_{i+1}} \leq ... \leq \nu_{r_{i+k}},\quad k \geq 2 \text{ and } \nu_{r_{i+k}} - \nu_{r_i} > \varepsilon \quad (16.4)$$

When condition (16.4) holds only inequality (16.3a) and not inequality

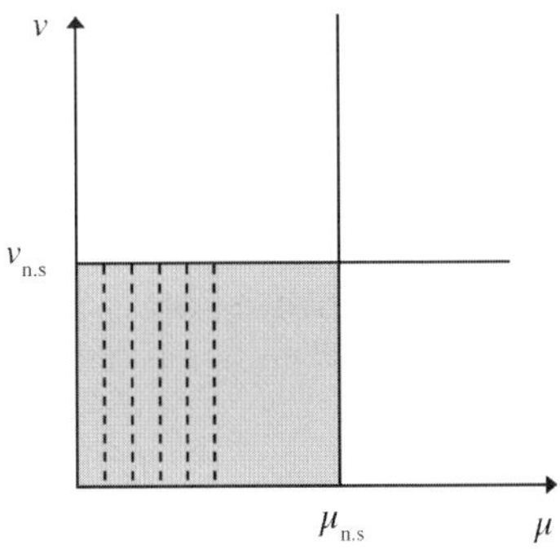

Fig. 16.2a Inequalities (16.2a)

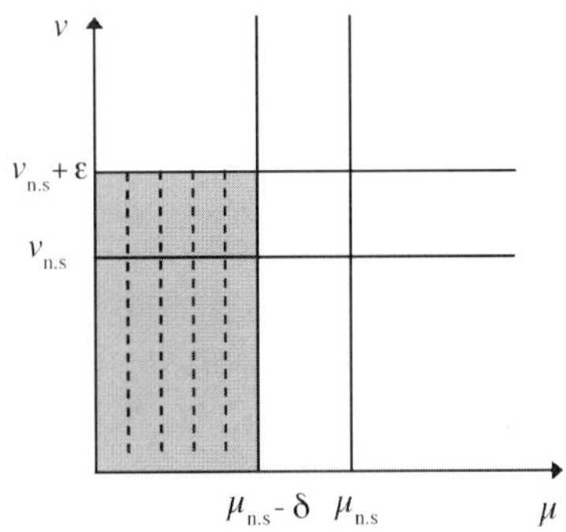

Fig. 16.2b Inequalities (16.2b)

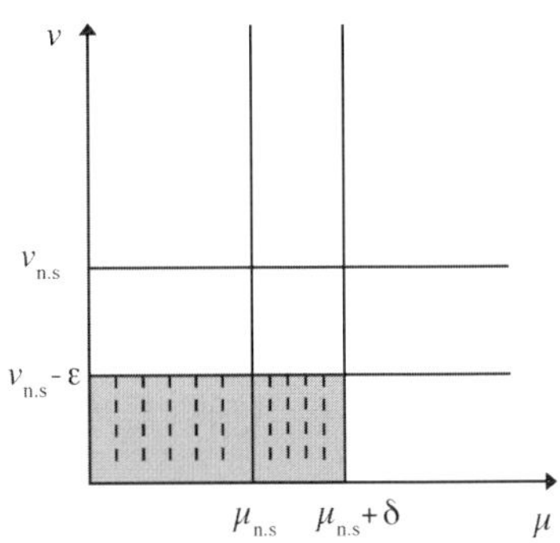

Fig. 16.2c Inequalities (16.2c)

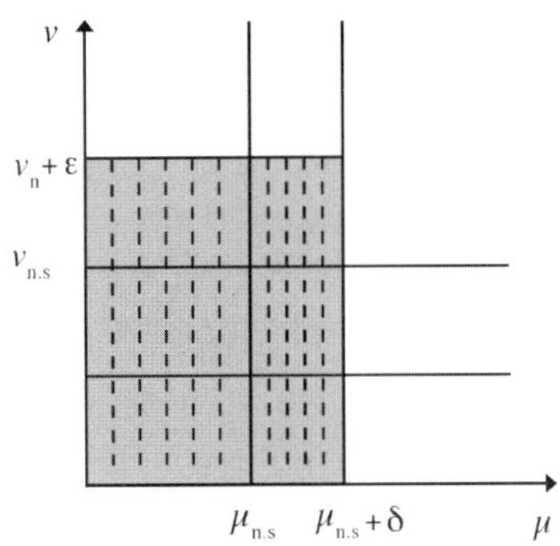

Fig. 16.2d Inequalities (16.3)

(16.3b) is valid. Condition (16.4) implies that the augmented fuzzy sets representing the impacts or variables are such that it is only possible to describe the relationship of $\mu_{n.s}$ and all succeeding terms or $\nu_{n.s}$ and all succeeding terms, but not both.

16.2.4 Augmented Fuzzy Set Difference Operator

Another useful operation for evaluating the effects of different strategies on impact reduction is the augmented fuzzy set difference operator. This can be used to obtain the reduction in impact with its associated degree of certainty for a given reduction strategy. It is defined as

$$(\mu_1, \nu_1) - (\mu_2, \nu_2) = (\max(0, \mu_1 - \mu_2), \min(\nu_1, \nu_2)) \tag{16.5}$$

Maximisation of the difference and zero is required to ensure that the resulting value of μ is positive. The minimum of the two values of ν is taken, as the degree of certainty of the difference is at most equal to each individual certainty acting on its own and combining the two certainties in this way gives the intersection or minimum.

16.2.5 Example: Strategically Targeted Impact Reduction

The example in section 15.6 showed that each stage of the computer life cycle has high environmental impacts, so that measures are urgently required to reduce these impacts. For the greatest effectiveness, a targeted reduction strategy will be required to order the impacts according to their values and degrees of certainty. This can be done using the algorithm in section 16.2.2. The augmented fuzzy sets for this problem are given in the example in section 15.6.

Since there are eight different variables and six life cycle stages, this gives 48 ordered pairs or membership grades and associated values of certainty. However these 48 different ordered pairs have only the following ten different values:

$(\mu_1, \nu_1) = (1, 0.50)$, $(\mu_2, \nu_2) = (0, 1)$,
$(\mu_3, \nu_3) = (0.66, 0.50)$, $(\mu_4, \nu_4) = (0, 0.75)$,
$(\mu_5, \nu_5) = (0.33, 0.50)$, $(\mu_6, \nu_6) = (0, 0.50)$,
$(\mu_7, \nu_7) = (1, 0.75)$, $(\mu_8, \nu_8) = (0.66, 0.75)$,
$(\mu_9, \nu_9) = (1, 1)$, $(\mu_{10}, \nu_{10}) = (1, 0.25)$

Taking $\eta = 0.33$, so μ_1, μ_3, μ_5 and $\mu_7 - \mu_{10} \geq \eta$ and $\mu_2, \mu_4, \mu_6 < \eta$
$v'_i = v_i$, for i =1,3,5,7-10, $v'_2 = 0$, $v'_4 = 0.25$, $v'_4 = 0.50$, so

$(\mu_{1.1}, v'_{1.1}) = (1,0.50)$, $(\mu_{1.2}, v'_{1.2}) = (0,0)$,
$(\mu_{1.3}, v'_{1.3}) = (0.66,0.50)$, $(\mu_{1.4}, v'_{1.4}) = (0,0.25)$,
$(\mu_{1.5}, v'_{1.5}) = (0.33,0.50)$, $(\mu_{1.6}, v'_{1.6}) = (0,0.50)$,
$(\mu_{1.7}, v'_{1.7}) = (1,0.75)$, $(\mu_{1.8}, v'_{1.8}) = (0.66,0.75)$,
$(\mu_{1.9}, v'_{1.9}) = (1,1)$, $(\mu_{1.10}, v'_{1.10}) = (1,0.25)$.

Using the augmented fuzzy set union operator in section 15.2.1,
$(\mu_{10.1}, v'_{10.1}) = (1,1) = (\mu_{10.1}, v_{10.1})$ as $\mu_{10.1} > \eta$

Reorder as
$(\mu_{2.2}, v'_{2.2}) = (1,0.50)$, $(\mu_{2.3}, v'_{2.3}) = (0,0)$,
$(\mu_{2.4}, v'_{2.4}) = (0.66,0.50)$, $(\mu_{2.5}, v'_{2.5}) = (0,0.25)$,
$(\mu_{2.6}, v'_{2.6}) = (0.33,0.50)$, $(\mu_{2.7}, v'_{2.7}) = (0,0.50)$,
$(\mu_{2.8}, v'_{2.8}) = (1,0.75)$, $(\mu_{2.9}, v'_{2.9}) = (0.66,0.75)$,
$(\mu_{2.10}, v'_{2.10}) = (1,0.25)$.

Then $(\mu_{10.2}, v'_{10.2}) = (1,0.75) = (\mu_{10.2}, v_{10.2})$ as $\mu_{10.2} > \eta$

Reorder as
$(\mu_{3.3}, v'_{3.3}) = (1,0.50)$, $(\mu_{3.4}, v'_{3.4}) = (0,0)$,
$(\mu_{3.5}, v'_{3.5}) = (0.66,0.50)$ $(\mu_{3.6}, v'_{3,6}) = (0,0.25)$,
$(\mu_{3,7}, v'_{3,7}) = (0.33,0.50)$, $(\mu_{3.8}, v'_{3.8}) = (0,0.50)$,
$(\mu_{3.9}, v'_{3.9}) = (0.66,0.75)$, $(\mu_{3.10}, v'_{3.10}) = (1,0.25)$

Then $(\mu_{10.3}, v'_{10.3}) = (1, 0.5) = (\mu_{10.3}, v_{10.3})$ as $\mu_{10.3} > \eta$

Then for $\delta = 0.34$, $\varepsilon \leq 0.25$ and $\eta = 0.33$ it can be easily be shown that

$(\mu_{10.5}, v'_{10.5}) = (0.66, 0.75)$
$(\mu_{10.5}, v'_{10.5}) = (0.66, 0.5)$
$(\mu_{10.6}, v'_{10.6}) = (1,0.25)$
$(\mu_{10.7}, v'_{10.7}) = (0.33, 0.50)$
$(\mu_{10.8}, v'_{10.8}) = (0, 0.5)$
$(\mu_{10.9}, v'_{10.9}) = (0,0.25)$
$(\mu_{10.10}, v'_{10.10}) = (0,0)$

The rather large value of δ has been chosen on account of the large spacing between the values. Only $v'_{10.9}$ and $v'_{10.10} < \eta$ so, using the values of the first three terms already obtained,

$(\mu_{10.1}, \nu_{10.1}) = (1,1)$
$(\mu_{10.2}, \nu_{10.2}) = (1, 0.75)$
$(\mu_{10.3}, \nu_{10.3}) = (1, 0.50)$
$(\mu_{10.4}, \nu_{10.4}) = (0.66, 0.75)$
$(\mu_{10.5}, \nu_{10.5}) = (0.66, 0.50)$
$(\mu_{10.6}, \nu_{10.6}) = (1,0.25)$
$(\mu_{10.7}, \nu_{10.7}) = (0.33, 0.50)$
$(\mu_{10.8}, \nu_{10.8}) = (0, 0.50)$
$(\mu_{10.9}, \nu_{10.9}) = (0,0.75)$
$(\mu_{10.10}, \nu_{10.10}) = (0,1)$

This ordering can then be translated back to give the reduction of specific impacts at particular life cycle stages. This gives the following targeted strategy for reducing impacts (only the first three stages of which are given here):

1. The energy consumption and waste generation of the use and reuse life cycle stage - augmented fuzzy set (1, 1).
2. The local air impacts of the manufacture and transport and distribution life cycle stages; the resource consumption and forest destruction of the use and reuse life cycle stage; the atmospheric impacts of the transport and distribution life cycle stage; and the resource consumption and landfill exhaustion impacts of the disposal and recycling life cycle stages - augmented fuzzy set (1, 0.75).
3. The energy consumption, resource consumption, local air impacts and waste generation of the material acquisition life cycle stage; the energy consumption and resource consumption of the material processing and manufacture life cycle stages; the local air impacts of the use and reuse life cycle stages; and the energy consumption & waste generation of the transport & distribution life cycle stage - augmented fuzzy set (1, 0.50).

16.2.6 Example: Evaluation of a Strategy for Impact Reduction

Since the use and reuse (fifth) life cycle stage of the example in section 15.6 has the highest environmental impacts, the effect of a reduction strategy on this life cycle stage will be considered. For ease of reference the table for the augmented fuzzy set for the use and reuse life cycle stage will be restated as table 16.1.

It will be assumed that a set of measures has been applied to reduce the environmental impacts of the computer life cycle and that the new augmented fuzzy set for the use and reuse stage of the life cycle is given in table 16.2.

Table 16.1 Impacts at use and reuse stage of computer life cycle

X	x_1	x_2	x_3	x_4	x_5	x_6	x_7	x_8
μ_5	1	1	0.66	1	0	0.66	1	1
ν_5	1	0.75	0.75	0.50	1	0.50	1	0.75

Table 16.2 Reduced impacts at use and reuse stage of computer life cycle

X	x_1	x_2	x_3	x_4	x_5	x_6	x_7	x_8
μ_{n5}	0.8	0.8	0.5	0.7	0	0.3	0.8	0.7
ν_{n5}	0.4	0.3	0.3	0.2	1	0.5	0.2	0.5

First the total reduced impact over the fifth life cycle stage will be obtained by combining the impacts in table 16.2. Using the definitions in section 15.2.1:

$$\mu_{5h} = 0.8, \quad \nu_{5h} = 0.4 \qquad \mu_{5m} = 0, \quad \nu_{5m} = 1 \qquad \mu_{5u} = 0.7, \quad \nu_{5u} = 0.5$$

For $\delta = 0$, $\varepsilon = 0$, conditions (a) and (b) in section 15.2.1 are not satisfied and $\mu_{5sup} = 0.8$, $\nu_{5sup} = 0.4$

For $\delta = 0.1$, $\varepsilon = 0.1$, conditions (a) in section 15.2.1 are satisfied and $\mu_{5sup} = 0.7$, $\nu_{5sup} = 0.5$

More generally
$$\begin{aligned} &\mu_{5sup} = 0.7, &&\nu_{5sup} = 0.5 &&\text{for } \delta \geq 0.1, \ \varepsilon \leq 0.1 \\ &\mu_{5sup} = 0, &&\nu_{5sup} = 1 &&\text{for } \delta \geq 0.8, \ \varepsilon \leq 0.6 \\ &\mu_{5sup} = 0.8, &&\nu_{5sup} = 0.4 &&\text{otherwise} \end{aligned}$$

Both the augmented fuzzy sets (0.7, 0.5) and (0.8, 0.4) give reasonable measures of the overall impact and which one is considered more appropriate will depend on the values chosen for δ and ε. (0, 1) does not give a useful measure of the overall impact (which is clearly non-zero), as indicated by the very high value of δ.

The difference operation will now be used to give augmented fuzzy sets for the reductions in impact in table 16.3.

Table 16.3 Augmented fuzzy set for the impact reductions

X	x_1	x_2	x_3	x_4	x_5	x_6	x_7	x_8
μ_{r5}	0.2	0.2	0.16	0.3	0	0.36	0.2	0.3
ν_{r5}	0.4	0.3	0.3	0.2	1	0.5	0.2	0.5

The overall reduction in impact at this life stage can be obtained by using the augmented fuzzy set intersection rather than the union operator. The intersection is used, since the overall reduction in impact is a measure of the reduction of each variable rather than the maximum reduction.

Using the definitions in section 15.2.2

$$\mu_y = \mu_n = 0.2,\ \ \nu_y = \nu_n = 0.2, \qquad \mu_g = 0,\ \ \nu_g = 1$$

Therefore, taking $\delta = 0.2$ gives an overall reduction of (0.2, 0.2). This shows that this strategy would only be successful to a limited extent, and that further reduction strategies would still be required.

16.3 Third Order Augmented Fuzzy Sets

16.3.1 Introduction and Definition

In many cases not all the information in a problem is of equal significance or relevance. For instance in a sustainable life cycle design problem it may be considered more important to reduce energy consumption than waste generation or equally important to reduce energy and resource consumption, with reduction of both energy and resource consumption considered more important than noise reduction. In some cases there will be a hierarchy of priorities. Therefore there is clearly a role for techniques which allow consideration of the *relative importance* or *priorities* attached to different variables or determination of the sensitivity of the resulting design, decision or other problem outputs to the priorities given to different factors and the effects of changing these priorities.

One relatively simple technique which allows representation of the priorities or significance of variables in addition to their values and the degree of certainty with which they are known is the *third order augmented fuzzy set*. This can be obtained by adding a third element to the second order augmented fuzzy set to represent the significance or priority of the information. Thus a third order augmented fuzzy set is an ordered triple, which is defined as follows:

$$f: X \rightarrow I \times I \times I \quad (16.6a)$$
$$f(x) = (\mu, \nu, \pi), \quad \forall x \in X, \; \mu,\nu,\pi \in I \quad (16.6b)$$

The first and third elements μ and π of the ordered triple (μ, ν, π) are membership functions, which represent the extent to which a particular (imprecise) property holds and its degree of priority or importance. The second element ν can again be expressed in either of the following two forms:

- A membership function which represents the degree of certainty with which the information is known
- A probability which represents the probability that the membership function has the value μ.

The third order augmented fuzzy set is particularly useful in, for instance, (sustainable) decision making and the comparison of the social and environmental impacts of different policy options. It allows different degrees of importance or priority to be assigned to different variables or impacts. It also allows investigation of the effect on the resulting decision of changes in the relative priorities given to the different variables. When X is finite, the third order augmented fuzzy set can be described in tabular form as shown in table 16.4.

Table 16.4 Third order augmented fuzzy set

X	x_1	x_2	...	x_n
μ	μ_1	μ_2	...	μ_n
ν	ν_1	ν_2	...	ν_n
π	π_1	π_2	...	π_n

As in the second order case, treating ν as a membership function or a probability gives two different expressions for the set operations of union and intersection for third order augmented fuzzy sets. In addition, for each of these approaches the third element can be treated in the following two ways, giving four expressions in total.

- The third element π is used to weight the first element μ.
- The third element π is treated analogously (symmetrically) to the other two elements

In principle extension is possible to classes of third order fuzzy sets which satisfy particular conditions, analogously to the classes of t-norms and t-conorms, but this will not be considered here. Only the first case will be discussed here, since the membership function treatment of ν gives simpler expressions, is probably more appropriate from a theoretical point of view and weighting the data gives a clear measure of its relative priority. To simplify the notation the argument x will be omitted.

16.3.2 Union and Intersection

As discussed previously, the definitions of the union operator for the fuzzy and second order augmented fuzzy sets are extensions of the standard crisp set union operator. The definition of *union* for 'weighted' *third order augmented fuzzy sets* can probably more appropriately be considered as a generalisation than an extension, as it can be obtained from the union operator for second order augmented fuzzy sets by replacing the first term μ of the second order augmented fuzzy set by its weighted value $\mu' = \mu\pi$. Therefore the algorithm in section 15.2.1 can be used to find (μ'_{sup}, ν_{sup}). π_{sup} is then put equal to the associated value of π and μ_{sup} is obtained from μ'_{sup} as μ'_{sup}/π_{sup}. For instance, if $(\mu'_{sup}, \nu_{sup}) = (\mu'_3, \nu_3)$, then $\pi_{sup} = \pi_3$ and $\mu_{sup} = \mu'_{sup}/\pi_{sup} = \mu'_3/\pi_3$. Similar comments can be made about the constants δ and ε to those in the second order case.

Therefore the union of n fuzzy sets with membership functions $f_1 = (\mu_1, \nu_1, \pi_1) \ldots f_n = (\mu_n, \nu_n, \pi_n)$ has membership function

$$f_{sup} = \bigvee_{i=1}^{n} f_i : X \to I \times I \times I \tag{16.7a}$$

$$f_{sup}(x) = (\mu_{sup}(x), \nu_{sup}(x), \pi_{sup}(x)), \ \forall x \in X, \mu_{sup}, \nu_{sup}, \pi_{sup} \in I \tag{16.7b}$$

with

$$\mu_{sup} = \mu'_u/\pi_u, \quad \nu_{sup} = \nu_u, \quad \pi_{sup} = \pi_u$$

if (ai) $\mu'_u + \delta \geq \mu'_h$, (aii) $\nu_u - \varepsilon \geq \nu_h$ or (aiii) $\mu'_u = \mu'_h$, $\nu_u = \nu_h$

$$\mu_{sup} = \mu'_m/\pi_m, \quad \nu_{sup} = \nu_m, \quad \pi_{sup} = \pi_m,$$

if (ai) but not (aii) is satisfied, (bi) $\mu'_m + \delta \geq \mu'_h$, (bii) $\nu_m - \varepsilon \geq \nu_h$

(c) $\mu_{sup} = \mu'_h/\pi_h, \quad \nu_{sup} = \nu_h, \quad \pi_{sup} = \pi_h$ otherwise

where δ and ε are chosen by the user and

$$h = \min j \in J,\ J = \{j:\ \nu_j = \max_{k \in K} \nu_k\},\ K = \{k:\ \mu'_k = \max_{i=1\ldots n} \mu'_i\} \tag{16.8a}$$

$$m = \min p \in P,\ P = \{p:\ \mu'_p = \max_{q \in Q} \mu'_q\},\ Q = \{q:\ \nu_q = \max_{i=1\ldots n} \nu_i\} \tag{16.8b}$$

$$u = \min w \in W,\ W = \{w:\ \mu'_w = \max_{z \in Z} \mu'_z\},$$

$$Z = \{z:\ \mu'_z \nu_z = \max_{i=1\ldots n} (\mu'_i \nu_i)\} \tag{16.8c}$$

$$\mu'_i = \mu_i \pi_i \quad i=1\ldots n \tag{16.8d}$$

The intersection of n fuzzy sets with membership functions $f_1 = (\mu_1,\ \nu_1,\ \pi_1)$ ... $f_n = (\mu_n,\ \nu_n,\ \pi_n)$ has membership function

$$f_{inf} = \bigwedge_{i=1}^{n} f_i : X \to I \times I \times I \tag{16.9a}$$

$$f_{inf}(x) = (\mu_{inf}(x),\ \nu_{inf}(x),\ \pi_{inf}(x)),\ \forall\, x \in X,\ \mu_{inf}, \nu_{inf}, \pi_{inf} \in I \tag{16.9b}$$

where, omitting the argument x to simplify notation,

$\mu_{inf} = \mu'_y/\pi_y,\quad \nu_{inf} = \nu_y,\quad \pi_{inf} = \pi_y$

if (ai) $\mu'_y - \delta \le \mu'_g$, (aii) $\nu_y + \varepsilon \le \nu_g$ or (aiii) $\mu'_y = \mu_g$, $\nu_y = \nu_g$

$\mu_{inf} = \mu'_n/\pi_n,\quad \nu_{inf} = \nu_n,\quad \pi_{inf} = \pi_n$

if (ai) but not (aii) is satisfied, (bi) $\mu'_n - \delta \le \mu'_g$, (bii) $\nu_m + \varepsilon \le \nu_g$

$\mu_{inf} = \mu'_g/\pi_g,\quad \nu_{inf} = \nu_g,\quad \pi_{inf} = \pi_g$ otherwise

where δ and ε are chosen by the user and

$$g = \min j \in J,\ J = \{j:\ \nu_j = \min_{k \in K} \nu_k\},\ K = \{k:\ \mu'_k = \min_{i=1\ldots n} \mu'_i\} \tag{16.10a}$$

$$n = \min p \in P,\ P = \{p:\ \mu'_p = \min_{q \in Q} \mu'_q\},\ Q = \{q:\ \nu_q = \min_{i=1\ldots n} \nu_i\} \tag{16.10b}$$

$$y = \min w \in W, \qquad W = \{w: \mu'_w = \min_{z \in Z} \mu'_z\}, \tag{16.10c}$$

$$Z = \{z : (\mu'_z, v_z) = \min_{i \in V}(\mu'_i \, v_i)\} \text{ if } \min_{i \in V}(\mu'_i \, v_i) \neq 0$$

$$\text{or } \min_{i \in V}(\mu'_i \, v_i) = 0 \text{ and } \mu'_z = \mu'_g, v_z = 0, \text{ or } \mu'_z = 0, v_z = v_n\}$$

$$Z = \{z: \ (\mu'_z, v_z) = \min_{i \in V}(\mu'_i + v_i)\} \text{ otherwise} \tag{16.10d}$$

$$\mu'_i = \mu_i \pi_i \quad i=1...n \tag{16.10e}$$

16.4 Augmented Fuzzy Sets with Negative Coefficients

There are a number of circumstances in which it is useful to be able to represent variables taking positive and negative values. Such circumstances include decision making, where both positive and negative impacts, generally referred to as benefits and costs, have to be considered. Such signed data may also be imprecise, uncertain, inconsistent and/or qualitative, making the use of fuzzy techniques appropriate. However, standard fuzzy techniques allow fuzzy and augmented fuzzy sets to take only positive values, generally between 0 and 1. There are a number of different approaches to representing positive and negative factors, all of which have advantages and disadvantages. One possibility is to divide the unit interval, for instance with values greater than 0.5 representing benefits and those less than 0.5 costs. This has the disadvantage of giving a reduced range of variation and therefore less discrimination unless values are taken to two decimal places. Another approach, which will be presented here, is to modify the definition of the augmented fuzzy set to allow the first term to take values between -1 and +1, with costs or disadvantages generally taking values between -1 and 0 and benefits values between 0 and +1.

This can be motivated in terms of the combination of the fuzzy set with a given beneficial property and the fuzzy set with the opposite disadvantageous property. To avoid adding another variable, membership of the extended fuzzy set consisting of the fuzzy set with the given property and its opposite property is represented by allowing μ to take values between -1 and 1. For instance, in the case of emissions from an industrial process or car exhaust, any increase in emissions is the negative property and any decrease in emissions is the positive property. In some cases the variable may act in one direction only, so that the opposite property is purely notional.

This gives the following modified definition of a third order augmented fuzzy set:

$$f: X \to I_{\pm} \times I \times I \quad (16.11a)$$
$$f(x) = (\mu(x), \nu(x), \pi(x)), \quad \forall x \in X, \; \mu \in I_{\pm}, \nu,\pi \in I, \text{ where } I_{\pm} = [-1, 1] \quad (16.11b)$$

Thus, for instance, if positive values of the first element $\mu(.)$ represent the fuzzy set *significant increases in emissions of a given toxic substance*, then negative values $\mu(.)$ represent the fuzzy set *significant decreases in emissions of the toxic substance*. It should be noted that not all (augmented) fuzzy sets can be meaningfully extended to allow negative membership grades. Whether or not this is possible will depend on the specific (augmented) fuzzy set.

Modified expressions are required for the union, intersection and complement, to take account of the sign of the first element in an appropriate way and produce a result with an appropriate sign. However only the expression for the union will be presented here. In order to obtain an appropriately signed answer the augmented fuzzy set operations are applied separately to the fuzzy sets with positive and negative first elements, after the modulus has been taken of the negative first element. Taking the modulus converts the augmented fuzzy set with a negative first element to one with a positive first element of the same magnitude, so that the augmented fuzzy set aggregation techniques in section 16.3.2 can be used.

This then gives two sets f_+ and f_- of aggregated 'benefits' and 'costs' which can be combined after the sign has been replaced in the first element of f_-, so that it is again negative. The two sets can then be combined by taking a weighted average, with the weighting coefficients chosen according to the relative importance of the 'costs' and 'benefits'. The algorithm is stated in the following section.

16.4.1 Union Algorithm

The *algorithm* for obtaining the *union* of *n third order augmented fuzzy sets* with *negative coefficients* and membership functions $f_i = (\mu_i, \nu_i, \pi_i)$ $(i=1...n)$ involves the following three stages:

1. Aggregate the third order augmented fuzzy sets with positive first elements using the algorithm in section 16.3.2 to give the fuzzy set $f_+ = (\mu_+, \nu_+, \pi'_+)$.

2. Aggregate the third order augmented fuzzy sets, with negative first elements to give the fuzzy set $f_- = (\mu_-, \nu_-, \pi'_-)$. This involves the following three steps:

   i. Form the fuzzy sets $f_j^* = (\mu_j^*, \nu_j, \pi_j)$, where $\mu_j^* = -\mu_j$ for each third order augmented fuzzy sets with negative first element.
   ii. Aggregate the third order augmented fuzzy sets $f_j^* = (\mu_j^*, \nu_j, \pi_j)$ using the algorithm in section 16.3.2 to give the fuzzy set $f_* = (\mu^*, \nu_-, \pi'_-)$
   iii. Obtain $f_- = (\mu_-, \nu_-, \pi'_-) = (-\mu^*, \nu_-, \pi'_-)$

3. Combine the two fuzzy sets f_+ and f_- by taking a weighted average with weighting coefficients π_+ and π_-, where the weighting coefficients π_+ and π_- determine the relative importance of positive and negative signed data or costs and benefits.

The algorithm for the union operator for augmented fuzzy sets with negative coefficients can be expressed mathematically as follows:

$$f_{sup} = f_+ \vee f_-, \; X \to I_{\pm} x I x I \tag{16.12a}$$

$$f_{sup} = (\mu_{sup}(x), \nu_{sup}(x), \pi_{sup}(x)), \; \forall x \in X, \mu_{sup} \in I_{\pm}, \nu_{sup}, \pi_{sup} \in I \tag{16.12b}$$

with (*step 3*)

$$\mu_{sup} = \frac{\mu_+\pi_+ + \mu_-\pi_-}{\pi_+ + \pi_-} \tag{16.13a}$$

$$\nu_{sup} = \frac{\nu_+\pi_+ + \nu_-\pi_-}{\pi_+ + \pi_-} \tag{16.13b}$$

where (μ_+, ν_+) is obtained from (*step 1*)

$$f_+ = \bigvee_{j \in J_+} f_j : X \to I \, x \, I \, x \, I \; \text{ for } J_+ = \{j : \mu_j \geq 0\} \tag{16.14a}$$

$$f_+ = (\mu_+(x), \nu_+(x), \pi'_+(x)) = \left(\underset{j \in J_+}{\mu_{sup}}(x), \underset{j \in J_+}{\nu_{sup}}(x), \underset{j \in J_+}{\pi'_{sup}}(x) \right)$$

$$\forall x \in X, \mu_+, \nu_+, \pi'_+ \in I \tag{16.14b}$$

and (μ_-, ν_-) is obtained from (*step 2*)

$$f\_(x) = \bigvee_{j \in J\_} f_j : X \to I \times I \times I \text{ for } J\_ = \{j : \mu_j < 0\} \quad (16.15a)$$

$$f\_(x) = (\mu\_(x), \nu\_(x), \pi'\_(x)) = \left(-\mu^*_{\substack{\sup \\ j \in J\_}}(x), \nu_{\substack{\sup \\ j \in J\_}}(x), \pi'_{\substack{\sup \\ j \in J\_}}(x) \right)$$

$$\forall x \in X, \mu\_ \in I\_, \nu\_, \pi'\_ \in I \text{ for } I\_ = [0, 1], \mu^* = -\mu\_ \quad (16.15b)$$

Thus $f_+ = (\mu_+, \nu_+, \pi_+)$ is the union of all the augmented fuzzy sets with positive first term and $f_- = (\mu_-, \nu_-, \pi_-)$ is the union of all the augmented fuzzy sets with negative first term. These expressions are obtained respectively from $\left(\mu_{\substack{\sup \\ j \in J_+}}, \nu_{\substack{\sup \\ j \in J_+}}, \pi_{\substack{\sup \\ j \in J_+}} \right)$ and $\left(-\mu^*_{\substack{\sup \\ j \in J\_}}, \nu_{\substack{\sup \\ j \in J\_}}, \pi_{\substack{\sup \\ j \in J\_}} \right)$, where these aggregated fuzzy sets are obtained by applying the union operator defined in section 16.3.2 to the augmented fuzzy sets with positive and negative first elements respectively.

Since the definition of the augmented fuzzy set union operator in section 16.3.2 is for positive μ, this operator is applied to the new fuzzy set (μ^*, ν, π), where $\mu^* = -\mu$. The fuzzy set f_- is then recovered by putting $\mu = -\mu^*$.

The priorities π_+ and π_- used in combining the fuzzy sets of positive and negative values or costs and benefits are chosen by the user and depend on their perceived relative importance. As indicated by the notation, π_+ and π_- are in general not equal to π'_+ and π'_-, the values of priority or weighting obtained from the union of the fuzzy sets with positive and negative first elements respectively. The value of π_{sup} has deliberately not been defined, as it is most appropriately chosen by the user if further aggregation is carried out, according to the priority or significance of the aggregated variable relative to the variables it is being combined with.

Thus the value given to the priority π in a particular fuzzy set is context dependent and may change, possibly even several times, within the same problem.

It should be noted that the ordering algorithm for augmented fuzzy sets presented in section 16.2 can easily be extended to third order augmented fuzzy sets and third order augmented fuzzy sets with negative coefficients. All this is required is replacing the union operator in loop step 1 of the algorithm (see section 16.2.2) by the union operator for third order augmented fuzzy sets or third order augmented fuzzy sets with negative coefficients.

16.5 Roundup of Chapter 16

Chapter 16 has continued the material in chapter 15 through the presentation of an algorithm for the ordering of augmented fuzzy sets and the introduction of third order augmented fuzzy sets. The ordering algorithm is required since augmented fuzzy sets cannot be compared directly using inequalities, due to the presence of two parameters. The ordering algorithm is based on the repeated application of the fuzzy set union or aggregation operator to obtain the 'maximum' augmented fuzzy set. This algorithm has a number of applications, including obtaining a strategy for the reduction of the environmental and social impacts of industrial processes.

Third order augmented fuzzy sets have three parameters rather than two and all three parameters take values between 0 and 1. The third parameter can be used to represent the significance or priority of a particular variable, such as a social or environmental impact, and is therefore of particular application in sustainable decision making. The third order augmented fuzzy set can further be extended by allowing the first parameter to take values between -1 and 1, for instance to represent costs and benefits or advantages and disadvantages in a sustainable development decision making problem.

The fuzzy set techniques presented in this and some of the earlier chapters will be illustrated by a case study of a decision problem on different strategies for achieving transport aims.

17 Case Study: Transport Decision Making

17.0 Learning Objectives

The chapter illustrates the techniques discussed in Part III by a case study of the application of augmented, intuitionistic and type 2 fuzzy sets to decision problems on different strategies for achieving transport aims.

Specific learning objectives include the following:

- Presentation of a case study of the application of third order augmented, intuitionistic and type 2 fuzzy sets to a decision problem on different strategies for achieving transport aims.
- Practice in applying fuzzy set techniques to both purely numerical and typical sustainable development problems.
- Increased understanding of the applications of the fuzzy set techniques considered in Part III to problems in sustainable development.

17.1 Case Study - The M74C: Presentation of the Problem

The M74C (also called the M74 northern extension) is a proposed urban motorway on the south side of Glasgow in Scotland, which is intended to link the city centre into the existing motorway network. The proposed route in shown in fig. 17.1 and again in fig. 17.3 at the end of section 17.3.2. It is supported by Glasgow City Council and some local businesses, particularly those involved in road freight or electronics. It is opposed by many residents throughout the city, particularly in the area the road will go through, environmental groups and some businesses. The planning process and limited initial public consultation concentrated on a particular route and did not investigate alternatives to building the road.

A local public inquiry was held at the end of 2003 and start of 2004 and its results released in early 2005 after an extended delay. Much of the land the road will pass through is contaminated with toxic waste, including chromium. Current treatment plans only extend to the road and not the

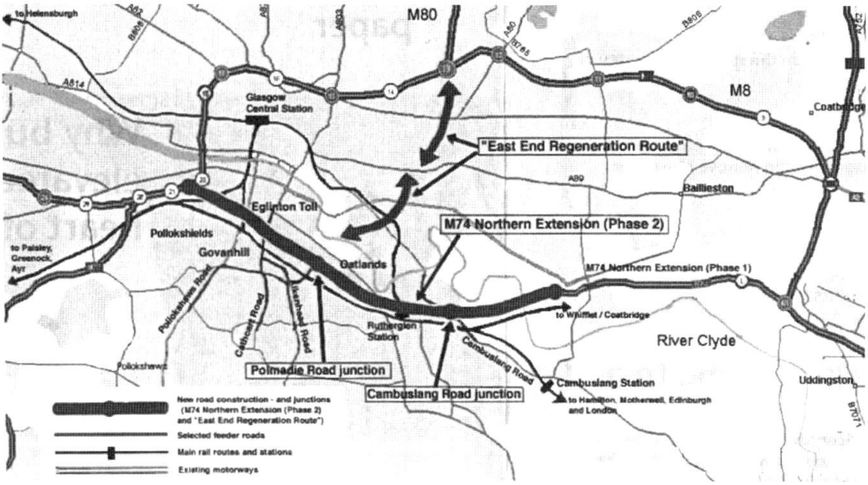

Fig. 17.1 Proposed route of M74C

surrounding area and do not include decontamination, but only covering the land in concrete. The inadequacy of the plans for the treatment of this contaminated land could provide grounds for a legal challenge to the construction of the road.

The decision problem will be analysed using third order augmented fuzzy sets with negative coefficients, intuitionistic fuzzy sets and type 2 fuzzy sets. To illustrate the techniques, three options which present a wide range of different strategies will be considered. These three options could be considered the first stage of a multi-stage decision process which allows details of the different options to be developed, after agreement in principle as to the most appropriate approach. The three options are as follows:

- Construction of the M74C.
- Maintaining the status quo (subsequently referred to as the status quo).
- An integrated package of measures, including the closure of urban motorways in the Glasgow conurbation and improvements in public transport facilities and the cycling network.

The integrated package of measures is intended to give a modal shift away from private car use to public transport, cycling and walking and from road freight to rail or mixed road-rail freight. Each of these options would be in addition to the 'route action plans' and the cycle network already agreed on by Glasgow City Council and in the process of being implemented. There may be some interaction between the two sets of measures. The use of augmented fuzzy sets is particularly appropriate, as available data is limited, uncertain, imprecise, largely qualitative and incomplete and numerical data is highly dependent on the assumptions made.

The space $I = \{x_1, ..., x_{12}\}$ of environmental and social impacts is:

x_1 - travel times
x_2 - accidents
x_3 - stress on the Kingston Bridge
x_4 - noise
x_5 - community separation
x_6 - energy consumption
x_7 - employment
x_8 - economic development
x_9 - equity
x_{10} - costs
x_{11} - traffic generation
x_{12} - emissions

Most of these environmental and social impacts are relevant to urban traffic planning in general, whereas the impact x_3 is specific to the Glasgow conurbation. This illustrates the fact that in many sustainable design and decision making problems there will be both impacts which are general and occur in other similar types of problems and impacts which are specific to the particular problem. The Kingston Bridge is a motorway bridge over the river Clyde which divides the northern and southern parts of Glasgow. It has suffered considerable structural damage as a result of excessive traffic volumes and has been subject to ongoing repair work for a number of years.

Many of the relevant impacts are the changes in the variables, as in the case of travel time savings (or increases), rather than their absolute values. This gives a mixture of positive and negative changes, frequently called benefits and costs. For each option the two life cycle stages of construction and operation will be considered. The total impact in each category for each option is equal to the combination (union) of the impacts for the two life cycle stages.

17.1.1 Assumptions Used in Deriving the Fuzzy Sets

Modelling generally requires some assumptions to be made, whatever the techniques used. The assumptions made in deriving the three types of fuzzy sets are as follows:

- In the M74C option major road building will lead to:

- Considerable traffic generation.
- Increases in emissions and energy consumption.
- No or only very limited positive effects on the local economy and employment.

- In the status quo option:
  - There will be some traffic growth, so that changes in traffic levels, emissions and energy consumption will not be zero.
  - This traffic growth will be reduced to a certain fairly limited extent by the cycle network and route action plans already decided on and being implemented.
  - Impacts for the construction stage are negligible compared to those for the operation stage of the life cycle.

- In the integrated package of measures option there will be:
  - A modal shift away from private car use and consequent traffic reduction.
  - Positive effects on the economy and employment.
  - Some uncertainty as to how successful the package of measures will be in achieving its objectives.

The assumptions in the status quo case state that some change will occur rather than the transport situation remaining totally unchanged, as measures already agreed on will still go ahead and current trends are for traffic growth to continue unless measures are taken to halt it. Since the integrated package of measures is intended to achieve a modal shift away from private car use, there is extensive empirical evidence that pedestrianisation can have a positive impact on urban areas (SACTRA 1999) and the assumptions include uncertainty of the effects of the package of measures, the assumptions in this case are also reasonable. The assumption of considerable traffic generation in the M74C case is backed by a body of research evidence, for instance (Pfleiderer and Dieterich 1995; SACTRA 1994). The growth in emissions and energy consumption will be a direct consequence of this traffic generation. The most controversial assumption is that the M74C will have no or very limited positive effects on the local economy and employment. However there is evidence that road building of this type facilitates movement out of, as well as into, an urban area (SACTRA 1999) and could therefore reduce or redistribute rather than increase local employment.

17.2 Case Study: Augmented Fuzzy Set Analysis

17.2.1 Augmented Fuzzy Sets for the Problem

The costs and benefits for the three options can be represented by the modification to (augmented) fuzzy set techniques discussed in section 16.4 to allow the variables to take values between -1 and 1, rather than just between 0 and 1. The fuzzy sets for the M74C, the status quo and the integrated package of measures are given below. Only values of the first two coefficients μ and ν are given, but not of the third coefficient π (which can be used to give the relative priority or importance of a given factor):

- To allow the effects of changing the value of the priority π on the decision reached to be investigated.
- Different values of π to be used when aggregating the two life cycle stages or the different categories of impact for each alternative.

As the context is the reduction of negative environmental and social impacts, these negative impacts or costs are given membership grades between 0 and +1 and positive impacts or benefits are given grades between -1 and 0. Therefore costs take positive values and benefits negative values and the overall signed result will give a measure of the total costs. The augmented fuzzy sets for the problem are given in tables 17.1.

Table 17.1a M74C: construction

X	x_1	x_2	x_3	x_4	x_5	x_6	x_7	x_8	x_9	x_{10}	x_{11}	x_{12}
μ	0.2	0.05	0.3	0.8	0.5	0.3	-0.1	-0.2	0.8	1	0.1	0.3
ν	0.3	0.2	0.2	0.7	0.5	0.3	0.3	0.5	0.7	1	0.3	0.4

Table 17.1b M74C: operation

X	x_1	x_2	x_3	x_4	x_5	x_6	x_7	x_8	x_9	x_{10}	x_{11}	x_{12}
μ	-0.2	-0.3	0.7	0.7	0.7	0.7	0.1	0.1	0.8	0.3	0.8	0.9
ν	0.2	0.3	0.3	0.6	0.6	0.5	0.2	0.2	0.4	0.5	0.5	0.7

Table 17.1c Status quo: construction

X	x_1	x_2	x_3	x_4	x_5	x_6	x_7	x_8	x_9	x_{10}	x_{11}	x_{12}
μ	0	0	0	0	0	0	0	0	0	0	0	0
ν	1	1	1	1	1	1	1	1	1	1	1	1

Table 17.1d Status quo: operation

X	x_1	x_2	x_3	x_4	x_5	x_6	x_7	x_8	x_9	x_{10}	x_{11}	x_{12}
μ	0.2	0.1	0.3	0.3	0	0.3	0	0	0.2	0.1	0.2	0.5
ν	0.2	0.2	0.3	0.5	0.8	0.3	0.7	0.7	0.5	0.5	0.5	0.5

Table 17.1e Integrated package of measures: construction

X	x_1	x_2	x_3	x_4	x_5	x_6	x_7	x_8	x_9	x_{10}	x_{11}	x_{12}
μ	0.1	0	0	0.1	0	0.1	-0.1	-0.1	-0.6	0.1	0	0
ν	0.3	0.3	0.4	0.2	1	0.7	0.4	0.5	0.6	0.4	0.8	0.7

Table 17.1f Integrated package of measures: operation

X	x_1	x_2	x_3	x_4	x_5	x_6	x_7	x_8	x_9	x_{10}	x_{11}	x_{12}
μ	-0.4	-0.6	-0.8	-0.8	-0.6	-0.7	-0.4	-0.4	-0.8	0.05	-0.8	-0.7
ν	0.2	0.4	0.5	0.5	0.5	0.5	0.3	0.3	0.5	0.6	0.5	0.5

17.2.2 Preliminary Analysis

The fuzzy sets for construction and operation can be combined to give an overall fuzzy set for each option. The effects of the operation life cycle stage will be experienced over a much longer period than those of the construction life cycle stage. Therefore values of $\pi = 0.3$ for construction and $\pi = 0.7$ for use have been chosen for the three options. Values of $\delta = \varepsilon = 0.1$ will be used. Details of the combination of the two life cycle stages will be given for the M74C case. First μ' is calculated, and added to tables 17.1a and b of values of x, μ and ν to give tables 17.2a and b respectively.

Table 17.2a M74C: construction

X	x_1	x_2	x_3	x_4	x_5	x_6	x_7	x_8	x_9	x_{10}	x_{11}	x_{12}
μ	0.2	0.05	0.3	0.8	0.5	0.3	-0.1	-0.2	0.8	1	0.1	0.3
μ'	0.06	0.015	0.09	0.24	0.15	0.09	-0.03	-0.06	0.24	0.3	0.03	0.09
ν	0.3	0.2	0.2	0.7	0.5	0.3	0.3	0.5	0.7	1	0.3	0.4

Table 17.2b M74C: operation

X	x_1	x_2	x_3	x_4	x_5	x_6	x_7	x_8	x_9	x_{10}	x_{11}	x_{12}
μ	-0.2	-0.3	0.7	0.7	0.7	0.7	0.1	0.1	0.8	0.3	0.8	0.9
μ'	-0.14	-0.21	0.49	0.49	0.49	0.49	0.07	0.07	0.56	0.21	0.56	0.63
ν	0.2	0.3	0.3	0.6	0.6	0.5	0.2	0.2	0.4	0.5	0.5	0.7

The values for x_3 to x_6 and x_9 to x_{12} are positive for both construction and operation and so can be combined using the union operator defined in section 16.3.2.

For x_3 $\quad \mu'_{3h} = \mu'_{3m} = \mu'_{3u} = 0.49, \qquad \nu_{3h} = \nu_{3m} = \nu_{3u} = 0.3$

Therefore condition (aiii) is satisfied and $\mu'_{sup} = 0.49$, $\nu_{sup} = 0.3$ and $\pi_{sup} = 0.7$, so $\mu_{sup} = 0.7$

For x_4 $\quad \mu'_{4h} = \mu'_{4u} = 0.49, \quad \nu_{4h} = \nu_{4u} = 0.6, \qquad \mu'_{4m} = 0.24, \quad \nu_{3m} = 0.7$

Therefore condition (aiii) is again satisfied and
$\mu'_{4sup} = 0.49$, $\nu_{4sup} = 0.6$ and $\pi_{sup} = 0.7$, so $\mu_{4sup} = 0.7$

It is appropriate to take values of π_+ and π_- equivalent to the values of π for the operation and construction stages. This gives $\pi_{1+} = \pi_{2+} = 0.3$ and $\pi_{1-} = \pi_{2-} = 0.7$ for x_1 and x_2, $\pi_{7+} = \pi_{8+} = 0.7$ and $\pi_{7-} = \pi_{8-} = 0.3$ for x_7 and x_8

Combination then gives $\mu_{1sup} = -0.08$, $\nu_{1sup} = 0.23$ and $\mu_{2sup} = -0.195$, $\nu_{2sup} = 0.27$.

The other variables can be treated in a similar way to give table 17.3 for the total impacts over both life cycle stages:

Table 17.3a M74C

X	x_1	x_2	x_3	x_4	x_5	x_6	x_7	x_8	x_9	x_{10}	x_{11}	x_{12}
μ_{sup}	-0.08	-0.195	0.7	0.7	0.7	0.7	0.04	0.01	0.8	1	0.8	0.9
ν	0.23	0.27	0.3	0.6	0.6	0.5	0.23	0.29	0.4	1	0.5	0.7

Table 17.3b Status quo

X	x_1	x_2	x_3	x_4	x_5	x_6	x_7	x_8	x_9	x_{10}	x_{11}	x_{12}
μ_{sup}	0.2	0.1	0.3	0.3	0	0.3	0	0	0.2	0.1	0.2	0.5
ν	0.2	0.2	0.3	0.5	1	0.3	1	1	0.5	0.5	0.5	0.5

Table 17.3c Integrated package of measures

X	x_1	x_2	x_3	x_4	x_5	x_6	x_7	x_8	x_9	x_{10}	x_{11}	x_{12}
μ_{sup}	-0.25	-0.6	-0.8	-0.53	-0.6	-0.46	-0.4	-0.4	-0.8	0.05	-0.8	-0.7
ν	0.23	0.4	0.5	0.41	0.5	0.56	0.3	0.3	0.5	0.6	0.5	0.5

17.2.3 Comparison of the Options

The data for the three options can be compared in a number of different ways, based on:

- The augmented fuzzy sets for the whole life cycle.
- Aggregation of the data to give, for instance, a smaller number of categories of impacts.
- Aggregate values of 'benefits' and 'costs'.
- An overall estimate of the impacts.

Aggregation simplifies the data and can facilitate comparisons, but can also reduce the quality of representation of the information. The fuzzy set for the first option, including the (relative) priorities or weights of the different impacts, is presented in table 17.4. These priorities are relative to the other impacts and are therefore different from the values previously used for the construction and operation life cycle stages.

Table 17.4 M74C

X	x_1	x_2	x_3	x_4	x_5	x_6	x_7	x_8	x_9	x_{10}	x_{11}	x_{12}
μ	-0.08	-0.195	0.7	0.7	0.7	0.7	0.04	0.01	0.8	1	0.8	0.9
ν	0.23	0.27	0.3	0.6	0.6	0.5	0.23	0.29	0.4	1	0.5	0.7
π	0.3	0.9	0.6	0.9	0.7	0.9	0.9	0.7	0.8	0.6	0.9	0.9

High values of priority have been given to all the impacts except travel time, although this is counter to the general practice of basing road decisions largely on the estimated travel time savings. However travel time savings have generally been considered for cars only and often been at the expense of cyclists, pedestrians and users of public transport. This gives rise to equity issues, as it is generally people on lower incomes, women, children and older people who are the main users of public transport and the main groups of pedestrians. Increased vehicle speeds have also allowed people to live further from their work and shopping and leisure facilities to be sited away from population centres. This has resulted in people travelling greater distances, with consequent increases in associated energy consumption and emissions, but only slight increases in journey times. For ease of evaluation values of $\mu_i' = \mu_i\pi_i$, ν_i, and π_i are given in table 17.5.

Table 17.5 M74C

X	x_1	x_2	x_3	x_4	x_5	x_6	x_7	x_8	x_9	x_{10}	x_{11}	x_{12}
μ'	-0.024	-0.176	0.42	0.63	0.49	0.63	0.036	0.007	0.64	0.6	0.72	0.81
ν	0.23	0.27	0.3	0.6	0.6	0.5	0.23	0.29	0.4	1	0.5	0.7
π	0.3	0.9	0.6	0.9	0.7	0.9	0.9	0.7	0.8	0.6	0.9	0.9

The first two variables x_1 and x_2 have negative values and therefore represent benefits. They can be combined to give (μ_-, ν_-). For all values of δ and ε, combination of these impacts gives $\mu'_- = -0.176$, $\nu_- = 0.27$, $\pi'_- = 0.9$, so that $\mu_- = -0.20$

The ten impacts x_3, ..., x_{12} with positive values can now be combined in a similar way. For x_3, ..., x_{12}

$$\mu'_h = \mu'_{12} = 0.81, \qquad \nu_h = \nu_{12} = 0.7, \qquad \pi_h = \pi_{12} = 0.9$$
$$\mu'_m = \mu'_u = \mu'_{10} = 0.6, \qquad \nu_m = \nu_u = \nu_{10} = 1, \qquad \pi_m = \pi_u = \pi_{10} = 0.6$$

Therefore for $\delta = \varepsilon = 0.1$, not all of the conditions (ai), (aii), (bi) and (bii) or (aiii) are satisfied and so

$$\mu'_+ = 0.81, \qquad \nu_+ = 0.7, \qquad \pi'_+ = 0.9, \text{ so that } \mu_+ = 0.9$$

When the positive and negative impacts are equally weighted i.e. $\pi_+ = \pi_- = 0.5$, combination of (μ_+,ν_+) and (μ_-,ν_-) gives $\mu_{sup} = 0.35$, $\nu_{sup} = 0.49$.

For a more accurate representation of the fact that ten of the impacts have positive values and only two negative values, values of π_+ greater than π_- should be used, for instance

$\pi_+ = 0.9$, $\pi_- = 0.1$ gives $\mu_{sup} = 0.79$, $\nu_{sup} = 0.66$
$\pi_+ = 0.7$, $\pi_- = 0.3$, giving $\mu_{sup} = 0.57$, $\nu_{sup} = 0.57$.

The impacts for the other two options can be combined in a similar way, using the same weights or priorities for each impact to give a fair comparison. Values of μ'_i, ν_i and π_i for the status quo and integrated package of measures options are given in table 17.6.

Table 17.6a Status quo

X	x_1	x_2	x_3	x_4	x_5	x_6	x_7	x_8	x_9	x_{10}	x_{11}	x_{12}
μ'	0.06	0.09	0.18	0.27	0	0.27	0	0	0.16	0.06	0.18	0.45
ν	0.2	0.2	0.3	0.5	1	0.3	1	1	0.5	0.5	0.5	0.5
π	0.3	0.9	0.6	0.9	0.7	0.9	0.9	0.7	0.8	0.6	0.9	0.9

Table 17.6b Integrated package of measures

X	x_1	x_2	x_3	x_4	x_5	x_6	x_7	x_8	x_9	x_{10}	x_{11}	x_{12}
μ'	-.075	-.54	-.48	-.48	-.42	-.41	-.36	-.28	-.64	.03	-.72	-.63
ν	0.23	0.4	0.5	0.41	0.5	0.56	0.3	0.3	0.5	0.6	0.5	0.5
π	0.3	0.9	0.6	0.9	0.7	0.9	0.9	0.7	0.8	0.6	0.9	0.9

For the status quo option, all values are positive and therefore

$\mu_- = 0$, $\nu_- = 1$

$\mu'_h = \mu'_u = \mu'_{12} = 0.45$, $\nu_h = \nu_u = \nu_{12} = 0.5$, $\pi_h = \pi_u = \pi_{12} = 0.9$

$\mu'_m = \mu'_7 = 0$, $\nu_m = \nu_7 = 1$, $\pi_u = \pi_7 = 0.9$

Therefore, for all values of δ and ε, condition (aiii) is satisfied and

$\mu'_+ = 0.45$, $\nu_+ = 0.5$, $\pi'_+ = 0.9$, so that $\mu_+ = 0.5$

For the integrated package of measures option, only x_{10} has a positive value, so that

$\mu'_+ = \mu'_{10} = 0.03$, $\nu_+ = \nu_{10} = 0.6$, $\pi'_+ = \pi_{10} = 0.6$, and $\mu_+ = 0.05$

The remaining eleven variables can now be aggregated, using $\mu^*_i = -\mu^*_I$ and $\mu'^*_i = -\mu_i\pi_i$

$\mu'^*_h = \mu'^*_u = \mu'^*_{11} = 0.72$, $\nu_h = \nu_u = \nu_{11} = 0.5$, $\pi_h = \pi_u = \pi_{11} = 0.9$

$\mu'^*_m = \mu'^*_6 = 0.41$ $\nu_m = \nu_6 = 0.56$, $\pi_m = \pi_6 = 0.9$

Therefore, for all values of δ and ε, condition (aiii) is satisfied and

$\mu'_{-} = -\mu'^{*}_{11} = -0.72$, $\nu_{-} = \nu_{u} = \nu_{11} = 0.5$, $\pi'_{-} = \pi_{11} = 0.9$, so that $\mu_{-} = -0.8$

When the positive and negative impacts are equally weighted

i.e. $\pi_{+} = \pi_{-} = 0.5$, combination of (μ_{+},ν_{+}) and (μ_{-},ν_{-}) then gives $\mu_{sup} = -0.38$, $\nu_{sup} = 0.55$.

However, an appropriate choice of weights to indicate that the majority of the impacts are negative, would have $\pi_{+} > \pi_{-}$, for instance $\pi_{+} = 0.1$, $\pi_{-} = 0.9$, giving $\mu_{sup} = -0.72$, $\nu_{sup} = 0.51$

Values of μ_{sup} and ν_{sup} are tabulated in table 17.7 for the three options of the M74C, the status quo and an integrated package of measures, denoted a, b and c, for six different pairs of values of π_{+} and π_{-}:

Table 17.7 Total impacts for the 3 alternatives for different values of π_{+} and π_{-}

Values of π_{+} and π_{-}	M74C $(\mu_{a.sup},\nu_{a.sup})$	Status Quo $(\mu_{b.sup}, \nu_{b.sup})$	Integ. Measures $(\mu_{c.sup}, \nu_{c.sup})$
π_{+}=0.9, π_{-}=0.1	(0.79, 0.66)	(0.45, 0.55)	(-0.04, 0.59)
π_{+}=0.7, π_{-}=0.3	(0.57, 0.57)	(0.35, 0.65)	(-0.21, 0.57)
π_{+}=0.5, π_{-}=0.5	(0.35, 0.49)	(0.25, 0.75)	(-0.38, 0.55)
π_{+}=0.3, π_{-}=0.7	(0.13, 0.40)	(0.15, 0.85)	(-0.55, 0.53)
π_{+}=0.1, π_{-}=0.9	(-0.09, 0.31)	(0.05, 0.95)	(-0.72, 0.51)
π_{+}=0, π_{-}=1	(-0.2, 0.27)	(0, 1)	(-0.8, 0.5)

In this case the analysis shows that the integrated package of measures clearly performs better than the other options for all values of π_{+} and π_{-} or relative weightings or 'costs' and 'benefits'. The option of maintaining the status quo is preferable to that of building the M74C for moderate to high values of π_{+}, and moderate to low values of π_{-}. It is less preferred for low values of π_{+}, and high values of π_{-} i.e. when little attention is paid to the disadvantages, such as the environmental costs. However this has often happened in practice, with the advantages of new roads being stressed and their environmental costs and negative equity changes being ignored. The consequences of inappropriate transport policies are illustrated in fig. 17.2. Further sensitivity analysis could be carried out to investigate the effect of the relative weights or priorities given to the different impacts.

Fig. 17.2 The consequences of inappropriate transport policies!

17.3 Case Study: Analysis Using Intuitionistic Fuzzy Sets

17.3.1 Intuitionistic Fuzzy Sets for the Problem

The first stage is to obtain the intuitionistic fuzzy sets for the construction and operation stages for the three options. These fuzzy sets are again derived using the assumptions in section 17.1.2 and therefore the augmented fuzzy sets presented in section 17.2.1 can be used to obtain the intuitionistic fuzzy sets for the three options. In this discussion to avoid confusion the membership functions in the augmented and intuitionistic fuzzy set cases will be denoted by $\mu_A(.)$ and $\mu_I(.)$ respectively.

In the augmented fuzzy set case the membership function $\mu_A(.)$ is able to take both positive and negative values to indicate the extent to which negative and positive impacts or a property and its opposite holds. In the intuitionistic fuzzy set case $\mu_I(.)$ and $\gamma(.)$ are membership and non-membership functions respectively. Therefore the positive and negative values of $\mu_A(.)$ in the augmented fuzzy set case correspond to values of $\mu_I(.)$ and $\gamma(.)$ respectively in the intuitionistic case. However the values of $\mu_I(.)$ and $\gamma(.)$ are required to satisfy the following inequalities

$$\mu_I(x) + \gamma(x) \leq 1 \quad \text{for all } x \in X \tag{17.1a}$$

$$\mu_I(x), \gamma(x) \geq 0 \tag{17.1b}$$

with the difference $1 - \mu_I(x) - \gamma(x)$ equal to the intuitionistic index $\pi(x)$. This represents the consistency of information with $\pi(x)$ equal to zero representing total consistency and $\pi(x)$ equal to one total inconsistency. Although consistency and certainty are not exactly the same, since there is no other information available on consistency, the consistency $[1- \pi(x)]$ will be taken equal to $\nu(x)$, giving

$$\pi(x) = 1 - \nu(x) \quad \text{for all } x \in X \tag{17.2}$$

The values of $\mu_I(x)$ and $\gamma(x)$ are required to satisfy

$$\mu_I(x) + \gamma(x) + \pi(x) = 1 \quad \text{for all } x \in X \tag{17.3}$$

and also to be related to the modulus of the positive and negative values, $\mu_I^+(x)$ and $\mu_I^-(x)$ respectively, of $\mu_I(x)$. The easiest way to meet all these requirements is to put

$$\begin{aligned} \mu_I(x) &= 0.5[1- \pi(x) + \mu_A^+(x)\{1- \pi(x)\}] \\ &= 0.5[1- \pi(x)][1 + \mu_A^+(x)] \end{aligned} \tag{17.4a}$$

$$\begin{aligned} \nu(x) &= 0.5[1- \pi(x) - \mu_A^+(x)\{1 - \pi(x)\}] \\ &= 0.5[1- \pi(x)][1 - \mu_A^+(x)] \quad \text{when } \mu_A(x) \geq 0 \end{aligned} \tag{17.4b}$$

$$\begin{aligned} \mu_I(x) &= 0.5[1- \pi(x) - \mu_A^-(x)\{1 - \pi(x)\}] \\ &= 0.5[1- \pi(x)][1 - \mu_A^-(x)] \end{aligned} \tag{17.5a}$$

$$\begin{aligned} \nu(x) &= 0.5[1- \pi(x) + \mu_A^-(x) \{1 - \pi(x)\}] \\ &= 0.5[1- \pi(x)][1 + \mu_A^+(x)] \quad \text{when } \mu_A(x) < 0 \end{aligned} \tag{17.5b}$$

Scaling of $\mu_I(x)$ by $1 - \pi(x)$ is required to ensure that the values of $\gamma(x)$ are positive, as required by (17.1b). If this scaling were to be omitted, negative values could be obtained. The details of the calculation are illustrated for variables x_1 to x_6 for the operation life cycle stage for the M74C option. The values of $\mu_A(x)$ and $\nu(x)$ presented in table 17.1 in section 17.2.1 are repeated in table 17.8 for ease of reference. The twelve variables in the space I of environmental and social impacts will also be repeated for reference:

x_1 - travel times
x_2 - accidents
x_3 - stress on the Kingston Bridge
x_4 - noise
x_5 - community separation
x_6 - energy consumption
x_7 - employment
x_8 - economic development
x_9 - equity
x_{10} - costs
x_{11} - traffic generation
x_{12} - emissions

Table 17.8a M74C: construction

X	x_1	x_2	x_3	x_4	x_5	x_6	x_7	x_8	x_9	x_1	x_{11}	x_{12}
μ_A	0.2	0.05	0.3	0.8	0.5	0.3	-0.1	-0.2	0.8	1	0.1	0.3
ν	0.3	0.2	0.2	0.7	0.5	0.3	0.3	0.5	0.7	1	0.3	0.4

Table 17.8b M74C: operation

X	x_1	x_2	x_3	x_4	x_5	x_6	x_7	x_8	x_9	x_{10}	x_{11}	x_{12}
μ_A	-0.2	-0.3	0.7	0.7	0.7	0.7	0.1	0.1	0.8	0.3	0.8	0.9
ν	0.2	0.3	0.3	0.6	0.6	0.5	0.2	0.2	0.4	0.5	0.5	0.7

Table 17.8c Status quo: construction

X	x_1	x_2	x_3	x_4	x_5	x_6	x_7	x_8	x_9	x_{10}	x_{11}	x_{12}
μ_A	0	0	0	0	0	0	0	0	0	0	0	0
ν	1	1	1	1	1	1	1	1	1	1	1	1

Table 17.8d Status quo: operation

X	x_1	x_2	x_3	x_4	x_5	x_6	x_7	x_8	x_9	x_{10}	x_{11}	x_{12}
μ_A	0.2	0.1	0.3	0.3	0	0.3	0	0	0.2	0.1	0.2	0.5
ν	0.2	0.2	0.3	0.5	0.8	0.3	0.7	0.7	0.5	0.5	0.5	0.5

Table 17.8e integrated package of measures: construction

X	x_1	x_2	x_3	x_4	x_5	x_6	x_7	x_8	x_9	x_{10}	x_{11}	x_{12}
μ_A	0.1	0	0	0.1	0	0.1	-0.1	-0.1	-0.6	0.1	0	0
ν	0.3	0.3	0.4	0.2	1	0.7	0.4	0.5	0.6	0.4	0.8	0.7

Table 17.8f Integrated package of measures: operation

X	x_1	x_2	x_3	x_4	x_5	x_6	x_7	x_8	x_9	x_{10}	x_{11}	x_{12}
μ_A	-0.4	-0.6	-0.8	-0.8	-0.6	-0.7	-0.4	-0.4	-0.8	0.05	-0.8	-0.7
ν	0.2	0.4	0.5	0.5	0.5	0.5	0.3	0.3	0.5	0.6	0.5	0.5

In the subsequent tables rounded values of μ and γ will be given to two decimal places. To ensure that μ, γ and π always sum to one, a '5' in the third decimal place will be rounded down for whichever of μ and γ has a lower value. From table 17.8b,

$$\mu_A(x_i) \leq 0 \text{ for } i=1,2$$
$$\mu_A(x_i) \geq 0 \text{ for } i=3...12$$

Using equation (17.2) together with equations (17.5) for the variables x_1 and x_2 and equation (17.2) together with equations (17.4) for variables x_3 to x_{12} gives

$$\pi(x_1) = 1 - \nu(x_1) = 0.8$$
$$\mu_I(x_1) = 0.5[1- \pi(x_1)][1 - \mu_A^-(x_1)] = 0.5[1 – 0.8][1 – 0.2] = 0.08$$
$$\nu(x_1) = 0.5[1- \pi(x_1)][1 + \mu_A^-(x_1)] = 0.5[1 – 0.8][1 + 0.2] = 0.12$$

$$\pi(x_2) = 1 - \nu(x_2) = 0.7$$
$$\mu_I(x_2) = 0.5[1- \pi(x_2)][1 - \mu_A^-(x_2)] = 0.5[1 – 0.7][1 – 0.3] = 0.10$$
$$\nu(x_2) = 0.5[1- \pi(x_2)][1 + \mu_A^-(x_2)] = 0.5[1 – 0.7][1 + 0.3] = 0.20$$

$$\pi(x_3) = 1 - \nu(x_3) = 0.7$$
$$\mu_I(x_3) = 0.5[1- \pi(x_3)][1 + \mu_A^+(x_3)] = 0.5[1 - 0.7][1 + 0.7] = 0.26$$
$$\nu(x_3) = 0.5[1- \pi(x_3)][1 - \mu_A^+(x_3)] = 0.5[1 – 0.7][1 - 0.7] = 0.04$$

$$\pi(x_4) = 1 - \nu(x_4) = 0.4$$
$$\mu_I(x_4) = 0.5[1- \pi(x_4)][1 + \mu_A^+(x_4)] = 0.5[1 - 0.4][1 + 0.7] = 0.51$$
$$\nu(x_4) = 0.5[1- \pi(x_4)][1 - \mu_A^+(x_4)] = 0.5[1 – 0.4][1 - 0.7] = 0.09$$

$$\pi(x_5) = 1 - \nu(x_5) = 0.4$$
$$\mu_I(x_5) = 0.5[1- \pi(x_5)][1 + \mu_A^+(x_5)] = 0.5[1 - 0.4][1 + 0.7] = 0.51$$
$$\nu(x_5) = 0.5[1- \pi(x_5)][1 - \mu_A^+(x_5)] = 0.5[1 - 0.4][1 - 0.7] = 0.09$$

$$\pi(x_6) = 1 - \nu(x_6) = 0.5$$
$$\mu_I(x_6) = 0.5[1- \pi(x_6)][1 + \mu_A^+(x_6)] = 0.5[1 - 0.5][1 + 0.7] = 0.43$$
$$\nu(x_6) = 0.5[1- \pi(x_6)][1 - \mu_A^+(x_6)] = 0.5[1 - 0.5][1 - 0.7] = 0.07$$

The remaining values can be calculated analogously and are presented in table 17.9. The subscript 'I' is omitted in table 17.9 and the subsequent analysis, since there is no longer any possibility of confusion with the augmented fuzzy set case.

Table 17.9a M74C: construction

X	x_1	x_2	x_3	x_4	x_5	x_6	x_7	x_8	x_9	x_{10}	x_{11}	x_{12}
μ	0.18	0.11	0.13	0.63	0.38	0.20	0.13	0.20	0.63	1	0.17	0.26
γ	0.12	0.09	0.07	0.07	0.12	0.10	0.17	0.30	0.07	0	0.13	0.14
π	0.7	0.8	0.8	0.3	0.5	0.7	0.7	0.5	0.3	0	0.7	0.6

Table 17.9b M74C: operation

X	x_1	x_2	x_3	x_4	x_5	x_6	x_7	x_8	x_9	x_{10}	x_{11}	x_{12}
μ	0.08	0.10	0.26	0.51	0.51	0.43	0.11	0.11	0.36	0.33	0.45	0.67
γ	0.12	0.20	0.04	0.09	0.09	0.07	0.09	0.09	0.04	0.17	0.05	0.03
π	0.8	0.7	0.7	0.4	0.4	0.5	0.8	0.8	0.6	0.5	0.5	0.3

Table 17.9c Status quo: construction

X	x_1	x_2	x_3	x_4	x_5	x_6	x_7	x_8	x_9	x_{10}	x_{11}	x_{12}
μ	0.5	0.5	0.5	0.5	0.5	0.5	0.5	0.5	0.5	0.5	0.5	0.5
γ	0.5	0.5	0.5	0.5	0.5	0.5	0.5	0.5	0.5	0.5	0.5	0.5
π	0	0	0	0	0	0	0	0	0	0	0	0

Table 17.9d Status quo: operation

X	x_1	x_2	x_3	x_4	x_5	x_6	x_7	x_8	x_9	x_{10}	x_{11}	x_{12}
μ	0.12	0.11	0.20	0.33	0.40	0.20	0.35	0.35	0.30	0.28	0.30	0.38
γ	0.08	0.09	0.10	0.17	0.40	0.10	0.35	0.35	0.20	0.22	0.20	0.12
π	0.8	0.8	0.7	0.5	0.2	0.7	0.3	0.3	0.5	0.5	0.5	0.5

Table 17.9e integrated package of measures: construction

X	x_1	x_2	x_3	x_4	x_5	x_6	x_7	x_8	x_9	x_{10}	x_{11}	x_{12}
μ	0.17	0.15	0.20	0.11	0.50	0.39	0.18	0.22	0.12	0.22	0.40	0.35
γ	0.13	0.15	0.20	0.09	0.50	0.31	0.22	0.28	0.48	0.18	0.40	0.35
π	0.7	0.7	0.6	0.8	0	0.3	0.6	0.5	0.4	0.6	0.2	0.3

Table 17.9f Integrated package of measures: operation

X	x_1	x_2	x_3	x_4	x_5	x_6	x_7	x_8	x_9	x_{10}	x_{11}	x_{12}
μ	0.06	0.08	0.05	0.05	0.10	0.07	0.09	0.09	0.05	0.32	0.05	0.07
γ	0.14	0.32	0.45	0.45	0.40	0.43	0.21	0.21	0.45	0.28	0.45	0.43
π	0.8	0.6	0.5	0.5	0.5	0.5	0.7	0.7	0.5	0.4	0.5	0.5

17.3.2 Analysis

The first stage in the analysis is the combination of the fuzzy sets for construction and operation using equations (15.13) in section 15.3.1 to give the total life cycle impact for each of the twelve variables for each option. Unlike in the augmented fuzzy set case, different weightings cannot be given to the construction and operation stages. The details of the calculation will be illustrated for variables x_1 and x_2 for the M74C option.

$$\text{For } x_1,\ \mu_{sup} = \max(0.18, 0.08) = 0.18 \quad \nu_{sup} = \min(0.12, 0.12) = 0.12$$
$$\text{For } x_2,\ \mu_{sup} = \max(0.11, 0.10) = 0.11 \quad \nu_{sup} = \min(0.09, 0.20) = 0.09$$

The other values can be calculated similarly and are presented in table 17.10.

Table 17.10a M74C

X	x_1	x_2	x_3	x_4	x_5	x_6	x_7	x_8	x_9	x_{10}	x_{11}	x_{12}
μ	0.18	0.11	0.26	0.63	0.51	0.43	0.13	0.20	0.63	1	0.45	0.67
γ	0.12	0.09	0.04	0.07	0.09	0.07	0.09	0.09	0.04	0	0.05	0.03
π	0.70	0.80	0.70	0.30	0.40	0.50	0.78	0.71	0.33	0	0.50	0.30

Table 17.10b Status quo

X	x_1	x_2	x_3	x_4	x_5	x_6	x_7	x_8	x_9	x_{10}	x_{11}	x_{12}
μ	0.50	0.50	0.50	0.50	0.50	0.50	0.50	0.50	0.50	0.50	0.50	0.50
γ	0.08	0.09	0.10	0.17	0.40	0.10	0.35	0.35	0.20	0.22	0.20	0.12
π	0.42	0.41	0.40	0.33	0.10	0.40	0.15	0.15	0.30	0.28	0.30	0.38

Table 17.10c Integrated package of measures

X	x_1	x_2	x_3	x_4	x_5	x_6	x_7	x_8	x_9	x_{10}	x_{11}	x_{12}
μ	0.17	0.15	0.20	0.11	0.50	0.39	0.18	0.22	0.12	0.32	0.40	0.35
γ	0.13	0.15	0.20	0.09	0.40	0.31	0.21	0.21	0.45	0.18	0.40	0.35
π	0.70	0.70	0.60	0.80	0.10	0.30	0.61	0.57	0.43	0.50	0.20	0.30

As is discussed in more detail in the augmented fuzzy set case in section 17.2.3, the data can be compared in a number of different ways. The total impacts for each option can be calculated using equation (15.13) in section 15.3.1. As an illustration the calculation for the integrated package of measures option is given below. The intuitionistic fuzzy sets for the other two options can be calculated analogously and the intuitionistic fuzzy sets for all three options are presented in table 17.11.

$$\mu_{sup} = \max(0.17,0.15,0.20,0.11,0.50,0.39,0.18,0.22,0.12,0.32,0.40,0.35) = 0.50$$

$$\gamma_{sup} = \min(0.13,0.15,0.20,0.09,0.40,0.31,0.21,0.21,0.45,0.18,0.40,0.35) = 0.09$$

$$\pi_{sup} = 1 - 0.50 - 0.09 = 0.41$$

Table 17.11 Total impacts

	M74C	Status Quo	Integrated Measures
μ	1.0	0.50	0.50
γ	0	0.08	0.09
π	0	0.42	0.41

The membership grade of an option in μ gives a measure of the extent of the total negative environmental and social impacts of that option and the membership grade in γ gives a measure of the extent to which these negative impacts do not hold or to which there are positive environmental and social impacts. Therefore, as in the augmented fuzzy set case, the option of building the M74C performs considerably less well in environmental and social terms than either maintaining the status quo or implementing an integrated package of measures. However, unlike in the augmented fuzzy set case, the performance of the integrated package of measures and the status quo are almost identical. This is probably due to the details of the method which gives greater weight to the membership

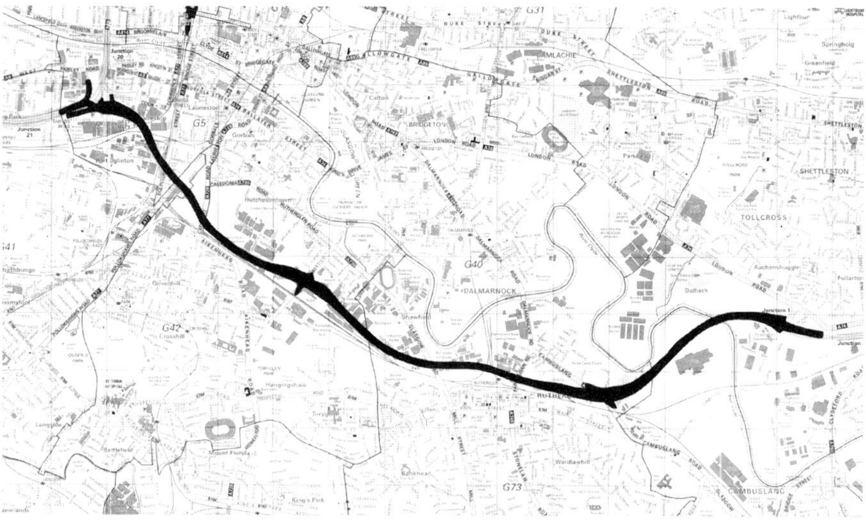

Fig 17.3 Proposed route of M74C

function μ(.) than the non-membership function γ(.). Repeating the analysis with μ(.) representing benefits i.e. positive environmental and social impacts would give greater weight to these benefits and therefore probably give a considerable difference between the performance of the status quo and integrated package of measures options. However this analysis will not be carried out here and is left to the reader as an exercise.

17.4 Case Study: Type 2 Fuzzy Sets

17.4.1 Obtaining Type 2 Fuzzy Sets

As in the intuitionistic fuzzy set case, the first stage is to obtain the type 2 fuzzy sets for the construction and operation stages for the three options. These fuzzy sets are derived using the assumptions in section 17.1.1 and therefore the augmented fuzzy sets for the two life cycle stages of the three options can be used to obtain the type 2 fuzzy sets.

The first stage in the calculation is obtaining linguistic descriptions for the type 1 fuzzy sets which replace membership functions for type 2 fuzzy sets. These linguistic descriptions can then be expressed as fuzzy sets. In the case of augmented fuzzy sets the degree of certainty is expressed by the second parameter ν. In the case of type 2 fuzzy sets the degree of certainty can be expressed by the spread of non-zero membership grades in the type 1 fuzzy sets, with increasing spread corresponding to a decreasing degree of certainty. Therefore the type 1 fuzzy sets can be expressed in terms of linguistic descriptions and modifiers which give the degree of spread of the non-zero membership grades. The linguistic descriptions and modifiers can be obtained from the parameters μ and ν respectively. These fuzzy sets are presented in tables 17.12. The twelve variables in the space I of environmental and social impacts will first be repeated for reference:

x_1 - travel times
x_2 - accidents
x_3 - stress on the Kingston Bridge
x_4 - noise
x_5 - community separation
x_6 - energy consumption
x_7 - employment
x_8 - economic development
x_9 - equity
x_{10} - costs
x_{11} - traffic generation
x_{12} - emissions

Table 17.12a M74C: construction

X	x_1	x_2	x_3	x_4	x_5	x_6	x_7	x_8	x_9	x_{10}	x_{11}	x_{12}
μ	pos low	pos non	pos low	prb hi	vps med	pos low	pos non	vps nglo	prb hi	vhi	pos non	pos low

Table 17.12b M74C: operation

X	x_1	x_2	x_3	x_4	x_5	x_6	x_7	x_8	x_9	x_{10}	x_{11}	x_{12}
μ	pos nlo	pos nlo	pos hi	vps hi	vps hi	vps hi	pos non	pos non	pos hi	vps lo	vps hi	prb vhi

Table 17.12c Status quo: construction

X	x_1	x_2	x_3	x_4	x_5	x_6	x_7	x_8	x_9	x_{10}	x_{11}	x_{12}
μ	non	non	non	non	non	non	non	non	non	non	non	non

Table 17.8d Status quo: operation

X	x_1	x_2	x_3	x_4	x_5	x_6	x_7	x_8	x_9	x_{10}	x_{11}	x_{12}
μ	pos low	pos non	pos low	vps low	prb non	pos low	prb non	prb non	vps low	vps non	vps low	vps med

Table 17.12e integrated package of measures: construction

X	x_1	x_2	x_3	x_4	x_5	x_6	x_7	x_8	x_9	x_{10}	x_{11}	x_{12}
μ	pos non	pos non	pos non	pos non	non	prb non	pos non	vps non	vps ngmd	pos non	prb non	prb non

Table 17.12f Integrated package of measures: operation

X	x_1	x_2	x_3	x_4	x_5	x_6	x_7	x_8	x_9	x_{10}	x_{11}	x_{12}
μ	pos ngmd	pos ngmd	vps nghi	vps nghi	vps ngmd	vps nghi	pos ngmd	pos ngmd	vps nghi	vps non	vps nghi	vps nghi

The type 1 fuzzy sets can be defined as follows:

- *negative high* (nghi): {(0.1, 1)}
- *negative medium* (ngmd): {(0.2, 1) + (0.3, 1)}
- *negative low* (nlo): {(0.4, 1)}
- *none* (non): {(0.5, 1)}
- *low* (lo): {(0.6, 1)}
- *medium* (med): {(0.7, 1) + (0.8, 1)}
- *high* (hi): {(0.9, 1)}
- *very high* (vhi): {(1, 1)}

The modifiers possibly (pos) very possibly (vps) and probably (prb) indicate the degree of certainty in the values with less certainty indicated by a greater spread of non-zero values. This is illustrated for the modified versions of the fuzzy set *none*, as follows:

- *possibly none*:
  {(0.2,0.4)+(0.3,0.6)+(0.4,0.8)+(0.5,1)+(0.6,0.8)+(0.7,0.6)+(0.8,0.4)}
- *very possibly none*:
  {(0.3,0.6)+(0.4,0.8)+(0.5,1)+(0.6,0.8)+(0.7,0.6)}
- probably none:
  {(0.4,0.8)+(0.5,1)+(0.6,0.8)}

The other modified fuzzy sets can be defined analogously.

17.4.2 Analysis

The type two fuzzy sets for the construction and operation stages for each option can now be combined using equations (15.23) in section 15.4. An outline of the calculation rather than full details is given for all variables for all the options.

Impacts for the M74C option

$f_{sup}(x_1)$ = possibly low V possibly negative low
= {(0.3,0.4)+(0.4,0.6)+(0.5,0.8)+(0.6,1)+(0.7,0.8)+(0.8,0.6)+(0.9,0.4)}
V {(0.1,0.4)+(0.2,0.6)+(0.3,0.8)+(0.4,1)+(0.5,0.8)+(0.6,0.6)+(0.7,0.4)}
= (0.3,0.4)+(0.4,0.6)+(0.5,0.8)+(0.6,1)+(0.7,0.8)+(0.8,0.6)+(0.9,0.4)

$f_{sup}(x_2)$ = possibly none V possibly negative low
= {(0.2,0.4)+ (0.3,0.6)+(0.4,0.8)+(0.5,1)+(0.6,0.8)+(0.7,0.6)+ (0.8,0.4)}
V {(0.1,0.4)+(0.2,0.6)+(0.3,0.8)+(0.4,1)+(0.5,0.8)+(0.6,0.6)+(0.7,0.4)}
= (0.2,0.4) + (0.3,0.6) + (0.4,0.8) + (0.5,1) + (0.6,0.8) +(0.7,0.6)+ (0.8,0.4)

$f_{sup}(x_3)$ = possibly low V possibly high
= {(0.3,0.4)+(0.4,0.6)+(0.5,0.8)+(0.6,1)+(0.7,0.8)+(0.8,0.6)+(0.9,0.4)}
V {(0.6,0.4)+(0.7,0.6)+(0.8,0.8)+(0.9,1)+(1,0.8)}
= (0.6,0.4) + (0.7,0.6) + (0.8,0.8) + (0.9,1) + (1,0.8)

$f_{sup}(x_4)$ = probably high V very possibly high
= {(0.8,0.8)+(0.9,1)+(1,0.8)}
V {(0.7,0.6)+(0.8,0.8)+(0.9,1)+(1,0.8)}
= (0.8,0.8) + (0.9,1) + (1,0.8)

$f_{sup}(x_5)$ = very possibly medium V very possibly high
= {(0.5,0.6)+(0.6,0.8)+(0.7,1)+(0.8,1)+(0.9,0.8)+(1,0.6)}
V {(0.7,0.6)+(0.8,0.8)+(0.9,1)+(1,0.8)}
= (0.7,0.6) + (0.8,0.8) + (0.9,1) + (1,0.8)

$f_{sup}(x_6)$ = possibly low V very possibly high
= {(0.3,0.4)+(0.4,0.6)+(0.5,0.8)+(0.6,1)+(0.7,0.8)+(0.8,0.6)+(0.9,0.4)}
V {(0.7,0.6)+(0.8,0.8)+(0.9,1)+(1,0.8)}
= (0.7,0.6) + (0.8,0.8) + (0.9,1) + (1,0.8)

$f_{sup}(x_7)$ = possibly none V possibly none
= {(0.2,0.4)+ (0.3,0.6)+(0.4,0.8)+(0.5,1)+(0.6,0.8)+(0.7,0.6)+(0.8,0.4)}

$f_{sup}(x_8)$ = very possibly negative low V possibly none
= {(0.2,0.6)+(0.3,0.8)+(0.4,1)+(0.5,0.8)+(0.6,0.6)}
V {(0.2,0.4)+ (0.3,0.6)+(0.4,0.8)+(0.5,1)+(0.6,0.8)+(0.7,0.6)+(0.8,0.4)}
= (0.2,0.4) + (0.3,0.6) + (0.4,0.8) +(0.5,1)+(0.6,0.8) +(0.7,0.6) +(0.8,0.4)

$f_{sup}(x_9)$ = probably high V possibly high
= {(0.8,0.8)+(0.9,1)+(1,0.8)}
V {(0.6,0.4)+(0.7,0.6)+(0.8,0.8)+(0.9,1)+(1,0.8)}
= (0.8,0.8) + (0.9,1) + (1,0.8)

$f_{sup}(x_{10})$ = very high V very possibly low
= {(1,1)} V {(0.4,0.6)+(0.5,0.8)+(0.6,1)+(0.7,0.8)+(0.8,0.6)}
= (1,1)

$f_{sup}(x_{11})$ = possibly none V very possibly high
= {(0.2,0.4)+ (0.3,0.6)+(0.4,0.8)+(0.5,1)+(0.6,0.8)+(0.7,0.6)+(0.8,0.4)}
V {(0.7,0.6)+(0.8,0.8)+(0.9,1)+(1,0.8)}
= (0.7,0.6) + (0.8,0.8) + (0.9,1) + (1,0.8)

$f_{sup}(x_{12})$ = possibly low V probably very high
= {(0.3,0.4)+(0.4,0.6)+(0.5,0.8)+(0.6,1)+(0.7,0.8)+(0.8,0.6)+(0.9,0.4)}
V {(0.9,0.8)+(1,1)}
= {(0.9,0.8)+(1,1)}

Impacts for the status quo option

$f_{sup}(x_1)$ = none V possibly low
{(0.5, 1)}
V {(0.3,0.4)+(0.4,0.6)+(0.5,0.8)+(0.6,1)+(0.7,0.8)+(0.8,0.6)+(0.9,0.4)}
= (0.5,0.8) + (0.6,1) + (0.7,0.8) + (0.8,0.6) + (0.9,0.4)

$f_{sup}(x_2)$ = none V possibly none
{(0.5,1)}
V {(0.2,0.4)+(0.3,0.6)+(0.4,0.8)+(0.5,1)+(0.6,0.8)+(0.7,0.6)+(0.8,0.4)}
= (0.5,1) + (0.6,0.8) + (0.7,0.6) + (0.8,0.4)

$f_{sup}(x_3) = f_{sup}(x_1)$
= (0.5,0.8) + (0.6,1) + (0.7,0.8) + (0.8,0.6) + (0.9,0.4)

$f_{sup}(x_4)$ = none V very possibly low

$\therefore f_{sup}(x_4) = \{(0.5,1)\} \vee \{(0.4,0.6)+(0.5,0.8)+(0.6,1)+(0.7,0.8)+(0.8,0.6)\}$
$= (0.5,0.8) + (0.6,1) + (0.7,0.8) + (0.8,0.6)\}$

$f_{sup}(x_5) =$ none V probably none
$\{(0.5, 1)\} \vee \{(0.4,0.8)+(0.5,1)+(0.6,0.8)\}$
$= (0.5,1) + (0.6,0.8)$

$f_{sup}(x_6) = f_{sup}(x_1)$
$= (0.5,0.8) + (0.6,1) + (0.7,0.8) + (0.8,0.6) + (0.9,0.4)$

$f_{sup}(x_7) = f_{sup}(x_5)$
$= (0.5,1) + (0.6,0.8)$

$f_{sup}(x_8) = f_{sup}(x_5)$
$= (0.5,1) + (0.6,0.8)$

$f_{sup}(x_9) = f_{sup}(x_4)$
$= (0.5,0.8) + (0.6,1) + (0.7,0.8) + (0.8,0.6)\}$

$f_{sup}(x_{10}) =$ none V very possibly none
$= \{(0.5,1)\} \vee \{(0.3,0.6)+(0.4,0.8)+(0.5,1)+(0.6,0.8)+(0.7,0.6)\}$
$= (0.5,1)+(0.6,0.8)+(0.7,0.6)$

$f_{sup}(x_{11}) = f_{sup}(x_4)$
$= (0.5,0.8) + (0.6,1) + (0.7,0.8) + (0.8,0.6)\}$

$f_{sup}(x_{12}) =$ none V very possibly medium
$\{(0.5, 1)\} \vee \{(0.5,0.6)+(0.6,0.8)+(0.7,1)+(0.8,1)+(0.9,0.8)+(1,0.6)\}$
$= (0.5,0.6)+(0.6,0.8)+(0.7,1)+(0.8,1)+(0.9,0.8)+(1,0.6)$

Impacts for the integrated package of measures option

$f_{sup}(x_1) =$ possibly none V possibly negative medium
$= \{(0.2,0.4)+ (0.3,0.6)+(0.4,0.8)+(0.5,1)+(0.6,0.8)+(0.7,0.6)+(0.8,0.4)\}$
$\vee \{(0,0.6)+(0.1,0.8)+(0.2,1)+(0.3,1)+(0.4,0.8)+(0.5,0.6)+(0.6,0.4)\}$
$= (0.2,0.4) + (0.3,0.6) +(0.4,0.8)+ (0.5,1) + (0.6,0.8) + (0.7,0.6) + (0.8,0.4)$

$f_{sup}(x_2) = f_{sup}(x_1)$
$= (0.2,0.4) + (0.3,0.6) +(0.4,0.8)+ (0.5,1) + (0.6,0.8) + (0.7,0.6) + (0.8,0.4)$

$f_{sup}(x_3)$ = possibly none V very possibly negative high
{(0.2,0.4)+ (0.3,0.6)+(0.4,0.8)+(0.5,1)+(0.6,0.8)+(0.7,0.6)+(0.8,0.4)}
V {(0,0.8)+(0.1,1)+(0.2,0.8)+(0.3,0.6)}
= (0.2,0.4)+(0.3,0.6)+(0.4,0.8)+(0.5,1)+(0.6,0.8)+(0.7,0.6) +(0.8,0.4)

$f_{sup}(x_4)$ = $f_{sup}(x_3)$
= (0.2,0.4)+ (0.3,0.6)+(0.4,0.8)+(0.5,1)+(0.6,0.8)+(0.7,0.6)+(0.8,0.4)

$f_{sup}(x_5)$ = none V very possibly negative medium
= {(0.5,1)}
V {(0,0.6)+(0.1,0.8)+(0.2,1)+(0.3,1)+(0.4,0.8)+(0.5,0.6)}
= (0.5,0.6)

$f_{sup}(x_6)$ = probably none V very possibly negative high
{(0.4,0.8)+(0.5,1)+(0.6,0.8)} V {(0,0.8)+(0.1,1)+(0.2,0.8)+(0.3,0.6)}
= (0.4,0.8)+(0.5,1)+(0.6,0.8)

$f_{sup}(x_7)$ = $f_{sup}(x_1)$
= (0.2,0.4) + (0.3,0.6) +(0.4,0.8)+ (0.5,1) + (0.6,0.8) + (0.7,0.6) + (0.8,0.4)

$f_{sup}(x_8)$ = very possibly none V possibly negative medium
= {(0.3,0.6)+(0.4,0.8)+(0.5,1)+(0.6,0.8)+(0.7,0.6)}
V {(0,0.6)+(0.1,0.8)+(0.2,1)+(0.3,1)+(0.4,0.8)+(0.5,0.6)+(0.6,0.4)}
= (0.3,0.6)+(0.4,0.8)+(0.5,1)+(0.6,0.8)+(0.7,0.6)

$f_{sup}(x_9)$ = very possibly negative medium V very possibly negative high
= {(0,0.6)+(0.1,0.8)+(0.2,1)+(0.3,1)+(0.4,0.8)+(0.5,0.6)}
V {(0,0.8)+(0.1,1)+(0.2,0.8)+(0.3,0.6)}
= (0,0.6)+(0.1,0.8)+(0.2,1)+(0.3,1)+(0.4,0.8)+(0.5,0.6)

$f_{sup}(x_{10})$ = possibly none V very possibly none
= {(0.2,0.4)+ (0.3,0.6)+(0.4,0.8)+(0.5,1)+(0.6,0.8)+(0.7,0.6)+(0.8,0.4)}
V {(0.3,0.6)+(0.4,0.8)+(0.5,1)+(0.6,0.8)+(0.7,0.6)}
= (0.3,0.6)+(0.4,0.8)+(0.5,1)+(0.6,0.8)+(0.7,0.6) +(0.8,0.4)

$f_{sup}(x_{11})$ = $f_{sup}(x_6)$
= (0.4,0.8)+(0.5,1)+(0.6,0.8)

$f_{sup}(x_{12})$ = $f_{sup}(x_6)$
= (0.4,0.8)+(0.5,1)+(0.6,0.8)

The total impacts for the three options can now be calculated. Equation (5.25) for the union of n type 2 fuzzy sets could be used. However it is probably easier to use equation (5.23) for the union of two type 2 fuzzy sets together with the commutativity and associativity properties to combine the sets pairwise. The details of the calculation will not be given. The type 2 fuzzy sets for the total impacts over the whole life cycle are as follows:

- M74C: (1,1)
- Status quo: (0.5,0.6)+(0.6,0.8)+(0.7,1)+(0.8,1)+(0.9,0.8)+(1,0.6)
- Integrated package of measures: (0.5,0.6)+(0.6,0.6)+(0.7,0.6)+(0.8,0.4)

As in the augmented and intuitionistic fuzzy set cases, the type 2 fuzzy set analysis shows that the performance of the M74C option is much worse than that of either the status quo or integrated package of measures. It also agrees with the augmented fuzzy set analysis and differs from the intuitionistic set analysis in differentiating between the performance of the latter two options, with the integrated package of measures performing significantly better than the status quo option.

17.5 Discussion

The case study has illustrated the application of three different fuzzy methods for representing and analysing imprecision and uncertainty in a transport decision making problem. The three methods considered were third order augmented fuzzy sets with negative coefficients, intuitionistic fuzzy sets and type 2 fuzzy sets. The analysis considered the impacts of twelve different environmental and social variables for the three transport planning options of the M74C (an urban motorway in the south of Glasgow in Scotland), maintaining the status quo and an integrated package of measures to achieve a modal shift away from private car use.

All three approaches found that the performance of the M74C option was much worse than that of either of the two other options with regards to the combination of negative environmental and social impacts and costs. This result is consistent with the findings of the Public Local Inquiry (Hickman and Watt 2004), though it should be noted that this inquiry did not investigate an option analogous to the integrated package of measures. The Inquiry reporters concluded that the construction of the M74C would have very serious adverse impacts on the environment of communities along the route during construction and operation, run counter to policies to promote social inclusion and environmental justice and impede the achievement of important Scottish Executive commitments to traffic reduction, public transport improvements and reduced carbon dioxide emissions and that the benefits of a temporary reduction in traffic congestion would progressively be lost due to traffic growth. There would be benefits to the local economy in some regions of Scotland, but at the expense of other regions.

However, despite the recommendation that the proposal for the construction of the M74C should not be authorised, the Scottish Executive has given its approval. A legal challenge has been lodged by Friends of the Earth Scotland and JAM74 (Joint Action Against the M74) on a number of grounds, including the failure by ministers to give adequate reasons for disagreeing with the conclusions of the Inquiry and the introduction of new evidence by ministers to support their rejection of these conclusions. There is likely to be direct action if the legal challenge proves unsuccessful.

The augmented fuzzy set and type 2 methods both found that the integrated package of measures was the best option overall. However, in the intuitionistic fuzzy set case the performance of the status quo and the integrated package of measures was almost identical, with the status quo option performing very slightly better, but the difference of no real significance.

The case study has illustrated the different ways in which the three approaches treat a combination of negative and positive impacts and uncertainty. The third order augmented fuzzy set approach has an additional parameter to give a measure of uncertainty and allows the measurement grade to take both positive and negative values to treat both positive and negative impacts.

The intuitionistic fuzzy set has two parameters and an intuitionistic index which gives a measure of consistency. The fact that the two parameters represent membership and non-membership in a particular set allows both negative and positive impacts to be treated by assigning one set to membership and the other to non-membership. However the approach is not symmetric in its treatment of the two types of impacts and therefore there will generally be some difference in the results according to whether positive or negative impacts are associated with membership rather than non-membership. The type 2 fuzzy set replaces membership grades by type 1 fuzzy sets and this additional level allows both positive and negative impacts and uncertainty to be handled.

The augmented fuzzy set approach is able to investigate the effect of giving different weightings to both the different impacts and the two life cycle stages. It is therefore probably the most appropriate method of the three to apply to problems of this type.

17.6 Tutorial Exercises

1. What is meant by the term 'fuzzy set'. Discuss briefly the advantages and disadvantages of fuzzy sets compared to crisp sets and probability theory in the context of potential applications in sustainable development.

2. Obtain the intersection, union and complements of the fuzzy sets with the following membership functions:

x	10	12	14	16	18	20
$\mu_1(x)$	0	0	0.1	0.2	0.8	1

x	10	12	14	16	18	20
$\mu_2(x)$	0.4	0.6	0.8	0.8	0.6	0.4

3. State four examples of environmental or social variables (not mentioned in this chapter) that could be appropriately represented using fuzzy techniques and derive fuzzy sets for them.

4. What are the main differences between augmented, intuitionistic and type 2 fuzzy sets. Give examples of the use of each of these techniques in the representation of environmental, social or developmental variables. Compare and contrast the advantages and disadvantages of the different representations.

5. Obtain the union, intersection and complement of the following augmented fuzzy sets:

x	10	12	14	16	18	20
$\mu_1(x)$	0.3	0.8	0.8	0.4	0.1	0.7
$v_1(x)$	0.7	0.7	0.3	0.8	0.5	0.5
$\mu_2(x)$	0.5	0.6	0.3	0.6	0.6	0.3
$v_2(x)$	0.6	0.5	0.8	0.5	0.4	0.4

$\mu_3(x)$	0.8	0.2	0.5	0.3	0.2	0.5
$v_3(x)$	0.4	0.2	0.6	0.6	0.8	0.2
$\mu_4(x)$	0.8	0.35	0.6	0.2	0.4	0.8
$v_4(x)$	0.3	0.1	0.5	0.7	0.7	0.4
x	10	12	14	16	18	20
$\mu_5(x)$	0.6	0.5	0.4	0.3	0.3	0.4
$v_5(x)$	0.7	0.3	0.6	0.5	0.1	0.8

6. Obtain a type 2 fuzzy set representation for the gear manufacturing process in section 15.5.

7. Obtain the total impacts across all categories for the different life cycle stages in the gear manufacturing process and computer life cycle examples in sections 15.5 and 15.6 respectively using:

   a. Second order augmented fuzzy sets
   b. Intuitionistic fuzzy sets
   c. Type 2 fuzzy sets

   Comment on these total impacts and any differences in the results obtained using the different techniques.

8. Use appropriate techniques to order the following augmented fuzzy sets for

   $\delta = \varepsilon = 0.2$ and $\eta = 0.33$

(0.7, 0.5)	(0.3, 0.4)	(0.5, 0.7)
(0.8, 0.4)	(0.4, 0.8)	(0.2, 0.7)
(0.3, 0.6)	(0.1, 0.9)	(0.5, 0.4)

9. Obtain the union of the following augmented fuzzy sets for appropriate values of δ and ε and investigate the dependence of the result on the values chosen for π_+ and π_-:

x	2	4	6	8
$\mu_1(x)$	-0.5	0.6	0.3	-0.7
$v_1(x)$	0.6	0.5	0.2	0.5
$\pi_1(x)$	0.9	0.4	0.5	0.8
$\mu_2(x)$	-0.7	0.4	-0.2	-0.6
$v_2(x)$	0.5	0.7	0.3	0.7

$\pi_2(x)$	0.8	0.6	0.5	0.5
$\mu_3(x)$	0.6	0.8	-0.4	0.6
$\nu_3(x)$	0.5	0.3	0.7	0.8
$\pi_3(x)$	0.4	0.3	0.6	0.5
x	2	4	6	8
$\mu_4(x)$	0.3	-0.3	0.7	0.4
$\nu_4(x)$	0.8	0.5	0.4	0.9
$\pi_4(x)$	0.4	0.8	0.6	0.8
$\mu_5(x)$	0.7	-0.6	-0.5	-0.5
$\nu_5(x)$	0.3	0.3	0.4	0.6
$\pi_5(x)$	0.5	0.6	0.7	0.9
$\mu_6(x)$	-0.5	0.5	0.5	0.3
$\nu_6(x)$	0.7	0.4	0.7	0.2
$\pi_6(x)$	0.3	0.7	0.4	0.6

10. a. Give three examples of sustainable design or decision making problems to which third order augmented fuzzy set representations could be applied and derive appropriate third order fuzzy set representations.

    b. Apply appropriate ordering techniques to obtain an impact reduction strategy or prioritise the different factors in the three problems you presented in the answers to question 10a.

    c. Apply an appropriate union operator to obtain the total impacts throughout the product or process life cycle or the total effects of the different factors over different decision making stages for the three problems presented in the answer to question 10a.

References and Additional Reading for Part III

References for Part III

Atanassov KT (1986) Intuitionistic fuzzy sets. Fuzzy Sets and Systems, 20:87-96

Atanassov KT (1989) More on intuitionistic fuzzy sets. Fuzzy Sets and Systems, 33:37-45

Atanassov KT (1999) Intuitionistic fuzzy sets: theory and applications. Physica Verlag, New York

Atanassov KT, Gargov G (1989) Interval valued intuitionistic fuzzy sets. Fuzzy Sets and Systems, 31:343-349

Bandemer H, Gottwald S (1995) Fuzzy sets, fuzzy logic, fuzzy methods with applications. John Wiley and Sons.

Blanchard N (1983) Fuzzy relations on fuzzy sets; fuzzy orderings; fuzzy similarities, two applications to ecology. Proc Int Conf on Systems, Man and Cybernetics, 132-135

Bustince H, Burillo P (1995) Correlation of interval-valued intuitionistic fuzzy sets. Fuzzy Sets and Systems, 74:237-244

Hersh MA (1998) A survey of green design with illustrations from the computer industry. IEEE Transactions on Systems, Man and Cybernetics, Part C, 28(4):428-540

Hersh MA (1999a) Fuzzy approaches to green design with examples from the computer industry. The 3rd Annual Conference of the Romanian Mathematical Sciences Society, Craiova, Romania

Hersh MA (1999b) The application of fuzzy methods to sustainable decision making. EUFIT '99, Aachen

Hersh MA, Hamburg I, Hamburg P (1998) The application of fuzzy methods to environmentally compatible production in manufacturing industry. Analele Univ. din Craiova, XXV:1-7.

Hickman RM, Watt D (2004) Roads (Scotland) Act 1984: Acquisition of Land (authorisation procedure) (Scotland) Act 1947, M74 special road (Fullarton Road to west of Kingston Bridge) orders report of public local inquiry into objections vol 1: main report. http://www.scotland.gov.uk/planning_appeals/seiru

Houghton JT, Ding Y, Griggs DJ, Noguer M, van der Linden PJ, Xiaosu D (2001) Climate change 2001: the scientific basis, contribution of working group I to the third assessment report of the Intergovernmental Panel on Climate Change (IPCC). Cambridge University Press

John R. (1998), Type 2 fuzzy sets: an appraisal of theory and applications. Int J of Uncertainty, Fuzziness and Knowledge-Based Systems, 6 (6):563-576

Karnik NN, Karnik JM (1998) Introduction to type-2 fuzzy logic systems. Proc IEEE Int Conf on Fuzzy Systems

Klir GJ, Yuan B (1995) Fuzzy sets and fuzzy logic theory and applications. Prentice Hall

Kruse R, Gebhardt J, Klawonn F (1994) Foundations of fuzzy systems. John Wiley and Sons

Leekwijck W van, Kerre EE (1999) Defuzzification: criteria and classification. Fuzzy Sets and Systems, 108:159-178

McCarthy JJ, Canziani OF, Leary NA, Dokken DJ, White KS (eds) (2001) Climate change 2001: impacts, adaptation and vulnerability, contribution of working group II to the third assessment report of the Intergovernmental Panel on Climate Change (IPCC). Cambridge University Press

Mendel JM (2001) Uncertain rule-based fuzzy logic systems: introduction and new directions. Prentice Hall

Mizumoto M, Tanaka K (1976) Some properties of fuzzy sets of type 2. Information and Control, 31:310-340.

Nakicenovic N, Swart R (2000) Special report on emissions scenarios. Cambridge University Press.

Newby HA, Popelka GA (1985) Audiology. 5th ed, Prentice Hall

O'Brien M (2000) Making better environmental decisions: an alternative to risk assessment. MIT Press, Cambridge MA

Pfleiderer RHH, Dieterich M (1995) New roads generate new traffic. World Transport Policy and Practice, 1(1):29-31.

Raffensperger C, Tickner J (eds) (1999). Protecting public health and the environment: implementing the precautionary principle. Island Press, Washington DC

SACTRA (1994) Trunk roads and the generation of traffic. Standing Advisory Committee on Trunk Road Assessment

SACTRA (1999) Transport and the economy. Standing Advisory Committee on Trunk Road Assessment, http://www.dft.gov.uk/stellent/groups/dft_econappr/documents/page/dft_econappr_022512.hcsp

Terano T, Asai K, Sugeno M (1987) Fuzzy Systems theory and its applications. Academic Press

Yager RR (1980) Fuzzy sets of type II in decisions. J of Cybernetics, 10:137-159

Zadeh LA (1965) Fuzzy sets. Information and Control, 8:338-353.

Zadeh LA (1975) The concept of a linguistic variable and its application to approximation reasoning III. Information Sciences, 9:43-79

Additional Reading for Part III

Atanassov KT, Gargov G (1998) Elements of intuitionistic fuzzy logic. part I. Fuzzy Sets and Systems, 95:39-52
Buckley JJ, Hayashi Y, Czogala E (1993) On the equivalence of neural nets and fuzzy expert systems. Fuzzy Sets and Systems, 53:124-129
Burillo P, Bustince H (1996) Entropy on intuitionistic fuzzy sets and on interval-value fuzzy sets. Fuzzy Sets and Systems, 78:305-316
Bustince H (2000) Construction of intuitionistic fuzzy sets with predetermined properties. Fuzzy Sets and Systems, 109:379-403
Bustince H, Burillo P (1995) Correlation of interval-valued intuitionistic fuzzy sets. Fuzzy Sets and Systems, 74:237-244
Bustince H, Burillo P (1996) Vague sets are intuitionistic fuzzy sets. Fuzzy Sets and Systems, 79:403-405
Cattaneo G (1997) Generalised rough sets. Studia Logica, 58:47-77.
Ciftcibasi and Altunay D (1998) Two-sided (intuitionistic) fuzzy reasoning. IEEE Trans on Systems, Man and Cybernetics, 28(5):662-677.
Demirci M (2000) Axiomatic theory of intuitionistic fuzzy sets. Fuzzy Sets and Systems, 110:253-266
Diamond P, Kloeden P (1990) Metric spaces of fuzzy sets. Fuzzy Sets and Systems, 35:241-249.
Dubois D, Prade H (1978) Operations on fuzzy numbers. Int J Systems Science, 9:613-626
Dubois D, Prade H (1987) Properties of measures of information in evidence and possibility theories. Fuzzy Sets and Systems, 24:161-182
Dubois D, Prade H (1990) The semantics of fuzzy 'if ... then ...' rules. In: Novak V, Ramik J, Mares M, Cherny M, Nekola J (eds) Fuzzy approach to reasoning and decision making. Kluwer academic publishers, pp 3-16
Dubois D, Prade H (1991) Fuzzy sets in approximate reasoning. Part 1: inference with possibility distributions. Fuzzy Sets and Systems, 40:143-202
Fishwick PA (1991) Fuzzy simulation: specifying and identifying qualitative models. Int J General Systems, 19:295-316
Geyer-Schulz A (1996) Fuzzy rule-based expert systems and genetic machine learning, 2nd edn. Physica-Verlag
George AV, Stoica MV, Barac A, Bărbăntan M (1985) Fuzzy models for energy demand management. Cybernetics and Systems: International J, 16:181-189
Giering EW III, Kandel A (1983) The application of fuzzy set theory to the modeling of competition in ecological systems. Fuzzy Sets and Systems, 9:103-127
Goguen JA (1967) L-fuzzy sets. J of Mathematical Analysis and Applications, 18:145-174
Gupta MM, Sanchez E (1982) Fuzzy information and decision processes. North-Holland

Hamburg I (1994) The Potential of fuzzy logic methods to support multicriteria group decision-making in the engineering process. In: Bouchon-Meunier & Yager (eds) Information processing and management of uncertainty in knowledge-based systems. Cité Internationale, Paris, pp 633-637

Hamburg I (1996) Using fuzzy logic for the modelling of product design and manufacture. In: Felix R (ed), EFDAN 96, pp 79-85

Hamburg P, Florescu L (1984) Topological structures in fuzzy lattices. J Mathematical Analysis and Applications, 101(2):475-490

Hersh MA, Hamburg I (1999a) Fuzzy logic and green design: a new approach. CIMCA'99, Vienna

Hersh MA, Hamburg I, Padeanu L (1999b), Integration of soft sustainable computing methods into engineering design environments. EDA '99, Vancouver

Hersh MA, Hamburg I, Padeanu L (2000) Integration of workflow management systems with other supporting software tools. HAAMAHA 2000, Krakow

Höhle U (1984) Fuzzy plausibility measures. In: Zimmermann HJ, Zadeh LA, Gaines BR (eds) Fuzzy Sets and Decision Analysis, North Holland, pp 83-96

John RI, Innocent PR, Barnes MR (1997) Type 2 fuzzy sets and neuro-fuzzy clustering of radiographic tibia images. Proc 6th IEEE Int Conf Fuzzy Systems

Kacprzyk J, Fedrizzi M (1992) Fuzzy regression analysis. Physica-Verlag

Kandel A (1986) Fuzzy mathematical techniques with applications. Addison-Wesley Publishing Company

Kawkernaak J (1978) Fuzzy random variables – I, definitions and theorems. Information Sciences, 15:1-29

Lee ES, Zhu Q (1995) Fuzzy and evidence reasoning. Physica-Verlag

Lin TY, Cercone N (1997) Rough sets and data mining: analysis for imprecise data. Kluwer Academic, Boston MA, London

Mabuchi S (1997) An interpretation of membership function and the properties of general probalistic operators as fuzzy set operators (II), extension to three-valued and interval-valued fuzzy sets. Fuzzy Sets and Systems, 92, 31-50

Mielczarski W (1998) Fuzzy logic techniques in power systems. Physica-Verlag

Negoită CV, Ralescu DA (1975) Applications of fuzzy sets to system analysis. Birkhäuser Verlag

Novák V, Ramík J, Mareš M, Cerny M, Nekola J (1992) Fuzzy Approaches to reasoning and decision making. Kluwer Academic Publishers

Oder C, Haasis H-D, Rentz O (1993) Analysis of the Lithuanian final energy consumption using fuzzy sets. Int J of Energy Research, 17:35-44

Orlowska E (1998) Incomplete information: rough set analysis. Physica-Verlag

Orlowski SA (1984) Fuzzy information in problems of resources allocation. IFAC 9th World Congress, Budapest

Pawlak Z (1982) Rough sets. Int J of Computer and Information Sciences, 11:341-356.

Pawlak Z (1997) Rough set approach to knowledge-based decision support. European J. of Operational Research, 99:48-57.

Pawlak Z, Slowinski R (1994) Rough set approach to multi-attribute decision analysis. European J of Operational Research, 72:443-459.

Sakawa M, Seo F (1983) Interactive multiobjective decision-making in environmental systems using the fuzzy sequential proxy optimization technique. Large Scale Systems Theory and Applications, 4(3): 223-243

Salksi A, Sperlbaum C (1990) Fuzzy logic approach to modeling in ecosystem research. Conf on Info Processing, 520-527

Sanchez E (1984) Fuzzy Information, knowledge representation and decision analysis. IFAC Proceedings Series, no 6, IFAC

Sato M, Sato Y, Jain LC, Jain LC (1997) Fuzzy clustering models and applications. Physica-Verlag

Shenoi S, Melton A (1989) Proximity relations in the fuzzy relational database model. Fuzzy Sets and Systems, 31:285-296

Slowinski R (1992) Intelligent decision support, handbook of application and advances of the rough sets theory. Kluwer Academic Publishers

Tarabishy MN, Grudzinski JJ (1996) An application of fuzzy logic to power generation control. Proc American Power Conf

Vila, MA, Kaprzyk, Verdegay JL (1994) Fuzzy Optimisation. Physica-Verlag

Von Mølbak T, Hammer L, Kristgeirsson J (1997) Einführung der Fuzzy-Control - Technik in der dänischen Kraftwirtschaft - praktische Erfahrungen, Probleme und Perspektiven. VGB Kraftwerkstechnik, 77:457-460

Yager RR (1992) Implementing fuzzy logic controllers using a neural network framework. Fuzzy Sets and Systems, 48:53-64

Zadeh LA (1974) The concept of a linguistic variable and its application to approximate reasoning I and II. Information Sciences, 8:199-249, 301-375

Zadeh LA (1978) Fuzzy sets as a basis for a theory of possibility. Fuzzy Sets and Systems, 1:3-28

Zadeh LA (1994) Fuzzy logic, neural networks and soft computing. Communications of the ACM, 37:77-82

Zadeh LA, Fu K-S, Tanaka K, Shimura M (1995) Fuzzy sets and their applications to cognitive and decision processes. Academic Press Inc

Zimmermann HJ (1995) Neuro + Fuzzy: Technologien – Anwendungen. VDI Verlag, Düsseldorf

Zimmerman HJ, Zadeh LA, Gaines BR (1984) Fuzzy Sets and decision analysis. North-Holland

Zhong N, Skowron A and Ohsuga S (1997) New Directions in rough sets, data mining, and granular-soft computing. Proc 7th International Workshop, RSFDGr '99, Springer Verlag, Yamaguchi

18 Looking Back and Moving Forward

18.1 Discussion of the Book's Contribution

This book has presented an innovative approach to analysing, increasing understanding of and starting to solve the problems of sustainable development. The approach is based on the use of mathematical and computational modelling techniques.

The technical material is organised into three parts which present the following topics:

- Part I: *systems theory and methodologies* for structuring complex sustainable development problems to make it easier to obtain a solution to them.
- Part II: *optimisation and decision making techniques* to support policy formulation and other decision applications.
- Part III: *fuzzy sets and techniques* for representing and analysing the mixture of quantitative and qualitative, uncertain and imprecise data occurring in environmental, developmental and social problems.

Each part includes the following:

- Introductory and more advanced theoretical material.
- Presentation of a number of methodologies and techniques.
- Illustration of the application of the methodologies and techniques to problems in sustainable development by a case study and a large number of examples. The very large number of examples and wide range of sustainable development topics covered is one of the distinguishing features of the book.
- Tutorial examples, references and additional reading materials.

These three parts are introduced by an *extended overview* of some of the main themes in sustainable development in Chapter 1. A brief discussion of the content and contribution of Chapter 1 and each of the three technical parts of the book will be given in the following subsections. This will be followed by a discussion of the way forward in section 18.2.

18.1.1 Chapter 1: Overview of Sustainable Development

Chapter 1 presents an overview of the main ideas in sustainable development as an introduction to the mathematical and computing techniques presented in subsequent chapters. Reflecting on the chapter shows just how unsustainable our world currently is and the dire social injustices and environmental crises we face. However the aim is to make readers aware of the urgent need for action and to encourage them to obtain more information, as well as to use the tools presented in the book.

The chapter stresses the importance of considering *both* the *social/developmental* and *environmental* aspects of sustainable development rather than focusing on one or the other and the *relationship* between the social and environmental aspects of unsustainability. It also highlights *inequalities* both *between and within countries* in access to resources, education, employment and income and the importance of the education of girls in particular for promoting development. Environmental problems, which include global warming and ozone depletion, are largely a result of the *unsustainable development paths* followed by the industrialised countries and will require them to act first, rather than expecting the 'developing' countries to pay the price in terms of restricted development while continuing to benefit from exploitation of the world's resources.

18.1.2 Part I: Systems Methodologies

Systems approaches can be used to *structure the very complex problems* that arise in sustainable development, with a view to increasing understanding, providing tools for analysis and either obtaining a solution to the problem or reformulating it to make it easier to solve with other techniques.

The theoretical material includes a discussion of the categorisation of systems approaches, for instance into 'soft' and 'hard' or engineering systems approaches, and a presentation of the *mathematical background* which underlies engineering systems approaches. This mathematical background is used in the derivation of state space system representations, which are illustrated by application to a model of the arms race between two (groups of) nations. The *other methodologies and techniques* include the SIMILAR Method and other engineering approaches, such as the program analysis and review technique (PERT), interaction matrices and tree representations, as well as soft systems approaches, such as Checkland's Soft Systems Methodology and the Systems Failures Method.

The SIMILAR Method and Checkland's Soft System Methodology are illustrated by *application* to the problem of waste reduction in Glasgow and the Clyde Valley in Scotland. *Other examples* include the environmental impacts at the different computer life cycle stages, population growth, regulation of traffic and the control of water outflow temperatures through the use of feedback and feedforward, cities and self-sufficient rural communities as examples of open and closed systems, a prison as an example of an open system, a model of a regional system and a model of the development of policy for sustainable development.

There are unfortunately a number of attitudinal, political and other *barriers to sustainable development*. Some of these barricrs are discussed using mental models and possible approaches to overcoming these barriers are presented in terms of multi-loop adaptive learning. The discussion is illustrated by an analysis of the catastrophic accident at the Bhopal plant in India and measurements of the ozone layer.

18.1.3 Part II: Optimisation and Multi-Criteria Decision Making

Policy formulation, regulatory measures and other action to promote sustainable development requires *choices* to be made about appropriate policies and measures, giving rise to a need for decision making and optimisation tools. These tools form the subject of Part II of the book. The *mathematical background* to the subject includes a presentation of the main concepts in optimisation, such as conditions for the minimum of a function of one and n variables and the use of test functions. Other background topics considered include an introduction to binary and preference relations, which are used in the mathematical theory of preferences, an introduction to utility theory and a categorisation of the different multi-criteria decision support methodologies. Decision problem formulation is also discussed and the main components, such as attributes, objectives and criteria are considered. *Modelling of the decision process* is also considered. The four stage model of intelligence, design, choice and review is presented and the differences between naturalistic and classical approaches to decision making are discussed.

An overview of methods for minimising functions of n variables is presented, as well as a more detailed discussion of *multi-criteria decision support methods*. The methods presented are categorized as outranking methods, including ELECTRE and PROMETHEE, aggregation of criteria methods, including P/G% and goal programming, the Analytical Hierarchy Process and multi-attribute utility theory. It should be noted that these methods can only be used to support the third, choice, stage in the four-stage model of decision making.

The multi-criteria decision support methods are illustrated by *application to decision making* between six alternatives for managing municipal waste to take account of resource consumption and environmental as well as cost factors. Other examples in Part III include environmental and social factors to be considered in purchasing decisions, the environmental impacts of energy generation, criteria for maintaining biodiversity and species preservation, attribute levels for *noise*, decision making about the impacts of coal fired power stations, the environmental and social impacts of an industrial process, the use of decision trees in decisions on energy generation and conservation and the ranking of hydroelectric power station projects.

18.1.4 Part III: Fuzzy Sets and Techniques

Problems in sustainable development frequent involve a mixture of quantitative and qualitative data, measured in different units and scales, as well as data which is inaccurate, imprecise and/or uncertain. There is therefore a need for *data handling techniques* for this type of mixed and poor quality data. One of the approaches to doing this, which is considered in Part III of the book, is the use of fuzzy sets and techniques.

The *mathematical background* to the subject includes an introduction to fuzzy set theory and the theory of aggregation operations, t-norms and t-conorms, which can be used to give intersections and unions for fuzzy sets. The theory of fuzzy complements is also considered. *Further theory* relates to the use of second and third order augmented fuzzy sets, intuitionistic fuzzy sets and type 2 fuzzy sets, all of which can be used to represent data which is both imprecise and uncertain, and their intersections, unions and complements. This is in distinction to ordinary fuzzy sets, sometimes called type 1 fuzzy sets, which can only be used to represent imprecision.

The techniques and methodologies presented are based on *augmented, intuitionistic and type 2 fuzzy sets*. They include an ordering algorithm for augmented fuzzy sets which can be used to obtain a targeted impact reduction strategy and techniques for applying third order augmented fuzzy sets with negative coefficients to decision making. There are also approaches based on the use of aggregation operations to obtain a measure of the total impact at a particular life cycle stage or the total effect of a particular environmental or social impact over all the life cycle stages.

Methods based on fuzzy relations and conditional statements can be used in modelling and are applied to examples relating to global climate change.

Augmented, intuitionistic and type 1 fuzzy set techniques are illustrated by *application* to a transport planning case study. Other examples include classification of hearing impairment, the emissions of industrial processes, printer noise levels and power consumption, and the environmental impacts of manufacturing processes for producing gears and the computer industry.

18.2 The Way Forward

This book has presented a number of mathematical and computational techniques for application to problems in sustainable development and an extended overview of some of the issues in sustainable development to facilitate their application. One very important area of application of this book is (higher) education. The *interdisciplinary nature* of sustainable development means that promoting it will require contributions from many different disciplines. Since it it is so important, education at all levels and in all disciplines should include material on sustainable development.

This book can make a contribution in supporting teaching about sustainable development at the undergraduate, masters and doctoral levels, as well as supporting specialist researchers. It will be of particular relevance in mathematically based disciplines, such as engineering and the sciences, but may also be of value in other areas.

Topics that could usefully be covered in education on sustainable development include (but are not limited) to the following:

- The *ethical and social responsibilities* resulting from work in a specific discipline, particularly with regards to sustainable development. Ethics in an educational context is often considered to be limited to issues such as avoiding plagiarism (unlawful and unattributed copying) and not falsifying data. However ethical responsibilities are much wider than this and should be considered to include taking responsibility for promoting sustainable development, for instance through choices about projects and deciding against projects which could have unsustainable consequences.
- The *interrelationships* between the social or developmental and environmental aspects of sustainable development.
- The *responsibility of the high income countries* for the current crises of unsustainability and consequently for taking action to promote sustainable development.

- *Specific topics* in sustainable development. Since it is such a wide topic choices will have to be made about specific topics and the focus and depth of coverage.

Further work should also include the development of a research agenda, focusing on the development and dissemination of the methodologies and their application areas. The development side has three main components:

- *Further development of the three areas covered in the book.* These are systems, optimization and decision making, and fuzzy sets. This would focus on topics within these areas not covered or only touched on briefly in the book and might include areas in which there is little existing research, such as the development of ordering algorithms and the fuzzification and defuzzification of intuitionistic and type 2 fuzzy sets.
- *Development of theory and methodologies in additional subject areas*, such as probability, stochastic processes, possibility theory and control theory.
- *Development of the application areas.* This would include both application areas discussed in the book, such as global climate change, waste disposal and transport planning, and application areas which have received limited attention in the book, such as nuclear power. This may become increasingly important, if the reduced availability of fossil fuels tempts governments to ignore the precautionary principle and public opinion to develop additional nuclear generation capacity rather than looking to renewable energy sources.

Supporting activities for both the educational and research agendas would include:

- A *software toolbox* to facilitate the application of the methodologies.
- A *book of case studies* of the application of the techniques to a range of problems in sustainable development.

Dissemination activities would initially focus on academic conferences, but should eventually target government and other decision makers. Relevant academic conferences include those on sustainable development, sustainable development for a particular discipline, sustainable development and education, and disciplinary education. Particular examples include the international conference series on Engineering Education in Sustainable Development. Dissemination activities would also involve making and further developing existing links with professional institutes and other international bodies, such as the Engineering for a Sustainable Future Professional Network of the

Institution of Electrical Engineers and the Supplementary Ways of Improving International Stability Technical Committee of the International Federation of Automatic Control.

Acknowledgements

Particular thanks are due to Michael Johnson for reading the manuscript several times, making numerous useful suggestions and helping me to make the book more user friendly. I am also very grateful to Arnold Heemnik for reading the manuscript, making some helpful suggestions and giving me encouragement and Dirk-Jan Peet for reading part of the manuscript, meeting up with me to discuss the book and giving me useful feedback on the book at the Conference on Engineering Education in Sustainable Development in Barcelona. I would also like to thank Karel Mulder for his support and encouragement. I am also grateful to Peter Hamburg for many useful comments on Part III of the book and Ileana Hamburg for her contributions and support.

I would like to thank Peter McKenna for turning my incomprehensible doodles into professional looking graphics. Without his help the book would look much less attractive. I would also like to thank Vi Romanes for general support during the preparation of the manuscript, Morag Grieg for help in obtaining reference details and Tom O'Hara for help with some of the formatting software.

Grateful thanks are also due to the following for permission to use figures:

- USAID and PSI for photo of a youth alert peer education team visits a secondary school in Africa.
- USAID and Management Sciences for photo of health voluntary HIV counselling and testing centre in Africa.
- Perkins School for the Blind for photo of deafblind children at Kwale. School for the Deaf in Kenya planting trees.
- Africa Recovery, United Nations for photo of a Tanzanian girl at school.
- Solar Engineering Services for fig. 1.2.
- Kip Smith of Weathersmith for photo of truck emitting diesel soot.
- Selena Travel Marketing for photos of Siberian white crane and swans on Agaya Lake.

- Wiley for figs. 6.1 and 6.3 from Systems Engineering by A.P. Sage (redrawn by P McKenna) and figures 6.4abc from Learning from Failure by J. Fortune and G. Peters.
- EndangeredSpecie.com for figs. 9.2 and 9.3.
- J P Brans and Ph Vincke for figs. 11.2a-f (redrawn by P McKenna).
- J P Brans, Ph Vincke and B Mareschal for fig. 11.3 (redrawn by P McKenna).
- Th Briggs, P L Kunsch and B Mareschal for fig. 11.4 (redrawn by P McKenna).
- JAM74 for figs. 17.1 and 17.3.

Biography of the Author

Dr Hersh is currently a Senior Lecturer in the Department of Electronics and Electrical Engineering at the University of Glasgow. Her research interests fall into the following three main areas:

- Assistive technology to support independence and social inclusion.
- The application of mathematical and computing techniques to sustainable design and decision making.
- Ethics and social responsibility issues in science and engineering.

This book has developed out of Dr Hersh's research and other activities in the area of sustainable development. Other work in this area includes teaching about sustainable development and design to MSc students and the development of fuzzy methods for analysing problems in sustainable development, design and decision making.

Although apparently distinct, these areas are interrelated. Truly sustainable development requires the involvement and integration of all sections of society, including disabled people and assistive technology can be an important tool for achieving this. Science and technology are often considered to be responsible for the current environmental crises. While it is true that some of the industrial, agricultural and other applications of science and technology have contributed to the current environmental and developmental crises the world faces, science and technology have an important role to play in supporting sustainable development. One way to ensure this is to apply ethical analysis techniques to science and technology projects. In this area Dr Hersh is developing different ethical analysis techniques, applying mathematical modelling techniques to problems in ethics and values and assessing and developing techniques for applying narrative ethics to support human centred technology developments.

Dr Hersh's research projects with students and colleagues in the assistive technology area include:

- A communication glove for deafblind people.
- An electronic abacus for blind students.
- Various devices to support leisure activities for blind people.

- Training modules for employers on employing disabled people and for (young) disabled people on assertiveness and their rights at work.
- An intelligent audiometer.
- Vibrating alarm clock for deafblind people.

She also has a successful conference series on assistive technology for vision and hearing impaired people with European Commission support and is publishing a series of books on assistive technology with Prof Michael Johnson. The first book in the series, Assistive Technology for the Hearing-Impaired, Deaf and Deafblind was published by Springer in 2003 and the next book, Assistive Technology for Blind and Visually Impaired People is expected to appear early in 2006.

Dr Hersh is a member of the Institute of Mathematics and its Applications, the Institution of Electrical Engineers and the Institute of Electronics and Electrical Engineers. She is also vice-chair of the International Federation of Automatic Control (IFAC) Technical Committee on Supplementary Ways of Implementing International Stability, Convenor of the IFAC Working Group on Ethics and a member on the IFAC Technical Committee on the Social Impact of Automation.

Dr Hersh also tries to implement the principles of sustainable development in her own life. While she admits to a serious lapse in terms of flying to conferences rather more frequently than she would like, she is vegan, commutes to work by bike, tries to minimise her personal consumption and redistributes or recycles as much as possible. She is also active in her trade union, particularly on equalities issues, and in a number of organisations campaigning for social justice.

Index

A **bold numeral** indicates the starting page number for a major section for a keyword entry or a definition. An *italic numeral* indicates a keyword entry in one of the summary sections at the end of a chapter.

Printing: Krips bv, Meppel
Binding: Stürtz, Würzburg